W9-AFC-583

techniques and materials in biology

techniques and materials in biology

Marjorie P. Behringer

Associate Professor of Biology
University of North Dakota

McGRAW-HILL BOOK COMPANY

New York St. Louis San Francisco Düsseldorf Johannesburg
Kuala Lumpur London Mexico Montreal New Delhi
Panama Rio de Janeiro Singapore Sydney Toronto

TECHNIQUES AND MATERIALS IN BIOLOGY

1234567890KPKP79876543

Library of Congress Cataloging in Publication Data

Behringer, Marjorie P 1912–
Techniques and materials in biology.

Includes bibliographies.
1. Biology—Technique. 2. Biology—Laboratory manuals. 3. Laboratory animals. I. Title.
QH324.B39 574′.028 73-5910
ISBN 0-07-004392-2

This book was set in Melior by Textbook Services, Inc.
The editors were James R. Young, Jr., and Carol First; the designer was Jo Jones; and the production supervisor was Joe Campanella.
The drawings were done by John Cordes, J & R Technical Services, Inc.

The printer and binder was Kingsport Press, Inc.

contents

preface

During twenty years of biology teaching I have been continually aware of a need for concise and accessible information about standard techniques and materials for the biology laboratory. Because of the need, I have found it necessary to collect books, manuals, and journals and to maintain an index file for ready access to the information. The system has been cumbersome, exasperating, and expensive. Far too many hours have been lost while trying to locate information for a restless student or a harried instructor waiting at my elbow. Perhaps other teachers and students, in both high school and college, have experienced the same frustrations, and perhaps they will gain some measure of assistance from this book. As well, the book may be useful in teacher-training courses, where all too often, it seems, the future biology teacher does not receive the sorely needed information about laboratory materials and techniques.

As a primary criterion for the selection of contents, I focused on the questions that I have encountered most often. As a result, the most pertinent problems to emerge were those concerned with establishing and maintaining a "living" laboratory. Therefore, the present volume is confined to procedures for obtaining and maintaining and for using living animals, plants, and microorganisms in the general biology laboratory.

I am indebted to many colleagues, both students and teachers, who have contributed to the book. I express my great appreciation to the biology faculty of the University of North Dakota, all of whom at one time or another have given me valuable assistance. In particular, I want to thank Dr. Vera Facey and Dr. Paul Kannowski, whose constant cooperation and encouragement allowed the completion of a tedious job.

Also I wish to acknowledge the assistance of professors in other institutions who read particular chapters and provided many helpful suggestions. They are Dr. Richard P. Aulie, biology, Chicago State College; Dr. Harold C. Bold, botany, University of Texas; Dr. Leland McClung, microbiology, Indiana University; Dr. Ingreth D. Olsen, zoology, University of Washington; and Dr. Willis Johnson, protozoology, Wabash College. Additionally, a number of research consultants associated with commercial suppliers have furnished pertinent ideas about the selection and care of appropriate laboratory organisms. In certain instances throughout the book, I have named commercial suppliers for particular organisms when only one source was known. This practice represents a deviation from the usual restraint of authors to name commercial suppliers, an unfortunate practice, it seems, as teachers are then denied access to important and necessary information. In no case do my references to suppliers represent an endorsement or guarantee of their products.

I am especially indebted to Mrs. Nelda Hennessy, who devoted many hours of careful attention to the organization and typing of the manuscript. Photographs not credited in the text were done by the author and Mr. Akey Hung, Biology Department, University of North Dakota. Financial support for the development of the book was provided by the University of North Dakota in three consecutive Faculty Research Grants.

I assume complete responsibility for errors and shortcomings of the book. Suggestions and corrections from teachers and students are important and are invited.

Marjorie P. Behringer

PART

one

ANIMALS IN THE LABORATORY

Part 1 emphasizes the importance of living animals in the biology teaching laboratory. Although most teaching personnel recognize that living animals provide more effective and more stimulating learning experiences than preserved animals, much laboratory teaching still revolves around dead organisms. It is true that living material is not appropriate for all studies; additionally, preserved materials often can serve as a valuable supplement. With most animal studies, however, a distorted, nonreacting, and dead organism can in no way replace a live animal. Although methods for preserving animals are included briefly in this section, the descriptions of techniques for obtaining and culturing may provide a greater ease in handling live animals and thus may encourage their greater use within the teaching laboratory.

Implicit in the use of live animals is a definite responsibility for their proper care and treatment. In this regard the following guidelines were prepared by the National Society for Medical Research at the request of the Science Clubs of America. The statements were approved by the National Society for Medical Research, the Institute of Laboratory Animal Resources (National Research Council), and the American Association for Laboratory Animal Science (1968).

GUIDING PRINCIPLES IN THE USE OF ANIMALS[1]

1. The basic aims of scientific studies involving animals are to achieve an understanding of life and to advance our knowledge of life processes. Such studies lead to respect for life.

2. Insects, other invertebrates and protozoa are materials of choice for many experiments. They offer opportunities for exploration of biological principles and extension of established ones. Their wide variety and the feasibility of using larger numbers than is usually possible with vertebrates makes them especially suitable for illustrating principles.

3. A qualified adult supervisor must assume primary responsibility for the purposes and conditions of any experiment that involves living animals.

4. No experiment should be undertaken that involves anesthetic drugs, surgical procedures, pathogenic organisms, toxicological products, carcinogens, or ionizing radiation unless a trained life scientist, physician, dentist or veterinarian directly supervises the experiment.

5. Any experiment must be performed with the animal under appropriate anesthesia if pain is involved.

6. The comfort of the animal used in any study shall be a prime concern of the student investigator. Gentle handling, proper feeding, and provision of appropriate sanitary quarters shall be strictly observed. Any experiment in nutritional deficiency may proceed only to the point where symptoms of the deficiency ap-

[1]From the National Society for Medical Research, Washington, D.C. By permission.

pear. Appropriate measures shall then be taken to correct the deficiency, if such action is feasible, or the animal(s) shall be killed by a humane method.

In addition to the proper treatment of laboratory animals, a strict regard for good conservation practices is incumbent upon all biology workers. Game and wildlife legislation places restrictions on the collection of most birds and many kinds of fish, reptiles, and mammals. The regulations vary among state and local governments, and a person must become familiar with all restrictive provisions by contacting local and state conservation offices. As a rule, no special permit is required for keeping protected species within an educational institution. Usually, however, individuals collecting and transporting these species must have a permit whether or not they are employees of an educational institution. Regardless of legislation, good conservation practices require that only the necessary number of animals be collected, that natural habitats not be unduly disturbed, and that wild animals be returned to their natural habitats when studies are completed.

I

OBTAINING AND HOUSING ANIMALS

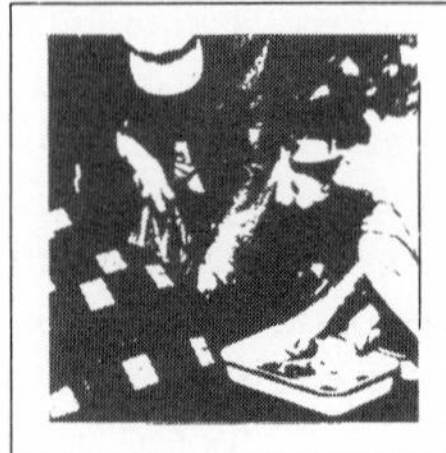

The first portion of Chap. 1 deals with some general techniques and equipment for collecting animals in the field. Specific techniques for individual animal groups may be found in Chaps. 2 to 4. A second portion discusses pertinent aspects of purchasing animals. A third and major portion of the chapter is concerned with the facilities and equipment for housing laboratory animals. A list of commercial sources for animals and equipment is given in Appendix B.

Before animals are brought into a laboratory, adequate arrangements should be made for their maintenance. As important, but perhaps less often recognized, is the disposal of animals when studies are completed. Ideally, wild animals are returned to their natural habitats. If sufficient facilities and assistance are available, a person may wish to continue the maintenance of various types of animals. Often, the animals are preserved for future studies or for displays. A most desirable practice is to establish outlets with local users, particularly with teachers of high school and elementary science courses.

GENERAL EQUIPMENT AND TECHNIQUES FOR FIELD COLLECTING

Collections may be transported in various types of containers that are clean and of ample size. Do not put saltwater animals into metal containers. Plastic containers

—bags, boxes, or pails—are commonly used for both aquatic and land animals. Reptiles and mammals may be carried in cloth bags or cages. Take care to isolate pugnacious or predatory forms and not to crowd animals. During transport, give animals ample ventilation and protect them against extreme temperatures of heat or cold.

Aquatic invertebrates and gill-breathing vertebrates may be closed for 24 to 48 h in sealed vessels, half-filled with the natural water and containing the maximum amount of trapped air above the water. During longer transport periods, open the container every 24 h. Aerate the water by stirring or by pumping air into the water with an aquarium pump or hand pump (a tire pump may be used). In hot climates, transport animals in portable iceboxes or Styrofoam receptacles containing bags of ice. Fishing-supply stores sell bags of chemicals that remain frozen within a carrying box for about 12 h. The bags are reusable, and the chemicals may be frozen in a home or laboratory freezer before traveling to the collecting site.

COLLECTING AQUATIC ANIMALS

Plankton towing net. The standard towing net consists of a 35-in-long conical bag of silk bolting cloth, attached to a 9½-in ring of bronze or stainless steel, with a collecting vial fastened into a ¾-in opening at the tapered end of the bag (see Fig. 1-1). The metal ring is suspended by three chain leaders from a metal swivel, from which extends a stout cord, 20 to 40 ft long. The net is thrown or carried into the water and drawn slowly across the surface (see Fig. 1-2). With large bodies of water, the net is towed from a moving boat. As water strains through the net, plankton collect in the vial and are then transferred to a larger container for transport to the laboratory. For quantitative studies, the amount of strained water is standardized by the towing distance and the number of tows.

The number and mesh (number of apertures per linear inch) of silk bolting cloth vary with the size of plankton to be collected, as indicated below.

Number	Mesh	Use
0	38	Very coarse; strains only largest organisms
6	74	Coarse mesh; strains the larger plankton
12	125	A general purpose net; strains water readily and retains most organisms
20	173	Fine mesh; smallest apertures for routine collecting
25	200	Very fine mesh for smallest plankton

Dredge net. Heavy nets are used to collect organisms from the deeper waters and the bottom surface (see Fig. 1-3). The size of net depends on the depth of water and method of collecting. Smaller nets, 1 to 2 ft wide and 10 to 15 in deep, are secured to poles for gathering bottom organisms by hand from the shore or boat. The same type of net may be attached to a long rope for hand towing. Deep-

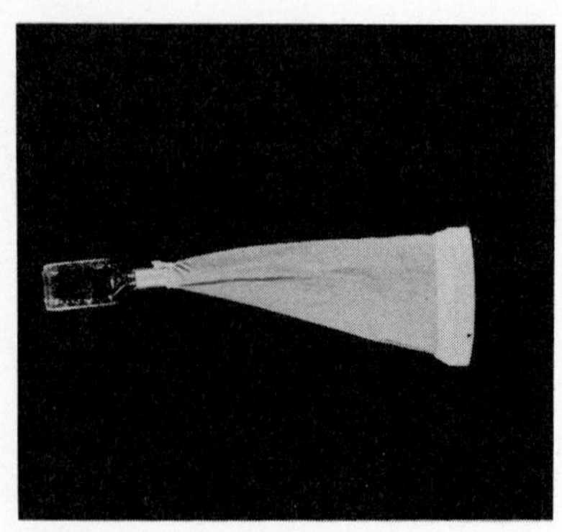

A

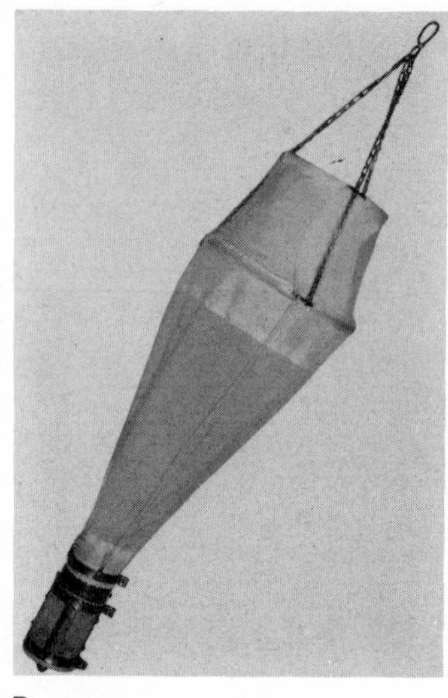

B

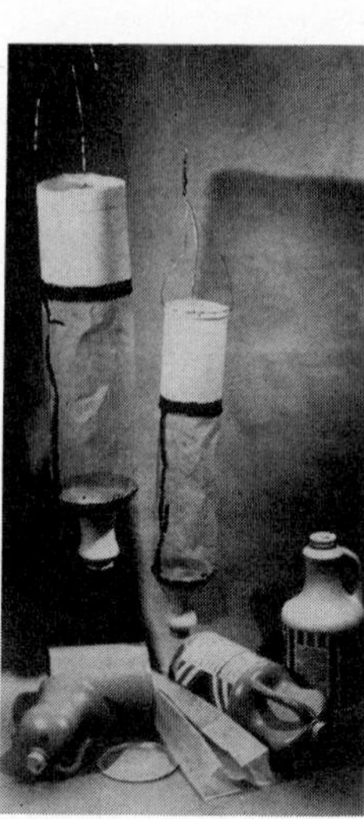

C

Figure 1-1 Equipment for collecting plankton. **A** Plankton towing net and collecting bottle from a commercial supplier. Net is attached to a cord and drawn horizontally through water by hand or boat. **B** Plankton net and collecting bucket from a commercial supplier; used for measuring plankton volume in a perpendicular column of water. Usually the net and bucket are lowered to a predetermined depth and are then pulled up. **C** Student-constructed net. The bottom and top portions of a plastic bottle are removed by cutting with scissors, leaving the cylinder shown at top of net. Water-resistant glue is used to seal together the edges of a strip of bolting cloth and to attach the cloth to two portions of the plastic bottle, as shown. After towing, the lower screw-on cap is removed to wash the collected plankton into a vessel. *(B Courtesy of Wildlife Supply Co. C Courtesy of T. F. Pletcher, Vancouver City College.)*

water collecting requires a larger net on a rope, towed by a boat, and pulled to the boat or shore with a winch. These nets are made of heavy nylon or wire mesh with an outer cover of heavy wire or rope screening to protect against snagging and tearing.

Bottom samplers and scrapers. Simple equipment may be constructed for collecting in shallow water. When quantitative sampling is not the purpose, a can

Figure 1-2 Methods of towing with plankton nets. **A** Towing by hand across a freshwater pond. **B** Towing by boat at Yaquina Bay, Oregon.

A

B

A

B

C

Figure 1-3 Collecting at bottom of water. **A** Bottom dredge net that is pulled by hand, or, in deep water, behind a boat. **B** Bottom sampler for qualitative sampling of insects, insect larvae, and other macroscopic forms in shallow streams. **C** Student-constructed bottom sampler, a simple device useful for collecting bottom scrapings in shallow water. The closed can is lowered to the bottom where the lid is opened to allow collecting of bottom sediment. The can is closed before bringing it to the surface. The lid is operated by a stiff wire with a handle fashioned at the top end. *(A and B Courtesy of Wildlife Supply Co.)*

attached to the end of a long pole is sufficient. A better scraping of the water bottom is accomplished if the can is attached to a short pole which is then hinged to a longer pole. For quantitative studies, a standard bottom sampler is required. The names of suppliers are listed in Appendix B. For less sophisticated work, a simple type of equipment may be constructed by attaching a can with a hinged lid to the end of a pole. A wire running from the lid to the upper end of the pole allows the lid to be opened when the container reaches the desired depth and then to be closed before the container is drawn to the surface. Several types of equipment are shown in Figs. 1-3 and 1-4.

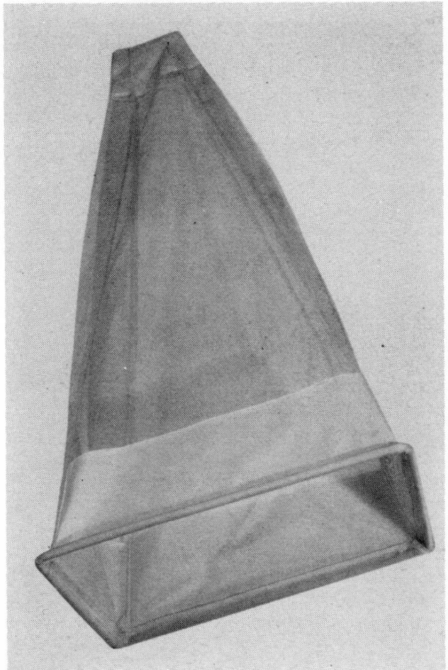

Figure 1-4 Stream drift net held in place with iron stakes driven into stream bed. A number of net frames may be set in a row across a stream. *(Courtesy of Wildlife Supply Co.)*

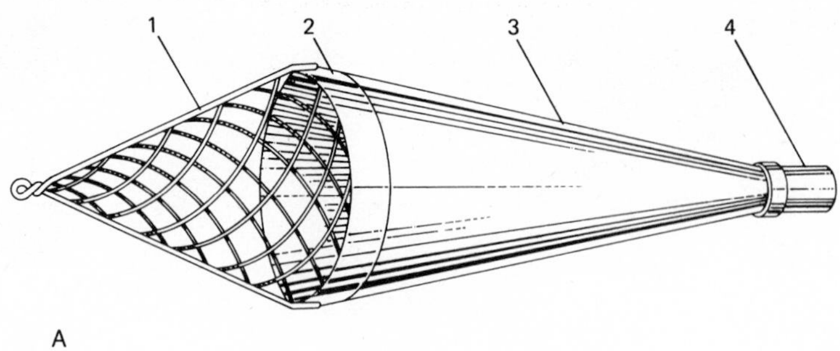

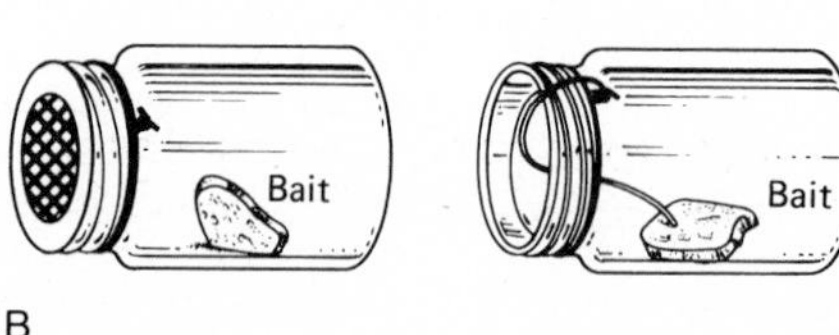

Figure 1-5 Traps for collecting aquatic animals. **A** Funnel net trap: 1, wire-mesh cone; 2, double metal ring for attachment of wire cone to 3, cloth net; 4, receiving vial. **B** Simple traps for collecting aquatic invertebrates, e.g., crayfish and planarians. Raw meat or other bait may be placed in a jar that rests at the bottom of water. If only small invertebrates are wanted, a mesh cover is fastened over the jar opening. **C** Fish trap from a commercial supplier. The trap may be left in the water overnight. If fish are not abundant, bait can be placed inside the trap. The two sections are pulled apart to remove fish. **D** Student-constructed trap, modeled from trap shown in **C.** Fish are removed through opening at center top after cover flap is turned back as shown.

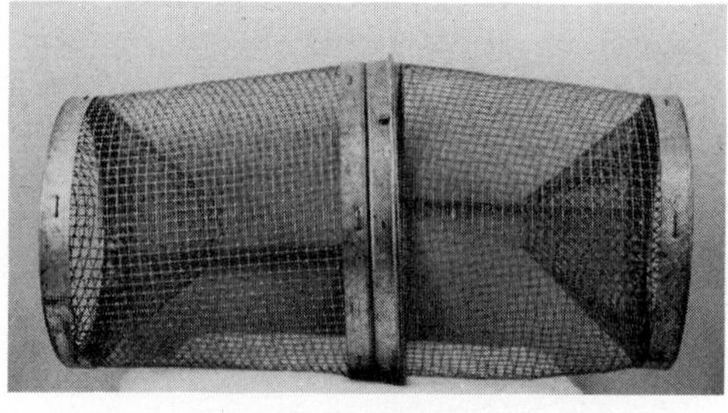

C

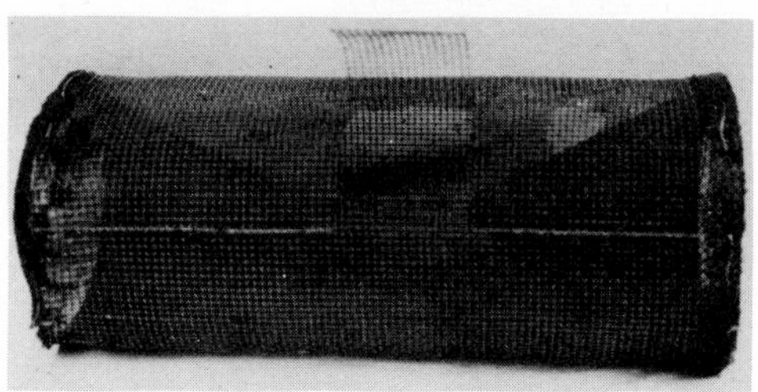

D

Dip nets and seines. Dip nets are used for obtaining small invertebrates at the water edge or from a boat. As a rule, the commercial equipment is more satisfactory, although adequate nets may be made from nylon netting purchased at a dry goods department store. Seines of various sizes are obtained from biology supply houses or from local fishing-supply companies. Seines and nets should be washed in clean, fresh water and placed flat or hung from a line in the open to dry completely before they are folded for storage.

Traps. Often aquatic invertebrates and small fish are collected with baited traps that are left in position for 24 h or longer. A simple trap consists of a jar equipped with a screen top and left at the bottom of the water overnight. A chunk of raw meat is used for bait. Various types of funnel-net traps are also used (see Fig. 1-5).

COLLECTING IN THE INTERTIDAL ZONE

The intertidal zone provides a unique hunting area because of the changing tidal waters. Generally a rocky zone is the most rewarding, as an abundance of animals live on the rocks and among the marine algae. Collecting is done during low tide (see Fig. 1-6). Tides reach a high and low level twice daily, with the extreme levels about 15 min later on each succeeding day of the lunar month. Tide timetables are published in local papers of coastal towns and are furnished by the weather bureau.

A collector must observe several precautions when following the outgoing tide. One must be alert to the passage of time to avoid getting trapped on rocks when the tide turns and rough waters begin moving toward the shore. Hip boots or rough-soled shoes are required for walking and climbing over the slick rocks and algae. Because of the possibility of falling, do not carry glass containers and sharp-pointed blades onto the rocks.

Sufficient equipment consists of several plastic bags and a dull-edged blade with a rounded end. A strong light is required for night collecting; many persons use a miner's headlight. Only an experienced person should collect alone at night. Knudsen (1966) recommends carrying a strong gasoline lantern in place of a battery-operated flashlight and placing a second gasoline lantern on shore above the tide level as an emergency guide in case the carried lantern becomes nonfunctional.

Take care to collect only the number of animals that is needed and to disturb the area no more than is necessary. Separate the larger animals into individual bags, with a small portion of algae for moisture. Most animals are transported more satisfactorily when no water is added to the container, although fresh seawater should be carried in a separate container to flush through a bag of animals every hour or so during transport.

COLLECTING LAND ANIMALS

Many of the laboratory mammals and birds are purchased from commercial suppliers. Methods for collecting amphibians and reptiles are described in Chap. 4

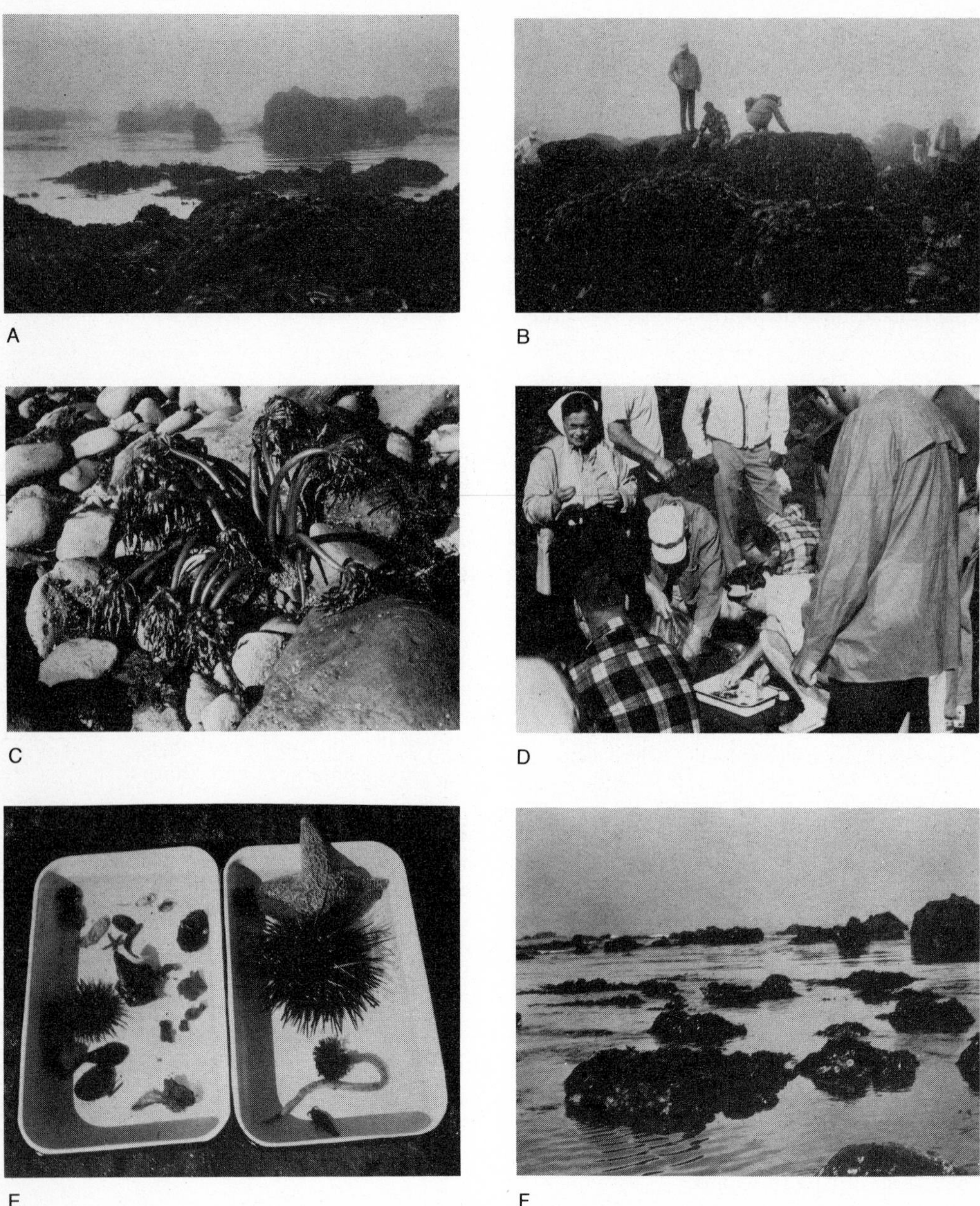

Figure 1-6 Collecting in intertidal zone at Cape Arago, Oregon. **A** Rocky coastline at early dawn during outgoing tide. **B** Collecting on rocks of intertidal zone. **C** The sea palm (*Postelsia palmaeformis*), a brown alga, flattened on rocks when tide is out. The hollow, rubbery, stemlike structure (the stipe) allows the alga to grow on rocks exposed to violent lashing of ocean waves. **D** A class examines collections upon returning to shore from the intertidal zone. **E** Animals collected by class in intertidal zone. **F** Incoming tide, several hours later than shown in **A**.

with the discussion of each group. Generally, wild birds and mammals are not captured for routine class studies.

Berlese funnel for soil arthropods. Both the Berlese funnel, described below, and the Baermann funnel, described in Chap. 2, are used to separate small animals from soil. Several kinds of Berlese funnels are available from suppliers. A simple and inexpensive type may be constructed as shown in Fig. 1-7. Note that the collecting bottle below the funnel may be closed with a stopper or a cotton plug to prevent animals from escaping. Place litter and coarse soil directly on the screen; put fine-grained soil on several thicknesses of cheesecloth that is laid over the screen. Heat from a 25- to 40-W lamp cause animals to move down the funnel into the bottle of water. Avoid overheating the soil, thus killing the animals before they move down into the bottle. If animals are to be preserved, remove the collecting bottle and add 70% alcohol to the water. Our experience has shown that if the alcohol is added to the bottle before separation, the animals are repelled by the fumes and may not move out of the soil.

Collecting insects. Use a lightweight aerial net to capture flying insects. Preferably, the net should be one that can be folded for carrying. Collect plant insects by brushing heavy sweeping or beating nets through plants, with the net opening turned upward to catch the falling insects. Alternatively, shake the insects from plants onto a paper or cloth spread below.

Although nets may be purchased, adequate equipment can be constructed rather easily as shown in Fig. 1-8. Fasten a bag, made from thin cotton or nylon cloth or from strong, fine-meshed netting, to a metal ring by means of a stout reinforcing-cloth tube sewn over the top hem. Then attach the two ends of the wire hook to a wooden handle. The pattern and size of the net are not critical. In general, the diameter of the metal ring for an aerial net is 12 to 15 in, and for a sweeping net, 8 to 10 in. The circumference of the bag is approximately three times the ring diameter, and the length of the bag about two times the diameter. The handle may be a 3 to 6 ft length, depending upon individual preference. Generally, a handle 5 or 6 ft long is desirable for a sweeping net.

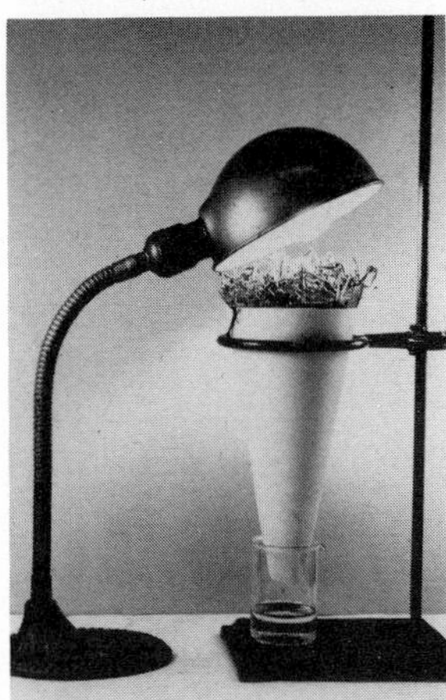

Figure 1-7 Student-constructed Berlese funnel consisting of a sheet of heavy, slick paper stapled at edges.

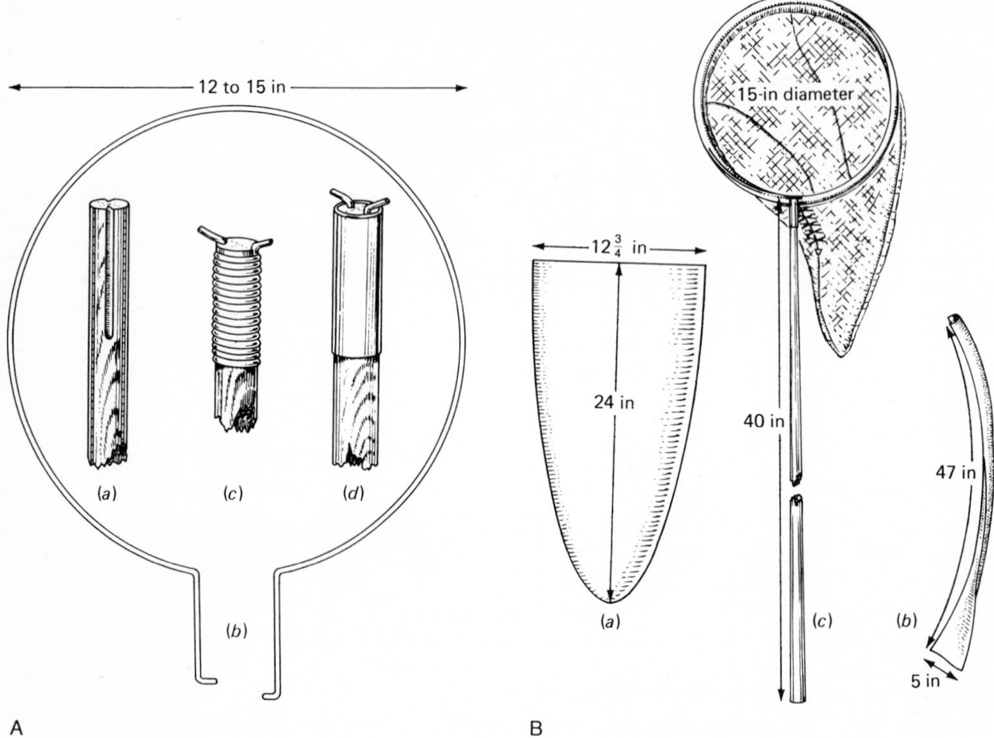

Figure 1-8 Constructing an insect net. **A** Net loop and handle. The short grooves, cut opposite each other at (*a*), end in holes through the handle that receive the hooks of the loop arms (*b*). The loop may be permanently bound to the handle (*c*), or a removable joint may be effected with a metal ferrule that can be slipped up and down (*d*). **B** Bag and completed net. The bag is cut from four pieces, each shaped as in (*a*). The top edge of the bag is bound with a narrow strip of stout muslin or light canvas (*b*), by which the bag is attached to the loop (*c*). After the bag is on the loop, the back vent may be closed with a string lacing, as shown in (*c*). The handle in (*c*) is the removable type shown in **A**. *(A and B Courtesy of the Illinois State Natural History Survey.)*

After an insect is netted, twist the top of the bag to prevent escape, and transfer the insect to a collecting jar or cage. If an insect is to be preserved, transfer it to a killing bottle. Prepare an *insect-killing jar* as shown in Fig. 1-9. Break pieces of potassium cyanide into small chunks under a fume hood. Put several chunks in the bottom of a large-mouthed bottle, 4 to 5 in in diameter. *Fumes are extremely toxic. Do not inhale fumes.* Place a ½-in layer of dry plaster over the potassium cyanide. To this layer, add a ½-in layer of wet plaster. Leave the open bottle under the fume hood until the plaster is dry. If bubbles form in the top layer, break them open and pour wet plaster into the spaces. When dry, add a circle of blotter over the top layer. Keep jar tightly closed except when adding insects. Do not inhale fumes accidentally when putting insects in the jar, and do not leave jar in strong sunlight. Because the fumes may discolor the insects, the animals should

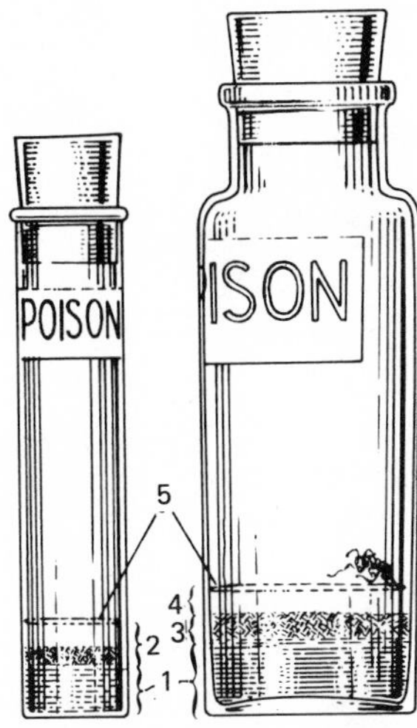

Figure 1-9 Constructing an insect-killing bottle. Potassium cyanide or calcium cyanide, at 1, is covered with a layer of dry plaster or cotton, 2, or sawdust, 3, and a layer of wet plaster, 4. When plaster is dry, a circle of blotter, 5, is added. *(Courtesy of the Illinois State Natural History Survey.)*

be removed from the bottle soon after they have died. Wipe accumulated moisture from the inner walls of the bottle with a cloth or paper that is then carefully disposed. Do not allow the chemical to contact the eyes, mouth, or cuts in the skin. Wash hands thoroughly after working with the chemical. The fume supply will be exhausted after one or two seasons. Dispose of the chemical in such a way that it will not become available to children or animals.

An *aspirator tube* is used for collecting small, fragile insects in quantity. Several types of commercial mouth aspirators are available, each consisting of a vial equipped with two tubes, a collecting tube, and a suction tube. Insects are drawn through the collecting tube into the vial by applying suction at the end of the suction tube. A screen over the inside opening of the suction tube helps to prevent drawing the animals and debris into the collector's mouth. Because a mouth aspirator may allow inhalation of dust and sand, a bulb type is more desirable. Figure 1-10 shows a bulb aspirator designed and recommended by Wheeler and Wheeler (1963).

Trap *night-flying insects* by suspending a light at the top edge of a white cloth,

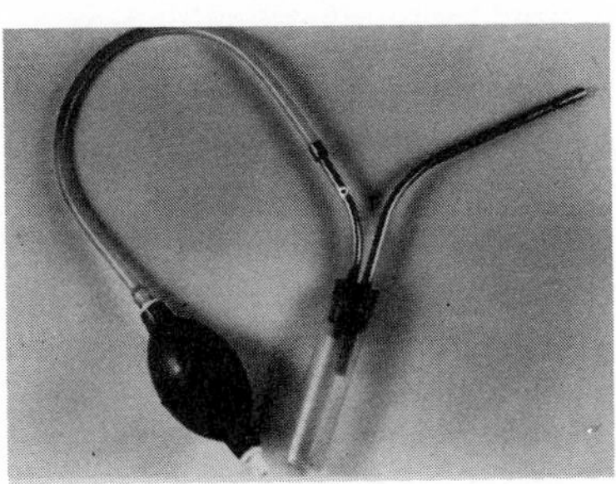

Figure 1-10 Insect aspirator for collecting small insects by bulb suction.

Figure 1-11 Insect sifter box. Soil and humus are sifted over white paper or cloth. The sifter, not more than half full, is shaken gently at first and then violently. Debris that does not pass through the mesh is emptied onto paper, and larger animals are separated out. *(Courtesy of the Illinois State Natural History Survey.)*

approximately 6 by 8 ft (for example, a bed sheet), that is fastened to an outdoor wall or between poles or trees. About 2 ft of the lower edge of the cloth should extend over the ground surface. A low-intensity light is best, such as a 15-W black-light neon tube. The same effect can be achieved with a yellow, blue, or green light bulb, or by painting the glass of a lantern or a flashlight with a dark water paint. The paint must dry completely before the light is used. To obtain live insects, one must remain at the collecting site and immediately transfer live animals to a collecting jar or cage. Killing jars and containers of preservative solutions may be devised as traps at the base of the white cloth.

A *sifter box* (see Fig. 1-11) is excellent for removing insects and other arthropods from soil and dried leaves, particularly during fall and winter months. The sifter consists of a wooden box, approximately 12 in square and 4 to 6 in deep, with a wire mesh bottom of 8, 10, or 12 meshes to an inch.

Various kinds of *insect traps* may be devised. A simple type is a small can containing food bait, which is sunk into the ground with the opening slightly below the ground surface. To prevent larger animals from taking the bait, cover the can with a board or rock, leaving a small opening for the entrance of insects.

PURCHASING ANIMALS

Ordering

Considerable information, other than cost, is required for purchasing live animals. Additionally, the procedure involves more than merely presenting a request form to the school purchasing office. The following aspects should be considered, primarily by carefully reading suppliers' catalogs and their other literature.

1. A supplier's specific instructions for writing a catalog order

2. Governmental requirements and restrictions for shipping certain animals and supplies

3. The seasons when certain animals are available

4. The time required by a supplier to fill an order and, thus, how far in advance of need an order must be placed

5. A supplier's policies for returning or exchanging an order

6. A supplier's policies regarding notification and claims for damaged shipments and diseased or dead animals

7. Procedures for special orders, such as those for unique animals or for the construction of equipment according to a buyer's design

8. The desired date for receiving a shipment and the name of the person who will receive shipments of live animals

Receiving Shipments

As with ordering, the policies of the supplier must be understood when receiving shipments. Pertinent considerations are listed here.

1. Immediately upon arrival of live animals, remove them from shipping cartons and transfer to appropriate housing conditions. Because animals require this immediate attention, the date of expected arrival must be made known to the supplier, who in turn must advise the buyer about any shipping delays and about the flight schedule of less hardy animals that must be transported immediately from an airport to a laboratory.

2. Save shipping tags and enclosed packing slips for any future returns or exchanges that may be necessary.

3. Read and observe any instructions enclosed with shipments for establishing animals in the laboratory.

4. Isolate and examine animals for disease before placing them with other animals.

5. Immediately report to the supplier any faulty shipment or diseased or dead animal. If animals are unique or costly, notify the supplier by telephone or telegram. In case the animal is rare or very costly, the supplier may require that the dead animal (or a portion of a large animal) be returned. Therefore, these animals or their body portions should be held in a freezer until the matter is resolved.

6. If the damage appears to have occurred during shipment, save any damaged cartons and immediately request an inspection and a written report from the carrier. Probably the supplier will ask for the carrier's report.

7. Aerate aquatic animals as soon as they arrive. If an air pump is not available at once, pump air with a medicine dropper into small bottles or vials, and stir the water of a large container.

8. Gradually accustom animals to new housing, as described for individual groups in Chaps. 2 to 4.

Assistance from Suppliers

As a rule, manufacturers and animal breeders make strong efforts to educate buyers about their products. On the basis of their experience, they are able to provide valuable information to *sincere* buyers through their literature or through direct correspondence. As well, many companies that sell primarily to educational organizations maintain an education office through which they regularly issue newsletters and other free literature. A partial list of these publications is given at the end of this chapter. A person must use the school letterhead paper when requesting that his name be placed on a newsletter mailing list or when asking for other free literature. Take all possible precautions to avoid indiscriminate and invalid requests, including those from enthusiastic but uninformed students. In reality, a supplier's catalog often may serve as a valuable educational device for both the teacher and student. However, the supplier must be protected from abuse of his services. Animal breeders and manufacturers are justified in attempting to furnish information to qualified buyers only.

HOUSING LABORATORY ANIMALS

The general requirements for housing laboratory animals are discussed here. Special housing is included in the following chapters with the discussions of animal groups. Wild animals often have specific requirements which differ greatly from those of animals reared in the laboratory. In these cases, a person must study the literature, often in journals, that pertains to individual animal types.

The Animal Room

Although animals must be accessible to students, they also must be protected against disturbance and transmission of disease. For this reason, a separate animal room is desirable, open only to qualified personnel. Ideally, aquariums and terrariums are located within the laboratory for formal or informal observations. Most mammals, birds, and reptiles require a greater privacy, however, as the ordinary laboratory activities may disturb their normal behavior, particularly their feeding and reproduction. Additionally, the regular chores of feeding and cleaning can be accomplished more easily in a separate animal room that adjoins the laboratory. Nevertheless, in many teaching situations where additional space is not available, all animals can be maintained in quiet corners of the laboratory.

When only a few animals are involved, a small animal room may be quite adequate, with facilities much the same as those of the laboratory. For those situations where many animals are kept on a permanent basis, the construction of the animal room merits special attention. Flooring, walls, and ceiling must be of material that is resistant to chemicals and easily cleaned. Preferably, the room can be cleaned by hosing, and a floor drain is installed toward which the floor slopes. The doors are close-fitting, with no more than a ¼-in gap. Wood doors must have a metal cover over the lower portion for protection against escaped, gnawing animals. Windows are not necessary if the room is air-conditioned.

Adequate lighting is provided for good working conditions in all areas of the room; light fixtures are of a design that permits easy cleaning and protects the lamps from water splashing. One or more sinks are installed, of adequate size for washing cages and with trap drains that catch solid wastes and prevent the entry of wild rats and mice. The room contains adequate cabinet counters with chemical-resistant tops. A sufficient number of outlets is provided for electricity, gas, air, and cold and hot water. The electrical fittings and wiring are waterproofed and protected against gnawing animals.

The tiers of cages are spaced so that air circulates through the entire room without drafts and so that cages can be serviced easily. Normally, the temperature is controlled at 20 to 22°C; humidity is regulated at 45 to 50%. Both are controlled more satisfactorily with mechanical air conditioning, particularly when many animals are housed in a large room. The absence of an unpleasant odor is a test for adequate cleaning and ventilation.

Breeding and experimental animals may require a regulated photoperiod, which can be provided by means of a time clock attached to the light switch. Timing devices can be obtained from hardware departments and biology supply houses.

Cages for Small Mammals

Mammals for the teaching laboratory include mice, rats, hamsters, guinea pigs, and rabbits. The equipment described in this section is suitable for all these groups, although the size of cage varies with the size of animal. Many suggestions for a large variety of equipment are found in published literature and in the catalogs of supply houses. Often the equipment can be constructed by the personnel of the biology department. Since living animals are desirable as a regular part of a biology course, a person is well advised to give serious consideration to the durability of equipment and to the ease of servicing cages for food and cleaning.

Opinions differ concerning the type of cage floor. Some persons believe the most desirable cages are those with a wire-mesh floor and a lower waste tray, both removable for cleaning. A flat board of adequate size can be put in the cage for animals to rest on. Resting boards seem unnecessary, however, and add greatly to the cleaning job as they must be scrubbed regularly and allowed to dry, preferably in the sun, before they are returned to a cage. Other persons prefer solid-bottom cages with animals placed on shavings. Sometimes wild mammals are kept on dirt which provides a dust bath for maintenance of pelage. In either case, bottom trays of stainless steel withstand erosion and wear longer than trays of other materials.

Metal cages. Galvanized iron and steel are used commonly for animal cages (see Fig. 1-12). During the galvanizing process the metal is coated with zinc, either by dipping in molten zinc or by an electrolytic method. The latter method is not suitable for animal cages, as only a thin coat of zinc is deposited on the inner metal. For best protection, the metal is galvanized after the cage is manufactured, thereby permitting a thorough coating of all joints and seams. Orders for commer-

A

B

Figure 1-12 Metal cages. **A** Transfer or carrying cage of galvanized steel. **B** Small metal cages. Food hopper is slotted; an opening is provided for drinking tube. **C** Metal cages in tiers. Paper from the roll at one side catches the excreta and is withdrawn at other side. **D** Rabbit cage unit, each cage with a feeder, drinking bottle, and waste pan. **E** Cage unit for guinea pigs. *(A–E Courtesy of Wahmann Manufacturing Co.)*

C

cial metal cages should specify this requirement. Although zinc is much more resistant to erosion than iron or steel, eventually spots of the zinc will wear away and holes will wear through the inner metal if the cage is not regalvanized. An aluminum laquer finish will give some additional protection, but only thorough and regular cleanings will ensure a long use of a cage. *Aluminum cages* are not practical, as the metal is soft and is attacked by alkalis of ordinary cleaning compounds. *Stainless steel cages* are practically indestructible but are expensive initially. *Enamel-coated wire cages*, such as those at pet stores, are designed for light duty only and are not satisfactory for more than temporary use.

D

E

Wood cages. Wood has one advantage in that it is warmer than metal. Otherwise, the disadvantages are many. Cleaning and drying the wood are arduous tasks; animal and food wastes quickly deteriorate the wood; gnawing animals destroy the construction; and wood frames are clumsy for handling and do not withstand rough treatment. A wood preservative will retard urine damage, but care must be taken not to select a preservative containing toxic chemicals. Occasionally, small cages are constructed from resin-treated plyboard, but the disadvantages of cleaning and wear are reduced only slightly.

Plastic cages. Both permanent and disposable plastic cages are desirable for small animals, such as mice, particularly when many cages must be cleaned regularly (see Fig. 1-13). *Polycarbonate* is almost indestructible and has the advantage of transparency, thereby allowing easy observation of animals. The material is resistant to weak acids and alkalis and is attacked slowly by strong concentrations of either; it can be autoclaved and withstands rough treatment. *Tyril and polypropylene plastics* are about one-third the cost of polycarbonate and are equally resistant to acids and alkalis. However, tyril cannot be autoclaved, and polypropylene is either translucent or opaque. *Disposable polystyrene cages* cost considerably less than other plastic cages. Although styrene is much less resistant to heat and chemicals, these cages can be used for extended periods when cleaned often. Plastic shoe boxes from department stores are inexpensive and quite satisfactory when covered with wire mesh. Generally, all types of plastic cages are less costly than those of other materials.

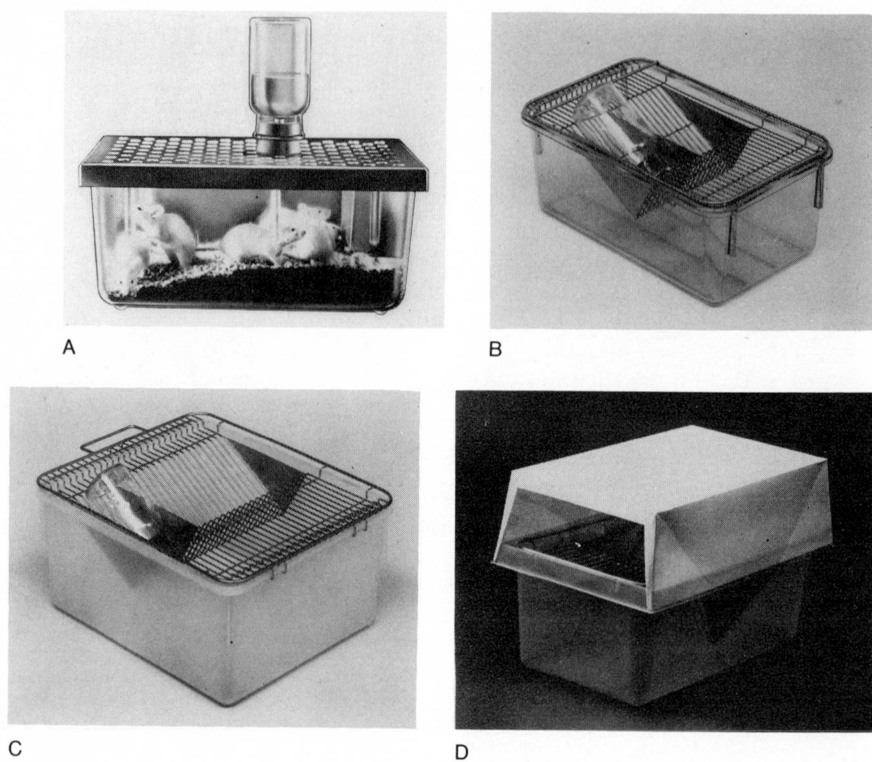

Figure 1-13 Plastic cages. **A** and **B** Clear plastic cages. **C** Translucent plastic cage. **D** Disposable mouse cage with a cover that allows isolation of animals. *(A–C Courtesy of Unifab Corp. D Courtesy of Carworth.)*

Litter for mammal cages. Litter is put in the bottom tray or inside a solid-bottom cage to absorb animal wastes, thereby retarding cage deterioration and protecting animals from their excreta. One-half to one inch of litter is sufficient; it is changed once or twice weekly, depending on the number and size of animals in a cage. The practice of using more litter for longer periods leads to excessive dampness and erosion of the cage, as well as a malodorous animal room. The litter may be pine shavings (used most commonly), cedar shavings, sawdust, or peat moss. Cedar shavings and cat litter from a pet store are desirable because of their deodorizing quality, but both are more expensive than soft-wood shavings. A combination of 1-part cedar shavings with 3-parts pine shavings has adequate deodorizing quality; additionally, the cedar helps to control skin mites and lice. Sawdust is used somewhat commonly in bottom trays, although it tends to clump when damp and becomes difficult to scrape from a tray. Sawdust is not desirable inside a cage, as the dampened material clings to animal skin and may cause chilling. Peat moss is used occasionally because of its great absorbency and because it is thought to be a deodorizer. Nevertheless, peat moss must be changed as regularly as other litter, and generally it is more expensive. Both peat moss and sawdust are more likely to become breeding places for insects than are the less compact wood shavings.

A number of litter materials are produced commercially. Not all commercial litters can be burned, however, and the disposal of wastes can become a considerable problem. Most companies will furnish a sample to prospective users. Several suppliers are listed in Appendix B.

Bedding for mammal cages. Bedding is furnished to breeding animals for building nests. Clean, shredded paper or torn newspaper is used most often for bedding and is quite satisfactory. Many laboratory workers believe that litter material in a solid-bottom cage is sufficient for nesting material as well. Breeding animals placed on mesh wire must be given bedding to prevent the young from slipping through the mesh and becoming chilled. Some persons like to provide nesting material to all animals, regardless of the type of cage floor and breeding periods. This procedure may be more a luxury than a necessity, however, since bedding greatly increases the cleaning chores. Bedding should be changed when it becomes soiled, generally once or twice weekly, except during the several days before birth of young and 12 to 14 days following birth. Unless the nest becomes badly soiled, many persons recommend that it not be disturbed until the young are weaned.

Water devices for mammal cages. The most common water devices are inverted bottles made of glass or plastic. Half-pint milk bottles are quite satisfactory. These are equipped with a rubber stopper or a screw-on cap through which a metal or glass drinking tube is inserted. The maximum capacity of a water container is about 500 ml, as a greater volume increases the likelihood of a sudden emptying of the bottle, due to the increasing volume of air as the animal drinks. Several bottles are used when more water is required. Water leakage and spontaneous emptying are the greatest hazards of the inverted bottle. The diameter of the drinking tube is of primary importance in this regard. A drinking tube with an inside diameter of 6 to 9 mm is desirable, with the distal or sucking end constructed to an inside diameter of 2 to 3 mm. A smaller tube may become air-locked, and a larger tube may lose water too readily. Glass drinking tubes are less likely to produce leakage than are metal drinking tubes. Bottle caps must fit tightly, since any air leakage may produce water leakage and a sudden emptying of the container. Not only is the deprivation of drinking water a concern when leakage occurs, but a sudden flooding may cause serious chilling of animals and a drowning of young.

Many types of bottles and large test tubes can be converted to water containers, and a great variety is available commercially. A bottle with a long, straight drinking tube can be located on the top of a cage, or one with a bent drinking tube can be fastened to the side of a cage. A short loop of coiled spring will secure a bottle firmly against the side of a cage. Several types of water bottles are shown in Fig. 1-14.

Food dispensers for mammal cages. Food containers are fastened to the side of a cage at a level low enough for animals to reach the food but high enough to prevent animals from depositing their wastes within the food dispenser. Various kinds of devices are available or can be constructed. Aluminum, stainless steel, and plastic dispensers are more satisfactory than glass; tin cans rust too quickly

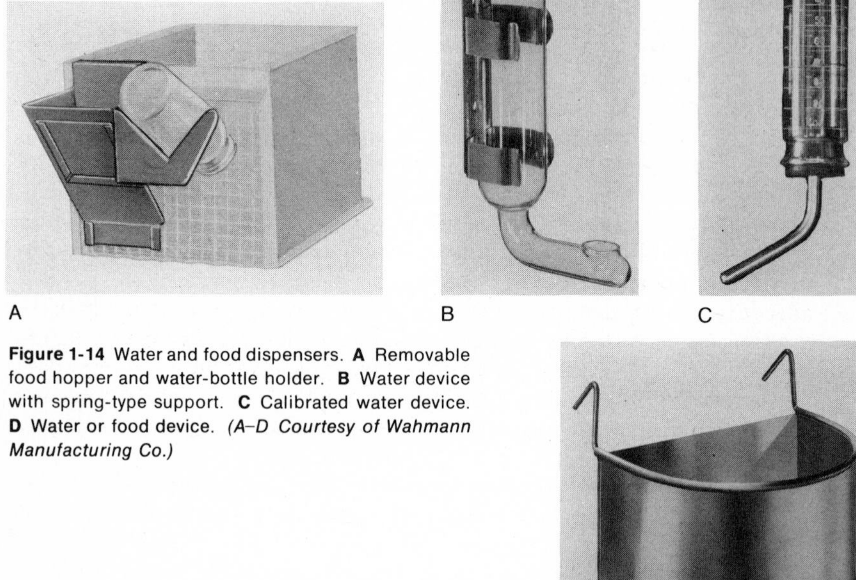

Figure 1-14 Water and food dispensers. **A** Removable food hopper and water-bottle holder. **B** Water device with spring-type support. **C** Calibrated water device. **D** Water or food device. *(A–D Courtesy of Wahmann Manufacturing Co.)*

and may have sharp edges. Dispensers attached outside the cage with an inside, bottom opening are highly desirable, both because of ease in adding food and because of protection from urine and fecal matter. Figure 1-14 illustrates various types of food dispensers.

Routine cleaning. The cleaning schedule varies with the number and kind of animals. A schedule with a minimum number of cleaning periods is described here. The ceiling and walls are vacuum-cleaned and mopped or hosed at least twice a year. Floors are washed with soapy water and hosed weekly. To control odor, particularly in mice and rat cages, litter is changed once or twice weekly. Wood, rubber, or plastic scrapers are used to remove litter from metal trays, as metal scrapers will damage a tray and thus promote erosion. The wastes and litter are deposited in a closed container and disposed in a manner that prevents the spread of infectious diseases and animal parasites. Bottom trays and wire floors are removed each week, washed with detergent water, rinsed thoroughly, and dried completely before they are returned to the cage. The entire cage is cleaned in hot, soapy water every 2 or 3 months. Disinfectants can be toxic to animals and should be avoided unless necessary for the control of disease, in which case the cages must be rinsed at length under running water and sun-dried when possible. Ordinarily, sterilization of cages is not required except when infectious diseases appear within a unit.

Each week, water bottles and drinking tubes are washed with detergent water and rinsed thoroughly. A pipe cleaner makes an excellent brush for the tubes, as the cleaners are flexible and can be pulled through a bent tube. Bottles are emptied at midweek, rinsed, and refilled with fresh water. On other days, water is added to fill bottles.

Because dry pellets make up much of the food for mammals, the outside dispensers require cleaning only at irregular intervals, unless the food becomes damp and molded. Inside-wall feeders require cleaning every 2 or 3 weeks; dishes placed on the cage floor must be emptied and scrubbed daily. Feeders are washed with detergent water, followed with a thorough rinsing.

Marine Aquarium Systems

Although marine animals greatly exceed freshwater forms in representative phyla, often these animals are neglected in general biology courses. Nevertheless, recent developments in marine aquarium systems allow maintenance of marine animals within both the coastal and inland laboratory. Because living marine animals can be a significant part of biology studies, and because many biology teachers have had little experience with saltwater aquariums, a detailed discussion is provided here.

The system may be one of three types: (1) a running ocean-water system, restricted to a coastal region; (2) a closed system where conditions remain near balance for an extended time, as with public aquariums; and (3) an open system where animals are removed and exchanged, with water conditions regulated as necessary. In reality, a closed system cannot be maintained indefinitely, even in public aquarium houses, and an open system is more appropriate for a teaching laboratory. The term "open" refers to the transfer and exchange of animals. It does not indicate an uncovered tank, since covers are used with all systems to reduce water evaporation and to prevent dust contamination.

Requirements are somewhat stringent for maintaining a saltwater aquarium, as water conditions must be controlled within fairly narrow bounds of temperature, pH, specific density, aeration, and ammonia content. Although algae are included with most shipments from supply houses, the plants are not good oxygenators; they are difficult to culture, and the dying plants foul the water. Therefore, it seems better to retain the living algae for a short observation period, followed by their preservation or disposal. Most animals of the Pacific and Atlantic waters require a refrigerated unit for successful laboratory maintenance, whereas many tropical and subtropical forms can be kept at room temperatures of 21 to 23°C. Before bringing animals into the laboratory, all supplies must be obtained, and the aquarium system must be operated until proper water conditions are established and controlled.

A beginner is advised to start with a small tank and a few animals. The literature should be reviewed, including science and science-teaching journals, and the publications of marine suppliers. See Appendix B for names of suppliers. When possible, a marine system should be observed in operation within another laboratory or at a commercial exhibit. A great variety of equipment is available; before making purchases, a person will be wise to survey the field to determine which types best fit his needs and his laboratory. Not to be overlooked is the consultant service provided by commercial suppliers.

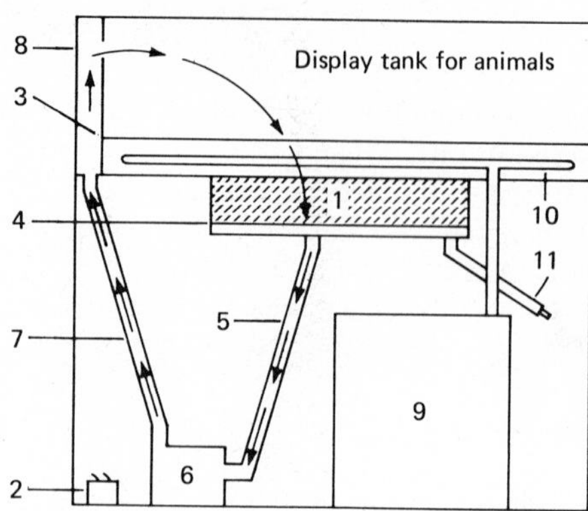

Figure 1-15 Scheme of a complete marine aquarium (Dayno Aqua-Labs). Operation: Filter box (1) with calcite filter. Water is initially filled 2 in from top of display unit. Pump and refrigeration turned on (2). Water passes through holes in removable shelf (3), through calcite (1), fine screen (4), tube (5), to pump (6), tube (7), inlet box (8), to display tank. Refrigeration compressor (9), refrigerator coil (10), and drain hose (11).

In a complete marine system, water is regulated by circulation through units for aeration, filtration, and refrigeration or heating. Figure 1-15 illustrates a scheme of a complete aquarium system. Several types of complete marine aquarium systems are marketed and are highly desirable for a long-term operation; however, adequate systems have been constructed in the laboratory by skillful persons. Because refrigerated units are costly, a number of reports have been published describing modifications for cooling the water; several of these reports are listed at the end of this chapter (Gallagher, 1967; O'Neill, 1968).

It is important to note that hardy marine animals can be maintained for several weeks, or even months, at room temperature without filtration and aeration. Accounts of short-term maintenance under these conditions have been reported by a number of persons, who have used either natural or synthetic seawater. Although techniques will vary with local situations, the method offers a means for enhancing student knowledge, particularly when a variety of animals are collected or purchased in small quantity throughout the school year. Indeed, experimentation for the establishment and control of marine cultures offers much opportunity for worthwhile research.

Containers for Marine Animals: Types and Care

In addition to commercial tanks, suitable containers are glass battery jars, large-mouthed 1- or 2-gal jars, and glass or plastic tanks constructed in the laboratory. Rectangular or square tanks are more desirable than globular ones, as they furnish a better view of the interior and a relatively greater surface area for gas exchange. A container of at least 20-gal capacity is recommended for a long-term system (see Fig. 1-16). Single animals, or a small group, can be maintained in smaller containers when other requirements are met. Plexiglas or Lucite containers are more durable than glass, although the plastic materials scratch easily. Plastic containers may be purchased, or they may be constructed from sheets of rigid plastic (3/8 to 1/2 in thickness), cut to desired size, and sealed with a solvent such as ethylene dichloride or with a caulking rubber compound such as silicone sealants available from local hardware stores or from General Electric Company, Waterford, New York. Figure 1-17 shows the Plexiglas tanks installed in the Marine Science Center of the University of Southern California at Avalon, California.

A

B

Figure 1-16 Refrigerated marine system. **A** 150-gal tank. **B** Close-up view of animals in tank.

A

C

B

Figure 1-17 Marine Science Center, University of Southern California. **A** View of Marine Center at Avalon, California. **B** Plexiglas tanks constructed and installed at the Marine Center. A plastic hose carries natural seawater into tanks from water outlets above. **C** Close-up view of a tank showing overflow outlet and sealed joints at corners and side walls. *(B–C Courtesy of Marine Science Center, University of Southern California.)*

The cement must not be toxic to animals; thus it is best to soak a new tank with seawater for several days to "season" the cement and to remove toxic substances. Plastic containers are placed on a solid surface and not in a frame, as the weight of the water may cause an unsupported tank to warp and curve. Because metal ions can be toxic to animals, the water must not contact metal at any point within the system. A metal lamp reflector above the tank must be shielded with glass or plastic to prevent condensed water from dripping into the tank. Detergents and strong chemicals are to be avoided for cleaning marine tanks, and ordinarily they are not necessary. If they are required, several changes of fresh water and a last change of salt water must be circulated through the system and removed, before the tank is filled and set into operation.

Several cautions are prescribed for storing and handling tanks that have cemented joints. Although new plastic cements are described as nondrying and nonhardening, a person may be well advised to continue the practice of keeping cemented tanks at least half-filled with water except during short periods of transport. The water seeps to the top of joints, thus preventing the drying of cement and a possible leakage. Tanks should be empty or nearly so when moved about, and should be picked up and carried by placing the hands under the ends of the container to avoid twisting the side walls. Grasping a tank at the top edge may displace the walls; a full tank may be set down too quickly, thus jarring and displacing the side walls. Only small containers molded from one piece of glass or plastic can be tilted for emptying; other tanks are emptied by suction. A soft, synthetic sponge is used to clean the walls and to soak up the last portion of water.

Seawater

At one time natural seawater was required for the marine aquarium. Later it was found that synthetic seawater was satisfactory when combined with a small proportion of natural ocean water. Often, however, the natural seawater could not be regulated within a laboratory situation because of decaying organic components and other toxic materials. Today relatively inexpensive synthetic sea salts are available which allow an easier maintenance of optimal water conditions.

Most companies furnish instructions for the preparation of synthetic seawater. In general, the salts are combined with distilled water to make a salinity of 33.5 parts per thousand or a specific gravity of 1.025. For delicate animals it may be necessary to measure the specific gravity periodically with a hydrometer that measures specific gravity in the range of 1.025. Hydrometers are available from most suppliers of biological equipment, and instructions for their use are included with the instrument. Add distilled water for that lost by evaporation to regain a specific gravity of 1.025. In most cases, however, the system can be controlled simply by replacing approximately one-fourth of the volume with fresh synthetic seawater every 1 or 2 weeks, depending on the number and kind of animals.

Water Circulation, Aeration, and Filtration

Water circulation is powered with a plastic centrifugal pump, controlled for overload protection and with a water turnover rate of at least one change each hour. Fay (1970) recommends a turnover rate of four times each hour. The water is aerated when dispersed into bubbles as it reenters the tank. In general, marine animals require more oxygen than freshwater animals, although some animals will survive with only one water change every 2 to 3 h at lowered temperatures of 13 to 15°C. Sessile or slow-moving animals may survive a week or longer with no water turnover if a relatively large water surface area is provided.

Filtration is accomplished with a filter unit installed in the water circulation system. Two types of filter units are in general use, and enthusiasts can be found for each type. These are (1) a bottom plastic plate covered with calcareous gravel or crushed shells and (2) a plastic basket, inside or outside the tank, which contains activated charcoal covered with glass wool. When the second type of filter is used, calcareous materials must be added to the filter basket or the tank. In general, the bottom filter seems the more satisfactory, particularly with tanks of 20 gal or more volume.

Filters require a periodic cleaning. Although excessive sedimentation can be removed with a bulb or mouth suction hose, the bottom plastic plate and calcareous materials must be removed at intervals and washed in running tap water, followed by a saltwater rinse. Basket filters are removed and cleaned in the same manner, with occasional replacement of the charcoal and glass wool. Fresh charcoal must be cured by soaking in salt water for about 24 h before use.

Proper aeration and filtration of the water are required for adequate oxygenation and for maintaining a pH of 8.0 to 8.2, as well as a low ammonia content. Both the removal of carbon dioxide during aeration and the leaching of calcareous materials contribute to an alkaline condition. Accumulation of ammonia in the water is a primary reason for animal deaths, not only because ammonia is toxic to animals but also because the accumulation of ammonia lowers the pH level and reduces the water capacity for oxygen absorption. Most marine animals excrete ammonia directly into the water; other organic wastes are converted partially to ammonia. The maximum ammonia tolerance for many marine animals is 0.05 ppm, although some animals such as the octopus can survive at much higher levels. Bacterial action converts much of the ammonia to nitrites and then to nitrates, which are less toxic to animals. Operating the system for one or more weeks before adding the animals becomes, then, a primary requisite for the establishment of a microbial flora in the filtrant materials and for the adjustment and control of water at a pH of 8.0 to 8.2.

Fay (1970) recommends that an aquarium system be soaked and recirculated for a minimum of 2 weeks before the tank is refilled with seawater and animals are added. Strickland and Parsons (1965) recommend periodic chemical tests of the water for ammonia, nitrites, and nitrates. An alternative method is to replace one-fourth to one-third of the water when the alkalinity drops below pH 8.0, and to remove excessive sedimentation and deposits on the walls at regular intervals. Dead animals must be removed at once, and uneaten food must not be left in the water.

Temperature Control

Most marine animals of the Atlantic and Pacific coastal waters require a temperature of approximately 15°C. The water is cooled with a refrigerating unit connected into the water-circulating tubes. In general, a ¾ hp unit is satisfactory for each 150 gal of water, although requirements vary with the surrounding temperature and with the number and kinds of animals. For long-term operation, the unit should include thermostatic and adjustable temperature controls, with a range from 10°C to ambient (surrounding) temperatures. A temperature above optimum is more dangerous than a lower temperature. Warm-water marine animals require a temperature range of approximately 21 to 24°C. If the room temperature does not permit this range, the tank must be equipped with one or more heaters with thermostatic controls.

Animals for the Marine Aquarium

A general rule for determining the number or size of slow-moving animals is to use 1 lb of animal for each 100 gal of water. Sessile animals may require only 40 gal of water/lb, and very active animals may require as much as 450 gal/lb. In any case, the animals should not be crowded, and it is a good policy to begin with a few animals and to increase the number gradually while keeping a check on water conditions.

Most animals will not survive abrupt water changes and require a "run-in" period. An adjustment from natural to synthetic seawater is achieved by gradually adding the synthetic water to the shipping or collecting container in which the animals arrive. An adjustment to a change in temperature is made by enclosing the animals in a plastic bag with their natural water and floating the bag in the aquarium for several hours. Occasionally, animals will be received in a depressed state, particularly if their transport is delayed, in which case they may require an immediate gentle aeration within the transport container. Sometimes animals can be revived more promptly if the original container of animals with an attached aerator pump is placed in a refrigerator or a cold room. Upon arrival, all animals should be inspected for signs of disease. Those with symptoms should be isolated in individual containers for an observation period.

The selection of animal types is related to the size and purpose of the aquarium, with an additional regard for animal compatibility. Because the animals from different coastal regions show great variations in their requirements and compatibility, no useful generalizations can be given here for the selection of animal groups. The marine suppliers who collect and maintain animals are in the best position to provide information concerning the animals shipped from their particular regions. Many of these suppliers furnish printed information about the selection and care of animals. A person who has had no experience with marine animals will need to obtain the literature of suppliers to learn what animals are available and the requirements for their feeding and maintenance. The names of several suppliers are given in Appendix B.

The animal groups described below are compatible under normal circumstances, with each group containing an appropriate number of animals for a designated volume of water. The groups are described here only as initial assistance to the person who does not know where and how to begin selecting animals. The lists cannot be used for placing orders—a current catalog must be obtained from a supplier.

Animals of the Atlantic Coast

Live Marine Animals for 10-gal Aquarium

4 small sea anemones *(Metridium)*

4 red stars *(Henricia)*

4 small starfish *(Asterias)*

4 sea urchins *(Arbacia and Strongylocentrotus)*

3 sea cucumbers *(Thyone)*

7 snails *(Nassa and Littorina)*

3 small mussels *(Mytilus)*

5 barnacles *(Balanus)*

4 hermit crabs *(Pagurus)*

2 clam worms *(Nereis)*

2 small horseshoe crabs *(Limulus)*

Green algae *(Ulva)*

2 chitons *(Chaetopleura)*

Animals of the Pacific Coast[1]

Laboratory tidepool. This is a diversified collection of hardy, live marine invertebrates and algae, designed to supplement either the introductory course in biology or a specialized course in invertebrate zoology. The collection consists of a minimum of 30 representative species of the following phyla and classes:

> Coelenterata (Anthozoa, 2 sp.); Sipunculoidea (1 sp.); Annelida (Polychaeta, 3 to 4 sp.); Arthropoda (Cirripedia, 1 sp.); Decapoda (4 to 5 sp.); Mollusca (Amphineura, 1 sp.; Gastropoda, 3 to 6 sp.; Pelecypoda, 2 sp.); Echinodermata (Echinoidea, 1 or 2 sp.; Asteroidea, 1 sp.; Ophiuroidea, 1 sp.); Chordata (Tunicata, 1 sp.); Algae (2 or 3 sp.).

The collection should be separated into two aquariums of 15 to 25 gal each, or otherwise segregated to separate the predators from potential prey, e.g., Asteroidea from Pelecypoda, or Decapoda from Sipunculoidea and Annelida.

Feeding marine animals. Herbivores will accept benthic (bottom) algae, lettuce, spinach, etc.; omnivores will feed on hard-boiled eggs or shrimp. Carnivores will accept frozen shrimp or fish or bits of clam flesh. Worms are good, and small goldfish may also be accepted by carnivores. Suspension-feeding animals will accept brine shrimp, suspended algae, and yeast. Animals may be isolated to a glass dish for feeding-habit studies. Sea anemones may be fed by placing chopped meat in the oral disc for ingestion. Starfish may be fed small clams or oysters in their shells. Avoid overfeeding, and remove uneaten food from the aquarium at the earliest opportunity. Daily feeding is not recommended, except for suspension feeders and fish. Most invertebrates will survive better if they are fed once each week.

---

[1]Adapted from Rimmon C. Fay: 1970, *Living Marine Organisms for Teaching and Research*, pp. 19 and 34, Pacific Bio-Marine Supply Company, Venice, Calif. By permission.

Animals of the Florida Coast[1]

Specimens of Broad Phylogenetic Representation (20–30 gal aquarium)

Porifera:	Globose orange sponge	*Tethya diploderma*
Coelenterata:	Small sea whip	*Pterogorgia* sp.
	Lump gorgonian	*Briareum asbestinum*
	Collared anemone	*Phyllactis conchilega*
	Rose coral	*Manicina areolata*
	Eyed coral	*Oculina diffusa*
	Green zoanthid colony	*Zoanthus sociatus*
Platyhelmia:	Polyclad	*Thysanozooan nigrum* or
		Pseudoceros sp.
Annelida:	Candy-striped fringe worm	Terebellidae—*Amphitrite* sp.
	Fan worms in *Halimeda* alga	Sabellidae—*Sabella* sp., *Branchiomma* sp.
Arthropoda:	Banded coral shrimp	*Stenopus hispidus*
	Arrow crab	*Stenorhynchus seticornis*
Mollusca:	West Indian chiton	*Ischnochiton floridanus*
	Flame fringed clam	*Lima scabra*
	Keyhole limpet	*Luccapina sowerbii*
	Nudibranch	*Tridachea crispata*
Echinodermata:	Thorny starfish	*Echinaster* sp.
	Smooth brittle star	*Ophioderma* sp.
	Black rock urchin	*Echinometra lucunter*
	2 yellow sea cucumbers	*Holothuria parvula*
Chordata:	Compound ascidian	*Botrylloides nigrum*
	Solitary black tunicate	*Ascidia nigra*
Marine algae:	1 green, 1 red, 1 brown	Various species

[1]Adapted from *Three Suggested Community Groupings for Use as Teaching Aids*, pp. 1–4, Tropical Atlantic Marine Specimens, Big Pine Key, Fla., 1968. By permission.

2 living rocks—Small sections of hard substrate with a natural population of sessile and near-sessile faunal and algal forms together with functional microorganisms and the organic sediments which support them.

Fish:	2 high hats	*Equetus pulcher*
	1 cardinal	*Apogon maculatus*

Feeding schedule. Daily: (1) A hatch of ⅛ teaspoon of *Artemia* eggs; (2) a quantity of chopped flesh, such as shrimp, lobster, crayfish, lean beef, or chicken, equivalent to a ½-in cube. Twice weekly: (1) feed anemone, directly in oral disc, a ⅜-in cube of any flesh suggested above; (2) feed a 1-in square of finely chopped fresh lettuce, spinach, turnip, or beet greens if a natural algal growth is not present.

The Freshwater Aquarium

Although the term "balanced aquarium" is encountered frequently, in reality a balanced condition is difficult to establish, and probably no aquarium remains balanced for an indefinite period. Therefore, attempts to establish a balanced aquarium in a teaching laboratory may be impractical as well as inappropriate. It seems very desirable for the aquarium to be "open," in the sense that organisms are added when they become available and are removed regularly for laboratory studies. Although, at times, animals must be purchased, generally native aquatic animals are more desirable (see Fig. 1-18).

Containers

The types and the care of aquarium containers are discussed in the preceding section on marine aquarium systems. That discussion is equally pertinent to the matter of selecting and handling freshwater tanks. In one respect, however, the care of freshwater containers is simpler, as freshwater animals are not as sensitive as saltwater animals to chemicals within the water. Consequently, detergents and

Figure 1-18 Freshwater aquarium.

Courtesy of Carolina Biological Supply Company

other chemicals, which must be avoided if possible with marine systems, can be used with less danger for cleaning freshwater containers. This becomes an advantage when removing oil or calcium deposits from the tank walls.

Clean a new or used container with detergent water, followed by several rinses. To remove lime deposit, fill the container with water to which is added about 100 ml of dilute hydrochloric acid for each 10 gal of water. Allow the container to stand until the deposit is dissolved. A heavy lime deposit may require a stronger concentration of acid and a soaking period of 24 to 48 h. Remove oil or fatty deposits by substituting ammonium hydroxide for the acid. To control algae growth or infectious animal diseases, empty the tank and refill with water to which is added enough crystals of potassium permanganate to make a pink color. Soak the container for 24 h. *Potassium permanganate is a poison;* care must be taken to keep the hands out of the water and the poison away from the eyes and mouth. After using any of these chemicals, rinse the tank with running water for several hours to soak the chemicals from the cemented joints.

Often the deposits can be removed by wiping the inside walls with paper towels or soft sponges. Plastic walls will scratch easily, even with paper towels, and a soft sponge or cloth is recommended for cleaning both the outside and inside walls of plastic containers. A scraping blade is not recommended for removing deposits from tank walls.

Aeration and Filtration

Tanks of a greater volume than 10 to 15 gal require aeration and filtration unless water is flowing through the container. Generally, aeration is not required for smaller containers except when the plants and animals are not in good proportion or when a tank contains only animals. Numerous types of equipment are available commercially, and a person is advised to observe the systems in operation before making expensive purchases.

When pressured air is piped into a laboratory, plastic tubing is run from the air outlet into the tank, where it is weighted with an air breaker that is inserted into the tube opening. The air breaker, most often of porous stone, disperses the air into bubbles, thus increasing the rate of aeration. A simpler method, requiring no stone breaker, is to plug the open end of the tube with a cork and to punch holes in the tube above the cork. Weight the tube by tying a stone to the end, or tape the tube to the wall of the tank above the water to hold the tube end within the water. Air pumps are employed when air outlets are not provided in a laboratory. "T" connectors and regulating valves allow air to be dispensed to a series of containers from one source (Fig. 1-19).

Two types of filter systems are available: (1) a bottom filter consisting of a plastic plate that is placed under the sand and gravel and (2) a basket filter, inside or outside the tank, which contains a bottom layer of activated charcoal and a top layer of glass wool. With both types, the water is circulated through the filter and returned to the tank; both require periodic cleaning, with the time interval dependent on the number of animals and plants.

Temperature

Most freshwater animals can be maintained at a room temperature of 20 to 24°C and can survive at 2 to 3° above and below this range if the temperature change is slow. Guppies tolerate temperatures from 18 to 29°C; goldfish live successfully in temperatures ranging from 10 to 26°C when fluctuations occur slowly. Aquarium heaters with thermostatic controls can be obtained from supply houses for situations where great fluctuations in room temperature occur during a 24-h period. Heated tanks are kept covered to maintain even temperatures and to reduce evaporation.

Location and Lighting

Before adding materials, place the tank in a permanent location under proper lighting and away from heat registers or drafty windows. A strong, diffused light from a north or east exposure is desirable, since a more direct sunlight encourages excessive growth of algae. A complete light spectrum can be obtained by combining two 40-W daylight fluorescent lamps with one 150-W incandescent lamp. Aquarium supply houses sell a GroLux lamp, manufactured by the Sylvania Lighting Products Company, which serves adequately. On the other hand, aquariums have been maintained for lengthy periods with only incandescent lamps that are adjusted at a proper distance to avoid overheating the water. The size of the tank and the number of plants determine the required intensity of light, a matter that is best resolved by observing the condition of aquarium plants and the accumulation of undesired algal growth. The light intensity may be regulated by adjusting the number of lamps or the distance between the tank and the light source, or by covering a portion of a tank when the intensity of natural light is too great.

Substratum Materials

Coarse sand or fine gravel is spread to a depth of about 2 in over the bottom of the tank. Soil and fine sand are not as satisfactory since the movement of animals may produce murky water, and both materials may pack and smother plant roots. Several kinds of plants, however, will take root only in soil, as with *Marsilea*, Chap. 8, in which case the plants can be grown in submerged pots of soil, thus preventing a disturbance by fish. Before being placed in the tank, the sand and gravel are washed in running water until it runs clear.

Figure 1-19 Methods for connecting a pump to one or more aquariums. *(From Enjoy Your Aquarium. By permission of Pet Library, Ltd.)* 1, Piston pump; 2, bottom filter; 3, charcoal; 4, glass wool; 5, porous air stone; 6, air lift tubes for under-gravel filter; 7, under-gravel or sub-sand filter; 8, ball-type air stone; 9, outside filter; 10, metal air release; 11, excess air.

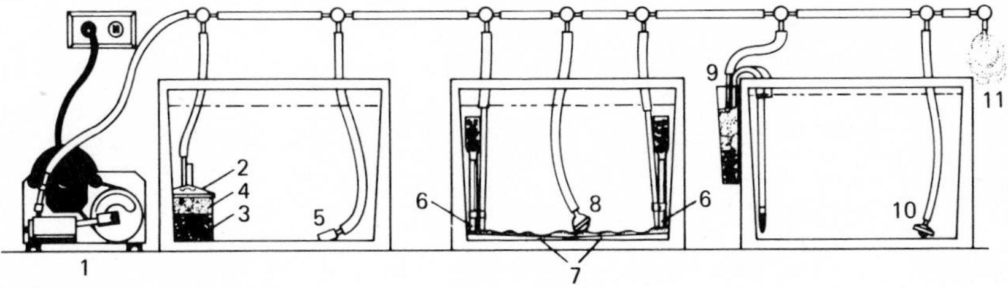

Water

Clear pond water is desirable when available. Tap water is satisfactory if allowed to stand several days in an open container to let chlorine gas escape into the air. Alternatively, tap water may be dechlorinated by adding 0.05 g sodium thiosulfate for each gallon of water, a treatment that allows immediate use of the water. The chemical can be obtained from aquarium or biology supply houses.

A tank is first half filled by pouring water on a sheet of plastic film or on a paper placed over the substratum. The film rises with the water and prevents a disturbance of the substratum. Plants are added, and the water level is brought to several inches below the top of the tank. Animals are not put in the tank until the rooted plants have started growing, generally a period of 1 or 2 weeks.

Freshwater Aquarium Plants

It is well to plant sparingly at first, as plant growth may crowd the tank in a short time. Spacing of plants depends on their expected growth, although a general rule is to allow about 4 in around all plants except *Vallisneria* (eel grass). The latter can be used as a filler plant at first and removed later as it multiplies and as other plants increase in size.

As a rule, tank-grown plants from aquarium houses are cultured easily, although occasionally the smaller varieties of floating or shallow-rooted plants, from a pond or a slow-moving stream, can be grown as successfully. *Great caution must be taken when floating plants* are placed in a tank that has water running through it. The water flow must be held to a very slow stream and the drain must be surrounded with a tightly secured and sturdy wire net or basket, placed several inches out of the drain but completely around it. The wire net prevents the plants from clogging the drain and producing a flood in the room.

Figure 1-20 contains photographs of many aquatic plants suitable for a freshwater aquarium. All can be obtained from biology and aquarium supply companies; many of the plants can be collected in various areas of the United States. Several references at the end of the chapter provide more detailed descriptions of aquatic plants and will assist with collecting local plants (Axelrod, 1967; Eyles and Robertson, 1963; Fassett, 1957; and Winterringer and Lopinot, 1966).

All plants, both purchased and collected, are rinsed in running tap water to remove algae and other contaminants. Some workers recommend soaking the plants in a potassium permanganate solution (light pink color—about one grain of powder per gallon of water) for 1 h before adding them to the tank. Taller plants are placed at the back of the tank, with lower plants spaced in front.

Freshwater Aquarium Animals

Since most aquarium difficulties result from overstocking the animals, the following discussion merits a close attention. The quantity of animals depends on the size of the tank, and two rules are used commonly: (1) one linear inch of fish (excluding the tail fin) for each 24 in^2 of water surface area, when the water is not aerated; or (2) two linear inches of fish for each 24 in^2 of surface or for each gallon of water when tank is aerated. (Calculate the volume of a tank from its cubic measurements, with 231 in^3 equal to 1 gal of water.) Both formulas apply only when the water temperature is 20 to 24°C. At a lower temperature, the quantity of fish can be increased somewhat; at higher temperatures, the quantity must be reduced. This is because water absorbs more oxygen at lower temperatures, and also because animal metabolic rate (and oxygen need) increases at higher temperatures.

The selection of animals depends on the laboratory studies and on the kinds of animals collected during field trips. In general, all freshwater animals can be grown in an aquarium. The adults of larger fish cannot be maintained easily, and wildlife legislation restricts the collection of many native species. Young animals and those that normally live in quiet streams or still ponds will adapt to aquarium life more successfully. As a rule, several small animals are preferable to one large animal. Predatory animals can be set up in a series of small, individual containers. Several snails for each gallon of water will act as scavengers of sedimentary wastes and will clear algae from the side walls. Both egg-laying and viviparous snails are desirable. Clams and newts can be added in the proportion of several 1-in animals for each 10 gal of water. Since these animals are slow-moving, their presence in the suggested number does not affect the formula for the number of fish.

Although small native fish are considered more preferable than goldfish, there are times when only the latter can be obtained. In spite of a somewhat deprecatory attitude toward the presence of goldfish in a laboratory, the fantails, the black mollies, and the ordinary goldfish can be useful and attractive when a careful selection is made for color and pattern. It must be noted, however, that these fish will eat the aquarium plants, perhaps even when sufficient food is provided. Also, goldfish are more subject to disease than are most other fish.

Because a large variety of aquatic animals are required in a teaching laboratory, the discussion of their care is included in the following chapters with the descriptions of animal groups. Suggested groupings of animals can be found in the catalogs of most biology and aquarium supply houses. The following plants and animals are suggested as only one kind of group for a 10-gal tank that is unaerated and unfiltered.

Plants	Animals
8 *Anacharis*	15 small snails or 6 large snails
8 *Myriophyllum, Ceratophyllum,* or *Cabomba*	10 one-inch fish or a like proportion of larger fish
8 *Vallisneria*	4 tadpoles
4 Dwarf *Sagittaria*	1 newt or larval salamander
4 *Ludwigia*	1 mussel
8 *Nitella* or *Riccia*	
1 small portion of duckweed	

A

B *Courtesy of Carolina Biological Supply Company*

C

D

E

F *Courtesy of Carolina Biological Supply Company*

G *Courtesy of Carolina Biological Supply Company*

H *Courtesy of Carolina Biological Supply Company*

I *Courtesy of Carolina Biological Supply Company*

J

K

L

Figure 1-20 Aquatic plants suitable for the laboratory aquarium. **A** *Ceratophyllum* sp. or coontail (*Vallisneria* in background). **B** *Myriophyllum* sp. or milfoil or parrot feather. **C** *Sagittaria* sp. **D** *Cabomba* sp. or fanwort. **E** *Anacharis* sp. **F** *Cryptocoryne willisii.* **G** *Pistia stratiotes* or water lettuce. **H** *Nymphoides* sp. or underwater banana. **I** *Vallisneria* sp. or corkscrew eelgrass. **J** *Lemna* sp. or duckweed. **K** *Ludwigia* sp. or false loosestrife. **L** *Echindorus* sp. or amazon swordplant, showing parent plant at rear and stolons with plantlets at front, which occasionally are formed in abundance. The swordplant is one of the most beautiful plantings in the laboratory aquarium.

Because no aeration and filtration are provided for the above sample group, about one-fourth of the volume of water should be replaced with fresh dechlorinated water at least once weekly. If organisms are not removed for laboratory studies, this proportion of plants and animals may reach a stage of near-balance, at which time water need not be replaced except at irregular intervals of 1 or 2 months, or when necessary due to evaporation loss. For larger tanks, both aeration and filtration are recommended, with an occasional removal of wastes and bottom sediment by siphoning. Probably the most satisfactory aquarium is one equipped with an inflow of tap water and an overflow drain. With a tank of 25 to 50 gal and a small incoming stream, no dechlorination is required and no aeration or filtration is necessary, although sedimentation must be removed occasionally.

The Terrestrial Vivarium

The housing of terrestrial animals, other than mammals, varies considerably according to the natural requirements of particular animal groups. In all instances, laboratory situations are established that resemble the outdoor habitat so far as is possible and feasible. These may be desert, woodland, or semiaquatic situations set up in boxes, tubs, pans, old aquariums, bowls, or jars. The containers must be durable and of ample size to allow sufficient space for animal movement. They must contain devices for food and water, insurance against animal escape, and an easy access to the animals. If desired, appropriate living plants can be added to produce a more natural and informative situation. However, plants are not required for the satisfactory maintenance of land animals, and their inclusion will increase the regular chores of caring for the vivarium. See Fig. 1-21.

Modified and Converted Types of Housing

Many times self-made cages and conversions of other containers are quite satisfactory and are less expensive than commercial equipment. For the laboratory with a restricted budget, the "do-it-yourself" plan seems very desirable. Even with an ample budget, the less expensive—and as satisfactory—substitutions appear to be good economy, since the plan permits the purchase of additional equipment for which substitutions cannot be made. This does not indicate that a disorderly and unattractive laboratory is necessary or desirable. Converted containers include

Figure 1-21 Desert terrarium. Cactus is in a plant pot. A desert pocket mouse (the male) can be seen in the foreground between two sticks where he will soon dig into the sand and become buried. The female pocket mouse has buried herself in the plant pot at the base of the cactus and is not visible.

discarded aquariums, battery jars, large-mouthed household jars, and kitchen plastic ware such as dishpans and icebox dishes. Additionally, discarded plastic containers of household products, trimmed to shape with scissors, can be substituted for various kinds of culture bowls and small cages. On one occasion a child's plastic wading pool was transformed into an oversize laboratory culture bowl. The BSCS publication *Innovations in Equipment and Techniques for the Biology Teaching Laboratory* (Barthelemy, Dawson, and Lee, 1964) contains excellent and imaginative ideas for constructing animal cages.

REFERENCES

Axelrod, Herbert R.: 1967, *Aquarium Plants in Color*, T.F.H. Publications, Inc., 245 Cornelison Ave., Jersey City, N.J. 07302, 32 pp.

Barthelemy, Richard E., James R. Dawson, Jr., Addison E. Lee: 1964, *Innovations in Equipment and Techniques for the Biology Teaching Laboratory*, pp. 22–35, D.C. Heath and Company, Boston.

Eyles, Don E., and J. Lynne Robertson, Jr.: 1963, *A Guide and Key to the Aquatic Plants of the Southeastern United States*, Public Health Bulletin 286, Bureau of Sport Fisheries and Wildlife, 151 pp. (Issued by U.S. Public Health Service, 1944; reprinted without change, 1963.)

Fassett, Norman C.: 1957, *A Manual of Aquatic Plants*, University of Wisconsin Press, Madison, 405 pp.

Fay, Rimmon C.: 1970, *Living Marine Organisms for Teaching and Research*, Pacific Bio-Marine Supply Co., P.O. Box 536, Venice, Calif. 90291, 41 pp. A catalog of organisms with prices and instructions for maintenance of a marine system. Free to biology teachers.

Gallagher, John L.: 1967, "An Economical Cooling System for Aquaria," *Am. Biol. Teacher*, **29** (7): 535–536. A modification is described in which the laboratory refrigerator is used for cooling the water of a marine aquarium.

Knudsen, Jens W.: 1966, *Biological Techniques*, Harper & Row, Publishers, Incorporated, New York, 525 pp.

O'Neill, Robert V.: 1968, "A Simple Marine Aquarium System," *Turtox News*, **46** (5): 166–168.

Strickland, J. D. H., and T. R. Parsons: 1965, *A Manual of Seawater Analysis*, Bulletin 125, 2d ed., Fisheries Research Board of Canada, Ottawa, 203 pp.

Wheeler, G. C., and J. Wheeler: 1963, *The Ants of North Dakota*, University of North Dakota Press, Grand Forks, 326 pp.

Winterringer, Glen S., and A. C. Lopinot: 1966, *Aquatic Plants of Illinois*, Illinois State Museum, Popular Science Series, vol. VI, Springfield, Ill., 141 pp.

Other Literature

Atz, James W.: 1964, *Some Principles and Practices of Water Management for Marine Aquariums*, Research Report 63, Fish and Wildlife Services, Bureau of Sport Fisheries and Wildlife, Washington, D.C.

Bay, Ernest C.: 1967, "An Inexpensive Filter-Aquarium for Rearing and Experimenting with Aquatic Invertebrates," *Turtox News*, **45** (6): 146–148.

Collins, H. H.: 1959, *Complete Field Guide to American Wildlife*, Harper & Row, Publishers, Incorporated, New York. Contains over 2,000 illustrations.

Conalty, M. L. (ed.): 1967, *Husbandry of Laboratory Animals*, Academic Press, Inc., New York.

Diehl, Fred A., J. B. Feeley, and D. G. Gibson: 1971, *Experiment Using Marine Animals*, Aquarium Systems, Inc., 33208 Lakeland Blvd., Eastlake, Ohio 44094, 92 pp. $4.50. Contains 29 laboratory exercises suitable for college or high school. Includes methods for setting up a marine aquarium and instructions for ordering specimens.

Guberlet, Muriel L.: 1956, *Seaweeds at Ebb Tide*, University of Washington Press, Seattle, 182 pp. $2.95.

Masters, Charles O.: 1967, "Requirements for the Culturing of Sea Animals," *Am. Biol. Teacher*, **29** (7): 537–540.

Masters, Charles O.: 1968, "The Aquarium as a Living Organism," *Carolina Tips*, **31** (7): 17–18.

O'Connell, R. F.: 1969, *The Marine Aquarium*, Great Outdoors Publishing Co., St. Petersburg, Fla., 159 pp. $6.95.

Ricketts, E. F., and J. Calvin: 1968, *Between Pacific Tides*, 4th ed., revised by J. W. Hedgpeth, Stanford University Press, Stanford, Calif.

Ross, H. H.: 1966, *How to Collect and Preserve Insects*, Circular 39, Ill. State Natural History Survey, Urbana, Ill., 71 pp.

Rudloe, Jack J.: 1971, *The Erotic Ocean: A Handbook for Beachcombers*, The World Publishing Company, New York, $15.00. Includes how to set up and maintain a saltwater aquarium, and chapters on collecting, culturing and preserving selected invertebrate groups found along the Atlantic and Gulf shores.

Sainsbury, David: 1967, *Animal Health and Housing*, The Williams & Wilkins Company, Baltimore, 329 pp.

Short, D. J., and D. P. Woodnott (eds.): 1969, *A.T.A. Manual of Laboratory Animal Practice and Techniques*, 2d ed., Charles C Thomas, Springfield, Ill., 350 pp. $14.00. An excellent reference for general housing and care of laboratory animals.

Spotte, Stephen H.: 1971, *Fish and Invertebrate Culture: Water Management in Closed Systems*, Interscience Publishers, John Wiley & Sons, Inc., New York. $8.95. The author is Director, Aquarium of Niagara Falls, Inc. Written for fish culturists, experimental biologists, and amateur or professional aquarists.

Waters, John and Barbara Waters: 1967, *Salt-Water Aquariums*, Holiday House, Inc., New York.

Free Literature from Biological Suppliers

Carolina Tips. Newsletter from Carolina Biological Supply Co., Burlington, N.C. 27215. Issued once or twice monthly.

Laboratory Animal Digest. Newsletter from Ralston Purina Co., Checkerboard Square, St. Louis, Mo. 63188. Issued four times a year.

S-E-A Scope. Newsletter from Aquarium Systems, Inc., 33208 Lakeland Blvd., Eastlake, Ohio 44094.

Turtox News. Newsletter from General Biological, Inc., 8200 S. Hoyne Ave., Chicago, Ill. 60620.

Turtox Service Leaflets. A set of 60 leaflets from General Biological, Inc.

Ward's Bulletin. Newsletter from Ward's Natural Science Establishment, Inc., P.O. Box 1712, Rochester, N.Y. 14603.

Ward's Culture Leaflets. A set of 25 leaflets from Ward's Natural Science Establishment.

2

LABORATORY CARE OF LOWER INVERTEBRATES

Included in this chapter are the lower invertebrates considered most useful for the biology teaching laboratory. For each group, special information is given concerning (1) uses in a teaching laboratory, (2) descriptions of distinguishing traits, (3) sources and collecting techniques, and (4) culturing the animals within the laboratory. For marine animals, reference is made to Chap. 1 where the installation of a marine aquarium and the general treatment of saltwater animals are described. Chapter 1 also contains general information about collecting and purchasing animals. In particular, a person should refer to Chap. 1 for pertinent information on handling animals when they are received in the laboratory. A list of commercial sources for animals will be found in Appendix B. The nomenclature for animals of this chapter is based on Hyman, vol. I (1940), vol. II (1951), vol. III (1951), vol. VI (1967); and on Pennak (1953).

PORIFERA: THE SPONGES

Sponges are difficult to maintain in the laboratory, as they generally begin to disintegrate within several weeks. Possibly the lack of successful culture is due to the highly specific and variable requirements among species such as water pH and mineral content, as well as lack of sufficient knowledge regarding suitable food. Although recent reports describe success with some of these problems, the subject seems to provide opportunity for much further research.

Freshwater Sponges

USES: For studies of (1) cell differentiation, (2) evolutionary divergence, (3) regeneration, (4) gemmule reproduction, and (5) reaggregation. See Whitten (1969).

DESCRIPTION: All are characterized by a skeleton containing both silicon spicules and varying amounts of fibrous spongin, and by the formation of gemmules. Representatives of only one family, the Spongillidae (see Fig. 2-1), are recorded for fresh waters, and about 30 species in this family have been located in the United States. Pennak (1953) lists as the most common American species: *Spongilla lacustris*, *S. fragilis*, *Meyenia mülleri*, and *Heteromeyenia tubisperma*.

Sponges become encrusted on rocks, logs, or other substratum in clean, slow-moving water at depths of 2 m or less. The mature growth may vary from several square centimeters to approximately 40 m^2. Coloration may be dark brown or tan, gray or flesh-colored, and green if symbiotic algae are present. Many species mature and attain maximum seasonal growth by late autumn, at which time deterioration sets in with perhaps a total degeneration and death or, in the warmer climates, the formation of many small sponge remnants which resume growth the next spring.

Of particular interest are the small resistant and dormant gemmules, formed throughout the sponge body during the growing season. These are microscopic spheroids, ranging in diameter from 150 to 1,000 microns and varying in color from light beige to yellow or brown. A single gemmule consists of a spicule covering, beneath which are several layers of tough, dead cells and a center portion of amoeboid, mesenchymal cells (see Fig. 2-2). The germination of gemmules offers interesting possibilities for individual student study.

Figure 2-1 Freshwater sponges. **A** *Spongilla* sp., a living sponge encrusted on a twig in the water. **B** *Spongilla* sp. on wall of aquarium tank.

A

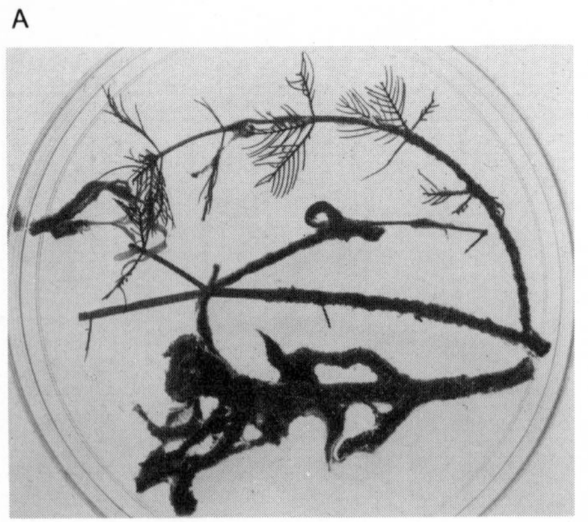

Courtesy of Carolina Biological Supply Company

B

Courtesy of Carolina Biological Supply Company

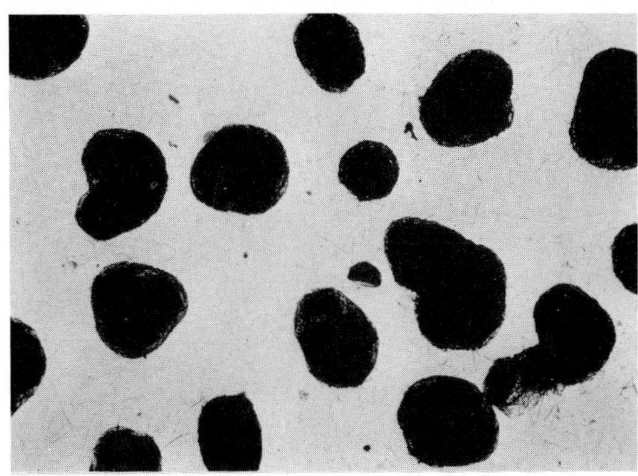

A *Courtesy of Carolina Biological Supply Company*

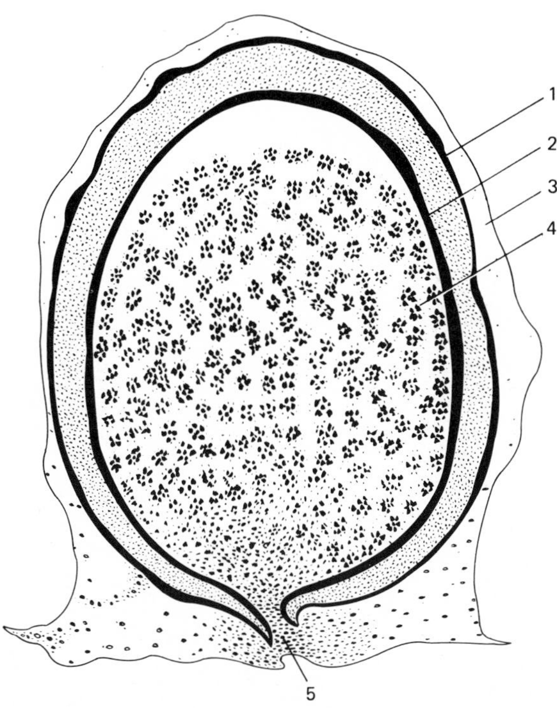

B

Figure 2-2 Gemmules of freshwater sponges. **A** Sponge gemmules. **B** A gemmule of the freshwater sponge, *Spongilla*, in process of hatching; 1, outer membrane of gemmule; 2, inner membrane of gemmule; 3, epidermis of young sponge; 4, food-laden archeocytes; 5, histoblasts streaming through opening of gemmule. *(B From The Invertebrates, vol. 1, Fig. 85 A, by Libbie Hyman, after Brien, 1932. Copyright 1940 by McGraw-Hill, Inc. Used with permission of the McGraw-Hill Book Company, New York.)*

SOURCES: Pick by hand from substratum in shallow water and by dredging in deeper water with a can, net, or rake. Freshwater sponges are extremely fragile and must be handled carefully. Transport in the natural pond water with algae and detritus from the collecting site. Several species can be purchased from biological supply companies listed in Appendix B.

CULTURE: Place the collection in an aerated container. The algae and bottom sediment serve as a food source. Generally, the animals degenerate within several weeks, although Pennak reports that *Meyenia fluviatilis* and species of *Spongilla* have been maintained for 5 to 6 months. Culturing has been more successful when small pieces of sponge (5 to 10 mm^2) are dropped onto a glass slide or other substratum where the cells may become attached within several hours. To do this, simply cut a sponge into small pieces or macerate it by pressing the body through a porous filtering cloth (for example, 25-mesh bolting cloth). Hold the slide in a dish of shallow water for several hours to allow the fragments to become attached. Then submerge the slide in water to a depth of about 5 cm and at a temperature of 15 to 23°C. Maintain the culture in running water, or change the water several times daily.

Reaggregation can be observed when small portions of freshly filtered cells are examined in drops of water on a slide under 100 to 400× magnification. Look for clumps of cells that move together and fuse. Small sponges may form in the dish within several weeks if the water is changed several times daily. The regeneration of a sponge can be observed under magnification during this time.

Saltwater Sponges

USES: For studies of (1) cell differentiation, (2) evolutionary divergence, (3) regeneration, and (4) aggregation of cells.

DESCRIPTION: Considerably more time has been devoted to the study of marine sponges than to freshwater forms, and approximately 5,000 species have been described. Smaller, suitable types for laboratory culture are *Scypha* (a sycon type formerly grouped with the *Grantia*), *Leucosolenia* (an ascon type), and *Haliclona* (finger sponge) (see Fig. 2-3).

SOURCES: A variety of small sponge types can be collected by hand from rocks, wharves, and other substrata of quiet water areas, or in the more turbulent intertidal zones. For inland laboratories, an adequate variety of hardy types is available from marine commercial suppliers (see Appendix B).

CULTURE: Maintenance of these animals is discussed in Chap. 1 under the title, *Marine Aquarium Systems.* See the section above on culture of freshwater sponges for techniques in observing regeneration and aggregation.

CNIDARIA (COELENTERATA)

Except for anemones and jellyfish polyps (scyphistoma), marine cnidarians are difficult to maintain. The freshwater hydra is used extensively in routine laboratory studies and in research.

A *Courtesy of Carolina Biological Supply Company*

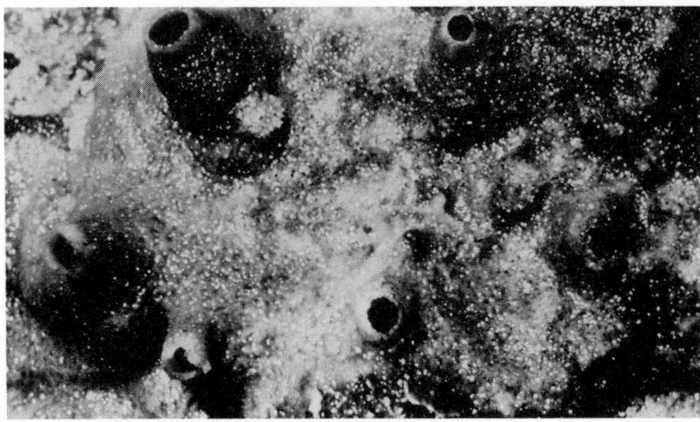

B *Courtesy of Carolina Biological Supply Company*

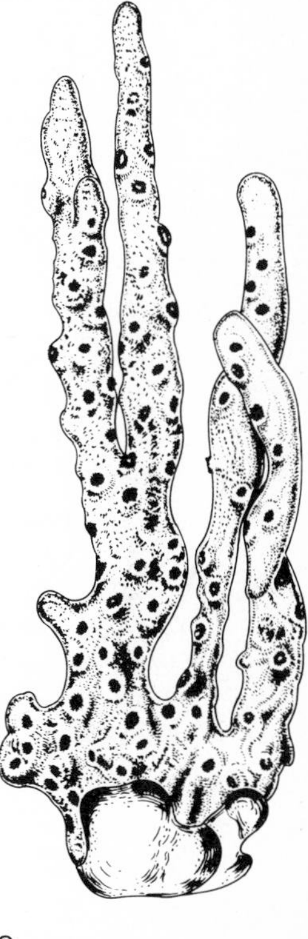

C

Figure 2-3 Saltwater sponges. **A** *Scypha* sp. **B** *Haliclona permollis* encrusted on rock. **C** *Haliclona* sp., dry specimen. *(C From The Invertebrates, vol. 1, Fig. 103 F, by Libbie Hyman. Copyright 1940 by McGraw-Hill, Inc. Used with permission of the McGraw-Hill Book Company, New York.)*

Hydrozoa: The Hydra

USES: (1) As food for larger aquatic invertebrates; (2) to observe locomotion, feeding, use of tentacles and nematocysts, budding, and regeneration. See Moog (1963), Chalkey and Park (1947), Loomis (1954), Loomis and Lenhoff (1956), and Galen (1969). In particular, see Lenhoff and Loomis (1961) for descriptions of various types of research with hydra.

DESCRIPTION: Species used commonly in the laboratory are *Pelmatohydra oligactis*, the "brown hydra," varying in color from light gray to brown and at times assuming the color of its food; and *Chlorohydra viridissima*, the "green hydra," which owes its color to the presence of symbiotic zoochlorellae. Both species are found in slow-moving streams or still ponds over most of the United States. A third species, *Hydra littoralis*, is an excellent organism for laboratory study but is more difficult to maintain. This species lives in swift-moving water, where it may form large masses of yellow or orange jelly on concrete or on the underside of rocks (see Fig. 2-4).

SOURCES: Look for hydra on plants, stones, logs, or debris in the water. Place the plants and other materials in a container of water taken from the collecting site. After 30 to 60 min, use a hand lens to look for expanded animals on the walls or bottom surface of the container. Also, they may be attached to the collected material and often will be seen hanging from the under surface of floating plants. Once a good collecting spot is found, generally animals can be collected at the same location year after year unless unusual conditions disturb the site. For short-time class studies, obtain animals 1 or 2 days before required, and retain the collections in the pond water or in a synthetic medium. Pure cultures are available from commercial suppliers. According to reports, commercial shipments sometimes arrive in a depressed condition and must be transferred immediately to fresh pond water or synthetic medium. We have not experienced this situation, however, and often retain commercial cultures for several days in the shipment bottles at 20 to 22°C.

Note: When hydra (and all other small aquatic animals) are received in the laboratory, remove the bottle cap immediately and gently aerate the culture by drawing the water in and out of the bottle with a medicine dropper. Cover loosely with the bottle cap or with punctured plastic film, thus giving the animals oxygen and at the same time retarding evaporation.

CULTURE: For successful culture, hydra require diligent care and considerable patience. They feed on live food, which must be maintained in sufficient quantity. Apparently all cultures eventually undergo a depression, when the animals shrink and disintegrate. Occasionally they can be saved if transferred to a fresh bowl of water. Possibly the depressions result from a rise in temperature, overfeeding, fouling of water, or aging of a clone. Sometimes, however, a depression occurs for no apparent reason. Conversely, animals sometimes become well established for an extended period, almost by chance it seems, in an aquarium containing plants and smaller invertebrates for food. The development of successful technique for culturing hydra remains a challenge.

Aquarium culture. Place animals in a small tank or widemouthed jar containing pond water; add aquatic plants and put under diffused light at 20 to 22°C. Aeration is debatable and may not be necessary if sufficient plants are present. Unless a large quantity of smaller invertebrates are growing in the container, add living animals as food at least three or four times weekly. The container must be small enough for hydras to come into contact with the food. Our success with aquarium cultures has been irregular and unpredictable, although we have observed situations where large numbers of hydra were maintained in synthetic medium for at least several months in unaerated, shallow, glass baking dishes. Brine shrimp were fed several times weekly.

Synthetic medium culture. Our experience indicates that the most reliable method is to put the animals in a synthetic medium and to maintain cultures of small invertebrates for food. Loomis (1953) describes a synthetic medium used in his laboratory for *Hydra littoralis*. The formula is sometimes named Littoralis me-

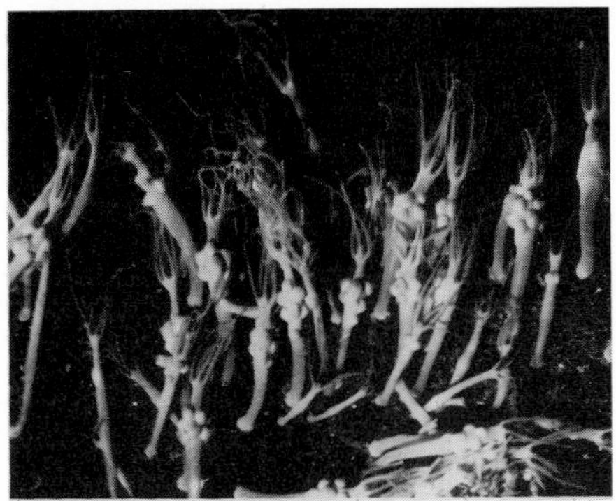

A *Courtesy of Carolina Biological Supply Company*

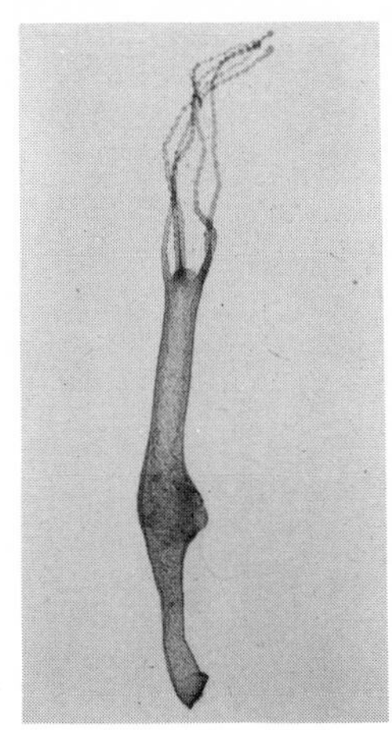

B *Courtesy of Carolina Biological Supply Company*

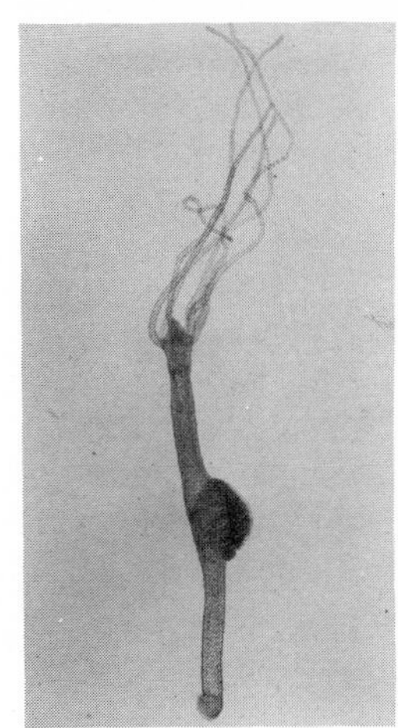

C *Courtesy of Carolina Biological Supply Company*

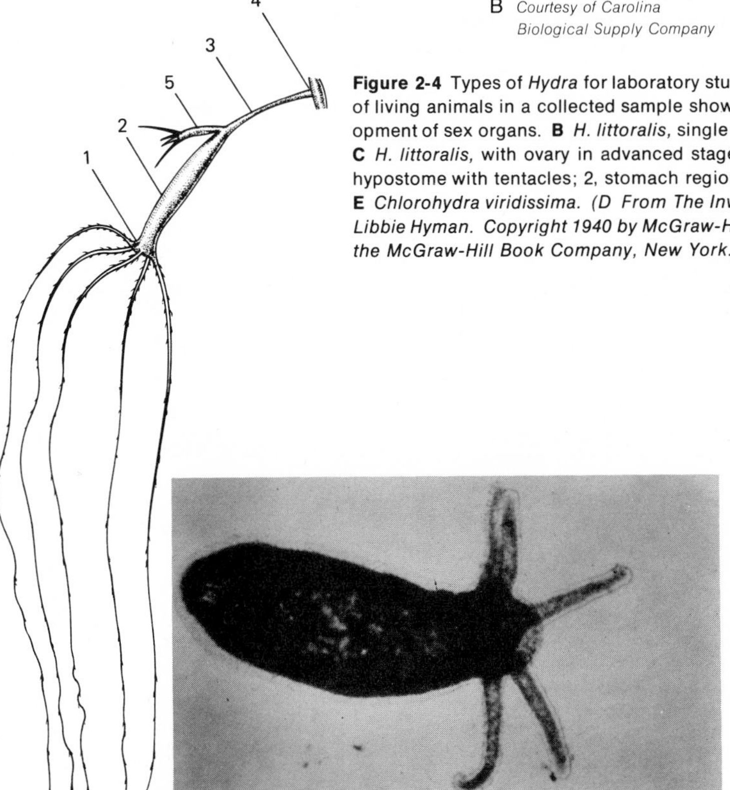

D

E *Courtesy of Carolina Biological Supply Company*

Figure 2-4 Types of *Hydra* for laboratory studies. **A** *Hydra littoralis,* a group of living animals in a collected sample showing a high percentage of development of sex organs. **B** *H. littoralis,* single animal with ovary in early stage. **C** *H. littoralis,* with ovary in advanced stage. **D** *Pelmatohydra oligactis;* 1, hypostome with tentacles; 2, stomach region; 3, stalk; 4, pedal disk; 5, bud. **E** *Chlorohydra viridissima.* *(D From The Invertebrates, vol. 1, Fig. 132 D, by Libbie Hyman. Copyright 1940 by McGraw-Hill, Inc. Used with permission of the McGraw-Hill Book Company, New York.)*

dium or L. medium. The following medium of Moog (1963) has always given us satisfactory results:

Stock Solution A	Stock Solution B
33.3 g $CaCl_2$	25 g $NaH_2PO_4 \cdot H_2O$
7.5 g KCl	112 g Na_2HPO_4
0.8 g KI	2 g disodium versonate
1 l distilled H_2O	1 l distilled H_2O

Add 1 ml of each stock solution (separately) to 1 l of distilled water. Do not combine solutions A and B before adding them to the water, as certain components will precipitate out. The two stock solutions will keep indefinitely if refrigerated to prevent mold growth.

Pour the medium into bowls to a depth of about 1 in and add hydra. Place the bowls in diffused light (or darkness) at a temperature of 20 to 22°C. For optimum growth, feed daily with *Daphnia*, *Aeolosoma*, or *Artemia*. Methods for culturing *Daphnia* and *Aeolosoma* are described later within their respective taxonomic groups. Techniques for growing *Artemia* (brine shrimp) are given below.

Culturing Artemia *for hydra food.* Add sodium chloride or uniodized table salt to a flat bowl containing about 1 in of dechlorinated tap water or distilled water. The salt concentration is not critical and can vary from 1 to 6%. Generally, about 1 tablespoon of salt is adequate for each pint of water. Stir the water until all salt dissolves. Sprinkle over the water surface a small quantity of dried *Artemia* eggs (from an aquarium or biology supply company). Stir the eggs gently to wet them completely. Do not move the bowl or stir the water roughly as eggs will wash onto the bowl wall and will not hatch. Hatching occurs in 24 to 48 h, at which time the small, orange-colored larvae (a nauplius) will be seen swimming at the top edge of the water and on the bottom surface. Two bowls of *Artemia* will feed 40 to 50 hydra, if fresh cultures are started in alternating bowls every 2 or 3 days. Generally, each bowl will furnish live larvae for several days.

To feed hydra, draw the brine shrimp into a medicine dropper and transfer them to a piece of fine-mesh cloth, such as a clean handkerchief, suspended over the opening of a small beaker. Run distilled water or the synthetic medium slowly over the cluster of shrimp and the surrounding areas to wash the salt water into the beaker below. Then place a finger under the cluster of shrimp and invert the cloth into the culture of hydra at several spots. After the hydra have fed, a period of 30 to 60 min, slowly drain all solution from the hydra bowls, taking care to retrieve any hydra that float out into the waste container. Most of the animals will cling to the bottom surface as the solution is removed. Wash the bowl and hydra with several changes of the synthetic medium until all shrimp are removed. A medicine dropper is effective for flushing the shrimp from the bowl. Although tedious at first, changing the solution and washing the bowl become simple operations after only a little practice.

Scyphozoa: The Aurelia Jellyfish

USES: Excellent for studying reproductive cycles of the asexual polyps and the sexual medusae (see Custance, 1967).

DESCRIPTION: *Aurelia aurita* (the "white sea jelly" or "moon jelly") is one of the most common jellyfish of the United States coastal waters (see Fig. 2-5). The dominant stage is the sexually reproducing medusa, with a flattened umbrella shape, with many short tentacles on the notched margin, and with the corners of the mouth extended into long, frilly oral arms that hang downward. Mature animals are 20 to 40 cm in diameter; the abundant nematocysts on the tentacles produce a painful sting to man. The *Aurelia* medusae are constant swimmers; most species stay in the upper waters. During summer months they may collect in great numbers at the water surface of bays, harbors, and estuaries.

Much variation in reproduction and development exists among jellyfish. In *Aurelia*, the medusa (jellyfish stage) is sexually separate. The sperm are carried by water currents into the female body where the eggs are fertilized and the larvae (planulas) develop. The planulas move to the outside where each may become attached to an object. In time, a planula develops into a polyp called a scyphistoma. Several scyphistomas may bud off of one central tube. Eventually, the scyphistoma may produce medusae by strobilation, a process in which the tubelike scyphistoma constricts into a number of saucer-shaped segments that develop and become separated into individual young jellyfish, each called an ephyra, which develops into a typical medusa. (See Fig. 2-6.)

SOURCES: Catch the small swimming medusae in the open water with a bucket

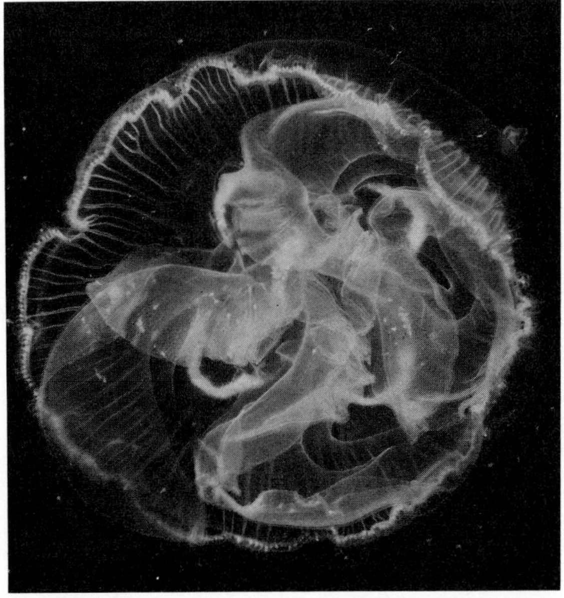

Figure 2-5 *Aurelia* sp., a mature medusa. *(By Robert Wolloch and Dorothy Spangenberg.)*

Figure 2-6 *Aurelia* sp. The polyp stage showing four scyphistomas budding from a lower, central tube. Advanced strobilation has occurred in two scyphistomas that have constricted into saucer-shaped segments, which break loose and develop into medusae. *(By Dorothy Spangenberg and Clarence M. Flaten.)*

or a hand net. Remove polyps from wharf pilings and from rocks of intertidal zones. Immediately place animals in a glass or plastic container of seawater. Because of their soft bodies, the animals must be handled gently; crowding in a container must be avoided. Transfer at once to a marine aquarium in the laboratory. Both the medusae and polyps can be purchased from commercial suppliers, although medusae do not ship well. (See Appendix B for commercial sources.)

CULTURE: Medusae are difficult to maintain, although small ones may live for several weeks under proper conditions. Polyps (the scyphistoma) are maintained rather easily and can provide valuable studies of asexual reproduction. Their maintenance is described in Chap. 1 within the discussion of the marine aquarium. Also see Spangenberg and Flaten (1967) and Custance (1967). Hill and Cather (1969) describe methods for culturing *Cassiopeia*, another Scyphozoa.

Anthozoa: Sea Anemones

USES: To study (1) regeneration, (2) feeding, (3) symbiotic algae, (4) extracellular digestion, and (5) firing of nematocysts. Exercises for these studies are described in the laboratory manual of Diehl et al. (1971). Also see Belcik (1968).

DESCRIPTION: In addition to anemones, Anthozoa include the corals, sea fans, sea feathers, and sea pens. All are marine polyps, either solitary or colonial, with one or more circles of tentacles on an oral disc. All are carnivorous. Suitable anemones for laboratory culture are the hardy cold-water *Metridium*, ubiquitous to the Atlantic and Pacific coastal waters; *Anthopleura*, found on intertidal rocks

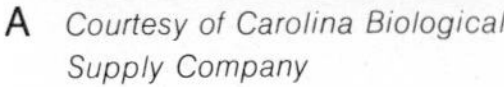

A Courtesy of Carolina Biological Supply Company

B

Figure 2-7 Anemones. **A** *Metridium marginatum*, from the Atlantic coastal waters. **B** *Anthopleura* sp., a Pacific tidepool anemone of green color due to symbiotic algae. *(B Courtesy of Marjorie Holstein.)*

of the Pacific coast; *Sagartia*, a burrowing sand anemone of both coasts; and *Condylactis* from the Florida and Gulf coasts (see Fig. 2-7).

SOURCES: Use a dull blade to remove an attached animal, or take portions of the substratum with the animal. Dig in sandy bottoms for burrowing anemones. Hardy types can be purchased from commercial suppliers (see Appendix B).

CULTURE: Small anemones can be maintained easily for long periods in natural or artificial seawater. See the discussion of the marine aquarium in Chap. 1 for their maintenance. Feed once or twice weekly by dropping onto the oral disc small portions of raw meat, such as fish, oysters, shrimp, beef, and earthworms. Do not overfeed. Remove all uneaten food within several hours.

PLATYHELMINTHES: TURBELLARIA (FREE-LIVING FLATWORMS)

The phylum Platyhelminthes consists of three groups: class Cestoda (tape worms), class Trematoda (flukes), and class Turbellaria (planarians and rhabdocoels). The first two classes are entirely parasitic; the last class is almost entirely free-living. The turbellarians include a great many marine and freshwater species and a few terrestrial species. The land planarians are not found often in the general biology laboratory, although they can serve as useful and interesting supplementary animals to the freshwater planarians.

Freshwater Planarians

USES: For studying (1) anatomy, (2) locomotion, (3) feeding (see Forrest, 1963), (4) regeneration (see Rose and McRae, 1969), and (5) tests for learning (see Jacobson and McConnell, 1962; and Jacobson, 1963).

DESCRIPTION: Most often freshwater planarians are identified simply as

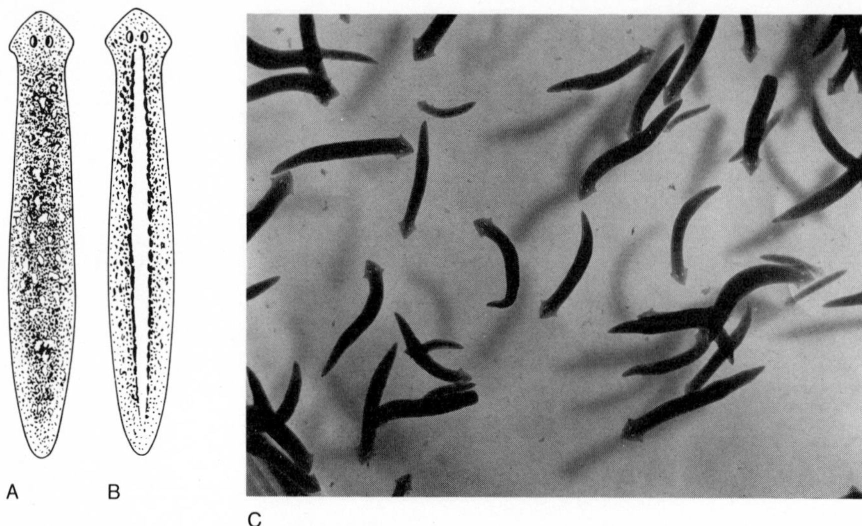

Figure 2-8 Planarians. **A** and **B** Color variants of *Dugesia tigrina*. **C** *Dugesia dorotocephala* in culture bowl. **D** *Polycelis coronata*, head region showing arrangement of eyes around dorsoanterior region. **E** *Phagocata gracilis* with multiple pharynges, from life, Pennsylvania. *(A, B, D, and E From The Invertebrates, vol. 2, by Libbie Hyman. Copyright 1951 by McGraw-Hill, Inc. Used with permission of the McGraw-Hill Book Company, New York.)*

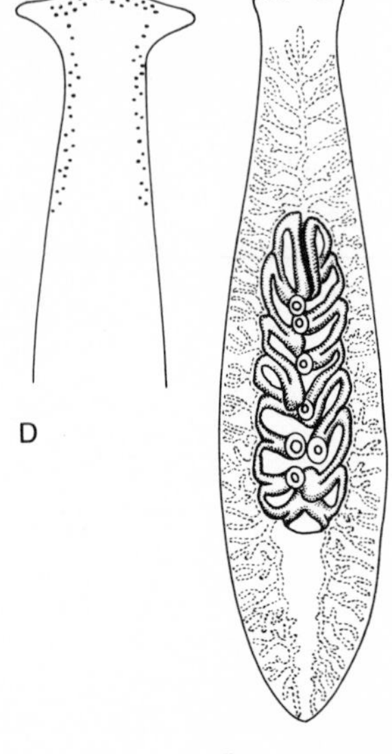

brown, black, or white types. Actually, the color and pigment pattern vary considerably, sometimes within the same species. Types used most often for laboratory studies include the following species (see Fig. 2-8):

1. *Dugesia tigrina*, the "brown" planarian; probably the most common species of the United States; color varies and may be gray, yellow, olive, or brown; may grow to 20 mm length; lives in slow-moving or still water.

2. *Dugesia dorotocephala*, the "black" planarian; most often in spring water; dark brown to black color; up to 30 mm long.

3. *Polycelis coronata* lives in mountain streams of probably most Western states; brown to black color; up to 20 mm long; many eyes scattered over the anterodorsal surface.

4. *Phagocata*. Many are white, including *P. morgani* in spring water of Eastern states and *P. oregonensis* in streams of Western coastal states (supplied by Carolina Biological Company from their Oregon office).

SOURCES: Obtain animals by taking bottom sediment and plants at the water edge or by looking for animals on rocks and other objects from shallow water. After 10 to 15 min, planarians can be seen moving on the glass wall of a container. Once a location is found, the animals often can be collected regularly from the same general area. Baiting planarians with small chunks of liver is described in much literature; we have had limited success with this method, and then only when searching in running water.

CULTURE: The animals are hardy and are cultured easily. Put them in a glass or enamel bowl containing natural pond water or dechlorinated tap water. (An easy method for conditioning tap water is to store it in an open jar which is refilled each time water is supplied to the culture bowl.) Cover the bowl loosely with a piece of cardboard to retard evaporation and to diffuse the light. If the bowl is left in open light, provide rocks for shelter. Optimum temperature is 20 to 22°C; if room temperature goes above 22°C, animals can be retained for 1 or 2 weeks on the top shelf of a refrigerator at 8 to 10°C. We have kept planarians in the refrigerator for as long as 3 weeks with 100 percent survival but with no growth. On several occasions, animals in the refrigerator have fragmented at an unusually high rate, with the fragments regenerating into small worms (an aspect that might well be explored).

For maximum growth, feed at least every other day. Planarians remain vigorous and healthy when fed only raw beef liver; the liver of most other animals is not suitable. Remove the liver within several hours and brush any attached worms back into the bowl. Rinse the bowl and animals with several changes of fresh water, taking care not to wash animals out of the bowl. Alternatively, animals can be sustained indefinitely when fed hard-boiled egg yolk three to four times weekly, supplemented occasionally with beef liver or liver meal from an aquarium supply company. Drain off water and egg yolk within several hours, and rinse until all egg particles are removed. Our experience indicates that dark-pigmented animals become a lighter color if retained on an egg diet for several days, a desirable feature when studying the internal structures.

At least once each week wipe the bottom and walls of the bowl to remove slime formed by animals. Detergents and other chemicals must be avoided. Aeration is not necessary if the water is changed every other day. As a rule, the animals will survive without food or a change of water for several weeks, although they shrink in size and eventually die if their care is not resumed.

Terrestrial Planarians

USES: Because of a relatively great length, the animal is particularly useful for studying regeneration gradients.

DESCRIPTION: These are the giants of planarians, with some tropical species reaching a length of 60 cm. When relaxed, the head assumes a broad, flat "cobra-snake" shape; when contracted, the head becomes a rounded anterior portion. Most species are tropical. *Bipalium kewense* (about 20 to 25 cm long) is a common species of the temperate zone (see Fig. 2-9). *Bipalium kewense* shows a high rate of regeneration, possibly higher than that of the more commonly used *Dugesia*.

SOURCES: The animals, commonly called "greenhouse worms," are found in rich soil of gardens and among flower pots within greenhouses. They occur in temperate climates over the United States. No commercial suppliers are known.

CULTURE: Put animals in a bowl, 3 to 4 in in diameter, containing about 1 in of rich garden soil, covered with a 1-in layer of dead leaves. Keep the soil and leaves moist but avoid an accumulation of standing water in the lower soil. Although animals feed on dead leaves and decaying humus material, it is well to feed several times weekly with oatmeal, cornflakes, or similar food, all soaked in milk. After 24 h, remove any uneaten portions to avoid mold growth.

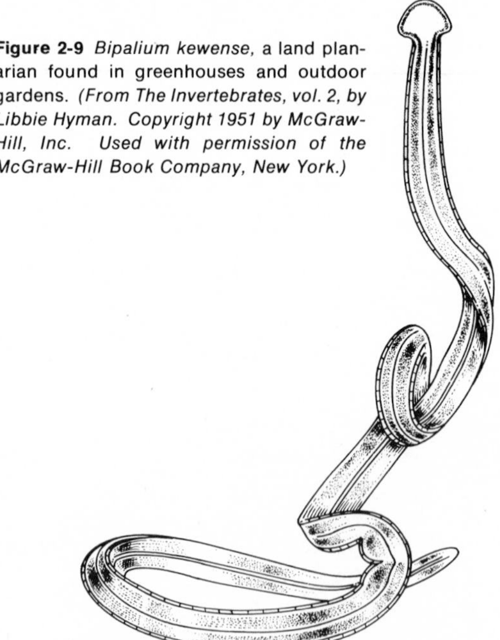

Figure 2-9 *Bipalium kewense,* a land planarian found in greenhouses and outdoor gardens. *(From The Invertebrates, vol. 2, by Libbie Hyman. Copyright 1951 by McGraw-Hill, Inc. Used with permission of the McGraw-Hill Book Company, New York.)*

THE FREE-LIVING NEMATODES (PHYLUM ASCHELMINTHES)

The class Nematoda and also other classes of the Aschelminthes are sometimes placed in separate phyla. Except for the insects, nematodes comprise perhaps the largest group of organisms in both number of individuals and number of species. They are among the most ubiquitous of all animals, occurring as plant and animal parasites and as free-living organisms in fresh water, marine water, and soil. Almost without exception, the free-living forms will appear in collections of algae and bottom sediment; rarely do they fail to develop in old hay infusions.

In spite of their abundance, little attention has been given in general biology courses to the culture and laboratory study of nematodes. The two types discussed here are obtained easily and may be used profitably for routine class observations or for individual student problems.

Turbatrix aceti (Vinegar Eel)

USES: Because of the transparency of its body, the animal is excellent for many kinds of studies, such as observing embryonic development and birth of young, digestion, and excretion. Other appropriate studies are adaptation to hydrogen-ion concentrations, reproductive rates, and taxic responses. See Behringer (1967) and Galen (1971).

DESCRIPTION: Elongated, cylindrical, and transparent body. Blunt anterior end; body tapers to a point at posterior end. Adult female is approximately 2 mm long; male is somewhat shorter. A rapid lashing movement is due to the presence of longitudinal muscles and no circular muscles. Ovoviviparous (see Fig. 2-10).

SOURCES: The animals may be found in tanks at a vinegar or pickling factory. Sometimes they appear in the sediment of cider vinegar that is untreated to inhibit bacteria and yeast. The worms feed on the sediment, often called "mother of vinegar." Generally, the simplest method of obtaining animals is to purchase a starting culture from a biology supply house.

CULTURE: Divide the starting culture into two bowls, each bowl containing about 1 l of nonpasteurized cider vinegar. The cheapest grade of pickling vinegar is satisfactory and can be purchased in large volume at many grocery stores. Cover the bowls with plastic film to retard evaporation and to reduce the vinegar odor. Punch 10 to 15 needle holes in the film. The population will become very dense within 2 to 3 weeks, at which time transfer a small portion of the culture with some of the bottom sediment to a fresh bowl of vinegar. The animals can be maintained indefinitely when vinegar is added to replace for evaporation and when fresh cultures are started every few weeks.

Soil Nematodes

USES: (1) Ecological studies of populations and adaptations. (2) Many of the same studies as for the vinegar eel, although soil nematodes are smaller and more difficult to observe. Also see Pramer (1964).

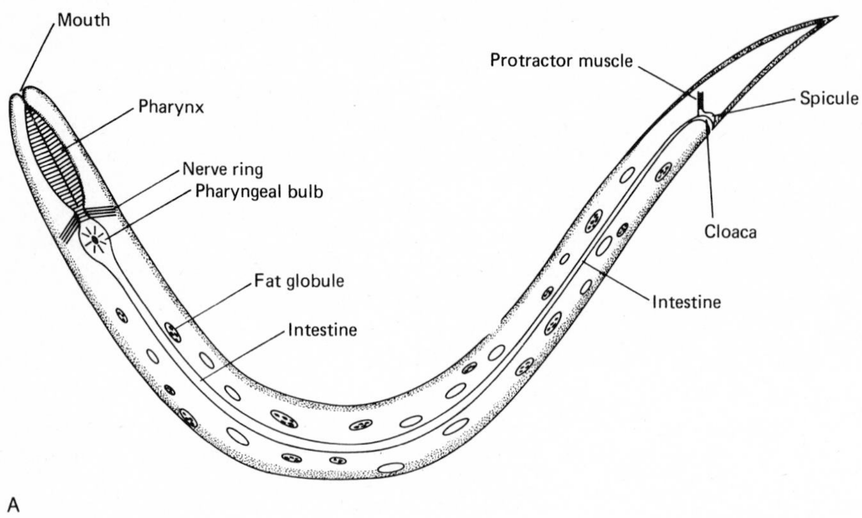

A

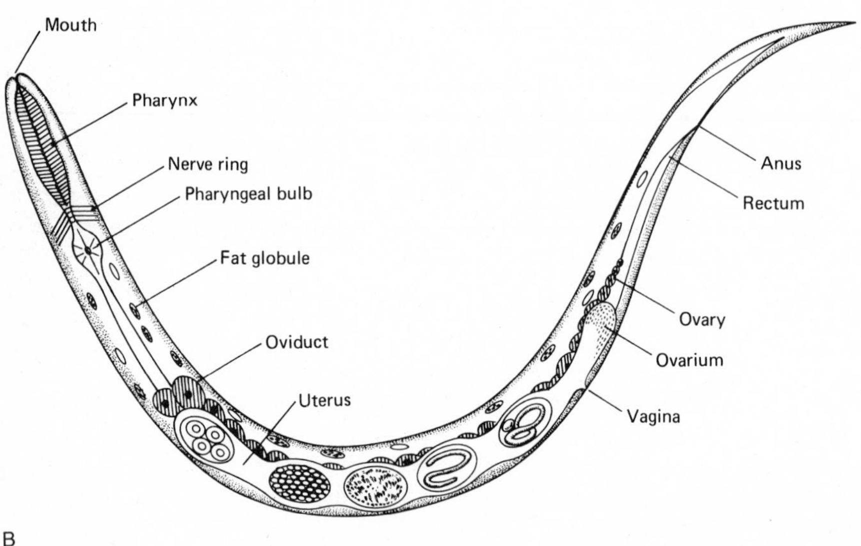

B

DESCRIPTION: Of same general body form as the vinegar eel; 0.2 to 0.5 mm long; transparent; sexually separate, and oviparous (see Fig. 2-11).

SOURCES: The organisms are found in soil containing decaying organic matter

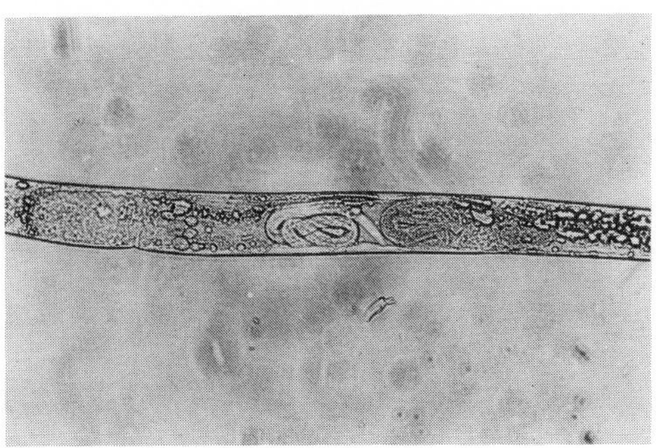

C

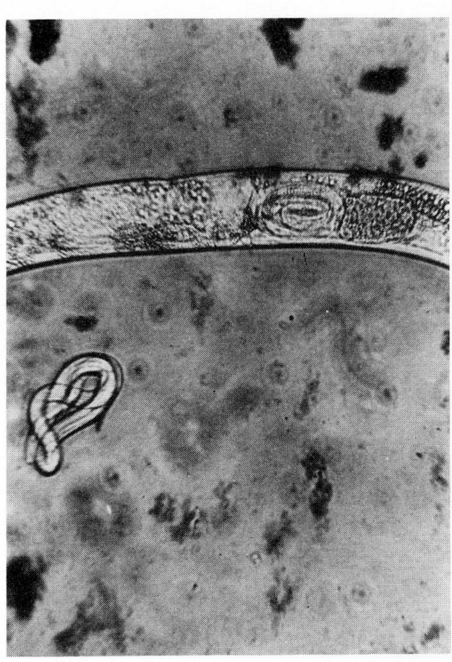

D

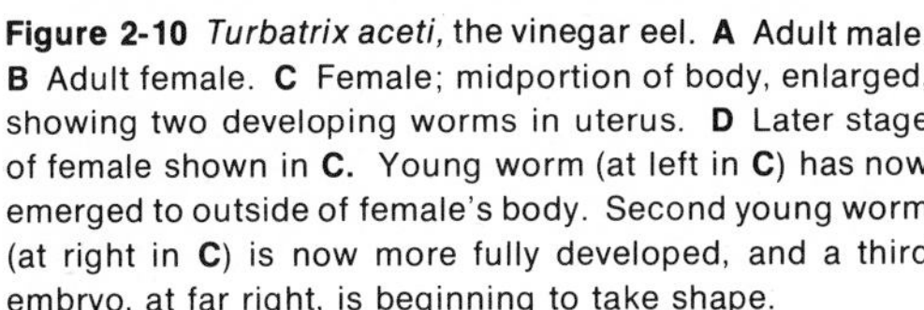

Figure 2-10 *Turbatrix aceti,* the vinegar eel. **A** Adult male. **B** Adult female. **C** Female; midportion of body, enlarged, showing two developing worms in uterus. **D** Later stage of female shown in **C.** Young worm (at left in **C**) has now emerged to outside of female's body. Second young worm (at right in **C**) is now more fully developed, and a third embryo, at far right, is beginning to take shape.

(manured soil is best) or in other decaying matter. Starting cultures of several genera can be obtained from biology supply houses (see Appendix B).

Separation from soil, Method 1. Construct a Baermann funnel as shown in Fig. 2-12. Place a small soil sample in a layer of gauze or in a fine sieve suspended over a beaker or test tube. Slowly add warm water over the soil; allow the soil to remain in place for several hours or longer until saturated. If nematodes are present, they wash through the gauze and collect at the bottom of the beaker. Examine drops of runoff water with the microscope. If many organisms are present, they may appear in the beaker water within minutes. Otherwise, allow the soil to remain undisturbed for at least 24 h before discarding the sample. Keep the soil moist by adding small amounts of water as necessary.

Separation from soil, Method 2. Soil nematodes can be obtained by burying a piece of potato, about 1½ inches square, several inches under rich soil. If present in the soil, nematodes will collect on the potato in 1 to 2 weeks and can be transferred to the potato medium described below.

CULTURE: Transfer several drops of nematodes to a piece of potato that has been sterilized in a petri dish or in a cotton-plugged test tube. Place at 20 to 25°C. A dense population should develop within 2 to 3 weeks. Alternatively, Pramer (1964) suggests placing several drops of organisms on salt agar prepared by adding 0.5 g NaCl and 1.5 g plain agar to each 100 ml tap water. The animal's movement and the internal organs can be observed when the plate is put under a microscope.

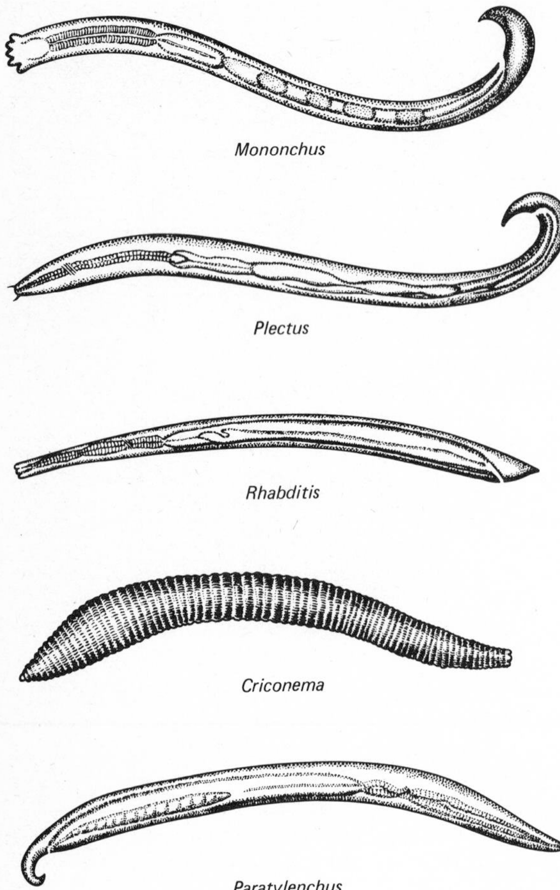

Figure 2-11 Typical soil nematodes. *(From David Pramer, Life in the Soil, Biological Sciences Curriculum Study.)*

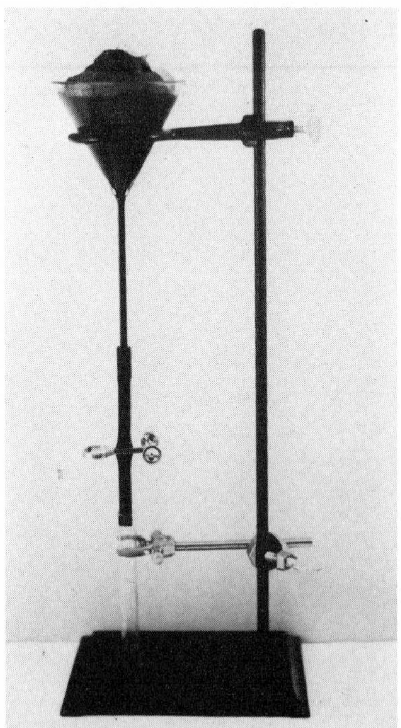

Figure 2-12 Baermann funnel for separation of nematodes from soil. A plastic tube empties into the test tube at bottom. Warm water poured over the soil sample on screen at top washes nematodes through screen into lower test tube.

ANNELIDA: THE SEGMENTED WORMS

Major classes of the phylum Annelida are Polychaeta, largely marine (*Nereis*); Oligochaeta, mostly freshwater and terrestrial (*Aeolosoma, Tubifex, Enchytraeus, Lumbricus*); and Hirudinea, the leeches, mostly freshwater parasites. The preserved *Lumbricus* long has been a classical laboratory organism; only recently has the living earthworm become recognized as a valuable organism for routine course work. Additionally, certain annelids are useful as food for other laboratory animals.

Aeolosoma (Freshwater Oligochaetes, "Spotted Worms")

USES: Particularly useful for studying fission of posterior segments. Transparency allows observation of internal organs. Feed to hydra by putting small worms or pieces of larger worms on hydra tentacles. Lonert (1967) describes regeneration studies suitable for general course work.

DESCRIPTION: Freshwater, segmented worm. Generally 1 to 3 mm long. Many have pigmented globules (red, yellow, or green) in epithelium. Ventral cilia and sometimes lateral cilia on prostomium. Four bundles of setae on most segments, two bundles at dorsolateral positions and two ventrolateral. Hermaphroditic. Generally reproduction is by budding from a posterior segment (see Fig. 2-13).

SOURCES: Found on bottom mud and plants or on floating plants of shallow water. Collect a portion of the bottom mud and plants with the pond water. Obtain pure cultures from a biology supply house.

CULTURE: *Method 1.* Place the natural substratum and several floating plants in a small container of pond water. Locate in a lighted spot during at least 8 to 10 h daily to allow aeration by the green plants. Replace evaporated water with pond water or conditioned tap water. *Method 2.* For pure strains, prepare a medium by boiling four to six grains of wheat, rice, or rye for 10 min in 300 ml of tap water. Add about one-half of a washed and shredded lettuce leaf. Allow medium to stand for 2 to 3 days for bacteria to accumulate before adding animals. Place in diffused light at 20 to 22°C. Gentle aeration is desirable unless the container provides a relatively large water surface. Each week add small portions (approximately 10 ml dry volume for each 500 ml of medium) of washed and shredded green plant food, such as lettuce, algae, or spinach. Allow plant material to remain in the water and to decay. Excessive decayed organic matter will be controlled if the medium is changed every 3 to 4 weeks. Change the water by slowly pouring off all but the bottom ¼- to ½-in layer and adding fresh medium prepared as before.

Tubifex (Freshwater Oligochaetes, "Tube Worms")

USES: Food for hydras, planarians, leeches, aquatic insect nymphs, and fish. Before feeding, wash mud from *Tubifex*, cut them into appropriate-sized portions, and place in contact with the animal. Hydra must be hand-fed by putting small pieces of *Tubifex* on their tentacles. Useful for studies of regeneration and adaptations for living in polluted waters.

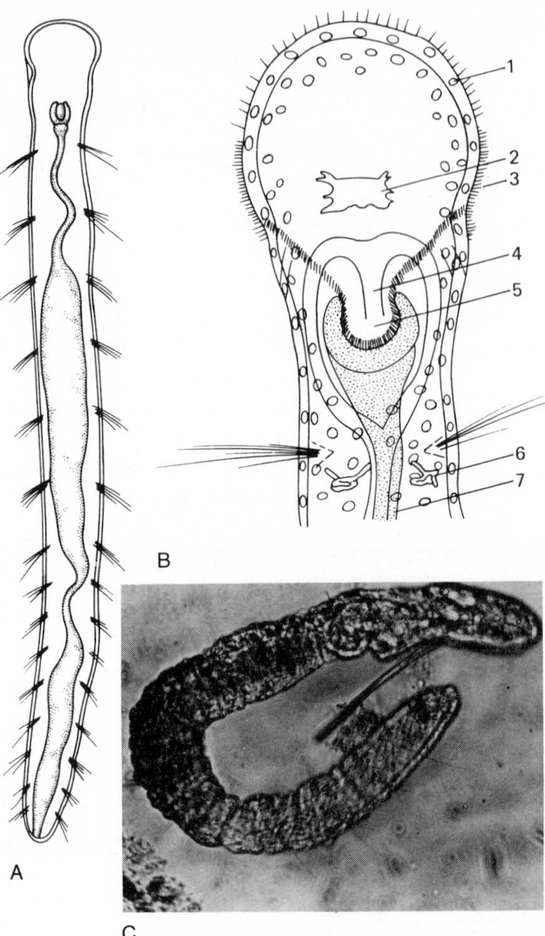

Figure 2-13 *Aeolosoma.* **A** *Aeolosoma leidyi,* dorsal view. Note formation of posterior zooid behind zone of fission. **B** *A. leidyi,* ventral view of anterior end; 1, pigment globules; 2, brain; 3, ciliated pit; 4, dorsal blood vessel; 5, mouth; 6, nephridium; 7, esophagus. **C** *Aeolosoma* sp., live animal swimming through water. **D** *Aeolosoma* sp., head region, showing pigment globules of live, swimming animal; spines are barely discernible. **E** *Aeolosoma* sp., posterior region of moving animal. *(A and B Based on Fresh-Water Invertebrates of the United States by Robert W. Pennak, modified from Kenk. Copyright 1953, The Ronald Press Company, New York.)*

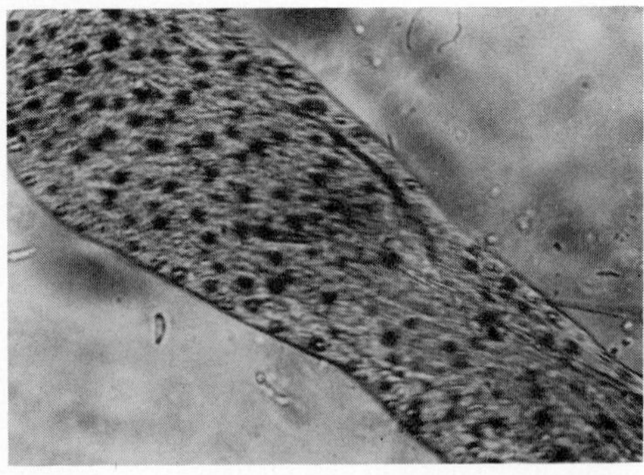

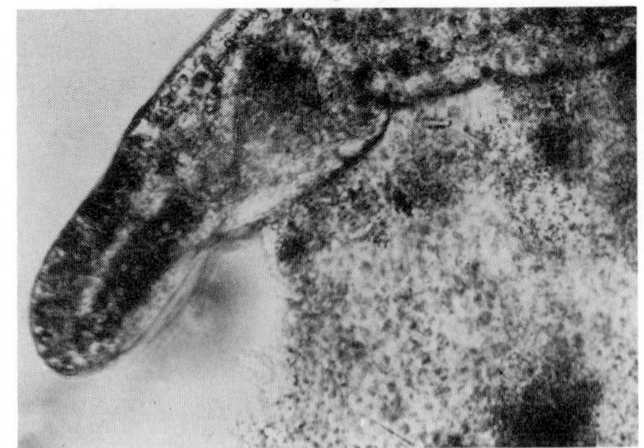

DESCRIPTION: Red or brown color; 30 to 100 mm long. Four bundles of setae on most segments, two bundles dorsolateral, two bundles ventrolateral. Most tubificids build mud tubes on substratum or in mud, where they bury anterior section at lower end of tube and wave posterior end out of tube opening for aeration (see Fig. 2-14). Hermaphroditic with cross-fertilization. Cocoons of developing embryos deposited on substratum during late summer and autumn.

SOURCES: Found in mud bottoms of rivers and lakes; sometimes very abundant in polluted waters and sewage tanks, where they may form thick, waving clumps. Obtain pure cultures from biology and aquarium supply houses. See Appendix B.

CULTURE: Put animals on a 1- to 2-in layer of mud of high organic content. Add pond water or conditioned tap water to depth of 4 to 6 in and leave at room temperature. The animals are hardy and survive under stagnant conditions with a relatively low oxygen concentration and at temperatures of 15 to 25°C. Every 1 to 2 weeks, add small portions of boiled lettuce or spinach leaves (boiling softens the leaves); bread, oatmeal, or other dry cereal soaked in milk; chopped, cooked meat; or boiled egg. Avoid excessive stagnation by subculturing every 4 to 6 weeks or by removing excess bottom debris and replacing a portion of the water.

Enchytraeus (Freshwater Oligochaetes, "White Worms")

USES: Food for fish, amphibians, lizards, small snakes, and turtles.

DESCRIPTION: White to pink color; 10 to 30 mm long. Four bundles of setae, two bundles dorsolateral, two bundles ventrolateral. Hermaphroditic with cross-fertilization; cocoons of developing young deposited on substratum during warm months outdoors and during most of the year indoors (see Fig. 2-15).

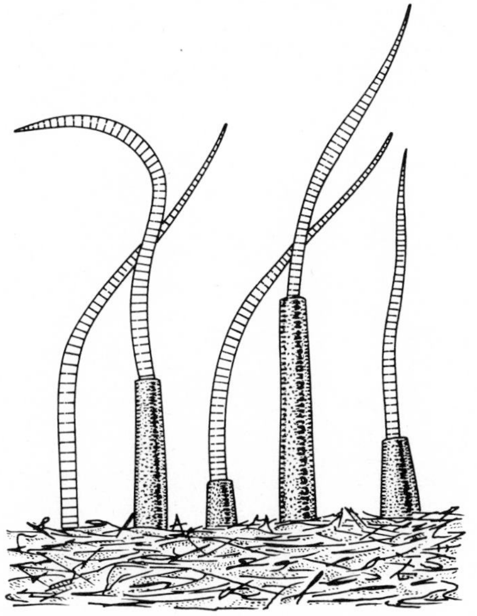

Figure 2-14 *Tubifex,* with posterior ends waving out of mud tubes built by the animals. *(Based on Fresh-Water Invertebrates of the United States by Robert W. Pennak. Copyright 1953, The Ronald Press Company, New York.)*

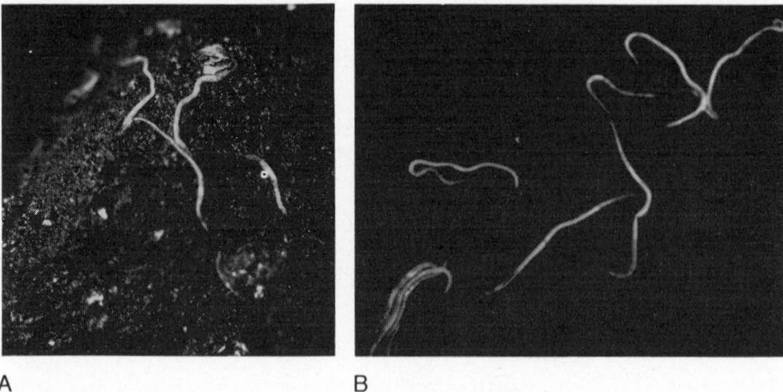

Figure 2-15 *Enchytraeus* sp. **A** "White worms" on mud in culture bowl. **B** The worms, removed from mud and placed in water for better view.

SOURCES: Probably distributed over most of the United States; in mud at the edge of rivers and in other very moist soil areas. Starting cultures can be purchased from commercial suppliers (see Appendix B).

CULTURE: Spread a 2- to 3-in layer of the collected mud with animals on the bottom of a flat bowl. Plastic dishpans may be used for large cultures. Place rocks, sticks, or pieces of glass on the soil. Add water to moisten throughout, but do not have water standing above the soil surface. If a culture is purchased, prepare the container in the same way using garden soil. Feed a variety of foods, such as boiled or baked potatoes, oatmeal, crushed crackers, bread, and cornflakes. Soak dry foods in milk, and vary the food from day to day. Put small masses of food together on the soil for the young, which develop from cocoons deposited on the food. Optimum temperature is 18 to 20°C. Animals require daily attention of feeding, sprinkling water lightly over the soil, and removing any uneaten food given the day before. Young develop in cycles; if worms are used as food for other animals, several cultures should be maintained. A dense growth results with proper care.

Lumbricus (Terrestrial Oligochaetes, the Common Earthworm)

USES: For studies of (1) physiology and anatomy, (2) behavior, (3) reproduction, and (4) embryonic development.

DESCRIPTION: Flesh-colored; flattened ventral surface; darker dorsal surface. Length of 13 to 30 cm. Usually more than 100 segments. Four pairs of setae on each of most segments, two pairs dorsolateral, two pairs ventrolateral. Hermaphroditic; cross-fertilization. Cocoons of developing embryos are deposited in soil (see Fig. 2-16).

SOURCES: Found in humus soil, near or above surface of wet soil, or at depth of 6 to 24 in in dry soil. Use a bright light to collect at surface of garden or lawn during early nighttime. Sometimes the animals may be collected in great numbers during the day or night, following a rain or when a plot of ground is soaked heavily. Obtain at fish-bait houses and biology supply houses, where the worms generally are packaged in cartons of damp peat moss. Cocoons (size of wheat grains) may be found throughout the peat moss. Occasionally, earthworm "farms" are located in suburban areas of larger cities.

A *Courtesy of Carolina Biological Supply Company*

B *Courtesy of Carolina Biological Supply Company*

Figure 2-16 *Lumbricus,* the common earthworm. **A** Adult worms. **B** Cocoons, each about 5 mm long.

CULTURE: Put worms in a container filled to a 12-in depth with leafy loam or rich soil (not clay). Keep soil moist but not wet. Lightly cover the container and put in diffused light at about 18°C. Forty to fifty large worms require about 12 in^3 of soil. Each week, sort through the soil and remove injured or dead animals. For food, mix crushed dead leaves into the soil and lay dead leaves over the surface. If leaves are not available, every 2 to 3 weeks spread moistened bread crumbs or crushed cornflakes over the soil and cover with about 1 in of soil. If the soil becomes sour or if water accumulates at the bottom of the box, immediately transfer worms to new containers. Remove any mold as soon as it appears.

Eisenia foetida, the fecal earthworm, is much hardier than *Lumbricus*. It is found in animal manure and maintained in containers of partly rotted cow or horse manure under the same conditions as for *Lumbricus*. *Eisenia foetida* produces a relatively large number of cocoons.

Freshwater Leeches (Hirudinea)

USES: For studies of (1) physiology, (2) adaptations, (3) behavior, and (4) ecology. Particularly useful for studying development of the young, which are produced abundantly in some species, for example, species of *Glossiphonia*.

DESCRIPTION: Most are flattened dorsoventrally. Oral and caudal suckers are used for attachment to substratum and to animal hosts. No setae; 33 segments and a prostomium. Segments are subdivided by external grooves. Length varies among species from 4 or 5 mm to 12 cm and longer. Much variation among and within species in color and markings. Highly contractile bodies. Hermaphroditic with cross-fertilization.

The relatively small *Glossiphonia*, *Helobdella*, and *Placobdella* adapt well to laboratory culture and produce many young which develop in a capsule fastened to the ventral body surface. The larger *Macrobdella* (the American medicinal leech) and *Philobdella* (the southern bloodsucker) feed on blood from man and

Figure 2-17 Ventral diagram of *Glossiphonia complanata,* showing reproductive system (stippled) and digestive system. Female organs shown only on the left; male organs shown only on the right; 1, mouth; 2, oral sucker; 3, proboscis; 4, salivary glands; 5, ovary; 6, testis; 7, ejaculatory duct; 8, gastric caecum (sixth); 9, seminal vesicle; 10, intestinal caecum (second); 11, rectum; 12, anus; 13, caudal sucker. *(Based on Fresh-Water Invertebrates of the United States by Robert W. Pennak, modified from Harding and Moore. Copyright 1953, The Ronald Press Company, New York.)*

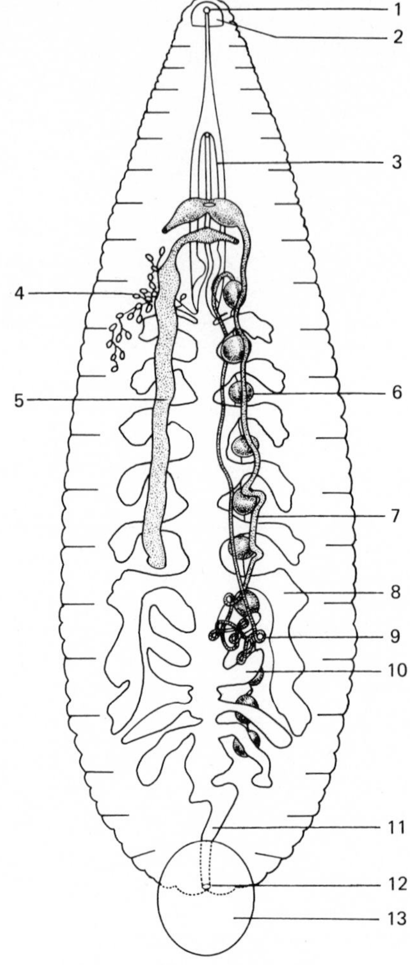

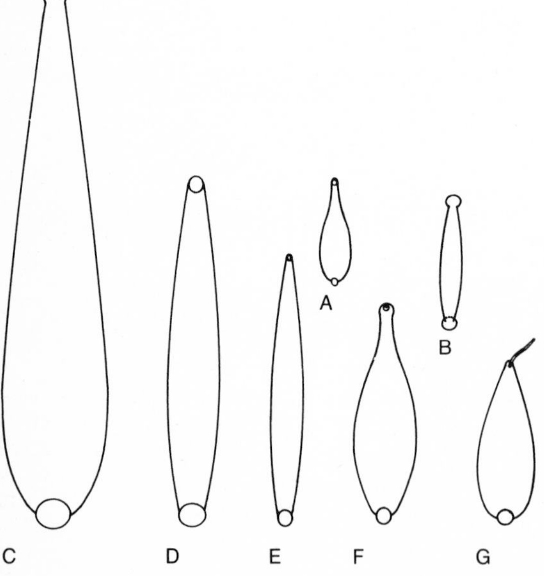

Figure 2-18 Relative shape and size of typical American leeches. **A** *Helobdella.* **B** *Piscicola.* **C** *Haemopis.* **D** *Macrobdella.* **E** *Erpobdella.* **F** *Placobdella.* **G** *Glossiphonia,* with extruded proboscis. *(A-G Based on Fresh-Water Invertebrates of the United States by Robert W. Pennak. Copyright 1953, The Ronald Press Company, New York.)*

other animals. The last two groups lay eggs in cocoons that are fastened to the substratum or buried in mud (see Figs. 2-17 and 2-18).

SOURCES: Animals are found in substratum and among floating algae of shallow water. Use a dip net for collecting larger swimming leeches, or pick them from rocks or sticks in the water. Often they may be picked from their hosts: fish, frogs, turtles, or in the mantle cavity of water snails. See Appendix B for commercial sources.

CULTURE: Most freshwater leeches are hardy and easy to culture. Put smaller leeches in bowls containing clean pond water or deionized water and aquatic plants. Place in diffused light at 18 to 20°C. Metal ions are toxic, particularly copper and chlorine. Animals do not survive in stagnant water. Remove decaying organic matter at once and change water if it becomes polluted. Every 1 or 2 weeks add several live water snails for food.

Put large leeches, such as *Macrobdella* and *Philobdella*, in bowls or tanks with sand at one end and shallow water at the other end. The animals can be kept for a long time on damp peat moss. These leeches are amphibious and tend to wander. Therefore, their containers should be tightly covered with fine-mesh screening. About every 6 months allow these leeches to suck blood from frogs or fish, or preferably from mice or rats.

MOLLUSCA: SNAILS AND SLUGS; CLAMS AND MUSSELS

The mollusks include more than 100,000 species of greatly diversified animals living in fresh or salt water or on land. General traits of most mollusks are: one or two calcareous shells secreted by a membrane, the mantle, which continues to form a lining within the shells; bilaterally symmetrical; unsegmented; coelomate; a visceral mass; mouth and sense organs in a head region; and a rasping organ, the *radula*, in the mouth or pharynx of most groups except the bivalves.

Principal classes of the phylum Mollusca are Amphineura (chitons); Gastropoda (snails, slugs, whelks, limpets); Scaphopoda (tusk shells); Pelecypoda (clams, mussels, oysters, scallops, shipworms, cockles); Cephalopoda (*Nautilus*, cuttlefish, squid, octopus).

Snails and Slugs

USES: For studies of (1) anatomy, (2) physiology, (3) feeding, (4) reproduction, (5) development of embryos in egg masses (see Davis, 1969), and (6) mitosis and meiosis (see Burch and Patterson, 1965). Snails are commonly used as scavengers in an aquarium. Slugs are interesting animals for a moist, woodland terrarium.

DESCRIPTION: *The Prosobranchia (gill) snails.* Most are marine; a few are large land or freshwater snails. All have gills, generally in the mantle cavity. Many have a thick-walled shell with an operculum which tightly covers the aperture when the body is drawn into the shell. Sexes are separate in most species. The group includes the abalones, limpets, large conchs, whelks, oyster drills, and periwinkles.

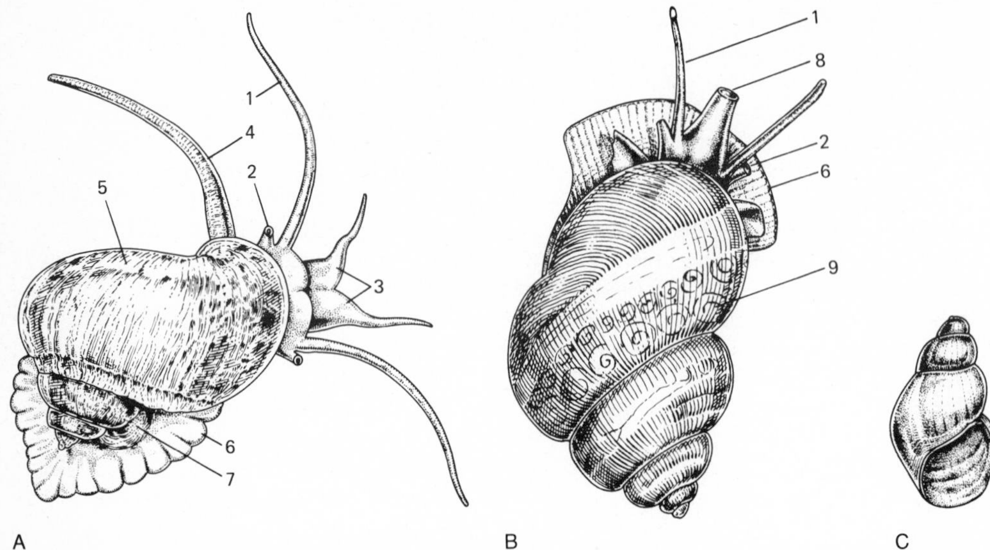

Figure 2-19 Prosobranch snails. **A** *Ampullarius,* extended. **B** *Viviparus* with embryos. **C** *Campeloma;* 1, tentacle; 2, eye; 3, labial palps; 4, siphon; 5, body whorl; 6, foot; 7, operculum; 8, snout; 9, embryos. *(A-C From The Invertebrates, vol. 6, by Libbie Hyman; A After Adams and Adams; B After Fischer. Copyright 1967 by McGraw-Hill, Inc. Used with permission of the McGraw-Hill Book Company, New York.)*

In addition to small marine forms, the following freshwater and land prosobranchs are suitable for the laboratory (see Fig. 2-19).

1. *Ampullarius.* Largest of freshwater snails. Native to Florida and Georgia. Both a gill and lung are present. Amphibious and lives in water or on muddy bank. Pink eggs, size of small peas, laid in clumps on substratum or on wall of aquarium. Feeds on aquarium plants if not given lettuce regularly.

2. *Viviparus.* Large snails with thin shells. Several species in lakes and rivers of central and eastern United States. Ovoviviparous. Feeds on algae. Good aquarium scavenger.

3. *Campeloma.* Similar to *Viviparus* in feeding and reproduction. Many species in middle and eastern United States. A thick, heavy shell. Good aquarium scavenger.

The Pulmonata (lung) snails and slugs. The pulmonates comprise a large group of snails and slugs distributed over all the United States. Most are terrestrial, although many are aquatic. No gills are present, and the name "Pulmonata" refers to the large, lunglike mantle cavity used as a respiratory organ. All are hermaphroditic, and in most species sperm are exchanged between two individuals. The shell of snails is well formed and whorled, and usually of a drab color; that of slugs is reduced to a thin, caplike, dorsal rudiment, covered by the mantle. One or two pairs of tentacles are present; a pair of eyes is located near or on the tentacles. Most species feed on vegetation. Among suitable, hardy animals for the laboratory are those described below (see Fig. 2-20).

1. *Lymnaea stagnalis.* A common pond snail with a long, cone-shaped shell (4 to 5 cm long). Lays eggs in large masses of clear jelly. A good aquarium scavenger.

2. *Planorbis rubra.* Red ramshorn snail. Flat, spiral shell 1 to 2 cm in diameter. Lays eggs in pink-colored jelly. A common aquarium snail and a good scavenger.

3. *Helisoma trivolvis.* The brown ramshorn snail. Flat, spiral shell 1 to 2 cm in diameter. Common in pond water. A good scavenger in aquariums. Feeds on algae and decaying vegetable matter. Clusters of eggs laid in clear jelly.

4. *Physa gyrina.* A pond snail with shell aperture on left side (most snails are "right-handed"). A fragile shell that must be handled carefully. A prolific aquarium animal that lays eggs in long, narrow masses of clear jelly. A good scavenger.

5. *Helix.* A common garden snail. Breeds well in a moist terrarium. Must be given lettuce regularly to prevent animal from feeding on terrarium plants. Eggs are deposited in the ground. *Helix pomatia* is the large, edible land snail of Europe, now introduced into the United States, where often it is a garden pest.

6. *Limax.* A slug with rudiment of shell that is covered with mantle. Nocturnal and can be found at night in damp places around houses, logs, and garbage cans. Maintain in a terrarium on bed of moist, dead leaves with rocks or twigs for the animals to crawl under during the day. Feed bread and dry cereal soaked in milk, raw potato and carrots, and leafy vegetables.

SOURCES: Collect slugs and land snails by hand from soil and under rocks and logs in moist areas of gardens and woods. Obtain aquatic snails by picking them from muddy banks, emergent vegetation, and shallow water, or by dredging with a net in deeper water. Most animals described above can be obtained from aquarium houses and biology supply companies (see Appendix B).

CULTURE: The following instructions are given for long-time maintenance. The animals can be kept for short periods under much less rigorous conditions so long as food and moisture are provided.

Put slugs and land snails in a terrarium or a small bowl containing a 1-in layer of small calcareous pebbles, a second layer of loose soil, and a top layer of dry leaves with several small dry twigs. Mix with the soil a small quantity of calcium carbonate or crushed shale to maintain an alkaline condition (about 1 g/l of soil). Keep all materials damp by sprinkling with water when necessary. Do not allow a bottom layer of water to accumulate. Twenty-five to thirty snails can be maintained in a 16 by 9 by 10 in receptacle. For breeding purposes, put several snails together in smaller receptacles prepared in the same manner. Place containers in diffused light at room temperature of 21 to 25°C. Cover securely with metal screen

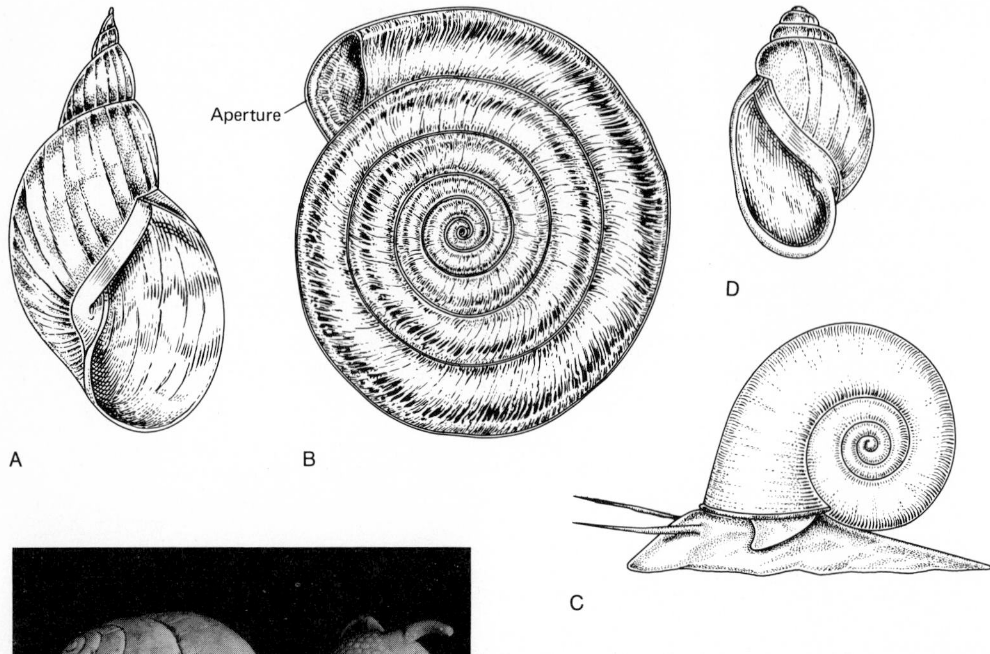

E *Courtesy of Carolina Biological Supply Company*

Figure 2-20 Pulmonate snails and slugs. **A** *Lymnaea stagnalis.* **B** *Planorbis* sp. **C** *Helisoma trivolvis.* **D** *Physa gyrina.* **E** *Helix* sp. **F** *Limax* sp. *(A, C, and D Based on Fresh-Water Invertebrates of the United States by Robert W. Pennak, C Modified from Baker. Copyright 1953, The Ronald Press Company, New York. B From The Invertebrates, vol. 6, by Libbie Hyman, after Harry and Hubendick. Copyright 1967 by McGraw-Hill, Inc. Used with permission of the McGraw-Hill Book Company, New York.)*

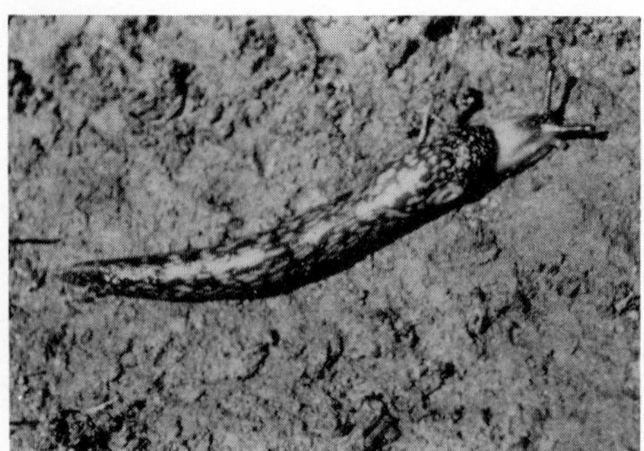

F *Courtesy of Carolina Biological Supply Company*

or cloth, or preferably with glass (leaving air spaces around the edge), or with punctured plastic film. A glass or plastic cover helps to maintain a higher humidity within the container. The decaying leaves furnish food for the animals. These should be supplemented several times weekly with lettuce leaves or other green leaves. Uncooked oatmeal, bran flakes, and bread (all soaked in milk) can be tried, as well as slices of raw potato and carrots. Remove food before it becomes molded; remove any mold that develops on the leaves or soil. If mold becomes excessive, transfer animals to a freshly prepared container.

Maintain aquatic snails in an aquarium or in culture bowls containing pond water or conditioned tap water. If the water is not aerated, it should be no more than 3 or 4 in deep. Many types of water snails are scavengers and help to keep an aquarium clean of decaying organic matter. Others feed on algae. Most water snails will eat aquarium plants if other food is scarce. Lettuce leaves, several times weekly, are suitable for most snails. The larger snails are ravenous feeders and produce much waste sediment, which must be removed regularly when many snails are in a container.

Freshwater Clams and Mussels

USES: (1) For studies of anatomy and physiology. (2) As aquarium animals.

DESCRIPTION: Bivalved, and the dorsally hinged shells are pulled tightly together by contraction of two short, thick muscles, the anterior and posterior adductors, fastened to the inner surface of each valve. A large anteroventral muscular foot is used for moving on or into substrate. No head, eyes, or tentacles. Posterior inhalant (lower) and exhalant (upper) siphons. Apparently, the terms "mussel" and "clam" often are used synonymously for freshwater bivalves, while only the term "clam" is used for saltwater forms. Almost all freshwater species belong to two families: Unionidae, large mussels of 20 to 250 mm length; and Sphaeriidae, small clams of 2 to 20 mm length (see Fig. 2-21).

SOURCES: Clams and mussels are found in unpolluted lakes, rivers, and

Figure 2-21 *Uniomerus* sp., freshwater mussel. **A** Lateral view. **B** Dorsal view.

A *Courtesy of Carolina Biological Supply Company*

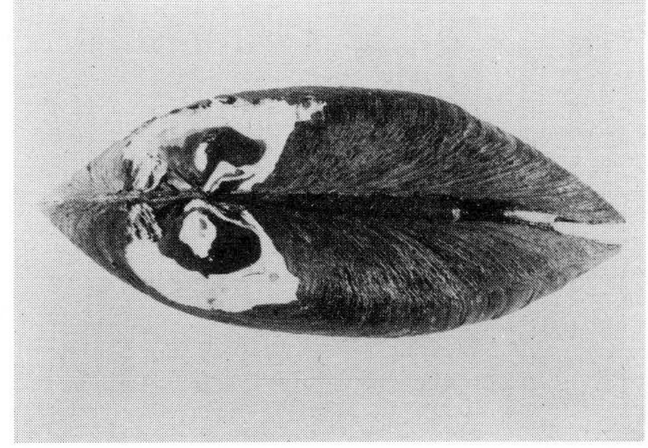

B *Courtesy of Carolina Biological Supply Company*

streams over all the United States. As a rule, they are most numerous in quiet areas of sand or gravel beds at depths of no more than 2 m, although some species can be found in abundance within the mud at the bottom of deep lakes. Generally, the animals are partially or totally submerged in the bottom materials. In shallow water, they can be obtained by hand picking or by pulling a rake through the substrate. In deeper water, a long-handled dipper or a dredge net is required. The animals can be transported for several days when wrapped in wet paper and kept at a cool temperature. See Appendix B for commercial sources.

CULTURE: Mussels and clams are not easily maintained for a long period. Put small animals (2 in or less in length) in an aquarium containing a bottom layer of 2 or 3 in of sand or gravel and plants that are already established. Maintain large mussels in large tanks of at least 20 gal, with slow-running water. The animals feed on algae, zooplankton, and organic waste.

REFERENCES

Behringer, M. P.: 1967, "Use of the Vinegar Eel in the Laboratory," *Am. Biol. Teacher,* **29** (7): 515–522.

Belcik, F. P.: 1968, "Metabolic Rate in Certain Sea Anemones," *Turtox News,* **46** (6): 178–181.

Burch, J. B., and C. M. Patterson: 1965, "A Land Snail for Demonstrating Mitosis and Meiosis," *Am. Biol. Teacher,* **27** (3): 203–207.

Chalkley, H. W., and H. D. Park: 1947, "Methods for Increasing the Value of Hydra as Material in Teaching and Research," *Science,* **105**: 553.

Custance, D. R. N.: 1967, "Studies on Strobilation in the Scyphozoa," *J. Biol. Educ.,* **6** (1): 79–81.

Davis, H. T.: 1969, "The Study of Aquatic Snail Embryos," *Am. Biol. Teacher,* **31** (3): 165–167.

Diehl, F. A., J. B. Feeley, and D. G. Gibson: 1971, *Experiments Using Marine Animals,* Aquarium Systems, Inc., Eastlake, Ohio.

Flichinger, Reed A.: 1959, "A Gradient of Protein Synthesis in Planaria and Reversal of Axial Polarity of Regenerates," *Growth,* **23** (3): 251–271.

Forrest, Helen: 1963, "Observing Planarians Feeding on Brine Shrimp," *Turtox News,* **41** (1): 34–35.

Galen, D. F.: 1969, "Culturing Methods for Hydra," *Am. Biol. Teacher,* **31** (3): 174–177.

Galen, D. F.: 1971, "Culturing and Using the Vinegar Eel," *Am. Biol. Teacher,* **33** (4): 237–238.

Hill, S. D., and James N. Cather: 1969, "A Simple Method for the Laboratory Culture of the Marine *Cassiopeia,*" *Turtox News,* **47** (8).

Hyman, L. H.: 1940, *The Invertebrates,* vol. I, *Protozoa Through Ctenophora,* McGraw-Hill Book Company, New York, 726 pp.

Hyman, L. H.: 1951, *The Invertebrates,* vol. II, *Platyhelminthes and Rhynchocoela,* McGraw-Hill Book Company, New York, 550 pp.

Hyman, L. H.: 1951, *The Invertebrates,* vol. III, *Acanthocephala, Aschelminthes, and Entoprocta,* McGraw-Hill Book Company, New York, 572 pp.

Hyman, L H.: 1967, *The Invertebrates, Mollusca I,* vol. VI, McGraw-Hill Book Company, New York, 792 pp.

Jacobson, A., and J. McConnell: 1962, "Research on Learning in the Planarian," *Carolina Tips*, **25** (7): 25–27.
Jacobson, A.: 1963, "Learning in Flatworms and Annelids," *Psychol. Bull.*, **60**: 74–94.
Lenhoff, H. M., and W. F. Loomis (eds.): 1961, *The Biology of the Hydra and of Some Other Coelenterates*, University of Miami Press, Coral Gables, Fla.
Lonert, A. C.: 1967, "Regeneration in *Aeolosoma*," *Turtox News*, **45** (11).
Loomis, W. F.: 1953, "The Cultivation of Hydra under Controlled Conditions," *Science*, **117**: 565–566.
Loomis, W. F.: 1954, "Environmental Factors Controlling Growth in Hydra," *J. Exptl. Zool.*, **126**: 223–234.
Loomis, W. F., and H. H. Lenhoff: 1956, "Growth and Sexual Differentiation of Hydra in Mass Cultures," *J. Exptl. Zool.*, **132**: 555–568.
Moog, Florence: 1963, *Animal Growth and Development*, BSCS Laboratory Block, D. C. Heath and Company, Boston. Student Manual, 83 pp.; Teacher's Supplement, 60 pp. $1.25 each.
Pennak, R. W.: 1953, *Fresh-Water Invertebrates of the United States*, The Ronald Press Company, New York, 769 pp.
Pramer, D.: 1964, *Life in the Soil*, BSCS Laboratory Block, D. C. Heath and Company, Boston. Student Manual, 62 pp.; Teacher's Supplement, 38 pp. $1.25 each.
Rose, S. L., and E. K. MacRae: 1969, "Planarian Regeneration," *Am. Biol. Teacher*, **31** (3): 168–172.
Spangenberg, D. B., and C. M. Flaten: 1967, "Growth and Metamorphosis in Jellyfish: A Picture Story," *Am. Biol. Teacher*, **29** (4): 306–309.
Whitten, R. H.: 1969, "The Fresh-Water Sponge," *Carolina Tips*, **32** (7): 25–26.

Other Literature

Berril, N. J.: 1957, "The Indestructible Hydra," *Sci. Am.*, **197** (6): 118–125.
Brondsted, H. V.: 1969, *Planarian Regeneration*, Pergamon Press, New York, 276 pp.
Buchsbaum, R.: 1948, *Animals without Backbones*, 2d ed., University of Chicago Press, Chicago. Many excellent photographs.
Burch, J. B.: 1962, *How to Know the Eastern Land Snails*, Wm. C. Brown Company Publishers, Dubuque, Iowa.
Burnett, A. L.: 1961, "The Growth Process in Hydra," *J. Exptl. Zool.*, **146** (1): 21–84.
Hardy, Sir Alister: 1965, *The Open Sea: Its Natural History, Part I: The World of Plankton, Part II: Fish and Fisheries*, Houghton Mifflin Company, Boston, 657 pp. A classic and fascinating account of the natural history of the open ocean. Contains numerous line drawings and color plates.
MacGinitie, G. E., and N. MacGinitie: 1968, *Natural History of Marine Animals*, 2d ed., McGraw-Hill Book Company, New York, 523 pp. An extremely interesting account of coastal water and intertidal animals. A good companion to Hardy's book on open-ocean animals.
Needham, James G.: 1937, *Culture Methods for Invertebrate Animals*, Comstock Publishing Associates, Cornell University Press, Ithaca, N. Y. (Dover Publications, Inc., New York, 1959.)
Ward, H. B., and G. C. Whipple: 1959, *Fresh-Water Biology*, 2d ed., W. T. Edmondson (ed.), John Wiley & Sons, Inc., New York.
Welsh, R. I., and J. H. Smith: 1960, *Laboratory Exercises in Invertebrate Physiology*, Burgess Publishing Company, Minneapolis.

LABORATORY CARE OF HIGHER INVERTEBRATES

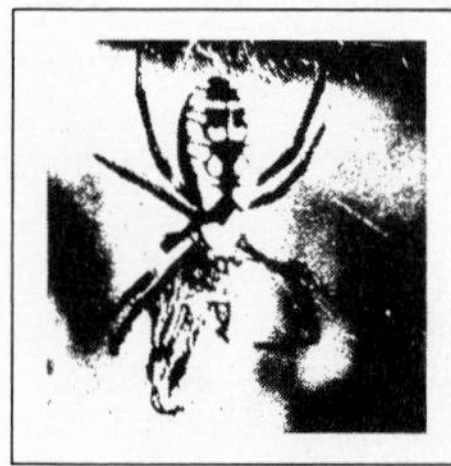

The Arthropoda and Echinodermata are discussed in this chapter. The animals described here are among those considered most appropriate for a teaching laboratory. The maintenance of marine forms and the collecting or purchasing of animals are discussed in Chap. 1. A list of commercial sources is given in Appendix B.

ARTHROPODA: CRUSTACEANS

Crustaceans comprise a large group of more than 26,000 species, most of them aquatic and gill-breathing, and with much diversity in structural adaptations. Distinctive traits are: (1) jointed appendages (characteristic of all arthropods), one pair on each of most segments and generally highly specialized for such functions as sensory perception, swimming, walking, feeding, reproduction, respiration, and protection; (2) a hard chitinous exoskeleton, often impregnated with calcium carbonate which produces a "crusty" appearance; and (3) two pairs of antennae, a distinction from other arthropods which have only one pair or no antennae.

Classification of Crustacea varies among authorities and remains inconclusive for some categories, particularly in the higher ranks. The following is a brief taxonomic outline for the animals described in this section.

Class Crustacea[1]

Subclass Branchiopoda

Order Anostraca: *Artemia* (brine shrimp). Sometimes all anostracans are called fairy shrimp.

Order Cladocera: *Daphnia* (water flea)

Subclass Copepoda

Order Cyclopoida: *Cyclops*

Subclass Malacostraca

Order Isopoda: *Armadillidium* (pill bug)

Order Amphipoda: *Gammarus*

Order Decapoda: *Cambarus* (crayfish)

Artemia (Brine Shrimp)

USES: The larval nauplius is used extensively as food for hydra, other small invertebrates, and small fish. The animal is excellent for studying (1) taxic responses, (2) larval development, (3) adult filter feeding, (4) swimming and feeding movements of larva and adult, and (5) adaptations to salt concentrations.

DESCRIPTION (ADULT): Body length, 1 to 2 cm. One pair of stalked compound eyes. Eleven pairs of thoracic appendages for swimming and feeding. No abdominal appendages. Elongated abdomen. Normally swims with ventral surface turned upward. Sexually separate; eggs carried by female in brood sac attached to ventral thorax. Young develop to adult through 12 to 14 instars within 3 to 6 weeks (see Fig. 3-1).

SOURCES: *Artemia* lives in salt lakes and salt basins and is abundant in the Great Salt Lake and the salt basins of Utah and California. Density and distribution may fluctuate. Collect animals with a dip net or a jar. In many areas, dried eggs must be purchased from aquarium houses or biology suppliers. Eggs remain viable for one or more years if stored under dry conditions.

CULTURE: Techniques for hatching nauplius from eggs are included with the discussion of *Hydra*, Chap. 2. Put adults in their natural salt water or in 6 to 10% sodium chloride–distilled water solution. Cover with glass or a punctured plastic film. Aerate gently. Mark the original water level and add distilled water (not salt water) for evaporation loss, bringing the water level to the original mark. Adults reach a larger size and mature more quickly in the higher salt concentrations. Add

---

[1]The taxonomy is based on W. D. Russell-Hunter: 1969, *A Biology of Higher Invertebrates*, Collier-Macmillan Canada, Ltd., Toronto.

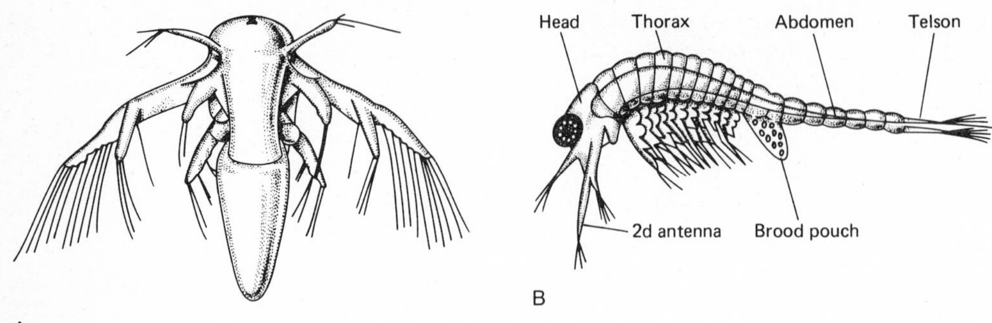

Figure 3-1 *Artemia salina,* the brine shrimp. **A** First instar (nauplius), the stage commonly used as food for small aquatic invertebrates. **B** Adult. *(A From Fresh-Water Invertebrates of the United States, by Robert Pennak. Copyright 1953, The Ronald Press Company, New York. B From The Science of Biology, 3d ed., by Paul B. Weisz. Copyright 1967 by McGraw-Hill, Inc. Used with permission of the McGraw-Hill Book Company, New York.)*

algae for food and provide yeast suspensions three or four times weekly. The shrimp will feed on flagellated, marine algae, for example, *Dunaliella,* which can be collected in salt lakes or warm coastal waters, or purchased from Carolina Biological Supply Co. Ordinarily, if algae are well established in the aquarium, no difficulty will be encountered in culturing the adult animal.

Daphnia (Water Flea)

USES: (1) Used extensively as food for aquatic invertebrates and small fish. (2) Excellent for a variety of class studies and individual research projects, such as comparison of reproduction among species; physiology, for example, effects of certain chemicals and various environmental factors on rate of heartbeat; and seasonal changes in morphology (cyclomorphosis). See Follansbee (1965) for laboratory tests with *Daphnia.*

DESCRIPTION: Cladocerans are mainly freshwater organisms that constitute an important component of aquatic food chains. General daphnid characteristics are: Body length, 2 to 4 mm. Trunk compressed inside a transparent carapace. A single compound eye. Five pairs of trunk appendages; and a sac-like heart in the dorsoanterior trunk (see Fig. 3-2A). The animal moves primarily with the second pair of antennae, which accounts for the irregular, jerky movement. Reproduction varies among species and apparently is related to ecological changes, although not all reproductive variation is understood. During much of the year, parthenogenetic eggs develop into female young within the dorsal brood chamber of the adult female (see Fig. 3-2B). During unfavorable conditions, reproduction seems to be primarily by means of fertilized eggs, which are resistant to extreme temperatures and drought conditions, and which may develop into both male and female young. Only a few fertilized eggs are present at one time in the brood chamber. The walls of the brood chamber thicken around the fertilized eggs to form a rough capsule, the ephippium, which is shed with the exoskeleton (see Fig. 3-2C). We

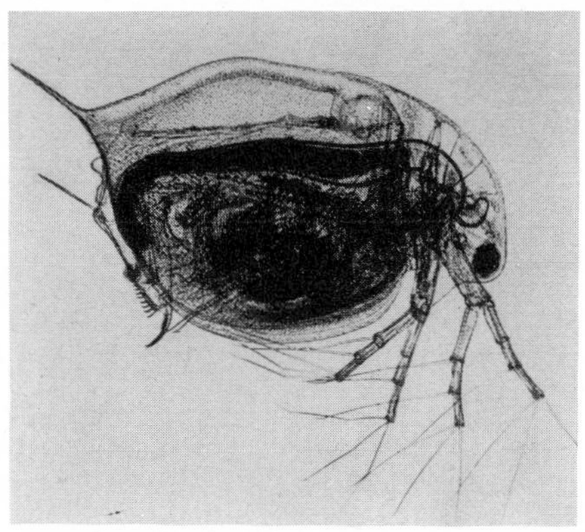

A *Courtesy of Carolina Biological Supply Company*

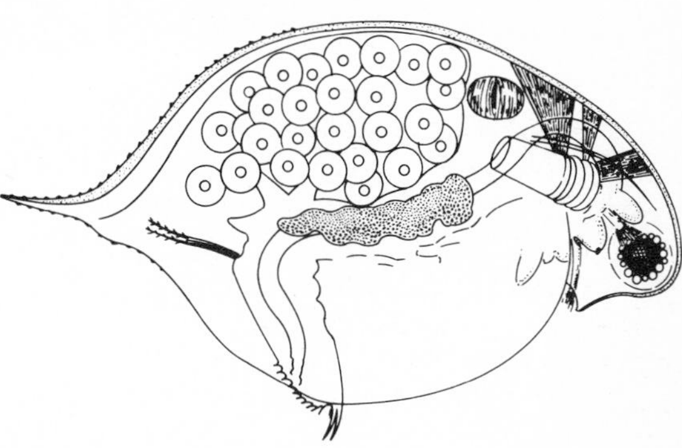

B

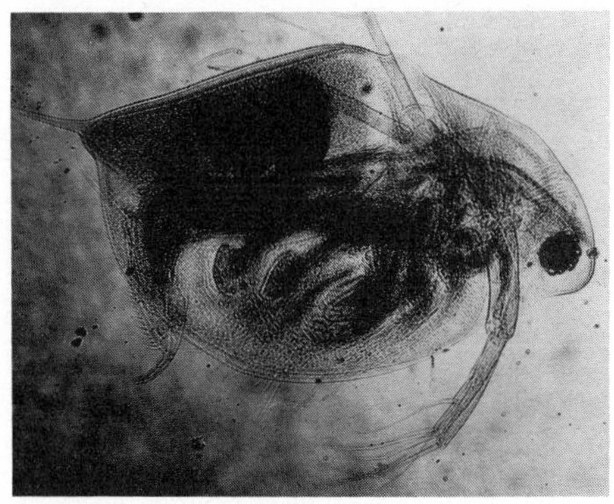

C

Figure 3-2 The *Daphnia*. **A** *Daphnia* sp., adult. **B** *D. pulex*, showing ovary (stippled) and parthenogenetic eggs in dorsal brood chamber. **C** *Daphnia* with three fertilized eggs in brood chamber. Brood chamber hardens to become an ephippium. *(B From Fresh-Water Invertebrates of the United States by Robert W. Pennak, modified from Ueno, copyright 1953, The Ronald Press Company, New York. C Courtesy of CCM: General Biological, Inc., Chicago.)*

have observed laboratory cultures of *Daphnia magna* which from early fall to the next late spring contained females with clusters of parthenogenetic eggs and other females with ephippia.

SOURCES: Depending on species, *Daphnia* is found at the surface or at the bottom of freshwater ponds and lakes. Many species make seasonal changes from the top surface in warm weather to the bottom water during cold weather. For collecting, use a dip net, a plankton tow net, or bottom dredge net of fine mesh. Once a productive collecting spot is located, often *Daphnia* can be collected regularly at the same site year after year. If animals are needed in great quantity, a commercial source may be more desirable.

CULTURE: Many methods have been described. Probably all are successful—and unsuccessful—at times. Although daphnids are considered delicate animals, we have had them to flourish with no attention for as long as 6 months within a covered aquarium containing algae and placed in a south window. Occasionally, they have flourished with no attention in large outdoor tubs containing aquatic plants. At other times, in spite of constant attention, cultures have been difficult to maintain.

For all methods described below, use pond or dechlorinated tap water, aerate gently, and put in diffused light at 20 to 24°C. Containers may be battery jars, widemouthed fruit jars, large beakers, or aquarium tanks. Eventually, the bottom sediment may contain encased, dormant, fertilized eggs. Therefore, the sediment should be examined before it is discarded.

Method 1. Probably the most successful and simplest procedure is to put *Daphnia* in their natural pond water, or in dechlorinated tap water, with about 50 animals for each liter of water. Feed hard-boiled egg yolk, prepared by crushing the yolk in water to make a thin emulsion. Add enough food to give the water a cloudy appearance. Repeat the feeding when the animals have consumed all the yolk and the water has become clear (about 4 to 5 days). Alternatively, a water suspension of yeast can be fed in the same manner.

Method 2. Add *Daphnia* to an established aquarium containing a dense growth of nonfilamentous algae. A dense green "soup" of *Chlorella* is excellent. The aquarium must not contain animals that feed on *Daphnia*. Add egg yolk or yeast as in method 1. Place in a location that receives direct sunlight several hours daily.

Method 3 (manure medium). This is an old standard method used successfully and exclusively by many persons. The *Daphnia* feed on colon bacteria which accumulate in the medium. We have had no greater success with this method than others, however, and the preparation of the medium involves more time and more materials.

To prepare a stock medium, combine 1 kg of garden soil (sandy loam) and 175 g of 10- to 12-day-old horse manure. Stir the mixture into 9.5 l of pond water. Allow to stand for 2 to 3 days at about 18°C. Remove any floating manure, and strain the liquid through a fine-mesh cloth, such as 130-mesh silk bolting cloth. Press enough sediment through the cloth to obtain a 1- to 2-mm layer of silt sediment in the container of strained liquid. Add 1 part of this stock medium (stirred well) to 2 to 4 parts of clear pond water or dechlorinated tap water. Proportions depend on density of stock medium; final medium should be a light-brown color. Add a small portion of stock medium to the animal container every 5 to 7 days to regain the light-brown color. The stock medium during early processing and the final medium have little or no odor and may be retained in the laboratory.

Land Isopods (Sowbug, Pillbug, Wood Louse)

USES: Although land isopods have been used little for laboratory work, they offer many possibilities for extremely interesting physiological studies, as suggested by Segal and Gross (1967).

DESCRIPTION: Body length, 5 to 20 mm. Dorsoventrally flattened and oblong to ovate (egg-shaped). Gray, brown, black, red, or yellow; may be mottled. A cephalothorax of head and first thoracic segment. Each of other seven thoracic segments with a pair of walking appendages. First five abdominal segments with a pair of greatly modified platelike appendages carrying gills or tracheae. Extension of sixth abdominal segment as a posteriorly pointed process between two uropod appendages.

Land isopods are the most successful of terrestrial crustaceans. Species of *Armadillidium* are known commonly as pillbugs because they can roll into a tight ball. *Porcellio* species, sometimes called sowbugs or wood lice, are able to flex the body only slightly and do not roll into a pill shape. Eggs are carried and hatched within a ventral thoracic brood pouch (marsupium) of females; the hatched young remain in the marsupium for a time. The light-colored swollen marsupium is easily observed (see Figs. 3-3 and 3-4).

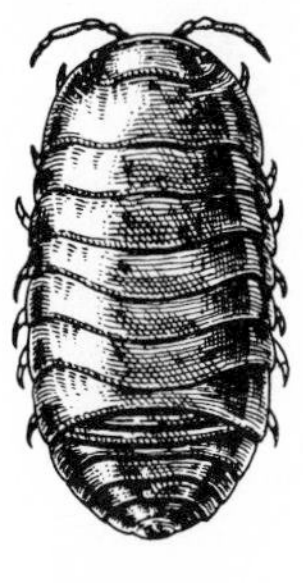

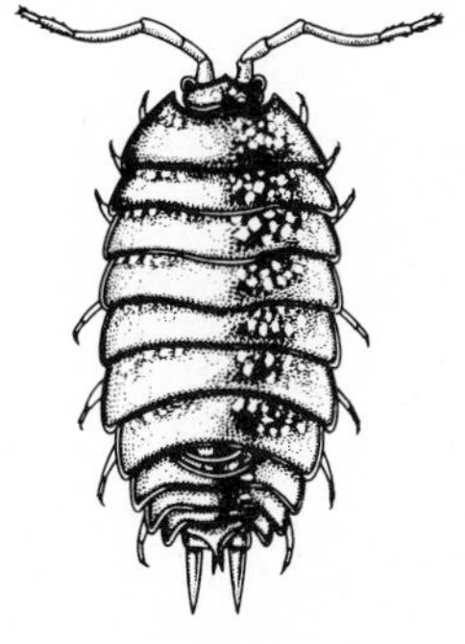

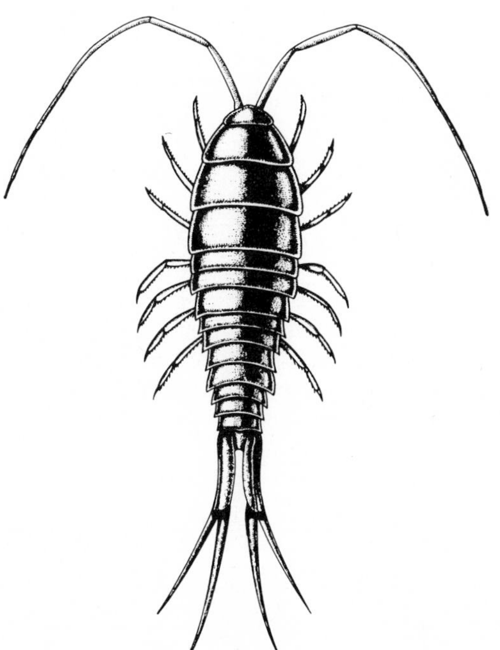

Figure 3-3 Land isopods. **A** *Armadillidum vulgare,* common pillbug, two views. Length of adult is about 16 mm. **B** *Porcellio laevis,* sowbug or wood louse, that can flex body slightly but does not roll into a pill shape. Length of adult is about 15 mm. **C** *Ligia exotica,* a large isopod (about 30 mm long in adult) found on saltwater beaches. *(A Courtesy of Illinois State Natural History Survey; B and C Biological Sceinces Curriculum Study.)*

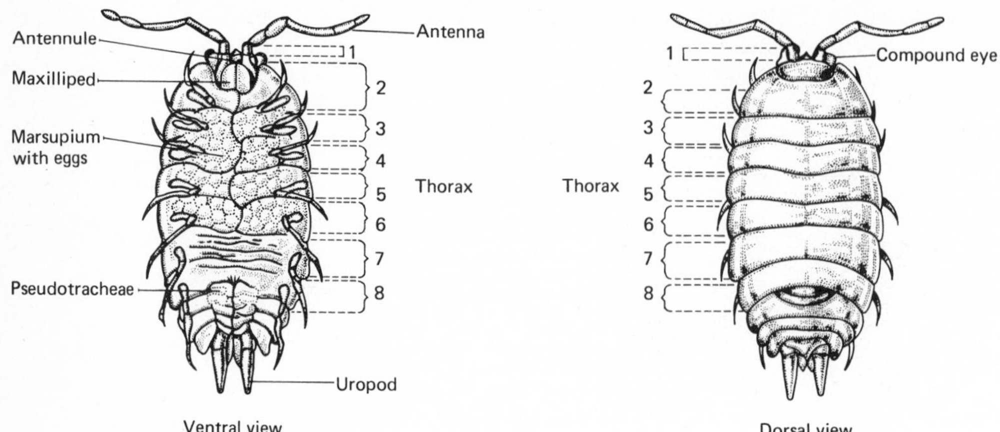

Figure 3-4 External structures of an isopod. *(Biological Sciences Curriculum Study.)*

SOURCES: *Armadillidium* and *Porcellio* occur in great abundance throughout the United States. Collect the animals in moist areas under rocks, logs, and leaves, or in damp soil under trees and at the side of buildings. Once a collecting spot is found, the animals can be trapped with a hollowed-out white potato placed under a thin layer of damp soil and leaves (see Fig. 3-5). After 24 to 48 h, many isopods may be found feeding inside the potato. *Ligia*, an amphibious, saltwater isopod, may be seen running along piers or wharfs (see Fig. 3-3C). Because of their swift movement, these are most easily collected by putting damp sacks or papers at sites where they are spotted. After 24 to 48 h, scoop up the sack or papers into a rounded mass and quickly transfer them to a closed container.

CULTURE: The animals are hardy and multiply quickly in culture. Prepare a layer of peat moss or dead leaves and twigs in any type of waterproof container (see Fig. 3-6). Keep substratum damp by sprinkling water over the bottom layer when necessary. Cover with plastic film to maintain a high humidity inside the container; punch air holes in the cover. Feed slices of potato or carrot every 2 to 3 days; remove old food to avoid molding. The animals are prolific, and many young will be found within the culture in several weeks.

Gammarus and *Hyalella* (Amphipods)

USES: Although not used extensively, amphipods can be effective substitutes for many of the studies usually conducted with other crustaceans. As a rule, these animals are cultured more easily than most other invertebrates. See Kuhn (1969).

DESCRIPTION: The adult body length of *Gammarus* varies from 15 to 25 mm. *Hyalella* is smaller (4 to 8 mm long). *Gammarus* is white or light yellow, and *Hyalella* is usually light brown or a pale green. Both are widespread and are commonly found in freshwater collections. Body is compressed laterally. Body regions consist of a cephalothorax containing the head and first thoracic segment,

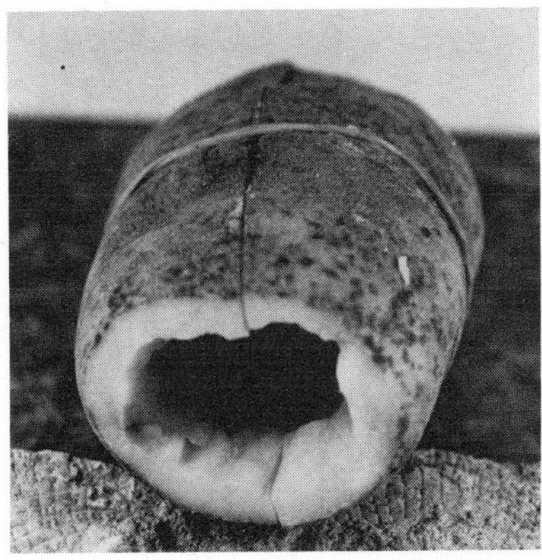

Figure 3-5 Potato trap for collecting land isopods.

seven additional free thoracic segments, six abdominal segments, and a small posterior telson. Much diversity is seen in adaptation of appendages.

When mating, the male carries the female on his back for as long as a week while the two animals swim and feed together. Eventually, the animals separate, the female sheds her exoskeleton and, shortly, a male transfers sperm to the marsupium on the ventral surface of the female. The two animals separate, eggs are released from the oviducts and pass into the marsupium where they are fertilized. The young develop within the marsupium during a period of 1 to 7 days and then are released to the outside. The number of instars varies from several to as many as eight or nine, depending on species. Amphipods are poor swimmers, are omnivorous in feeding, and are scavengers in mud and sediment (see Fig. 3-7).

SOURCES: Amphipods live in marine littoral zones, in estuaries, and in the

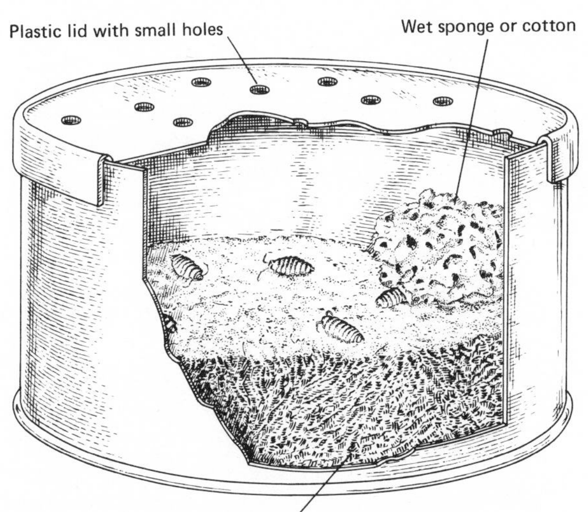

Figure 3-6 Isopod culture chamber. *(Biological Sciences Curriculum Study.)*

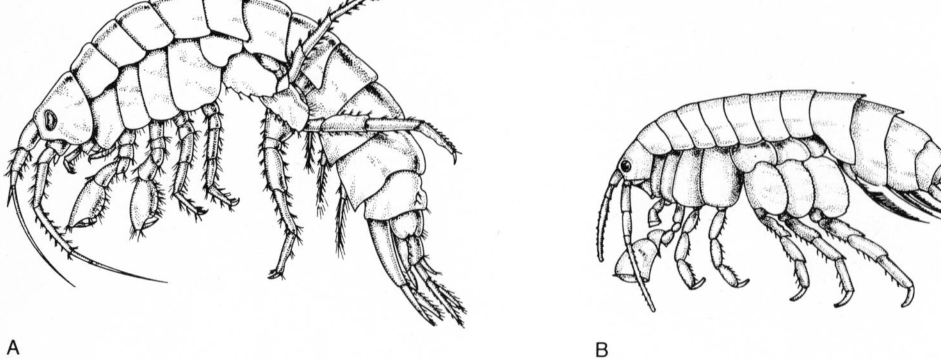

Figure 3-7 Amphipods or scuds. **A** *Gammarus fasciatus,* adult male. **B** *Hyalella knickerbockeri. (A From Fresh-Water Invertebrates of the United States by Robert W. Pennak, modified from Kunkel, The Ronald Press Company, New York.)*

upper meter of clear ponds and streams. Occasionally they will be found in great abundance. Collect the animals with dip nets and plankton tow nets. If abundant, they can be scooped into jars or other containers at the water edge.

CULTURE: The animals are hardy and can be cultured easily in an aerated aquarium that contains well-established plants and no animals that feed on amphipods, such as fish. Amphipods feed on dead plants and animals and other wastes, and on floating plankton. If the aquarium is well established with a dense, green "soup" of unicellular algae, no additional food is necessary. Alternatively, put the animals in a separate culture bowl and feed yeast or hard-boiled egg yolk as described for *Daphnia.* Larson (1971) reports that he has maintained *Hyalella* for a 4-yr period in an aquarium containing a dense culture of unicellular, green algae.

Crayfish *(Cambarus)*

USES: Crayfish are excellent for studies of movement, behavioral patterns, reproduction, and development of young. Their hardiness and ease of care make them especially useful in the laboratory. See Strayer (1969).

DESCRIPTION: Body length, 15 to 130 mm. Cephalothorax covered with a carapace; appendages, 19 pairs. Male usually with larger pincers than female on first pair of walking legs, and with larger first and second pairs of swimmerettes that are modified for sperm transfer. Color range of dark brown, tan, red, orange, and occasionally green-blue (see Fig. 3-8).

Females produce ten to several hundred eggs, generally in the spring. The eggs are glued to her swimmerettes, where they are fertilized by sperm previously transferred in spermatophores from a male into a ventral seminal receptacle or onto the ventral surface of the female. The hatched young remain attached to the female swimmerettes through the first two or three instars, a very fascinating observation for students. Spring-hatched young may pass through six to ten molts by autumn, at which time molting and growth come to a stop.

SOURCES: Crayfish are found throughout the United States in running streams and still bodies of water. Use a coarse dip net to collect in mud bottoms of shallow water or among plants and under rocks. Adults tend to hide during the day and are located by dislodging bottom debris or by waving a net through the plants.

A *Courtesy of Carolina Biological Supply Company*

B

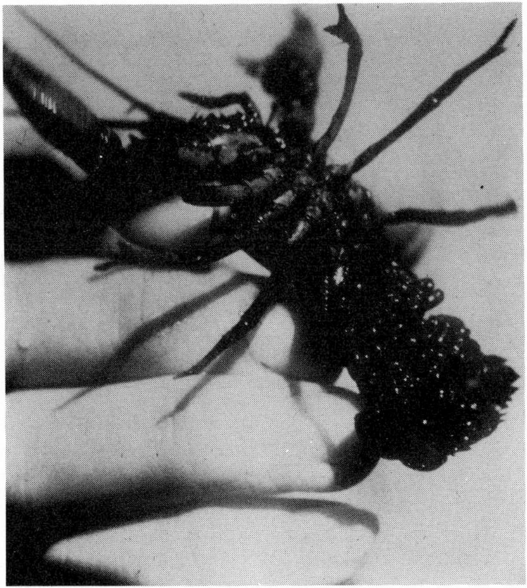

C *Courtesy of Carolina Biological Supply Company*

Figure 3-8 The crayfish. **A** *Cambarus* sp., large southern crayfish. **B** Adult male, ventral view, showing the large first and second pair of swimmerettes that extend anteriorly to between the walking legs. **C** Adult female, ventral view, showing eggs and young crayfish attached to swimmerettes.

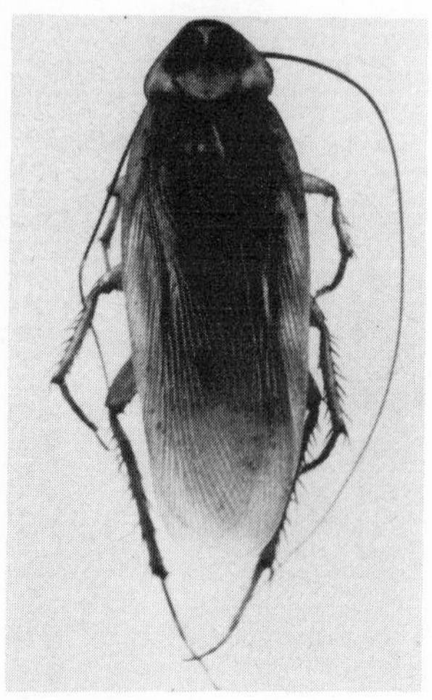

A

B

C

D

Figure 3-9 The cockroach. **A** *Periplaneta americana* or American cockroach. **B** *Blattella germanica* or German cockroach. **C** *Blatta orientalis* or oriental cockroach; left, male and right, female. **D** *Blattella germanica,* female with egg case partially extruded. *(D Permission of Ward's Natural Science Establishment, Inc. Rockester, N.Y.)*

Crayfish are obtained easily with traps or fish hooks baited with small chunks of raw pork or beef and left overnight in known collecting spots.

CULTURE: Crayfish are hardy animals, tolerating a fairly wide range of pH and temperature. Animals from still or slow-moving water adapt more readily to aquarium situations. The freshly collected animals should be adjusted to a change in water temperature by enclosing them in a small container of the natural habitat water which is then floated in the aquarium water until the two water temperatures are nearly the same. Provide a hiding place, with a loose stack of small rocks or other objects. In particular, young crayfish require seclusion during molting periods. Large crayfish may uproot aquarium plants. Feed small chunks of fresh or frozen raw meat, including fish. Remove uneaten food within 1 h after feeding.

ARTHROPODA: THE INSECTS

Culture techniques are described for the insect groups considered most useful in laboratory studies, and for several other groups that are often maintained as food for other animals. All insects of this section can be maintained with little difficulty when proper housing and feeding are provided. Special collecting techniques are given here; general techniques are discussed in Chap. 1. Nomenclature is based on Borror and DeLong (1971).

Cockroaches (Order Orthoptera; Family Blattidae)

USES: Cockroaches are especially useful for (1) physiological studies; and (2) as food for spiders, praying mantids, amphibians, and reptiles. For laboratory tests see Richards (1963), Segal and Gross (1967), and Yurkiewicz (1970).

DESCRIPTION: Incomplete (gradual) metamorphosis. Chewing mouth parts. Two pairs of wings (reduced in some species), first pair a tough cover and second pair for flying and folded fanlike when resting. Body flattened dorsoventrally. Rapid runner. Feeds on all types of organic material, including kitchen food and clothing. Suitable species for the laboratory: the large American cockroach, *Periplaneta americana*, 2 to 3.5 cm long and reddish brown; the smaller oriental cockroach or "water bug," *Blatta orientalis*, 2 to 2.5 cm long, and dark brown; and the German cockroach or crotonbug, *Blattella germanica*, about 1.5 cm long and a light-brown color (see Fig. 3-9).

SOURCES: The three species are common both indoors and outdoors in the warmer areas of the United States, and inside heated buildings in colder regions. Look for them in dark crevices of closets and cabinets. Around the exterior of dwellings, the insects may be found under boards, leaves, and trash, as well as around garbage cans and garbage dumps. Capture roaches by quickly laying a fine-mesh net or cloth over them, or by quickly inverting an open jar over the animal. See Appendix B for commercial sources.

CULTURE: Establish cultures in large-mouthed gallon jars or in any other similar type of container (see Fig. 3-10). Cover receptacles with a fine-mesh screen, fastened *securely* with a rubber band or other device. Both adult and young roaches escape easily through any small opening. A ring of Vaseline around the inside top wall will help to prevent escape. For experimental work, small bottles or vials are satisfactory containers when plugged with cotton or perforated cardboard caps. For breeding cages, add a bottom layer of wood shavings, shredded paper, or peat moss. Place the culture in a dark, warm area (21 to 25°C). Provide pieces of apple, potato, bread, or dog or rabbit food pellets. Add fresh lettuce leaves several times weekly. *Note:* Roaches must have a *constant* supply of water on heavily soaked

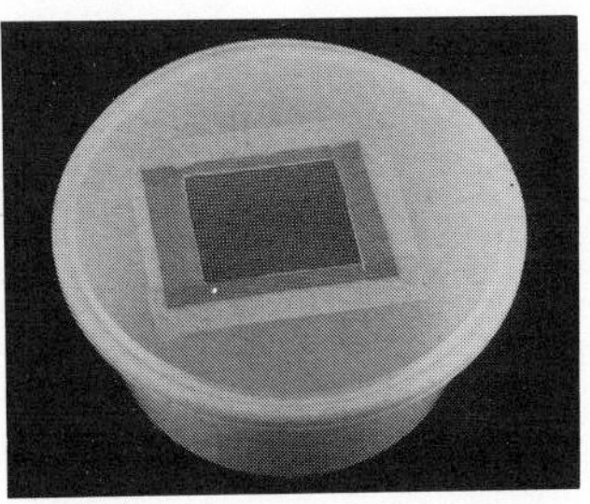

Figure 3-10 Insect culture chamber, devised from a plastic food container.

cotton pads in a small dish, or by other devices that protect animals from drowning in open water. Transfer cultures to clean cages with fresh bottom materials at least every 2 months, or more often to eliminate a foul odor when many roaches are caged. *Periplaneta* and *Blatta* deposit egg cases (ootheca) in bottom materials, each oothecum containing 15 to 40 eggs. *Blattella* carries the egg case partially extruded from the brood chamber until a short while before the eggs hatch. Other species may retain the oothecum in the brood chamber until after the eggs hatch. Young nymphs develop at a relatively slow rate, requiring several months to a year to reach maturity.

Crickets (Order Orthoptera; Family Gryllidae)

USES: Considerable research has been done concerning behavior, physiological adaptations, and speciation in crickets. See Webb (1968) and Alexander (1967) for fascinating accounts of crickets, and experiments that are suitable for the general biology laboratory.

DESCRIPTION: Incomplete (gradual) metamorphosis. Chewing mouth parts. Two pairs of wings, first pair thickened as a cover, second pair used for flying and folded fanlike when at rest. Suitable types for laboratory study: *Nemobius* (ground cricket), *Gryllus* (field cricket), and *Acheta* (house cricket; see Fig. 3-11). Females of these groups display a conspicuous ovipositor, used for depositing eggs in the ground or in crevices. The stridulations, or singing, of the male cricket (the female of most species does not stridulate) are produced when a scraper on

Figure 3-11 The cricket. **A** *Acheta* sp. or house cricket, female. **B** *Gryllus* sp. or field cricket, male. **C** *Gryllus* sp. or field cricket, female.

A

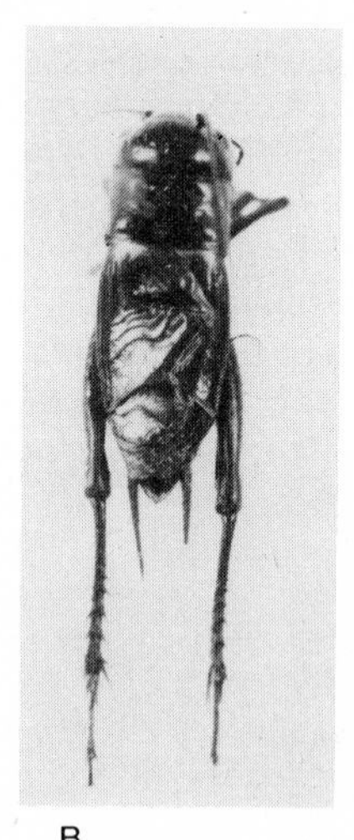

B

C

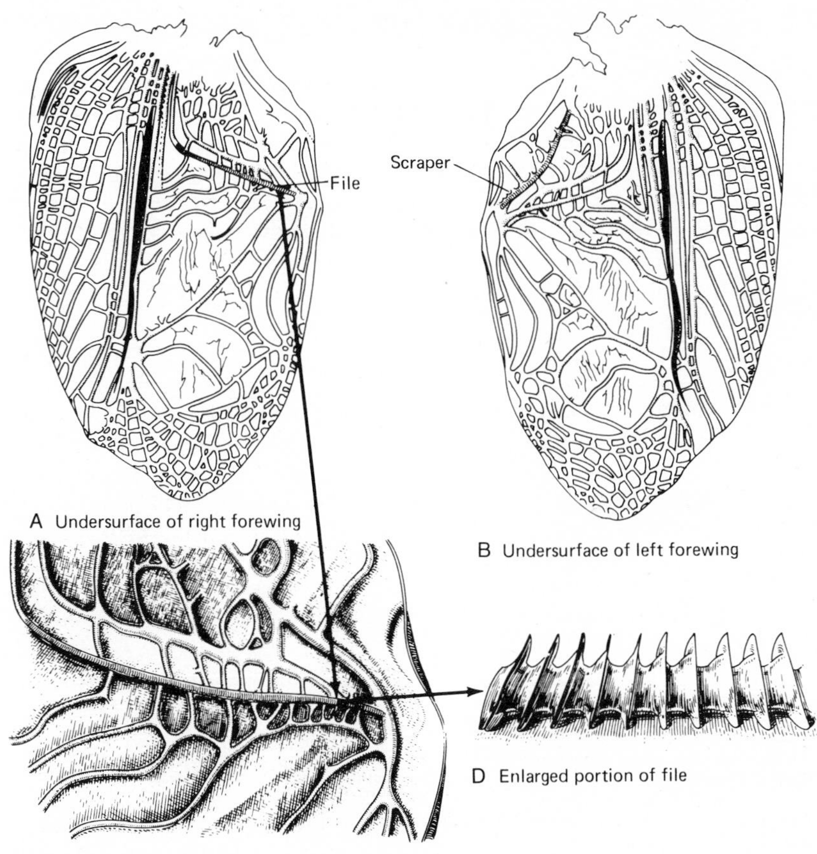

Figure 3-12 Stridulatory apparatus of a male field cricket. The scraper, a group of spines under the edge of the left forewing (**B**) is struck against the file, a row of fine teeth (**C** and **D**), located on the undersurface of the right forewing (**A**). Right and left wings appear to be in reversed locations, above, because the undersurface of each wing is shown. *(From Alexander, 1967.)*

the first wing is rubbed over a file on the femur, or when the two first wings (tegmina; tegmen, singular), are rubbed together (see Fig. 3-12). The fluttering movement of the tegmina can be observed easily during stridulation.

SOURCES: Collect ground and field crickets by passing a sweeping net over grass and shrubs. Use a net or jar to capture house crickets. The animals can be obtained from biological supply houses. Field crickets often can be obtained in large quantity at local fish-bait stores.

CULTURE: Put crickets in jars or other containers, covered securely with small-mesh screen or with a metal lid punctured to provide air (see Fig. 3-10). Crickets will eat through a cloth cover. Fish-bait stores often house the animals in large wooden boxes (about 3 by 4 by 2 ft) with a hinged lid. Also fish-bait stores sell cardboard cartons and small metal cages constructed specially for crickets. No bottom material is required in the cage. We have kept as many as 20 to 25 adults in a gallon container for 3 to 4 weeks with a low mortality. Crickets eat a variety of foods, including apple, potato, lettuce, and dry cereal. Chicken starter mash and crushed dog biscuits are easily supplied and furnish a well-balanced diet. Several times weekly, supplement the diet with lettuce leaves. Remove old, uneaten food before it molds. At all times, supply water on a heavily soaked cotton pad in a small dish, or by other devices that protect the crickets from drowning in open water.

Grasshoppers and Katydids (Order Orthoptera)

USES: Grasshoppers can be retained in quart jars for 1 or 2 weeks to observe incomplete metamorphosis—the shedding of exoskeletons, and emergence of nymphs.

DESCRIPTION: Incomplete (gradual) metamorphosis. Adult with two pairs of wings, first pair usually a hard cover, and second pair for flying. All with chewing mouth parts (see Fig. 3-13).

Short-horned grasshoppers (family Acrididae). Probably the most common and most destructive of grasshoppers. Abundant during the summer in grass and shrubs. Antennae much shorter than body. One tympanum (auditory organ) each side of first abdominal segment. Females have a short, clublike ovipositor.

Males sing during daytime by cracking hind wings together in flight or by scraping spines of the third leg against lower edge of front wing, producing a buzzing sound. Short-horn grasshoppers eat plants, at times causing great crop damage. The large southern lubber, *Romalea microptera,* and the western lubber, *Brachystola magna* (both 5 to 7 cm long and with reduced wings and flight), can be maintained successfully in the laboratory, although the large size makes housing and feeding more difficult. The female of most species inserts the posterior end of abdomen into the soil, forming a chamber into which 100 to 150 eggs are deposited, after which a protective covering is secreted over the chamber opening. Eggs hatch into nymphs the following spring.

Long-horned grasshoppers and katydids (family Tettigoniidae). Long, hairlike antennae. Most species with a tympanum on tibia of front leg. Females with flattened, elongated ovipositor, and males with stridulating organs on front wings.

These insects are strong singers and each species has a characteristic song. The katydid lives in trees and shrubs. Most species feed on plants; a few feed on other insects. Eggs are deposited in the ground or in the soft tissue of plants during the summer and fall; nymphs emerge the next spring.

SOURCES: Collect with a sweeping net from grass, shrubs, and trees. See Appendix B for commercial suppliers.

CULTURE: Put grasshoppers and katydids in covered gallon jars, cardboard cartons, or fine-mesh wire cages that have a solid bottom and a solid frame, several inches high, around the bottom of walls (see Fig. 3-10). A gallon container will

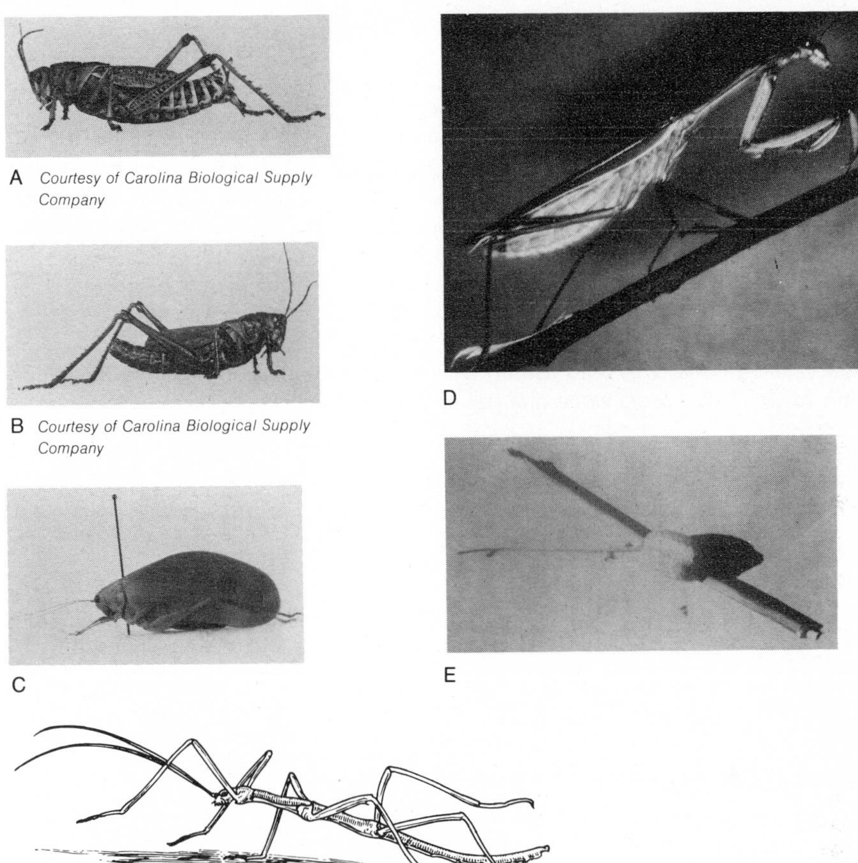

Figure 3-13 Grasshoppers, katydids, and mantids. **A** *Romalea* sp., female, short-horn southern lubber. **B** *Romalea* sp., male, short-horn southern lubber. **C** Katydid. **D** *Stagmomantis carolina,* praying mantis, common in central and southern United States. A harmless insect that feeds on roaches, spiders, and other insects. Generally a bright-green color, and may reach length of 4 in. An excellent laboratory insect and easy to maintain. **E** Egg case of praying mantis *(S. carolina),* green, gray, or brown. Collect on branches during late fall and winter. During spring, in the laboratory, each case may produce 20 or more bright-green, miniature mantids. **F** The walking stick, *Diapheromera* sp. Walking sticks have no wings and closely resemble sticks and twigs; they may grow to a length of 6 to 8 in. Harmless, and interesting because of mimicry. *(D and E Courtesy of CCM: General Biological, Inc. Chicago. F Courtesy of Illinois State Natural History Survey.)*

provide adequate space for 10 to 15 grasshoppers or katydids. When animals are held for only 1 to 2 weeks, no bottom material is necessary, although fresh green leaves should be furnished daily for food and for water sprinkled over the leaves. If animals are to be held for a longer time, spread several inches of dry, clean sand over the cage bottom. Keep sand moist throughout the time that egg cases are being deposited. To study the developing young, strain the egg cases from the sand, and transfer them to dishes of fresh, moist sand. Put animals in a well-ventilated and warm area (23 to 28°C). If retained longer than 2 to 3 weeks, expose animals for several hours daily to direct sunlight or to complete white light (incandescent plus daylight fluorescent lamps). Feed abundantly at all times, preferably with leaves gathered at the collecting site. Although most species feed on a wide variety of plants, some species exhibit a definite requirement for local plants. Other foods that can be tried are lettuce leaves, pieces of apple, grass cuttings, and grass planted in the bottom sand or in pots. Furnish water at all times on cotton-soaked pads or by daily sprinkling of plant leaves. Remove uneaten food before it decays or molds. For long-time maintenance, wash cages and replace bottom materials every 1 or 2 weeks to inhibit infection and parasites. Provide food and water for nymphs in the same manner as for adults.

Tenebrio (Mealworm; Order Coleoptera; Family Tenebrionidae)

USES: *Tenebrio molitor* is cultured widely for feeding the larvae (mealworms) to laboratory animals, including frogs, toads, salamanders, lizards, turtles, and snakes.

DESCRIPTION: Complete metamorphosis. Larvae and adults with chewing mouth parts. Eggs, nearly microscopic, 0.4 to 0.8 mm in diameter. Light-brown larvae (mealworms) to 25 to 30 mm long. Hard-crusted, light-brown pupae to 13 to 17 mm long. Adults, dark brown, and to 15 to 18 mm long, and with general beetle characteristics of a first pair of tough, covering wings, and a second pair of flying wings (reduced in some species). See Fig. 3-14.

SOURCES: Collect *Tenebrio* at flour mills, or obtain a beginning culture from a biology supply house.

Figure 3-14 The mealworm. *Tenbrio molitor.* **A** Larva; lateral view, at top; dorsal view, at bottom. **B** Pupa; ventral view, at top; lateral view, at bottom. **C** Adult.

CULTURE: Various types of containers are suitable, such as shallow pans, wood boxes, glass or earthen jars, and 5- to 10-gal metal cans. Line the bottom of the container with a sheet of wax paper and spread a layer of food over the paper. Probably wheat bran is used most often. Other satisfactory foods are coarse wheat millings (the first dark, coarse milling, sometimes called "red dog" flour), breakfast cornflakes (crushed), dry wheat cereal, and chicken starter mash. To eliminate infestation by other beetles, sterilize bran from wheat mills in an oven at 60 to 65°C for 6 h. Packaged, commercial food should not need sterilizing.

Spread a portion of the beginning culture of beetles over the bottom food layer, and add subsequent layers of food and animals. Finally, sprinkle on top a layer of shredded raw carrots. Cover the container with a securely fastened fine-mesh wire or cloth to prevent the escape of flying adults. With beginning cultures and with those kept in shallow pans, add shredded carrots over the top surface once each week. For deeper cultures, such as those in 5-gal cans, add layers of wheat food as the animals multiply, and provide shredded carrots at least twice weekly to the top surface. Although some persons sprinkle water over a culture each day, we do not find that water is necessary when sufficient shredded carrots are provided. Also, additional water encourages the growth of mold. At 20 to 23°C the life cycle is completed in about 6 months.[1]

Large Moths and Butterflies (Order Lepidoptera)

USES: For studying life histories and feeding adaptations.

DESCRIPTION: Complete metamorphosis; larvae (caterpillars) with chewing mouth parts; adults with a sucking tube, coiled under head when not in use. Two pairs of membranous wings. Body cover of scales, sometimes brightly colored and in striking patterns.

In general, moths are nocturnal; butterflies are diurnal. Adult moths have a fuzzy or velvety body appearance; butterflies have a smooth body surface. Many moths pupate in a parchmentlike cocoon, butterflies in a smooth, tough chrysalid case. Some adult moths (not all species and generally only males) have feathery antennae; butterflies have smooth antennae. Moths rest with wings flat on the body, butterflies with wings upright and nearly together.

Eggs are laid in clusters on plant leaves or stems during the spring or early summer. Occasionally, eggs can be found in the field and transferred to small covered dishes in the laboratory. Larvae hatch from eggs during late spring or summer and vary in length from 2 to 10 cm, according to species. Larvae generally feed on plant leaves, sometimes voraciously. Under natural conditions, most species pupate during late summer or early fall, at which time the larva becomes fastened to a twig, a fence, or some other object. The following spring, adults emerge with shrunken and crumpled wings, which very shortly become expanded as body

---

[1]Maurice Jasper, stockroom manager of the biology department, University of North Dakota, has regularly maintained cultures in 5-gal metal cans for indefinite periods by the above method. For food, he uses a 1:1 mixture of "red dog" flour and wheat bran or chicken starter mash.

fluid is pumped into the wings. As a rule, the adult life span is reduced to a period of mating and laying eggs. Some species use the long proboscis for sucking nectar from flowers; others do not feed and may have reduced mouth parts.

Practically all Lepidoptera are suitable for laboratory culture, although those with destructive larvae must be carefully controlled to prevent escape. The larger Lepidoptera include: swallowtail butterflies (family Papilionidae); white and sulfur butterflies (family Pieridae); monarch butterflies (family Danaidae); skippers (family Hesperiidae); sphinx or hawk moths, tobacco and tomato hornworms (family Sphingidae); giant silkworm moths (family Saturniidae); tent caterpillars (family Lasiocampidae). Of these, the giant silkworm moths are among the most beautiful, including the following: cecropia moth *(Hyalophora cecropia)*, 12 to 15 cm wingspread; promethea *(Callosamia promethea)*, resembles cecropia but is smaller; the very beautiful luna moth *(Actias luna)*, light green with long tails on back wings and about 10 to 15 cm length; polyphemus *(Antheraea polyphemus)*, 7

Figure 3-15 The butterfly. **A** *Papilio glaucus*, a swallowtail. **B** *Colias eurytheme*, a sulfur butterfly. **C** *Danaus plexippus*, monarch butterfly. **D** *Polites* sp., a skipper butterfly.

A

B

C

D

Figure 3-16 The moth. **A** *Celerio lineata,* the white-lined sphinx moth. **B** *Hyalophora cecropia,* the cecropia moth. **C** *Antheraea polyphemus.* **D** *Callosamia promethea.* *(C and D Courtesy of CCM: General Biological, Inc., Chicago.)*

to 12 cm long, a yellow to brown color and a large spot on each wing. See Figs. 3-15 and 3-16.

SOURCES: See Chap. 1 for collecting techniques. Commercial sources are listed in Appendix B.

CULTURE: Maintain larvae in covered jars, in insect cages of fine-mesh wire, or in other similar vessels, all covered with porous material. Some species are extremely restricted in the plant that is eaten. Therefore, when collecting animals, the surrounding plants should be noted, and so far as possible, leaves from these plants should be supplied as food in the laboratory. Leafy twigs in bottles of water will last longer, although the open bottle top must be covered around the twig to prevent insects from drowning. Sometimes lettuce and other leaves can be substituted when particular leaves are difficult to obtain. Most larvae eat a large quantity of food, and a fresh supply must be given every 1 or 2 days. Before adding fresh leaves, dip them in water both to wash the leaves and to add drinking water. When removing dead leaves, use a brush to gently transfer the larvae to fresh leaves.

Drinking water must be supplied daily by sprinkling a small amount over the leaves or by other means that prevent animals from drowning. A slightly dampened blotter or paper towel over the inside bottom surface of the container will give added moisture and will catch animal waste. Generally, no cleaning of the container is necessary except to remove and replace the bottom blotter.

In late summer or early fall, cocoons and chrysalids, still attached to twigs or other materials, may be transferred to the laboratory. Probably the simplest procedure for maintaining locally collected species is to put the caged pupae in protected outdoor areas. Otherwise, the natural temperature and humidity requirements must be provided within the laboratory. Retaining adults for breeding purposes is not done easily, although it may be successful if adults are allowed to emerge from pupae in a large cage or box with sufficient space for flying. A tight greenhouse is very desirable if necessary security can be established in regard to opening and closing of doors and the use of the plant house by other persons. Drinking water must be supplied by sprinkling plant leaves within the greenhouse.

Drosophila (Fruit Fly; Order Diptera; Family Drosophilidae)

USES: *Drosophila melanogaster* is used commonly for genetic studies. The flies, particularly the large *D. virilis*, are excellent food for frogs, toads, and lizards, when an open culture bottle of flies is placed in the animal cage. The animals feed on the adult flies as they emerge. The cage must be covered with fine muslin cloth to prevent escape of the flies. Generally mold does not develop if the bottles are laid horizontally inside the animal cage. Fresh cultures are added as needed.

DESCRIPTION: First pair of membranous wings; second pair reduced to knobbed, balancing organs, the *halteres*. Complete metamorphosis; larva with chewing mouth parts; adult with a lapping proboscis.

The life cycle from egg to adult requires 10 to 11 days at 25°C and 13 to 15 days at 20°C. Generally the female does not mate until 8 or 10 h after emerging from the pupa. Single eggs, about 0.5 mm long, are produced at a rapid rate during the first week following emergence and at a slower rate for perhaps a month. The ovoid egg, laid on ripe fruit in nature, is equipped with two anterior filaments that prevent the egg from sinking into the moist medium. Within 2 or 3 days the eggs become larvae that feed actively on the medium, producing a rough, sometimes crumbled texture (called "working" the medium). The "worked" medium is an indication of success in starting a culture. A larva passes through three instars (two moltings), reaching a final length of 4 or 5 mm within 3 to 5 days, at which time the larva crawls up the side of a culture bottle and changes into a pupa inside the third larval skin. During the next 5 or 6 days, while internal changes are occurring, the pupal covering changes from a soft, white skin to a dark-brown, hard case. The adult emerges with crumpled wings, which become expanded within the first several hours, during which time the body assumes a more rotund shape and the bands of the abdomen change to a darker color. The adult may live 2 or 3 months. Figure 3-17 shows the distinguishing traits of the male and female *D. melanogaster*.

SOURCES: Obtain pure strains from commercial suppliers. Collect wild flies in a ½-pt cream bottle or other small-mouthed vessel containing about 1 tablespoon of mashed banana or other ripe fruit. Lemon juice, squeezed over the fruit, provides an added attractant. Place the open bottles in areas where the flies may be found, for example, near a fruit stand or garbage can, or in the preparation room of a grocery market. Leave the bottle in place for several hours; when flies have entered the bottle, close it with a cotton plug or cloth cover. Transfer flies to fresh medium in the laboratory. Thompson (1967) has described an effective fly trap, shown in Fig. 3-18.

CULTURE: Various types of containers are suitable: ½-pt cream bottles or 125- to 250-ml erlenmeyer flasks, both plugged with nonabsorbent cotton (wrap the plug in a layer of cheesecloth) or sponge-rubber plugs; large test tubes with cotton plugs; and ice-cream cartons covered with punctured plastic film. We find that the small flasks are particularly useful because the large base and smaller opening help to prevent the loss of flies during transfers.

Many kinds of medium are suggested in the literature. The following medium has proved satisfactory because it is relatively inexpensive and solid enough that it does not become dislodged when the bottle is inverted.

540 ml H_2O

50 g cornmeal

25 g rolled oats (not quick-cooking)

50 g dextrose

5 g dried yeast

6 g agar

0.8 g mold-inhibitor powder dissolved in 5 ml 95% ethyl alcohol

Mix ingredients and heat until the medium has boiled several minutes, stirring constantly to avoid sticking and lumping. Pour hot medium into culture vessels to a depth of 1.5 to 2 cm. A folded strip of paper inserted part way into the soft medium is often recommended as a place for pupation. We consider the paper unnecessary, however, since pupation occurs as readily on the container wall above the medium. Plug or cap the containers and sterilize at 15 lb pressure for 15 min in an autoclave or a pressure cooker.

Before transferring flies to a bottle, place 2 or 3 drops of yeast, in water suspension, over the cooled medium and allow bottles to set for 24 h. The yeast helps to prevent drying and provides food for the larvae and adults. As the flies develop, add 2 or 3 drops of yeast suspension regularly once or twice weekly. Bottles of medium may be stored in the refrigerator for 2 to 3 weeks before use.

Maintaining cultures. Store cultures in a room or incubator at minimum-maximum temperatures of 20 to 25°C. Do not put cultures in the refrigerator, in direct sunlight, or near a cold window or heat source. Once or twice weekly, add several

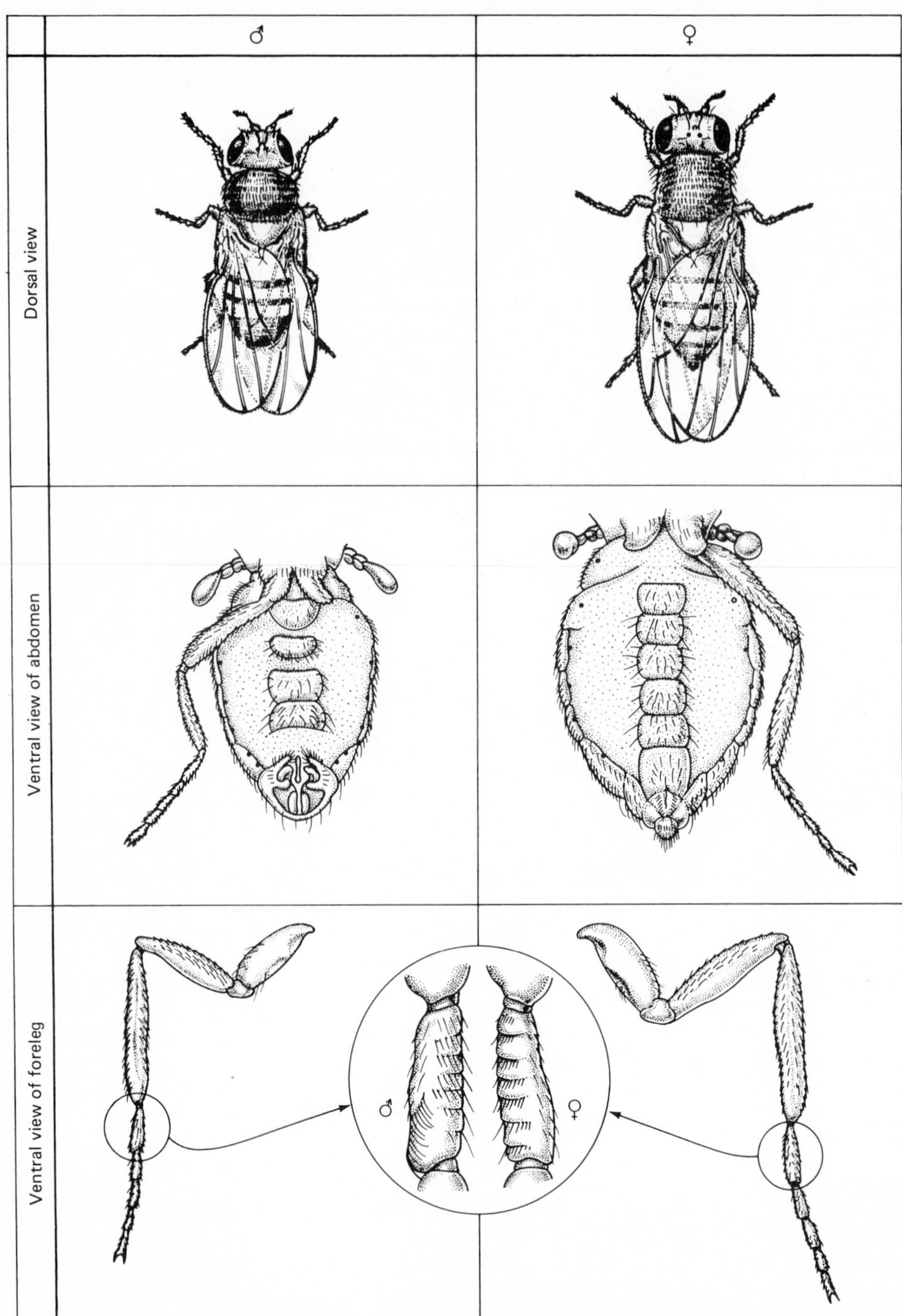

Figure 3-17 Distinguishing traits of male and female *Drosophila*. Left, male. Right, female. *(Biological Sciences Curriculum Study.)*

drops of yeast suspension. Maintain at least two cultures of each stock to avoid losing a strain because of poor growth, accident, or disease. Transfer stock cultures to fresh medium every 2 or 3 weeks.

If mold develops, it is advisable to discard the entire culture by overetherizing the flies and sterilizing the plugged bottle of flies for 15 min with 15 lb pressure before emptying the bottle. If efforts must be made to save a molded strain of flies, use a small brush to dislodge and transfer pupae to a small dish of potassium permanganate solution for 30 sec (one or two crystals in water to produce a light-red color), and then transfer the pupae to fresh medium. Wash and sterilize the brush before using it again. Several transfers may be required during subsequent generations before the mold is lost.

An infestation by tyroglyphid mites is a constant hazard and may become extremely difficult to eradicate. Generally the mites can be avoided if old culture bottles are cleaned immediately. The adult and early nymph of the mite feed on the fly medium, but a last nymph stage may become fastened to adult flies, sometimes in great numbers, causing a general debilitation or death. The mites may be transferred from bottle to bottle and at times may invade an entire laboratory, necessitating strenuous procedures for eradication. A continuous practice of cleanliness and constant observation for mites are extremely expedient. Take measures to prevent outdoor wild flies from entering the laboratory and carrying the mite to the stock cultures. Within the laboratory, trap wild flies in open bottles of medium to which are added several drops of lemon juice. Isolate all suspected cultures, and immediately sterilize them in an autoclave. Sometimes all cultures must be discarded in this manner. If mites invade the room, wash or spray the entire laboratory – walls, ceiling, floors, furniture, and all facilities – with a strong

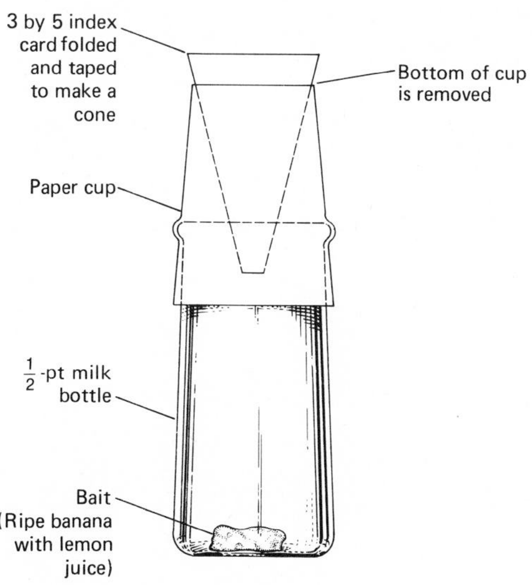

Figure 3-18 An inexpensive fruit fly trap. *(After Thompson, November 1967, Turtox News,* **45** *(11).)*

water solution of disinfectant, such as Lysol. (Use heavy rubber or plastic gloves.) On rare occasions, the services of a commercial exterminator may be required.

Transferring flies to fresh medium. If the food is soft within the bottle of flies, tap the bottom of the bottle against the palm of one hand or on a sponge, quickly remove the plug, and immediately invert an open bottle of fresh medium over the opening of the bottle of flies. Hold the two bottles tightly together while the flies move upward into the fresh container. If the food is not soft, a more efficient transfer can be done as follows. Tap the bottom of the bottle of flies several times against the palm of one hand to shake the flies to the lower area; quickly remove the cap of the fly bottle, and immediately invert this bottle over an open bottle of fresh medium. Hold the two bottles in tight position while gently tapping the lower bottle on a book or pad of cotton, thus causing the flies to drop into the lower bottle. Do not immediately discard the original culture bottles from commercial suppliers as they may contain early stages of flies. If the old medium is dry, add several drops of yeast suspension and store the container at 20 to 25°C. Label the fly strain and date of transfer on each new culture.

Anesthetizing flies. Because flies are killed easily with other anesthetics, such as chloroform, ethyl (diethyl) ether is used most often. (Anesthetics are discussed in Chap. 5.) Ethyl ether is extremely volatile and must be used with great care. Extinguish all flames and remove sources of static electricity. Administering ether from squeeze bottles or spray cans is not recommended. With beginning students or with large classes where supervision is difficult, a teacher may prefer to anesthetize the flies and distribute them to the students. Otherwise, the ether is dispensed to groups in small dropping bottles. At the end of the laboratory period, all remaining ether is returned to the storage can. Bottles of ether are not stored at work benches, particularly if the laboratory is used by other groups.

If the medium has been pulverized excessively by larvae, transfer adults to empty bottles or to bottles of fresh medium before anesthetizing. A satisfactory etherizer is shown in Fig. 3-19, where a strip of cotton cord is fastened around the outer surface of a funnel tube. Place several drops of ether on the cotton, and insert the funnel into a small flask (or bottle). Do not allow drops of ether to fall into the flask, as flies are killed by contact with liquid ether. Put a small pad of cotton inside the funnel to plug the top opening of the funnel tube; hold the funnel tightly against the flask for about 15 sec while ether fumes collect in the flask. Then remove the plug from the funnel, quickly tap the bottom of the fly bottle against one hand, remove the plug of the fly bottle, and immediately invert the container down tightly inside the funnel. Tap the fly container gently to cause flies to drop through the funnel into the etherizer. If moist or pulverized medium drops into the funnel tube, it may be necessary to stop the operation and remove the medium, a process that generally results in the escape of flies.

If flies are to be revived for later studies, they must be anesthetized lightly, a technique that requires practice to learn the amount of ether and the required time for a particular flask size. If fumes in the flask become too concentrated, the flies may die at once. Otherwise, the flies move about for several seconds and then fall to the bottom as they become anesthetized. When the flies stop moving, remove and restopper the fly bottle. Pour anesthetized flies on a white card or paper to observe with a hand lens or a stereomicroscope. Use a soft, tapered brush or a blunt probe (a lead pencil) to move the flies about under the lens. Dead flies are recognized by the position of their wings which are extended vertically above the body, whereas the living, anesthetized flies hold the wings in a horizontal position, parallel to the body.

Flies will remain anesthetized for several minutes and will begin to move about as they revive. To reetherize, brush the flies back into the anesthetizing flask, first adding ether to the cotton strip and plugging the top opening of the funnel tube as before. Alternatively, convert one-half of a petri plate into a reetherizer by taping a small square of cotton, covered with cheesecloth, against the inside top surface. (Flies may become entangled in uncovered cotton and will die if they contact liquid ether.) Put 1 or 2 drops of ether on the pad, and brush the reviving animals under the inverted plate as shown in Fig. 3-19. To prevent the escape of flies be certain to select a plate that lies flat against the table.

Note: When transferring anesthetized flies to a bottle of fresh medium, lay the bottle on its side and place flies on the lower surface, leaving the bottle in that position until the flies are revived. Do not put anesthetized flies on the medium, as they may sink into the moisture and suffocate.

Collecting virgin females. Ordinarily, the females do not mate until about 8 h after emerging from the pupal case. A simple procedure for collecting virgin females is to remove *all* adult flies from a bottle early in the morning, a task requiring *close observation.* As flies emerge during the day, the females are separated

A

Figure 3-19 Equipment for anesthetizing *Drosophila.* **A** Etherizing equipment. **B** Reetherizer.

B

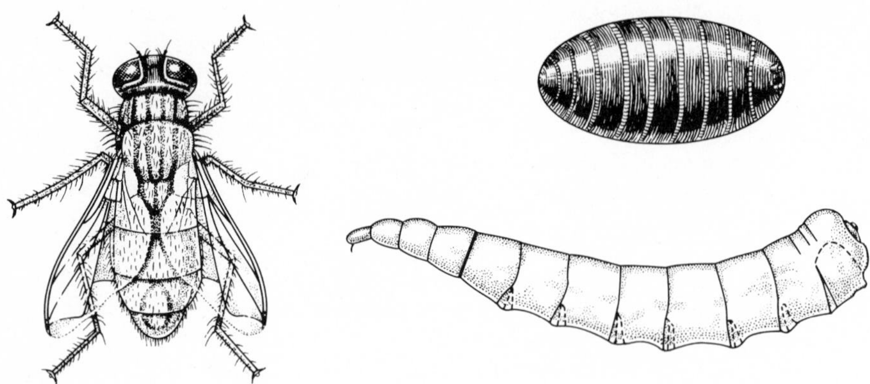

Figure 3-20 The house fly, *Musca domestica*. Adult fly, at left. Pupa, at top right. Larva, at lower right. *(Courtesy of Illinois State Natural History Survey.)*

and isolated into bottles of fresh medium. When adults first emerge, the bands on the abdomen are lightly colored and the abdomen is not yet filled out. To separate the sexes, look for the ventral, dark-brown genitalia of the male, shown in Fig. 3-17. These can be detected with the unaided eye, although a beginner should use a hand lens or a stereomicroscope.

The House Fly (Order Diptera; Family Muscidae)

USES: A versatile animal for routine or special studies such as reproduction, development, genetics, behavior, and effectiveness of insecticides. Use as live food for amphibians and reptiles.

DESCRIPTION: First pair of membranous wings; second pair reduced to knobbed, slender *halteres* used as balancing organs. Complete metamorphosis. A maggot larva. Adult with lapping mouth parts. Worldwide distribution and perhaps the most ubiquitous of insects (see Fig. 3-20).

SOURCES: Pure cultures of *Musca domestica* can be obtained from biology supply houses and, at times, from research laboratories of the U.S. Department of Agriculture. Wild house flies may be collected with a fly trap, such as that shown in Fig. 3-18.

CULTURE: Maintain 40 to 50 adult flies in a gallon, widemouthed jar, or a similar container. Glass or clear plastic containers allow better observation and easier manipulation of the flies. Around the mouth of the jar securely fasten a sleeve, fashioned as a tube from a 24-in square of cotton muslin with two opposite sides of the square sewed tightly together as shown in Fig. 3-21. A satisfactory sleeve can be made from ladies' hose or men's long socks, with the foot portion cut away. Loop the long sleeve into a knot to close the outer end, or close the end with a rubber band or string. When supplying food and water, shake the flies from the sleeve down into the jar and insert the supplies through the sleeve with the outer end draped tightly around the arm. Retain animals at 21 to 25°C, removed from direct sunlight, cold windows, or heat radiators.

Furnish water at all times on heavily soaked cotton pads placed in a small dish, or use some other watering device that protects flies from drowning. Provide a constant supply of 0.1 *M* sucrose solution on cotton pads (34.2 g sucrose in 866 ml water). Although adult flies can be sustained on sugar water alone, if egg-laying is desired, a protein food must be given. Probably milk is used most often, thus supplying both food and water. Either fresh or sour milk is satisfactory; it may be easier to store dried or condensed or evaporated milk, each mixed with an equal volume of water for feeding. Dispense the milk on heavily soaked cotton pads in a small dish. To avoid a disagreeable odor, give a new supply of milk in a clean dish at least every 2 days.

Eggs are deposited on the cotton pads, beginning about 4 days after adults emerge from pupal cases. A female may lay 100 or more single eggs daily during a period of 7 to 10 days. Clusters of the cream-colored, cylindrical eggs, each about 3 mm long, can be seen in crevices of the cotton pad. A hand lens or stereomicroscope will help to locate the egg clusters. Invert the cotton pads with eggs onto the surface of larva medium (described below) in half-filled quart jars or other containers. Allow the pads to remain on the medium for 2 days.

A variety of larva food can be used, including canned dogfood, or crushed dog biscuits moistened lightly with whole milk or with a 1:1 mixture of dried milk and water. The food may soon develop an unpleasant odor, and the culture must be removed to an isolated closet or room.

Generally, larvae hatch during the first day of transfer of cotton pads and move down into the medium. After 4 or 5 days of growth, the larvae begin to migrate toward the surface, where pupation occurs. Although not necessary, it is well to add a 2-cm layer of sand over the medium on about the fourth day after eggs are transferred to the medium. The larvae will move into the sand to pupate; and the

Figure 3-21 Fly culture chamber. A long, tubelike sleeve fashioned from a square of cotton muslin and taped to the mouth of a jar. The sleeve is tied into a knot at the outer end.

pupae are separated more easily from the sand than from the top region of the medium.

Flies are most easily transferred in the pupal stage, although the pupae must be handled gently. For stock cultures, transfer 40 to 50 pupae to each gallon jar equipped with a cloth sleeve. Depending on their later use, transfer other pupae to petri plates, small individual boxes, or test tubes. Store the pupae at about 23°C. If room humidity is low, maintain pupae on slightly dampened cotton. Adult flies emerge on the sixth or seventh day following pupation (the tenth or eleventh day after eggs are transferred to the larva medium). When feeding house flies to other laboratory animals, transfer the fly in the pupal stage. The animal cage must be covered with a fine muslin cloth to prevent the escape of flies as they emerge from the pupal cases.

The Blow Fly (Order Diptera; Family Calliphoridae)

USES: Same as for the house fly. See Dethier (1955).

DESCRIPTION: Blow flies have the same general characteristics as the house fly. Adults are about the same size or a little larger than house flies. Often they are metallic blue or green, distinguishing them from the flesh flies (Sarcophagidae) which are black with gray stripes on the thorax. The blow fly lays eggs on animal carcasses and other decaying organic matter, where the larvae (maggots) develop and feed.

Apparently all species can be grown without difficulty. During the 1930s *Phaenicia sericata* and *Phormia regina* were cultured extensively in medical research laboratories. The sterile-clean larvae were used to clear away suppurative tissue in postoperative treatment of osteomyelitis.

SOURCES: Obtain flies from biology supply houses or research laboratories. Collect the larvae and pupae from carrion, or with raw meat in jars or fly traps (a less disagreeable method) located near garbage cans and animal barns (see Fig. 3-18).

CULTURE: Transfer 40 to 50 adults or pupae to a gallon jar with an attached cloth sleeve as shown for house flies (Fig. 3-21). Furnish drinking water and cane sugar as described for house flies. For oviposition, provide 1-in squares of raw liver slightly moistened with several drops of water and loosely stacked in a dish, thus furnishing crevices where the eggs will be deposited. Replace with fresh liver every 24 h. If eggs are present, gently scrape them from the meat into a bottle of sterile medium prepared by the following instructions.

20 g agar — 100 g dry yeast

100 g powdered milk — 1 l water

2 g mold-inhibitor powder dissolved in 5 ml of 95% ethyl alcohol

Mix ingredients thoroughly. Heat and boil for several minutes, *stirring continuously* to avoid sticking and lumping. Pour 1-pt jars one-half full. Cover lightly with a screw-on, solid-top lid, and sterilize for 15 min at 15-lb pressure. Allow medium to cool, and add 2 or 3 drops of yeast in water suspension to the surface of the medium before transferring the eggs (25 eggs in a 1-pt bottle). Do not place the eggs in drops of moisture; if necessary, allow the capped bottles to set for 24 h before adding the eggs. After transferring the eggs, fill the bottle with sawdust or shredded paper sterilized in a cotton-plugged container within a dry oven for 2½ h at 110 to 115°C.

Although the larvae and pupae will develop without the strict sterile conditions described above, bacterial decay may produce an unpleasant odor, requiring the removal of cultures to an isolated location.

Maintain all stages at 21 to 25°C. Larvae move up into the sawdust to pupate. Pour sawdust and pupae out of bottles, remove pupae from sawdust, and transfer to adult-fly containers or to other containers for special uses. The approximate times for development at 23°C are: egg hatches within 2 days; larva pupates within 8 days; adult emerges from pupa within 4 days—a total of approximately 14 days for completion of development.

Mormoniella Wasp (*Nasonia;* Order Hymenoptera; Family Chalcididae)

USES: (1) For genetic studies. See Whiting and Caspari (1957); Whiting (1967); Darling (1968); and Walker and Pimentel (1966). (2) For observing parasitism of mormoniella on the pupae of the flesh fly, *Sarcophaga.*

DESCRIPTION: The chalcid wasps constitute a large group of important insects that usually are very small, 4 to 5 mm or less in length. Because of their small size, the wasps are often overlooked by nonspecialists in spite of their great number over the United States. Most species are parasitic on other insects, including many of the crop pests. Thus, the chalcids are important in the control of insect pests, and many species have been imported for this purpose. Whiting (in 1948) was among the first to develop methods for genetic studies of the mormoniella wasp (*Nasonia vitripennis*), and the host fly, *Sarcophaga bullata.* Since that time the minute wasp has come to occupy a permanent place in routine genetic studies, although it has in no way preempted the position of the *Drosophila.*

The adult mormoniella is 4 to 5 mm long. The female has dark antennae and long wings extending the length of the abdomen. The male has lighter-colored antennae and short wings that reach to about half the length of the abdomen (see Fig. 3-22).

Mormoniella deposit their eggs on the pupae of flies, including the flesh flies (Sarcophagidae) and blow flies (Calliphoridae). The flesh fly, *Sarcophaga bullata,* is used commonly as the host for laboratory work (see Fig. 3-23). During oviposition, the adult female wasp punctures the outer pupal cover (puparium) of the fly and deposits eggs on the surface of the fly pupa inside the puparium. At the same time, the female wasp makes punctures into the fly pupal body, forming permanent openings through which she feeds on the fly body. Larval wasps hatch between the outer puparium and the inner fly pupa, where they feed on the pupa surface. The larvae develop into naked wasp pupae which remain on the surface of the fly pupa while developing into the adult wasp. Adult wasps puncture the outer puparium, crawl to the outside, and mate immediately. The entire life cycle

A Courtesy of Carolina Biological Supply Company

B Courtesy of Carolina Biological Supply Company

Figure 3-22 Mormoniella wasp *(Nasonia vitripennis).* **A** White-eyed male. **B** Female depositing eggs into puparium of the flesh fly, *Sarcophaga bullata.*

A Courtesy of Carolina Biological Supply Company

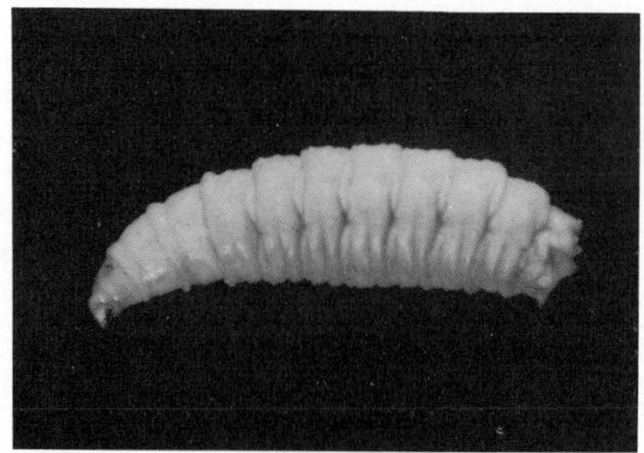

B Courtesy of Carolina Biological Supply Company

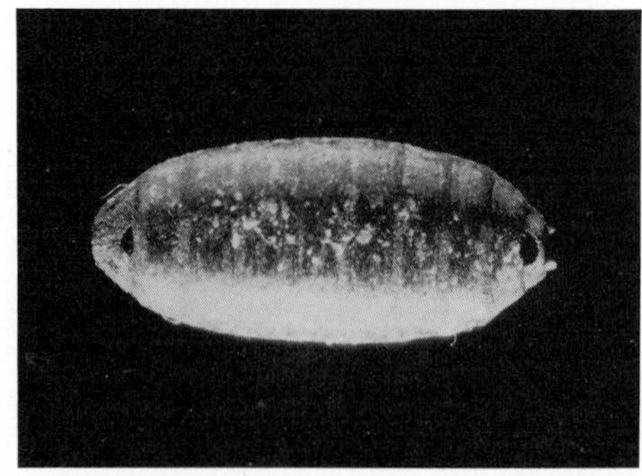

C Courtesy of Carolina Biological Supply Company

Figure 3-23 *Sarcophaga bullata,* flesh fly. **A** Mature female; **B** Larva; **C** Puparium, approximately 2.5 cm long. Note holes in wall through which adults of mormoniella have emerged.

requires about 10 to 12 days at 27°C and about 14 to 16 days at 22°C. Female wasps develop from fertilized eggs, and male wasps from the unfertilized eggs, as is true of most Hymenoptera.

SOURCES: Wild wasps can be collected in parasitized puparia of blow flies and flesh flies. This is not a simple nor pleasant task since it entails probing under and around the decaying flesh of dead animals or their excrement. Unless the research involves local mormoniella, the wasp is obtained from a research laboratory or from Carolina Biological Supply Co.

CULTURE: Several procedures can be followed for maintaining cultures to use in genetic studies.

1. *Purchasing Parasitized Sarcophaga Pupae.* The simplest and most expensive procedure (also the least interesting) is to purchase parasitized *Sarcophaga* puparia containing wasp pupae from adult wasps of the desired parent crosses. Upon receiving the fly puparia, place them in an incubator at 27°C or at room temperature. Normally, adult wasps emerge within a week, depending on the temperature and their stage of development when received in the laboratory. Etherize the adult wasps and observe them with a hand lens or stereomicroscope to determine the ratio of genetic traits.

2. *Purchasing Mormoniella Pupae and the Sarcophaga Puparia.* A more interesting procedure, which provides greater learning value, is to purchase several strains of mormoniella pupae (generally about 35 pupae of a strain are furnished in a vial) and a quantity of unparasitized fly puparia. Store the fly puparia until needed in the refrigerator at 8 to 10°C. The students sort the wasp pupae by sex and by special genetic traits visible through the pupa skin, such as eye or body color. The female is distinguished by the ovipositor, seen as a pale streak on the ventral abdomen. According to the desired crosses, put several wasp pupae of each sex in cotton-plugged vials and add two or three fly puparia. Store in an incubator at 27°C or at room temperature. Within several days, the adult wasps emerge and mate, and the females deposit eggs in the fly puparia. Remove the adult wasps on about the sixth day following the initial preparation of the vials, which is after the eggs are laid but before the F_1 wasps emerge. If the F_1 wasps are to be crossed, open the fly puparium on the seventh or eighth day (before emergence of adults), identify the traits and sex of pupae (F_1 wasps) with a stereomicroscope, and place six or eight, each, of males and females together in a vial according to desired crosses. Add several fly puparia. When the F_1 adult wasps emerge, they will mate immediately, and their eggs will be deposited in the puparia. The adult F_2 wasps will emerge in about 2 weeks, depending on the temperature.

3. *Maintaining stock cultures of flies and wasps.* If many flies and wasps are required on a regular basis, obtain initial cultures of unparasitized *Sarcophaga* puparia and various strains of wasps in fly puparia. Maintain stock cultures of the fly, using the same techniques as those described earlier for the blow fly. Maintain stock cultures of mormoniella on the fly puparia by adding fresh fly puparia before

wasps emerge, or by transferring wasp pupae (removed from fly puparia and sorted as desired) to fresh fly puparia. NOTE: Both the unparasitized and parasitized fly puparia can be stored in the refrigerator at 8 to 10°C for several weeks or longer.

Ants (Order Hymenoptera; Family Formicidae)

USES: For studies of social behavior.

DESCRIPTION: Social insects with nonreproductive, wingless female workers, the largest caste in number of individuals; and with reproductive males and females, generally winged (see Fig. 3-24). Chewing or lapping mouth parts. Ovipositor often modified into a stinger. First abdominal segment pinched-in to form a stalk, or *petiole*, between thorax and other portion of abdomen; a dorsal projection on petiole that distinguishes ants from antlike Hymenoptera. Most species omnivorous. Much variation in type of nest and number of ants in colony.

Periodically, winged males and females are produced in great numbers that leave the nest for a nuptial flight. The female is fertilized, and the male soon dies. The female finds a nest site, breaks off her wings, and seals herself into a hollow chamber where she lays eggs and secretes a food through her mouth for the subsequently developing larvae (grubs). The larvae pupate, and the first adults emerge as worker females that hunt and carry food to later broods of developing workers. The queen, fertilized only the one time, continues to lay eggs for several years. At regular intervals, once a year for many species, reproductive males and females are produced and additional colonies are established following the nuptial flights.

Most types of ants are cultured easily. The large, mound-building ants of the genus *Formica* are found over most of the United States and are among the most interesting for laboratory observation. The name derives from the ability of many species to spray formic acid as a defense mechanism. When closely confined in collecting bags, the ants may die from the fumes of their own secretion. Color varies among species and may be black, red and black, dark brown, reddish brown, reddish yellow, or yellow. Other interesting and easily maintained mound builders are the large, red harvester ants, *Pogonomyrmex*, found in the Central Plains and southwestern and western United States. A considerable disadvantage in handling these ants is their formidable sting.

SOURCES: The best collecting time is during the warmer months. Often the winged males and females can be found in nests during August and September. Use a trowel or spade to turn up the soil of a nest. Quick digging is required since many ants, when disturbed, immediately move deeper into the earth, including the winged males and females, and the workers carrying with them the brood. Separate the ants from the soil with tweezers or an aspirator (see Chap. 1), and place them in a covered container with a portion of soil. Alternatively, the soil can be scooped into a large container and carried to the laboratory for separation. For successful maintenance, a queen ant must be obtained with the colony.

CULTURE: A variety of culture chambers are suitable, including quart or gallon jars filled with the nest soil and covered with a porous material. Soil obscures

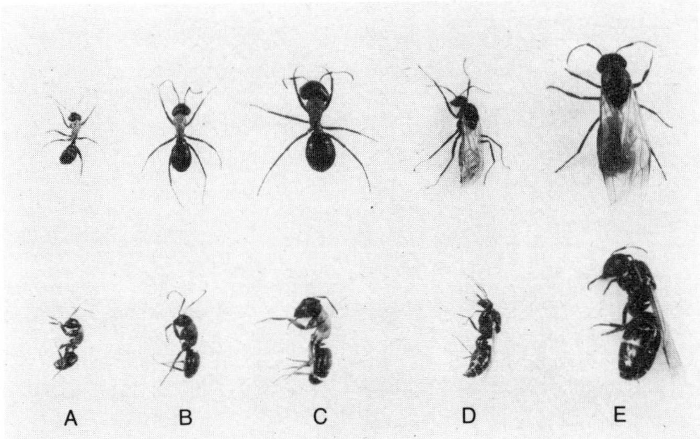

Figure 3-24 Castes of the carpenter ant, *Camponotus noveboracensis*. Top row, dorsal view; bottom row, lateral view. **A–C** Three subcastes of workers. **D** Drone. **E** Queen.

much observation and is not necessary unless one wishes to observe ant excavations. Commercial ant chambers (without ants) are sold by many supply companies. A portion of the chamber should be dark for a nesting area. Red light is not perceived by ants, and thus a red glass or red cellophane covering will permit observation of activity without disturbing the ants. Provide ventilation through cotton-plugged holes in the chamber walls. Maintain nests at a temperature of 21 to 25°C. According to Wheeler and Wheeler (1963) the best all-purpose nest is the one shown in Fig. 3-25. Forrest (1962) describes a nest consisting of three plastic chambers (ordinary sandwich boxes) connected by plastic tubing.

Supply water at all times on a sponge or cotton pad, or by sprinkling water over the soil. The type of food varies among species. Many will eat honey on bread, moistened sugar, raw or hard-boiled egg yolk, dry cereal flakes, mealworms, and larval and adult flies.

Figure 3-25 Artificial ant nest. Right chamber and top-left chamber are covered with clear glass. Lower-left chamber is covered with red glass. Water is added to small sponges inside the nest with a pipette inserted through stoppered openings at left.

Honey Bee (Order Hymenoptera; Family Apidae)

USES: For studying behavioral patterns of social insects. A window beehive allows students to observe the extremely fascinating dances of bees, as well as other social activities.

DESCRIPTION: The honey bee, *Apis mellifera*, is the most domesticated of all insects. Most colonies are located in domestic beehives; occasionally domestic colonies will escape and become "wild" colonies in hollow trees. The honey bee and the social bumble bee are among the most important of insect plant pollinators. As with most Hymenoptera, unfertilized eggs develop into haploid males and fertilized eggs into a queen or female workers (see Fig. 3-26). A female becomes a queen or a worker depending on the kind of food given the larva. When the colony population reaches a certain size, a new queen is developed, and either the new or old queen leaves the hive in a swarm with a group of workers to build a nest elsewhere. The new queen mates during a nuptial flight and thereafter spends all her time laying eggs. For descriptions of social behavior, see the literature listed at the end of the chapter.

A *Courtesy of Carolina Biological Supply Company*

B *Courtesy of Carolina Biological Supply Company*

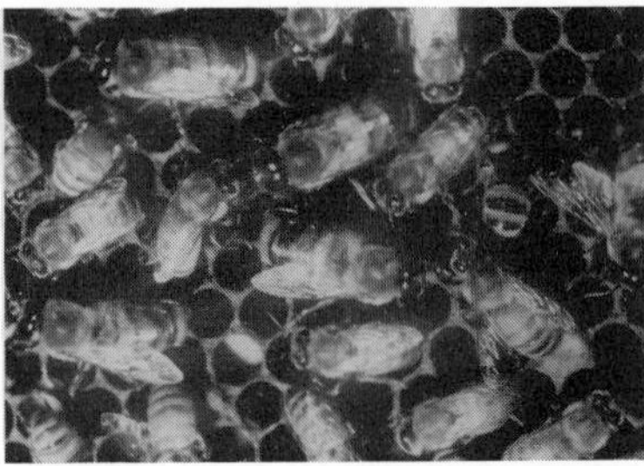

C *Courtesy of Carolina Biological Supply Company*

Figure 3-26 Castes of bees. **A** Queen, at center, laying eggs and attended by workers. **B** Worker bees on surface of honeycomb. **C** Drones with smaller workers.

Courtesy of Carolina Biological Supply Company

Figure 3-27 An observation beehive.

SOURCES: Collecting and transferring colonies of wild honey bees is best done by an experienced beekeeper. Wild colonies do not settle down easily to laboratory maintenance and often will swarm and depart in mass from a window hive. Additionally, wild bees may inflict a more painful sting than do domestic strains. Colonies of domestic bees can be obtained from biological supply houses or from local beekeepers.

CULTURE: Observation beehives are available from biology supply houses, with a colony of bees and instructions for maintaining them; or a person may wish to construct an observation hive similar to the one shown in Fig. 3-27. Place the hive in a window, with a hose leading to the outside for bees to use as an entrance and exit. Colonies are established most easily during late spring or summer months. During winter months, honey or diluted cane syrup must be supplied to the brood chambers.

ARTHROPODA: ARACHNIDA

Arachnids include the scorpions, spiders, ticks, mites, and the harvestman. General traits that distinguish arachnids from other arthropods are the four pairs of jointed legs, the absence of antennae, simple eyes and no compound eyes, and the first pair of appendages (mouth parts) modified into chelicerae, which may be pincers or fanglike claws and in spiders may contain poison glands. Scorpions and spiders are maintained with little difficulty. Mites and ticks are difficult to maintain since many require blood of specific animal hosts for food. Methods for culturing the last two groups can be found in Needham (1937).

Scorpions (Order Scorpionida)

USES: Interesting as a primitive, terrestrial animal. Excellent for studying courtship behavior, patterns of reproduction, and development. Use is restricted because of limited distribution over the United States, although the animals can be purchased from several commercial suppliers (see Appendix B).

A *Courtesy of Carolina Biological Supply Company*

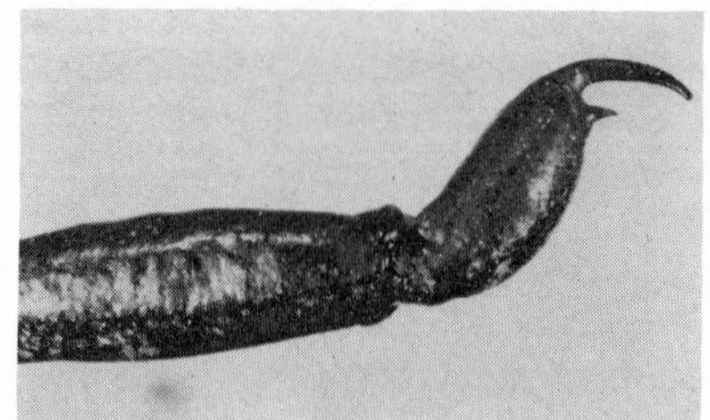

B *Courtesy of Carolina Biological Supply Company*

Figure 3-28 The scorpion. **A** Mature animal. **B** Close-up view of stinger.

DESCRIPTION: Mostly tropical and subtropical; distribution extends into central United States and throughout Pacific coastal region. Oldest known land arthropods; sometimes called "living fossils." Species of Gulf States from 2 to about 6.5 cm in length; a painful sting. Scorpions of Arizona and New Mexico to 4 in long; venom toxic to humans. A relatively short, square cephalothorax (prosoma) covered with a carapace. Four pairs of walking legs. One pair of raised eyes at middorsal carapace; several pairs of eyes at anterior margin of carapace. Pair of large anterior pedipalps (second pair of appendages), each with a greatly enlarged distal pincer. A primitive abdomen consisting of a thick seven-segmented preabdomen and a slender five-segmented postabdomen. Stinging barb and pair of poison glands in last segment. (See Fig. 3-28.)

Scorpions display an interesting courtship behavior which may last several days. The male deposits a spermatophore on the ground and then moves the female in such a way as to attach the spermatophore to her ventral abdomen. The young are produced alive and crawl upon the dorsal surface of the female where they remain for several days. Although reports vary regarding the first food of the young, the author has collected several females, each dead and with 20 to 25 young scorpions moving about on her back. After 5 to 6 days, when they were apparently feeding on the mother's body, the young scorpions began to move away.

SOURCES: Scorpions are nocturnal, and during the day may be found under leaves, rocks, and other debris. At night, the animals can be found in open, woody regions or desert areas. Often they come into houses, garages, or outbuildings. Collecting is generally a random matter. Occasionally a number of animals may be found in scattered locations within a building or at an outdoor collecting site. Collect by laying a net or a jar over the moving animal. If scorpions are to be handled, wear gloves to avoid a painful sting.

CULTURE: Put scorpions in any type of container covered with a punctured lid. Put soil or sand on the bottom surface, and add small stones, leaves, and twigs. Maintain at room temperature. Sprinkle water daily over the animals but avoid excess moisture. Keep a synthetic sponge or cotton pad in the cage, moistened daily with water. Once or twice weekly, feed several small mealworms or other small insects. Provide a constant supply of chick starter mash in a small dish or lid.

Spiders (Order Araneida)

USES: Study of behavioral patterns, including web spinning and the development of young.

DESCRIPTION: A large group of more than 30,000 species. Length from less than one millimeter to several centimeters. Cephalothorax and the unsegmented abdomen separated by a narrow "waist." Unique characteristic of producing silken threads and spinning webs with ventral abdominal spinnerets. Chelicerae (first pair of mouth appendages) with poison glands that empty through fangs.

All spiders "bite," and the poison from most spiders forms necrotic spots that heal slowly. The venom of the black widow spider *(Lactrodectus mactans)* is neurotoxic to man and causes severe muscular pain with a possibly fatal respiratory paralysis. In most species the female is larger than the male, and after mating, the female may kill and eat the male, for example, the black widow spider. Some species exhibit an elaborate courtship pattern. Most lay eggs in a silk sac, attached to various objects or carried by the female. The young resemble the adult except for size. Spiders prey on insects and other small animals. Figure 3-29 shows the types described here.

SOURCES: *Tarantulas.* Collect by laying a net or large pail over the spider as it walks over the ground. When disturbed, tarantulas may jump several feet, and they must be approached quietly. *Web spiders.* With as little disturbance as possible, cut the surrounding vegetation to which the web is attached. Transfer the spider in the web to an appropriate sized cardboard carton for transporting to the laboratory. Knudsen (1966) describes how he carries web spiders (without the webs) in paper bags and transfers them to a 10 by 14 in wooden frame that is hung by a wire from the ceiling of the laboratory. A paper cone is thumbtacked to the corner of the frame; the spider is placed within the cone, and the cone is plugged for several hours. He states that about 90 percent of the spiders will remain in the new location when the plug is removed, and that they spin new webs each night. *Black widow spiders.* These animals are distributed throughout the southern and western regions of the United States. During warm seasons they may be located in garages and other outbuildings. Because of the severe and perhaps fatal effects of its toxin, the black widow must not be handled. If a slow and careful approach is made, a jar can be clamped over the spider and it can be maneuvered into the jar which is then covered with a punctured lid. The spider will kill other small animals and must be placed alone in a container. Great care must be exercised to prevent the spider's escape. Kill the spider by overanesthesia once the living animal and the red hourglass mark on the ventral abdomen have been observed.

CULTURE: Most spiders can be maintained without difficulty in the laboratory. Put the animals in containers of appropriate size and with vegetation from their natural habitat. Sprinkle water daily over the substratum and the vegetation. Feed small mealworms, flies, cockroaches, and other small insects.

A Courtesy of Carolina Biological Supply Company

B Courtesy of Carolina Biological Supply Company

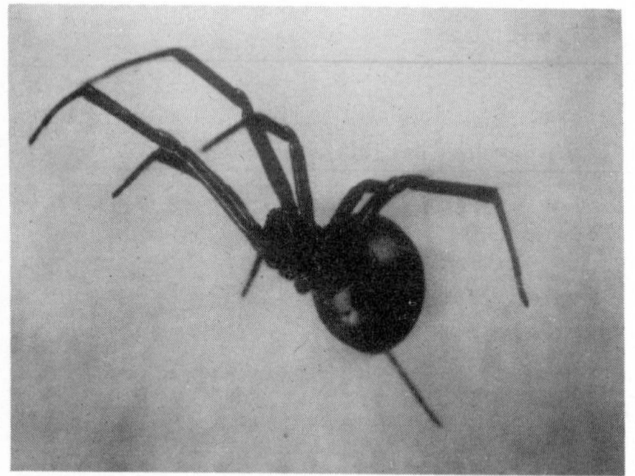

C Courtesy of Carolina Biological Supply Company

Figure 3-29 Spiders. **A** Tarantula, *Dugesiella hentzi.* **B** Garden spider, *Argiope aurantia,* feeding on prey trapped in web. Note the characteristic ladder in the web. **C** Black widow spider, *Lactrodectus mactans,* showing red hourglass marking on ventral surface of abdomen.

ARTHROPODA: MILLIPEDES AND CENTIPEDES

Probably little attention has been given to using millipedes and centipedes in a teaching laboratory. Because of their general distribution over the United States and their availability throughout the year in many regions, possibly both can provide effective demonstrations of physiological functions. Millipedes, in particular, are maintained easily.

Millipedes (Class Diplopoda; "Thousand-legged Worm")

USES: Studies of behavior and development of the young.

DESCRIPTION: A millipede, of course, is not a worm nor does it have a thousand legs. The two pairs of short, hairlike legs on most body segments probably account for the common name. All possess one pair of short antennae; the body is long and tubelike; the number of segments varies from about 25 to 100 or more. The southern millipede, *Spirobolus*, may grow to a length of 6 to 9 cm, has a crusty, brittle body covering, and when disturbed emits a malodorous fluid and draws the body into a flat, tight coil. The smaller *Julus*, of 1- to 2-in length, is distributed over most of the United States and may be seen slowly moving about during both day and night. Apparently millipedes are harmless to man. (See Fig. 3-30.)

Our own observations indicate that the female of at least several species provides considerable care for the eggs by laying them in a cavity "nest" which is fashioned either slightly below the soil surface or in a rounded crest that rises slightly above the surface. Buchsbaum (1948) states that the female guards the nest. The hatched young have a greatly reduced number of segments, with only one pair of legs on each segment. As the animal grows, pairs of abdominal segments fuse, resulting in two pairs of legs for most abdominal segments of the adult.

SOURCES: Look for millipedes among dead leaves and under rocks and logs in dark, damp areas of gardens and woods.

CULTURE: Maintain in a flat receptacle, 3 or 4 in deep, containing several inches of loose humus soil with a top layer of dead leaves. Cover the container with punctured plastic film. The animal is herbivorous and feeds on the decaying plant material or occasionally on green plants. For additional food, add portions of potato and carrot each week. Remove molded food immediately. Keep receptacle slightly moist by sprinkling water over materials several times weekly. Avoid excess moisture.

Centipedes (Class Chilopoda; "Hundred-legged Worm")

USES: For studies of behavior and development; especially interesting for studying positive thigmotaxis, demonstrated by their continuous running about in unobstructed areas and their cessation of movements when at least two sides of their bodies come in contact with solid surfaces.

Figure 3-30 *Spirobolus* sp., southern millipede, approximately 3½ in long.

Courtesy of Carolina Biological Supply Company

DESCRIPTION: Centipedes are not worms, and the number of legs varies among species. The body is flattened dorsoventrally; number of segments varies from about 15 to 170 or more. One pair of legs on each segment. The first pair is modified to a pair of poison claws. Centipedes move swiftly; most species are nocturnal; and apparently all are carnivorous, feeding mostly on worms and insects which they kill with their poison and crush with their strong mandibles. *Scolopendra*, of southwest United States and the tropics, may reach a length of 20 to 30 cm. Their poison can produce death in small animals and a violent reaction in man; all should be handled with forceps or gloved hands. The "house" centipedes *(Scutigera)* and small "garden" centipedes *(Lithobius)* are not harmful and may be beneficial, as they feed on undesirable insects. Apparently, most centipedes lay their eggs in the upper soil. Except for size, the hatched young resemble the adult. (See Fig. 3-31.)

SOURCES: During the day, collect centipedes under rocks and logs of desert regions, or in moist, dark areas. At night, look for animals in open desert regions or open woods and gardens.

CULTURE: Place in receptacles of 3- to 5-in depth, containing loose soil and humus covered with a layer of dead leaves. Close the receptacle with a tight cover of screen or other material punched with small openings. Keep container tightly closed except during treatments and studies, as centipedes will attempt to escape at all times and particularly during the night. Feed with live roaches and other insects of appropriate size.

ECHINODERMATA

USES: (1) Within inland laboratories for display of sea life. (2) For studies of regeneration, reproduction, and development. The sea urchin is especially useful for class study of the developing eggs. Techniques for this work are described below.

Figure 3-31 Centipedes. **A** *Scolopendra* sp., large centipede of southwest United States; approximately 4 in long. **B** *Scutigera* sp., a house centipede. *(B Courtesy of Illinois State Natural History Survey.)*

A *Courtesy of Carolina Biological Supply Company*

B

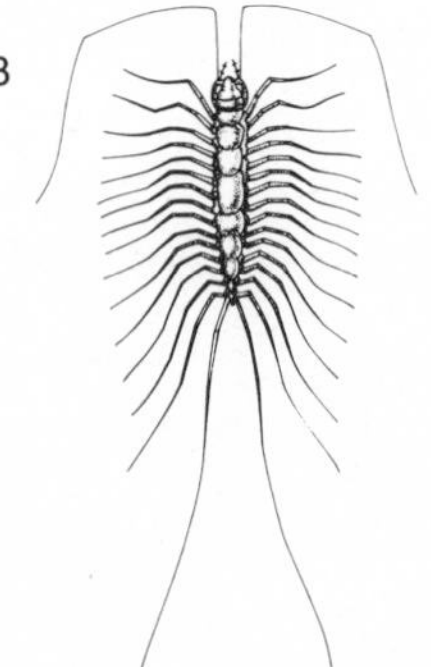

DESCRIPTION: About 6,000 species, all exclusively marine. Most with a bilaterally symmetrical and free-swimming larva that metamorphoses into a sessile or slow-moving, radially symmetrical adult. Most adults with a spiny skin and five radii from a short oral-aboral axis. All with endoskeleton of calcareous plates in dermis, and a water-vascular system of coelomic tubes leading to exterior, hollow tube feet.

Many echinoderms can be maintained without difficulty in natural or artificial seawater, including the sand dollar, starfish, brittle star, sea cucumber, and sea urchin (see Fig. 3-32). A marine aquarium is required; detailed instructions for installation of the aquarium are given in Chap. 1.

SOURCES: Collect from wharf pilings, rocks, and tide pools of intertidal zones, and in bottom sand of shallow water. Use dredge nets in deeper water. Knudsen (1966) describes a starfish drag consisting of several string mops attached to a heavy chain in which starfish and brittle stars become entangled as the mops are drawn across the water bottom. For transporting to the laboratory, wrap the sea stars and sand dollars loosely in moist seaweeds or in newspapers first dipped in ocean water. Put sea urchins into a plastic or glass container of seawater. Carry sea cucumbers in a separate container of seawater. Do not crowd the animals. Styrofoam receptacles are excellent; during warm weather the animals can be cooled with a bag of ice inside the Styrofoam container. A variety of echinoderms are available from commercial suppliers. See the sources listed in Appendix B.

CULTURE: Small-sized specimens are recommended for laboratory maintenance. Feed starfish with live snails, clams, and oysters; give small pieces of raw seafood or beef to sea urchins. More detailed instructions for feeding are given in Chap. 1 with the discussion of the marine aquarium.

Techniques for Laboratory Study of Sea Urchin Reproduction

We find that the techniques for induction of spawning and artificial fertilization often are simpler and more reliable for the sea urchin than for the more commonly used frog. Species are collected or purchased during the particular months in which they are fertile. Along the North Atlantic coast, the purple urchin *Arbacia* is fertile from April to August; and the green urchin *Strongylocentrotus* is fertile from February to late April. Before ordering animals, write for a bulletin of general instructions regarding shipments and prices. On the Pacific coast, *Strongylocentrotus purpuratus*, a purple urchin, is fertile from December through April, and *Lytechinus pictus* is ripe during the summer months. The Pacific Bio-Marine Supply Company, P.O. Box 536, Venice, Calif. 90291, specializes in sending a kit of the animals with necessary supplies.

Generally, we conduct the studies in March or April, at which time we order *Strongylocentrotus* from Pacific Bio-Marine Supply Company. Upon arrival at the airport, the shipment is transferred at once to the laboratory, where the animals are examined immediately. Any dead organisms are discarded, and the others are returned to the shipping container of ocean water. The open container is placed at a temperature of 15 to 20°C, with aeration provided by a plastic hose from an air outlet. The animals are hardy and survive for at least several days under these

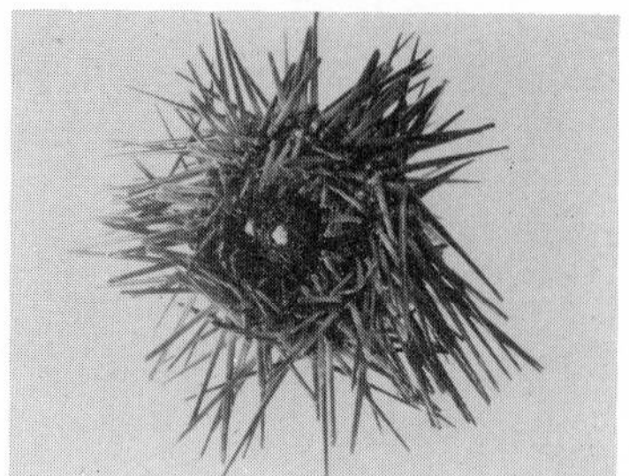

A *Courtesy of Carolina Biological Supply Company*

C *Courtesy of Carolina Biological Supply Company*

B *Courtesy of Carolina Biological Supply Company*

D

conditions. Little or no animal mortality has occurred during shipments. If more than one or two animals die, the living animals must be transferred at once to a fresh container of artificial seawater. (See Appendix B for suppliers of synthetic sea salts.) Since the sex cannot be determined by external observations, probably at least 18 to 24 animals should be ordered with each shipment, although we have obtained both sexes in orders of only one dozen animals.

To induce spawning, put an animal in a dry dish with the aboral surface upward; inject into the lantern-coelom 2.0 ml of 0.5 *M* KCl (37.3 g KCl in 1 l of distilled water). Generally within a few seconds the animal releases germinal fluid with eggs or sperm through the gonopores onto the aboral surface—a yellowish

Figure 3-32 Echinoderms. **A** *Abacia* sp., sea urchin of North Atlantic coast. **B** *Strongylocentrotus purpuratus,* sea urchin of Pacific coast. **C** *Asterias* sp., starfish, dorsal view. **D** *Asterias* sp., ventral view of live starfish on aquarium wall. **E** Starfish of Oregon Coast. **F** Sand dollar, at right. **G** Sea cucumber. *(E and F Courtesy of Marjorie Holstein.)*

F

G *Courtesy of Carolina Biological Supply Company*

secretion from a female and a white secretion from a male. Figure 3-33 illustrates the techniques.

For immediate studies of sperm and eggs, use the tip of a toothpick to transfer a small drop of either fluid to several drops of natural or synthetic ocean water on a concave slide. To observe fertilization and cleavage, both fluids are transferred to one slide where they are gently mixed in the seawater.

More often it is desirable to prepare and store eggs and sperm for later class studies. In this case, invert and suspend a female, after injection, on the top edge of a beaker so that the aboral surface is in contact with seawater in the container. During about a 15-min period, the eggs will be shed and will settle to the bottom of the beaker. Wash the eggs twice by slowly pouring off most of the seawater and adding fresh seawater. Leave the eggs in a last quantity of approximately 100 ml of

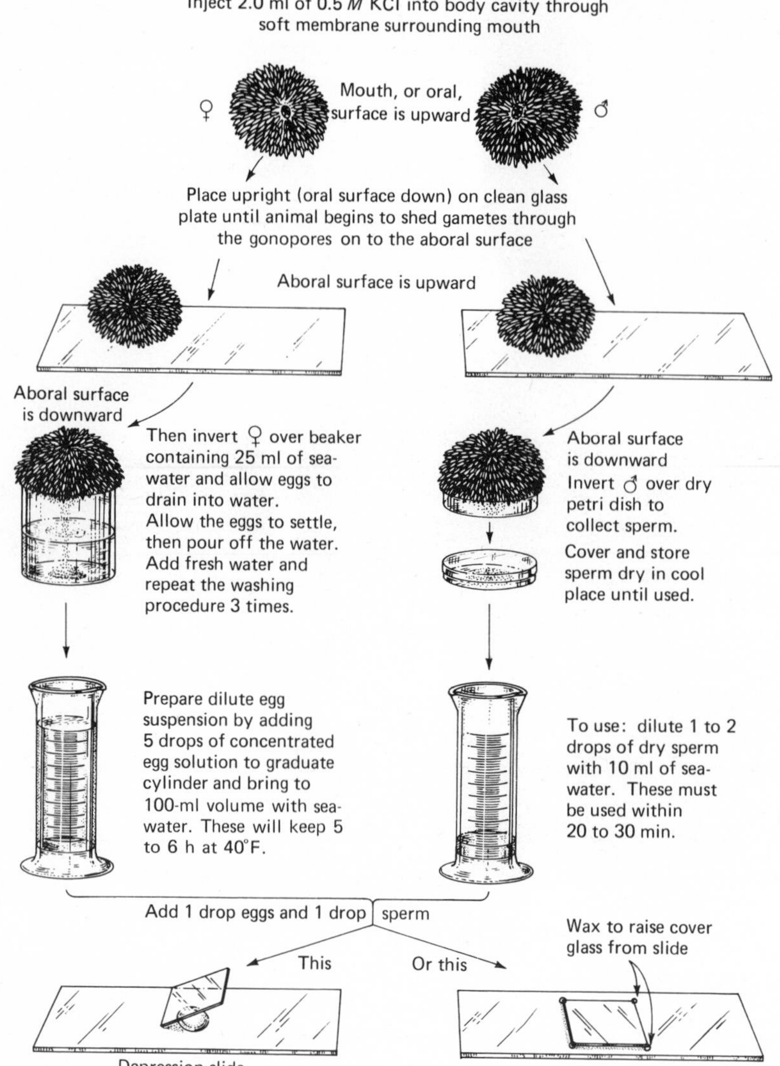

Figure 3-33 Techniques for artificial fertilization of sea urchin eggs. *(From An Experimental Approach to Biology by Peter Abramoff and Robert C. Thomson. W. H. Freeman and Company, San Francisco. Copyright 1966.)*

seawater. Cover the beaker and store on the top shelf of a refrigerator at 8 to 10°C for no longer than 5 or 6 h. Prepare the sperm for storage by transferring sperm fluid, after injection, to a small, dry container. Cover and store the container in the same manner as for the eggs. At time of use, transfer several small drops of sperm fluid to 10 ml of seawater; sperm become viable within about 10 min and remain viable for 10 to 20 min. At the same time, gently stir the suspension of eggs; transfer several drops of the suspension to drops of sperm fluid on a slide and study with a microscope. Depending on room temperature, the first cleavage of an egg occurs within 1 to 2 h, and can be observed in concave or deep-well slides. Alternatively, transfer several drops of the egg and sperm suspensions to test tubes of seawater; plug the tubes with cotton, and place them at 22 to 23°C for periodic studies of the developing eggs.

Note: Before use, all glassware and instruments must be thoroughly washed and rinsed with distilled water. Equipment for the eggs and sperm *must be kept separated* to avoid fertilization before the desired time.

REFERENCES

Alexander, R.D.: 1967, *Singing Insects*, Rand McNally & Company, Chicago, 86 pp.

Borror, D.J., and D.M. DeLong: 1971, *An Introduction to the Study of Insects*, 3d ed., Holt, Rinehart and Winston, Inc., New York, 812 pp.

Buchsbaum, Ralph: 1948, *Animals Without Backbones*, 2d ed., The University of Chicago Press, Chicago, 405 pp.

Darling, Don: 1968, "*Mormoniella* in the Genetics Laboratory," *Sci. Teacher*, **35** (1): 60–61.

Dethier, V.G.: 1955, "The Physiology and Histology of the Contact Chemo-receptors of the Blowfly," *Quar. Rev. Biol.*, **30**: 348–371.

Follansbee, Harper: 1965, *Animal Behavior*, D.C. Heath and Company, Boston. BSCS Laboratory Block. Student Manual, 70pp; Teacher's Supplement, 51 pp. $1.25 each. Contains excellent laboratory exercises for behavioral studies of *Daphnia*.

Forrest, Helen: 1962, "An Adaptable Ant Nest for Culture and Experiment," *Turtox News*, **4** (8): 186–188.

Hoste, R.: 1968, "The Use of *Tribolium* Beetles for Class Practical Work in Genetics," *J. Biol. Educ.*, **2** (4): 365–372.

Knudsen, Jens W.: 1966, *Biological Techniques*, Harper & Row, Publishers, Incorporated, New York, 525 pp.

Kuhn, D.J.: 1969, "The Gammarus," *Am. Biol. Teacher*, **31** (3): 160–161.

Larson, Omer: 1971, personal communication, University of North Dakota.

Needham, James G.: 1937, *Culture Methods for Invertebrate Animals*, Comstock Publishing Associates, Cornell University Press, Ithaca, N.Y. (Dover Publications, Inc., New York, 1959), 590 pp.

Richards, A. Glenn: 1963, *The Complementarity of Structure and Function*, BSCS Laboratory Block, D.C. Heath and Company, Boston. Student Manual, 78 pp.; Teacher's Supplement, 66 pp. $1.25 each.

Russell-Hunter, W.D.: 1969, *A Biology of Higher Invertebrates*, Collier-Macmillan Canada, Ltd., Toronto, 224 pp.

Segal, Earl, and W.J. Gross: 1967, *Physiological Adaptation*, BSCS Laboratory Block, D.C. Heath and Company, Boston. Student Manual, 109 pp.; Teacher's Supplement, 60 pp. $1.25 each.

Strayer, James L.: 1969, "Living Crayfish in the Laboratory," *Am. Biol. Teacher*, **31** (3): 162–164.

Thompson, James N., Jr.: 1967, "Inexpensive Fruit Fly Trap Adapted for Classroom Use," *Turtox News*, **45** (11): 282–283.

Walker, Ilse, and D. Pimental: 1966, "Correlation Between Maternal Longevity and Incidence of Diapause in *Nasonia vitripennis* Walker (Hymenoptera, Pteromalidae)," *Gerontologia*, **12**: 89–98.

Webb, S.D.: 1968, *Evolution*, BSCS Laboratory Block, D.C. Heath and Company, Boston. Student Manual, 53 pp.; Teacher's Supplement, 55 pp. $1.25 each.

Wheeler, George C., and Jeanette Wheeler: 1963, *The Ants of North Dakota*, University of North Dakota Press, Grand Forks, 326 pp.

Whiting, P.W., and S.B. Caspari: 1957, "Mormoniella Merry-Go-Rounds," *J. Heredity*, **48**: 31–35.

Whiting, P.W.: 1967, "*Mormoniella* Notes: Larval Diapause," *Carolina Tips*, **30** (8): 29.

Yurkiewicz, W.J.: 1970, "The Madeira Cockroach as an Experimental Animal," *Am. Biol. Teacher*, **32** (1): 39.

Other Literature

Alexander, Richard D., and Donald J. Borror: 1956, *The Songs of Insects* (12-in long-playing record), Cornell University Press, Ithaca, N.Y.

Arnett, Ross H., Jr.: 1968, *The Beetles of the United States: A Manual for Identification*, American Entomological Institute, Ann Arbor, Mich., 1112 pp.

Atkins, J.A., C.W. Wingo, W.A. Sodeman, and J.E. Flynn: 1958, "Necrotic Arachnidism," *Am. J. Trop. Med. Hyg.*, **7** (2): 165–184.

Borror, D.J., and R.E. White: 1970, *A Field Guide to the Insects of America North of Mexico*, Houghton Mifflin Company, Boston, 406 pp.

Bristowe, W.S.: 1958, *The World of Spiders*, William Collins Sons & Co., Ltd., New York, 304 pp.

Creighton, W.S.: 1950, *The Ants of North America, Museum Comp. Zool. Bull.* Harvard University, Cambridge, Mass., **104**: 1–585.

Demerec, M., and D.P. Kaufman: 1964, *Drosophila Guide*, 7th ed., Carnegie Institute of Washington, Washington, D.C.

Dethier, V.G.: 1963, *The Physiology of Insect Senses*, John Wiley & Sons, Inc., New York, 266 pp.

Ehrlich, P.R., and A.H. Ehrlich: 1961, *How to Know the Butterflies*, Wm. C. Brown Company Publishers, Dubuque, Iowa, 262 pp.

Glass, Bentley: 1965, *Genetic Continuity*, BSCS Laboratory Block, D.C. Heath, Boston. Student Manual, 154 pp.; Teacher's Supplement, 89 pp.

Helfer, Jacques R.: 1963, *How to Know the Grasshoppers, Cockroaches and Their Allies*, Wm. C. Brown Company Publishers, Dubuque, Iowa, 353 pp.

Hyman, L.H.: 1955, *The Invertebrates*, vol. IV, *Echinodermata*, McGraw-Hill Book Company, New York, 763 pp.

Johnson, D.L., and A.M. Wenner: 1966, "A Relationship Between Conditioning and Communication in Honey Bees," *Animal Behavior*, **14**: 261–265.

Klots, A.B.: 1965, *The World of Butterflies and Moths*, McGraw-Hill Book Company, New York, 207 pp.

Pennak, Robert W.: 1953, *Fresh-Water Invertebrates of the United States*, The Ronald Press Company, New York, 769 pp.

Ross, H.H.: 1966, *How to Collect and Preserve Insects*, Circular 39, Illinois State Natural History Survey, Urbana, Ill., 71 pp.

Stone, Alan, et al.: 1965, *A Catalogue of the Diptera of America North of Mexico*, USDA Handbook 276, 1969 pp.

von Frisch, Karl: 1956, *Bees: Their Vision, Chemical Senses, and Language*, Cornell University Press, Ithaca, N.Y., 119 pp.

von Frisch, Karl: 1967, *The Dance Language and Orientation of Bees*, The Belknap Press of Harvard University Press, Cambridge, Mass., 566 pp.

Wenner, A.M.: 1962, "Simple Conditioning in Honey Bees," *Animal Behavior*, **14**: 149–155.

Wingo, Curtis W.: 1960, *Poisonous Spiders*, University of Missouri Agricultural Extension Service, Bull. 738, 11 pp.

LABORATORY CARE OF VERTEBRATES

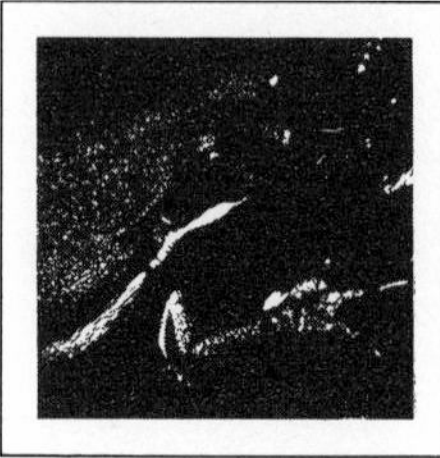

This chapter discusses the special uses, distinguishing traits, sources, and care of appropriate laboratory fish, amphibians, reptiles, birds, and mammals. A greater discussion of housing for these animals is presented in Chap. 1. Commercial sources for animals are given in Appendix B, and special handling techniques are discussed in Chap. 5. Nomenclature for the aquarium fish is based on Axelrod et al. (1967); nomenclature for other vertebrates is from Blair et al. (1968).

FISH

In general, small native fish are more desirable and less expensive than those purchased from commercial suppliers. Often fish-bait houses can supply small native fish in quantity and at a low cost. Check with county and state government offices to learn about state and local laws governing collecting fish. As a rule, no restrictions are placed on small fish used for educational purposes. Procedures for installing and maintaining freshwater and marine aquariums are discussed in Chap. 1.

General Instructions

1. When possible, obtain small fish of 1 to 2 in length. Mature fish do not always adapt successfully to new surroundings; breeding fish may be toward the end of reproductive age.

2. Fish from ponds and slow-moving streams adapt more readily to laboratory situations than do those from swift water.

3. If the tank is aerated, plants are unnecessary except for fish that lay eggs on plants or that use the plants for food.

4. Sand and gravel are not necessary except to hold rooted plants or to furnish materials for nests of certain fish.

5. Crowding of animals is the primary cause of debilitation and spread of disease. Without aeration, allow at least 24 in^2 of water surface for each inch of fish, including the tail fin. Provide more space for very active fish.

6. Although fish must receive adequate food, they should not be overfed, as judged by any remaining uneaten food, which must be removed 2 or 3 h after feeding.

7. Avoid abrupt changes in water temperature, for example, from abrupt changes in room temperature or from differences in water temperature when fish are transferred.

8. Use pond water or dechlorinated (conditioned) tap water. Distilled water does not contain the necessary minerals, although distilled water should be used to replace water lost by evaporation in order not to increase the concentration of minerals. Condition tap water by letting it stand in an open container for 48 h to allow chlorine gas to escape. Alternatively, tap water can be used at once when conditioned by adding 0.05 g sodium thiosulfate for each gallon of water. Aquarium stores and biological suppliers sell water-conditioning chemicals in tablet or powder form.

9. The young fry of all laboratory fish feed on dense cultures of ciliates, dried infusoria, or powdered fish food. These can be obtained from aquarium stores or from biological supply companies.

10. Live animals used for fish food include *Artemia, Daphnia, Tubifex, Enchytraeus*, and chopped earthworms. Probably *Artemia* (brine shrimp) is the most desirable type for small fish; the frozen *Artemia* from an aquarium store provides an easy method for storing and feeding fresh food. The latter is particularly desirable as a reward food in fish-learning tests, since the frozen shrimp do not swim away from the test area.

11. Handle fish with clean nets that are wetted in the aquarium water. If fish must be handled by hand, wash and rinse the hands thoroughly and dip them into the aquarium water before touching the fish. Dry hands may remove a protective mucous covering on the fish body, and thus allow a parasitic infestation.

12. Before adding new fish to a community aquarium, isolate them in a separate container for 1 or 2 weeks to observe for disease.

13. Introduce new fish into a community by floating them in a plastic bag of their collecting or shipping water, both to adapt them to a temperature change and to accustom the community of fish to the new animals.

Diseases of Fish

The prevention of disease is more important than the treatment. Unless a fish is costly or an unusual type, generally the most efficient procedure is to dispose of a diseased animal. Treatment may be tedious and unsuccessful. Several of the most common and most easily treated diseases are described below.

Ich. Ichthyophthiriasis, White Spots, or "Ich" is probably the most common disease and perhaps the most easily treated. The disease is caused by a protozoan parasite *Ichthyophthirius,* which encysts on the skin and produces small white patches on the fins, body, and gills. A first few white patches are followed by many patches over the entire body. Untreated animals die within several weeks. The protozoan spreads in a free-swimming stage to other fish. The parasite may be carried into the laboratory on the common goldfish from small aquarium counters of variety stores.

For treatment, transfer infested fish to bowls of fresh water containing *1 drop* of 0.75% malachite green solution in a gallon of water. (0.75% malachite green solution = 0.75 g malachite green powder in 100 ml water; store in refrigerator. Prepare fresh solution every 2 weeks.) The treatment is very successful, but an overdose will kill the fish almost immediately. Do not use this medication with young fry. Additionally, change the water of the aquarium, and to the fresh water add the 0.75% malachite green solution in the same proportion. If a charcoal filter is used, it must be removed during medication as the charcoal adsorbs and retains the medicine. Observe the aquarium for a week and isolate any other fish that show infestation. When no new infestation appears for a period of 1 week, replace the treated water with fresh water. A less effective medication, but safer for young fry or developing eggs, is 1 ml of 1% methylene blue solution for each gallon of water. (Prepare by adding 1 g methylene blue powder to 100 ml of 95% ethyl alcohol.) An older treatment, less effective but safe for young fish, is to put the fish in a 10% NaCl solution.

If the tank becomes heavily infested, dispose of the fish, plants, and sand. Soak the tank for several days with a pink solution of potassium permanganate (two or three crystals of $KMnO_4$ for each gallon of water). Following this, soak and rinse the tank with several changes of fresh water. Only rarely is the potassium permanganate treatment necessary and then generally because of extreme neglect.

Velvet or rust. The disease is caused by infestations of the protozoan *Oodinium;* it is often confused with Ich, but in place of white patches, a velvety coat of small, light-yellow granules forms over the skin. If untreated, the fish die within several weeks. The protozoan spreads in a free-swimming stage to other fish. The disease is more persistent than Ich, and several weeks may be required to rid the animals and the water of the parasite. Treat with the malachite green solution as described for Ich. An older, effective medication is to add 1 drop of supersaturated copper-sulfate solution to each 5 gal of water.

Mouth "fungus," tail rot, popeye, and other bacterial infections. Bacterial infections are uncommon in well-kept aquariums. Symptoms of these diseases are open sores and internal abscesses, both with a collection of mucus. Not all causative agents have been identified for specific infections. The pop-eye disease may be due to both bacterial infection and collection of mucus. Infecting organisms include slime bacteria (myxobacteria), *Pseudomonas, Aeromonas,* and *Haemophilus.*

Transfer diseased fish to separate containers and treat with antibiotics, which also are added to the aquarium water. Effective antibiotics are tetracycline hydrochloride (50 mg/gal of water) and Ampicillin, a form of penicillin (50 mg/gal of water).

Larger ectoparasites. Occasionally fish will carry on their skin small leeches and fish lice *(Argulus).* Remove the parasites with tweezers and treat the wound with Mercurochrome. Several types of small skin flukes produce a characteristic "shimmy" in swimming, and fish may be seen scraping their bodies against plants, gravel, and other objects. When infestation of flukes is heavy, mucus and blood appear on the skin and gills. The flukes can be seen if a fish is put under a stereoscopic microscope. For treatment, remove fish and dip them in formalin water prepared by adding 5 ml of 40% formalin to 1 gal of water. Return fish to the tank. Add to the aquarium water 1 drop of supersaturated copper-sulfate solution for each 5 gal of water.

EGG-LAYING FISH

Goldfish *(Carassius; Minnow Family)*

USES: These are hardy fish of versatile use for many studies, including ecology, genetics, physiology, and behavior.

DESCRIPTION: Fish breeders, particularly in Japan and China, have produced much variation in color and fins, including the black, white, red, yellow, and variegated colors with single, double, and long, flowing fins (see Fig. 4-1A). Aquarium fish may reach a length of 3 or 4 in; goldfish in outdoor pools may reach a much larger size. Goldfish breed at about 2 years of age, at which time the female becomes swollen with eggs and the male develops white spots on the gills. Usually the male pursues the female for several days before spawning time. Under proper conditions, spawning may occur several times a year.

The goldfish represents an exotic member of the minnow family (Cyprinidae). A number of native minnow species may be collected locally, for example, the fathead minnow, *Pimephales promelas,* a gray, heavy-bodied fish, 2 to 4 in long, and occurring in lakes and streams of central and western United States (Fig. 4-1B). See Blair et al. (1968) and Carlander (1969) for taxonomy and descriptions of the minnow family.

A

B

Figure 4-1 A Goldfish, *Carassius* sp. **B** Northern fathead minnow, *Pimephales promelas;* males showing three rows of tubercles on head that appear at spawning season.

SOURCES: Obtain from aquarium or biology supply houses. "Wild" goldfish may have accumulated in local rivers and streams as a result of their release by home aquarists.

CULTURE: Water must be clean and aerated; mineral content and pH are not critical. Probably the most satisfactory food is the commercial goldfish food obtained from aquarium houses. A daily feeding at about the same hour is recommended. Do not overfeed, as indicated by uneaten food after several hours, which should be removed.

Put breeding animals in a 5- to 10-gal tank containing a heavy growth of rooted plants. During a period of several hours, the female deposits 500 to 1,000 eggs (1 or 2 mm in diameter) over the plants, to which the eggs adhere. The male releases sperm among the eggs. Remove the adults when egg laying is finished. Depending on water temperature, the eggs hatch within 5 to 7 days. During the first week after hatching, feed dense cultures of ciliates to the young fish. The newly hatched fish also feed on each other, thus controlling the population density in the aquarium.

Three-spined Stickleback (*Gasterosteus aculeatus*)

USES: An excellent animal for studying courtship and reproduction. See Müller-Schwartze (1968), Marler (1968), and Tinbergen (1953).

DESCRIPTION: From 2 to 4 in long. Green to brown color. Male with orange or red belly at breeding time.

SOURCES: Distributed over most of the United States and common in many creeks, rivers, lakes, and ponds. Many sticklebacks, for example *Apeltes*, live in brackish and salt water.

CULTURE: Sticklebacks are pugnacious and must be separated from other fish. For optimum growth, keep water at 15 to 18°C. Feed brine shrimp, *Daphnia*, and *Tubifex*. During an occasional emergency, dry fish food may be accepted. The aquarium should be heavily planted with rooted, grasslike plants such as *Vallisneria*.

Under normal conditions the fish breed during spring months, at which time the male displays a brilliant coloration and begins a series of courting activities (see Fig. 4-2). At this time the animals should be separated into groups of one male and two or three females in a 10- to 15-gal tank containing well-rooted plants, filamen-

A

B

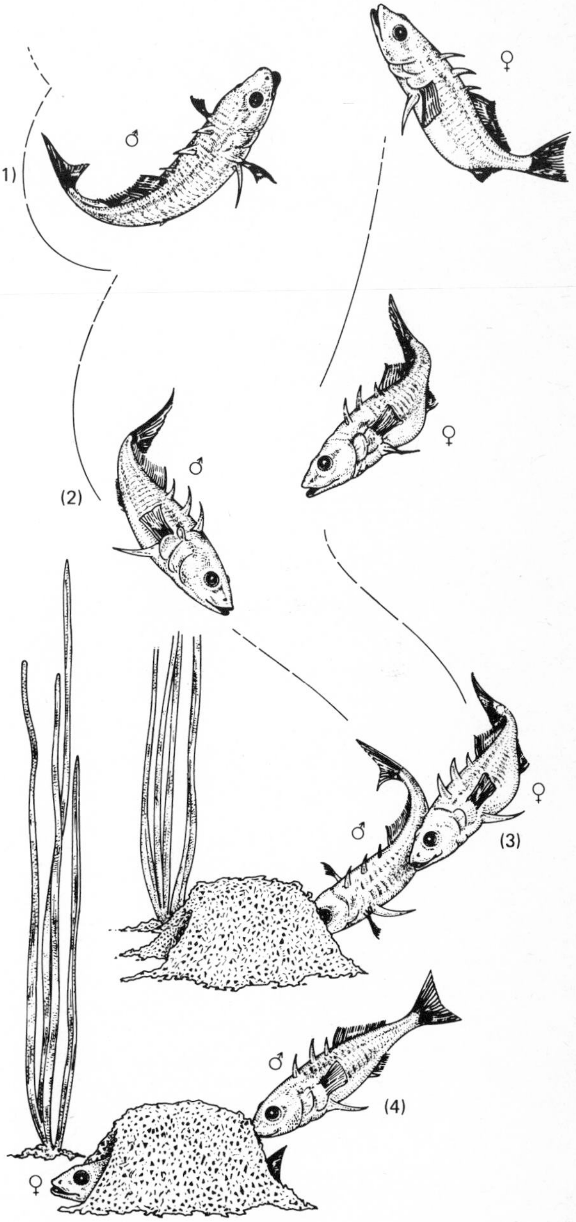

Figure 4-2 Sticklebacks. **A** *Eucalia* sp., five-spined or brook stickleback. (First three spines are erect; last two spines are folded against body and are faintly visible on dorsal line.) **B** Five-spined stickleback with all spines erect. **C** Courtship sequence of three-spined stickleback, *Gasterosteus aculeatus.* Following a series of courtship displays at (1), the male leads the female to the nest at (2). The male puts his snout into the nest (3); the female moves past him into the nest. The male prods her tail region, stimulating the female to lay eggs (4). The female leaves the nest and is chased away by the male. He may then attract several other females to the nest, after which he cares for the eggs and young fry. *(C From the Study of Instinct, by N. Tinbergen, Copyright 1951, Clarendon Press, Oxford.)*

C

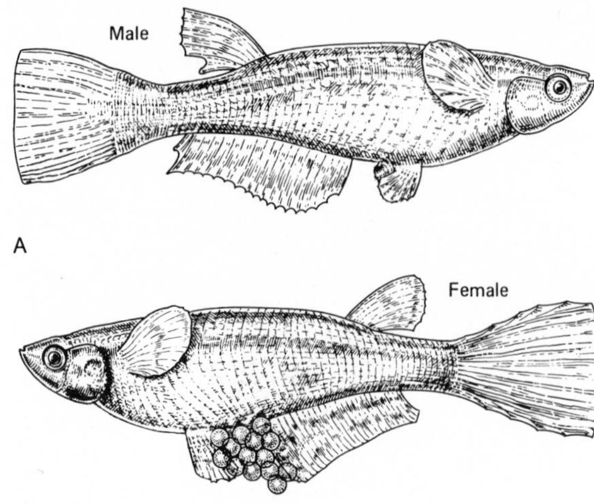

Figure 4-3 Japanese medaka, *Oryzias latipes.* **A** Male. **B** Female.

Courtesy of Carolina Biological Supply Company

tous algae, and string. The male uses the plants and string to build a ball-shaped nest with a passageway through the middle. After a series of courtship displays, the female deposits eggs in the nest and is then routed away by the male. Several more females may immediately add eggs to the nest. The male guards the eggs during the hatching period of 8 to 14 days and protects the young fry.

Japanese Medaka *(Oryzias latipes)*

USES: Excellent for studies of embryology and genetics. Not used extensively in the United States, but a very desirable fish for the teaching laboratory. See Kirchen and West (1969), Daniel (1966), and Briggs and Egami (1959).

DESCRIPTION: Adult females about ¾ to 1½ in long; males slightly smaller. Golden, blue-black, white, or variegated (see Fig. 4-3). Peaceful in communities and unusually hardy. Mature at 1½ to 2 months. Under good conditions, life span is 2 to 4 yr. Natural breeding period from April to October, but can be induced throughout the year. Found in great abundance in rice fields of Southeast Asia, where they feed on mosquito larvae. The fish has had much use in Japan for research and teaching, but is less known in the United States.

SOURCES: Native to Japan and Southeast Asia. Obtain from commercial suppliers.[1]

CULTURE: Maintain at 20 to 25°C in slightly acid pond or conditioned tap water (pH 5.5 to 6.8, but not critical). Forty to fifty fish can be put in a 10-gal tank. Artificial aeration is not required if the tank contains a strong growth of oxygenating plants, for example, *Vallisneria* and *Sagittaria.* Cover tank to prevent animals from jumping out. Do not subject to sudden temperature changes. Medakas accept all kinds of tropical fish food, as well as live brine shrimp, mosquito larvae,

[1]Carolina Biological Supply Co. distributes the fish with a Medaka Breeding Kit.

and *Enchytraeus*. Animals that are mating should be given live food once or twice weekly. Feed the young fry with freshly hatched brine shrimp or dense cultures of ciliates until they are several weeks old.

To induce breeding, place about four males and eight females in an aerated 5-gal tank (or fewer animals in a widemouthed gallon jar) under a daily photoperiod of 16 h light and 8 h dark. Generally, the female begins spawning in 8 to 10 days, producing from 1 to 70 eggs (generally 20 to 30) at one time during early morning. During a breeding season, one female may produce as many as 3,000 eggs. The male spreads milt over the eggs, which remain attached to the female vent for several hours, after which they drop and become attached to plants. Young fry hatch in 1 to 3 weeks depending on temperature. Because adults often eat the eggs after they drop from the female, remove the cluster from the female's body before 9 AM, and transfer to a small culture bowl of water, without plants if developing eggs are to be observed. To prevent mold growth on the developing eggs, add 1 ml of 1% methylene blue solution to each gallon of water distributed to the culture bowls (1 g methylene blue powder in 100 ml of 95% ethyl alcohol). The transparency of eggs and their rapid cleavage permit exceptionally good class studies and individual student research.

Siamese Fighting Fish *(Betta splendens)* and Other Anabantids

USES: Excellent for studying (1) visual communication and aggression, (2) reproduction, and (3) genetics of body color. See Southwick (1968a), Southwick (1968b), and Daugs (1967).

DESCRIPTION: Attractive fish with long, flowing fins and brilliant colors, most often red, green, blue, or white with red fins (see Fig. 4-4 A and B). Length, 2 to 2½ in. Anabantids possess an auxiliary respiratory organ, the labyrinth, an air-breathing organ in the gill cavity. They frequently come to water surface to gulp air. Males are very pugnacious and must be isolated, each in a separate chamber.

Figure 4-4 *Betta splendens*, Siamese fighting fish. **A** Male. **B** Male *Betta* in pursuit of smaller female *Betta*.

A

B

Females are peaceful together. As a rule, mature male anabantids are more brilliantly colored than females, and anal fins are longer and more pointed.

Other anabantids suitable for the laboratory are the paradise fish (*Macropodus opercularis*), and croaking and kissing gouramis (*Trichopsis, Colisa,* and *Trichogaster*) (see Fig. 4-5). Breeding activities, including a bubble nest, are much the same as those described below for *Betta splendens*. Paradise fish grow to a 3-in length, are semitropical, and require a water temperature of 21 to 25°C. The male is very pugnacious and must be isolated. Gouramis may grow to a 6-in length; generally they are peaceful but may attack smaller fish. Gouramis require a water temperature of 25 to 28°C for breeding. The "kissing" of some gouramis occurs, not as a part of courting, but more as a competitive activity where two fish slowly put their mouths together in an effort, it seems, to "stare each other down." The gouramis' large sucking lips are adapted to eating soft algae.

SOURCES: Native to India, southeastern Asia, the Philippines, and Formosa. Obtain from local aquarium suppliers or by airmail shipment from distributing companies. See Appendix B for suppliers.

CULTURE: Because of their air breathing, these fish do not require large containers. One male can be kept without aeration in a widemouthed 2-qt container. No plants or artificial aeration are required; pH is not critical (5.5 to 7.0); temperatures of 20 to 29°C are tolerated if there is no abrupt change. A temperature of 25 to 27°C is required for breeding. Feed tropical fish food, supplemented occasionally with live *Artemia, Daphnia, Tubifex,* or *Enchytraeus*. For adult fish, Daugs (1967) recommends Purina Trout Chow, number 1 size.[1]

Betta splendens matures at 6 to 8 weeks, at which time males must be isolated to prevent fighting. To induce breeding, put a male and female in a tank, with the female protected by a glass partition or by being placed in a screen-covered jar within the water. Add floating plants for attachment of the bubble nest. Cover the tank with screen to prevent fish from jumping out. Bring water temperature to 27°C. Give both animals live food several times daily. When ready to spawn, gen-

[1]From Ralston Purina Co., Checkerboard Square, St. Louis, Mo. 63102.

Figure 4-5 *Trichogaster* sp. Gourami showing the typical pair of long, threadlike ventral fins and long anal fin.

erally in 1 or 2 weeks, the female will be rounded and heavy; a small white tube, the ovipositor, will extend from the vent just anterior to the anal fin. The female will attempt to swim through the glass to the male. During this time, the male will display his colors to the female and will construct a nest of air bubbles (from a sticky saliva secretion) among the plant leaves at the water surface or occasionally in the open water. At this time, release the female and observe the fish for possible fighting. If fighting occurs, return the female to the enclosement for another 12 to 24 h. During spawning, the female approaches the nest and turns upside down as the male curves his body around her in an appearance of pressing out the eggs. A few eggs are expelled at one time; the male spreads milt over the eggs and then takes them into his mouth to put them in the bubble nest. The process is repeated many times during several hours, generally with a total of several hundred eggs produced. When spawning is completed, the male again begins to fight the female, and she must be removed at once. The male guards the nest during incubation, which at 27°C lasts for 24 to 36 h. Generally the male protects the young for another 24 h, but he soon begins to eat them and must be removed. Feed dense cultures of ciliates to the young fry.

To study the developing eggs, use a medicine dropper to transfer eggs from the nest to small dishes of aquarium water kept at 27°C. Place a single egg in several drops of water on a concave slide and observe with a stereomicroscope. To coordinate the studies with laboratory hours, slow the stages of development by lowering the water temperature.

Egyptian Mouthbreeder *(Haplochromis multicolor)* and Other Cichlids

USES: Especially interesting for studying the female behavior of carrying eggs in the mouth. See Axelrod et al. (1967) and Axelrod and Schultz (1955).

DESCRIPTION: Males up to 3 in long; females slightly smaller. Green to yellow color. Large head and mouth. Generally peaceful in a community. Most cichlids lay eggs on stones, plants, and occasionally on the glass aquarium wall, with both parents guarding the eggs. A few cichlids, for example, the Egyptian mouthbreeder *(Haplochromis multicolor)* and the West African mouthbreeder *(Tilapia melanopleura)*, carry the eggs in the mouth, foregoing any feeding until the young hatch.

The Egyptian mouthbreeder is used more often in the laboratory than other cichlids. The female lays 30 to 80 eggs in a depression made by the male. She then takes the fertilized eggs into her mouth, where they remain until hatching time, 8 to 12 days later. The mouth is greatly distended during this time, and the parent does not feed. Spawning and taking eggs into the mouth may occur quickly and often are not observed. Distention of the mouth may be the first evidence of spawning. After hatching, the young fry remain near the parent and at times, particularly when in danger, will return to inside the female's mouth.

The angel fish *(Pterophyllum scalare)*, a favorite with aquarists because of its beauty, is not often found in laboratories, perhaps because of cost. It may grow to 5 or 6 in in length, is slow-moving and gentle, and should not be placed with aggressive fish. A water temperature of 25 to 27°C is required for breeding. Female places eggs on leaves or stems, from which the newly hatched young hang for several hours. The animals exhibit an interesting and graceful swimming movement, using only the pectoral fins (see Fig. 4-6).

SOURCES: Native to African rivers. Obtain at aquarium houses.

CULTURE: Place in planted and aerated aquarium. The fish are peaceful and must be protected from pugnacious animals. They are heavy eaters and must be given an abundance of live food, frozen chopped meat, and lettuce leaves. Dry commercial food may be accepted. The optimum breeding temperature is 25 to 27°C.

Sunfish (*Lepomis*)

USES: For studying reproductive behavior and care of eggs by the male.

DESCRIPTION: Attractive aquarium animals, native to United States. Local collecting makes them relatively inexpensive. Although not pugnacious, they should not be placed with other kinds of fish because of size and hearty meat-eating habits. May reach a length of 6 to 8 in in the wild. Color varies but may be green, blue, or yellow. Markings vary. Small fish, collected in spring or summer, will adapt easily to aquarium conditions and may spawn the next spring. At spawning time, the male forms a depression in the sand and lures the female to the nest. After eggs are laid, the male chases the female away and guards the eggs until they hatch, generally in 2 or 3 days.

SOURCES: Found in streams and lakes over the United States. Collect with a net or seine.

CULTURE: Put fish in a well-planted and well-aerated aquarium containing pond or conditioned tap water at 20 to 23°C. Give daily feedings of fresh or frozen chopped meat, small live fish, or chopped earthworms. Transfer young fry to small bowls of aquarium water with floating plants or other aeration. Feed infusoria until young are large enough to eat chopped meat.

Figure 4-6 Angel fish, *Pterophyllum scalare*.

Figure 4-7 Guppy, *Lebistes reticulatus;* adult female at center.

LIVE-BEARING FISH

Guppies *(Lebistes reticulatus)*

USES: Of versatile use for many types of study. See Silvan (1966).

DESCRIPTION: Generally a gray body with dark markings; many color variations in mutants. Adult males about 1½ in long and with a pointed anal fin, the *gonopodium*, used to transfer sperm. Mature females about 2 in long and with a fan-shaped anal fin (Fig. 4-7). Hearty and prolific; live young every 4 weeks under good conditions. No special breeding time.

In a large group, males dart quickly at all times among immature and mature females, transferring sperm at random. The sperm, perhaps from several males, are stored in the female reproductive tract. First brood, with generally only a few young, is produced when the female is 10 to 12 weeks old. When under bright light and with a water temperature of approximately 25°C, female may produce a brood of about 60 young every 22 days. Although not definitely established, possibly the eggs of one brood are fertilized by sperm from several males.

SOURCES: Native to South America, Trinidad, and Barbados. Obtain from commercial aquarium suppliers.

CULTURE: Place in pond or conditioned tap water, with pH of 5.5 to 7.0 (not critical), and at room temperature. Added aeration is not necessary when tank contains rooted, oxygenating plants such as *Vallisneria*.

Put breeding animals under bright light and at 25°C. Transfer young to a separate container to prevent adults from eating them. A heavy plant growth helps to protect the young until they are transferred. Alternatively, carefully transfer females to a brood chamber just before breeding time, as indicated by the female's swollen body. The brood chamber may be any type of container, suspended in the water and holding the female within the chamber but allowing the young to escape. A suggested type is a plastic bowl, covered with screen or porous cloth and

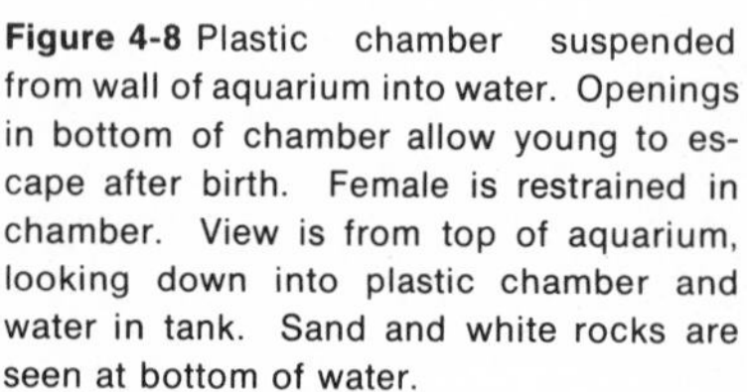

Figure 4-8 Plastic chamber suspended from wall of aquarium into water. Openings in bottom of chamber allow young to escape after birth. Female is restrained in chamber. View is from top of aquarium, looking down into plastic chamber and water in tank. Sand and white rocks are seen at bottom of water.

with holes punched in the bottom for escape of the young (see Fig. 4-8). Feed the fry with dense cultures of ciliates, powdered fish food, or brine shrimp. Try infusoria powder from an aquarium supply house.

Mosquito Fish or "Minnow" *(Gambusia)*

USES: For studying (1) aggressive behavior, (2) the genetics of color and markings, (3) physiology, and (4) reproduction. Inexpensive and extremely hardy, even under neglectful conditions.

DESCRIPTION: Males to 1 or 1½ in in length; females to 2 or 2½ in long. Color varies from almost black to gray. Pugnacious and scrappy with other types of fish; may kill snails. Very hardy and tolerates much variation in water conditions (see Fig. 4-9).

SOURCES: In shallow streams over most of the United States. *Gambusia affinis* is found in the Southwestern states and east to Alabama; *G. affinis holbrooki* is found along the eastern coast from New Jersey to Florida. Collect with hand nets, or purchase at fish-bait stores. Available from biological supply companies. See Appendix B for commercial sources.

CULTURE: Put in pond or conditioned tap water at 20 to 25°C. Aerate artificially or with a heavy growth of green plants. *Gambusia* are voracious eaters of most any type of food—commercial goldfish food, bread crumbs, small, live invertebrates, small guppies, and mosquito larvae. Treat breeding animals as described for guppies, separating the young from adults.

Figure 4-9 *Gambusia affinis.*

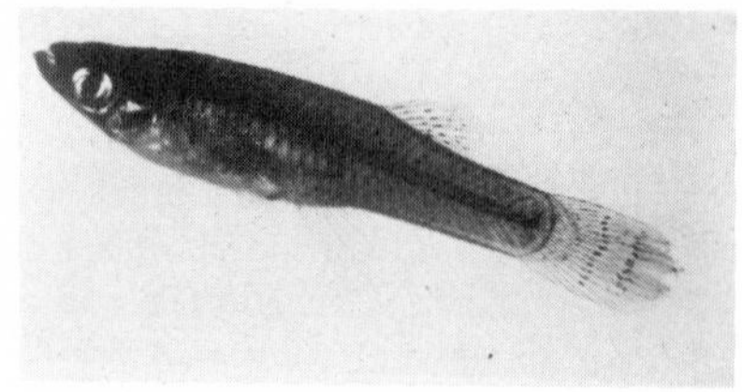

Courtesy of Carolina Biological Supply Company

AMPHIBIA

Amphibians include the salamanders and newts of the order Urodela (amphibians with tails) and the toads and frogs of the order Anura (amphibians without tails). Typically, amphibians are smooth-skinned tetrapods with an aquatic, gill-breathing larva (tadpole) that generally develops into a terrestrial, lung-breathing adult. Most amphibians reproduce and lay eggs in water. See Table 4-1 for information on the breeding habits of common amphibia. Although frogs are used more extensively in the teaching laboratory, other animals of this group can be maintained as easily and can serve as excellent and interesting organisms for a wide range of studies. See Bishop (1943), Blair (1958), Conant (1958), Goin and Goin (1971), Regal (1966), Rugh (1962), and Stebbins (1954).

Salamanders and Newts

USES: For various studies, including (1) ecology, (2) physiology, (3) reproduction and development, (4) regeneration of appendages, (5) anatomy, and (6) behavior.

DESCRIPTION: All with a naked skin and a tail; generally with two pairs of legs. The aquatic, gill-breathing larva resembles the adult in body form. Some retain the larval form and external gills throughout life; others metamorphose into lung-breathing adults that may become almost completely terrestrial or may spend much time in the water. All are nocturnal, and most return to water for breeding.

The sex of these animals cannot be established easily except during breeding seasons, when the male may display a more greatly swollen ridge surrounding the cloaca vent. Otherwise, the external sex traits vary greatly among species. When mating, the male places a packet of sperm (the spermatophore) in the female's cloaca or in the nearby water, where the female encloses the packet in her cloaca. Most animals of this group are commonly called salamanders. The name newt is used interchangeably with salamander, although newts are mostly aquatic whereas most salamanders become terrestrial. Some newts may pass through a subadult terrestrial stage for several years and then return to aquatic life, for example, the red-spotted eft, a terrestrial stage of *Notophthalmus viridescens*.

Hardy animals for the laboratory are the following (see Fig. 4-10):

Ambystoma. Distributed over most of United States. Generally a terrestrial and lung-breathing adult that returns to water for breeding. Some remain a larva and reproduce in the larval form called an *axolotol*, for example, *A. tigrinum*, the tiger salamander.

Necturus. The mudpuppies or waterdogs. Entirely aquatic and retain gills throughout life. The young of small types are maintained easily in the laboratory.

Notophthalmus (black-spotted newt, striped newt) and *Taricha* (red-bellied newt). Mostly aquatic. Develop into adults with lungs.

SOURCES: Use coarse nets and seines to collect swimming animals. Locate others in mud banks, under logs and stones, and near moist land areas. Pick up amphibians by quickly grasping them around the top, middle portion of body. Ordinarily, a tight grasp does not injure them. Occasionally at nighttime, particu-

TABLE 4-1. BREEDING HABITS OF SOME COMMON AMPHIBIA*

ANIMAL	POPULAR NAME	LOCALITY	BREEDING	NO. OF EGGS
Frogs				
Acris gryllus	Cricket frog	Central U.S.	May to July	Few
Hyla crucifer	Spring peeper	Eastern seaboard	April	1,000
Hyla versicolor	Tree frog	Eastern U.S., Canada	May and June	50
Pseudacris nigrita	Swamp tree frog	All U.S. except New England	March and April	500 to 1,500
Rana catesbeiana	Bullfrog	East of Rockies	May to August	6,000 to 20,000
Rana clamitans	Green frog	Eastern N. America	June to August	5,000
Rana pipiens	Leopard frog	Entire U.S.	March to May	5,000
Rana sylvatica	Wood frog	Entire U.S.	March to May	3,000
Toads				
Bufo americanus	American toad	Northeastern U.S.	April and May	6,000
Bufo fowleri	Fowler's toad	Central & Eastern U.S.	April to June	8,000
Salamanders				
Ambystoma jeffersonianum	Jefferson salamander	Eastern U.S., south Canada	Early Spring	300
A. opacum	Marble salamander	East & Middle West	September to October	100 to 250
A. maculatum	Spotted salamander	Eastern U.S.	January to May	100 to 200
A. tigrinum	Tiger salamander	U.S., Canada, Mexico	December to April	100
Plethodon cinereus	Red-backed salamander	Entire U.S.	June and July	14
Notophthalmus viridescens	Common newt	U.S., south Canada	April to June	20 to 30

*Adapted from Roberts Rugh: 1962, *Experimental Embryology*, 3d ed., Burgess Publishing Company, Minneapolis, pp. 31–32.

larly during rainy seasons of spring and fall, great numbers of land salamanders may be observed migrating to water for food and breeding. Transport aquatic forms in containers of water; put land animals in covered vessels with mud or

A

B *Courtesy of Carolina Biological Supply Company*

Figure 4-10 Salamanders and newts. **A** *Ambystoma tigrinum,* the tiger salamander. **B** *Notophthalmus viridescens,* the spotted newt. *(A Courtesy of Joe K. Neel.)*

moistened paper. Do not expose to extreme temperatures or direct sunlight while in transit. See Appendix B for commercial sources.

CULTURE: Young animals and smaller species are maintained more successfully than others. Put gill-breathing animals into an aquarium with plants. If the animals will metamorphose to lung-breathing adults, provide water at one end of the container and a damp soil area at the other end. Add several rocks or small logs for hiding places. Cover with a screen. Place salamanders at 21 to 25°C and newts at cooler temperatures of about 18 to 20°C. Newts may not survive if kept at higher temperatures for a prolonged period.

Feed adults once or twice weekly with chopped raw meat, earthworms, mealworms, and other insects. Some types are ravenous eaters and may require more frequent feedings. Avoid pollution from decaying food. When animals are first brought into the laboratory, force-feeding with forceps may be necessary. Large animals may nip off the toes of other animals and should be isolated.

Toads and Frogs

USES: For many types of studies, including reproduction and development, physiology, anatomy, ecology, and behavior.

DESCRIPTION: A tadpole larva with gills, long muscular tail, horny beaklike jaws, and coiled intestines. Adult with lungs, naked skin, no tail, and two pairs of limbs. Hind limbs are adapted for jumping. Larvae are generally herbivorous; adults carnivorous. Most are nocturnal.

The sex of adults may be identified by several general traits, which vary, nevertheless, among individuals of the same species. In general, the male is smaller in body size; the fingers are stouter and may have black horny deposits on the upper surfaces; the thumb and thumb pad are enlarged; and the throat region may be discolored with wrinkles or folds that give evidence of vocal pouches. These differences can be observed throughout the adult life but are more pronounced during breeding seasons.

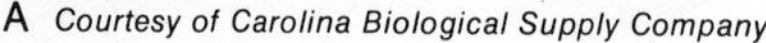

A *Courtesy of Carolina Biological Supply Company* B *Courtesy of Carolina Biological Supply Company*

Figure 4-11 Tree frogs. **A** *Hyla crucifer,* the spring peeper, well-camouflaged against surroundings although the inflated throat pouch is visible. **B** *Hyla cinerea,* the green tree frog, identified by light stripe on each side. *(B Permission of Ward's Natural Science Establishment, Inc., Rochester, N.Y.)*

Toads. Common in the United States are *Bufo americanus* of eastern United States, and *B. woodhousei* of the Great Plains and western United States. Rough, warty skin with numerous mucous glands. Large parotoid (shoulder) mucous glands. Mainly terrestrial but return to water for breeding.

Tree frogs. Examples are *Hyla crucifer,* the spring peeper of eastern half of United States, and *H. cinerea,* the green tree frog of eastern, central, and southern United States (see Fig.4-11). Relatively small size and generally a smooth skin. Body colors change from bright green to brown. Adapted to arboreal life, with long limbs and adhesive discs on the toes. Male with a relatively large throat pouch that inflates to a round balloonlike structure. Require a moist habitat.

Frogs. Common species, found over most of United States, are *Rana pipiens* (leopard frogs); *Rana catesbeiana* (bullfrogs); and *Rana clamitans* (green frogs). A smooth, moist skin. Live in or near water. Hind digits are webbed, front digits free. Males with large vocal pouches. See Figs. 4-12 and 4-13.

SOURCES: Look for frogs in shallow water or on land near the water edge. The most productive collecting is done at night with coarse nets or seines. During breeding or rainy seasons, the singing males can be spotted with a flashlight, which is extinguished when the singing stops, as a person approaches, and turned on when the singing resumes. Local and state laws must be observed relative to collecting animals with a light. Approach animals cautiously, watching for the glowing eyes. Because of their quick, leaping movements, netting is difficult and requires a slow approach. Often the animals can be blinded momentarily with a sudden bright light. Once the frog is spotted, quickly set a net over the animal. Transport frogs in closed plastic bags containing entrapped air, or in other covered containers that have small openings for ventilation. Surround animals loosely with dampened materials such as plants, peat moss, or shredded paper. Frogs can be retained in the container without food for several days if adequate ventilation is provided and if the surrounding materials are kept moist.

Courtesy of Carolina Biological Supply Company

Figure 4-12 The leopard frog, *Rana pipiens.*

Toads move farther from water than do frogs. During rainy or humid weather, look for them in gardens, on roadways, and around or under buildings. Because they move by hopping and not by leaping, they are more easily captured than frogs. Handle toads with caution, as the mucus from their epidermal glands may be strongly irritating to a person's skin, eyes, and mouth. Transport in a covered container that provides adequate ventilation. Moist packing around the toads is

Figure 4-13 The bull frog, *Rana catesbeiana.*

Courtesy of Carolina Biological Supply Company

not necessary unless they are to be retained in the container for more than 1 or 2 days.

Tree frogs are more difficult to capture because of their smaller size and because they are generally well concealed in plants. Trap them with small nets, and transport in the same manner as for frogs. See Appendix B for commercial sources.

CULTURE: Toads can be held without food for several weeks in most any type of container if ample ventilation and moisture are provided and if a cool temperature of about 18 to 20°C is maintained. For more permanent maintenance, put animals on gravel or coarse sand in an old aquarium or other container covered with screen. Add rocks or twigs for shelter. Provide a bowl of water, and maintain moist surroundings with damp, crumpled paper or with water sprinkled daily over the sand. Keep at a cool temperature.

Put tree frogs in covered containers with dense growths of green plants or fresh green leaves. Provide leafy twigs for climbing. Add a bowl of water, and sprinkle water daily over the container. Keep at a temperature comparable to the animal's natural habitat.

Put frogs in water with resting places, such as rocks, that rise above the water level, or with one side of the tank elevated so that a portion of the floor surface remains above water. Schmidt and Hudson (1969) describe their use of 10 by 6 by 4 in styrene boxes for small species and juveniles of larger species. They report that acetate boxes are lethal to the young. A tank with slow-running water is most desirable; otherwise, the water must be changed, and the container cleaned every few days.

Frogs can be held for several days in a widemouthed gallon jar, covered securely with screen or muslin cloth, and with about 1 in of water in the jar. Depending on animal size, put no more than three to five frogs in a gallon jar. Change the water daily and wash out the jar. Generally, this can be done without removing the frogs. If the jar temperature cannot be controlled below 24°C, put the container on a top shelf of the refrigerator at a temperature of 8 to 10°C. Alternatively, frogs can be held for several weeks (reportedly for longer periods) in containers of damp peat moss placed in the refrigerator at 8 to 10°C. Ample ventilation must be provided, and the moss must be kept damp.

Feed adult amphibians live mealworms, crickets, and flies. Chopped earthworms can be tried. Occasionally force-feeding is necessary, although frogs may survive without food for several weeks.

Maintaining tadpoles. During spring and early summer, masses of frog eggs can be found floating on the surface of quiet water. Toad eggs are deposited in strings of jelly on submerged plants or sticks. Dip the eggs from the water with a jar or bucket. Handle the eggs gently, and remove with them at least 1 or 2 qt of pond water containing floating algae. In the laboratory, place the container in diffused light (not direct sunlight) at 20 to 23°C. Tadpoles will develop within 2 or 3 days and will feed on the algae. Transfer them to an aquarium of pond water or dechlorinated tap water; add floating green plants and a small quantity of filamentous algae. About 200 tadpoles can be sustained in a 5-gal container if the water is aerated gently and continuously. Feed lettuce leaves two or three times weekly. Boil the lettuce for about 1 min to soften the tissue; cool the leaves before adding them to the aquarium. Remove uneaten lettuce after several days and add fresh, boiled leaves. Diatoms and algae that cling to the tank wall are good food also. A water suspension of boiled egg yolk is excellent food, but may produce difficulties, as an excess amount will foul the water within a 24-h period, thus necessitating a complete change of water and cleaning of tank with several water rinses. To avoid the latter, transfer tadpoles to a bowl of egg suspension for several hours of feeding. As tadpoles increase in size, more tank space must be provided. Before the animals metamorphose to lung-breathers, rocks or logs should be provided as resting surfaces above the water level.

Diseases. Amphibians are subject to few infections, and ordinarily these are avoided with proper housing and food. As a rule, the most satisfactory procedure is to dispose of seriously diseased animals. A common infection of frogs, however, is "red leg," which must be guarded against constantly and treated at once when it appears. Probably several organisms produce the disease (Gibbs et al., 1966), which appears most often as a severe hemorrhaging within the ventral skin of the hind legs and abdomen, followed by convulsions and coughing up of blood. If untreated, the animal dies within several days. All new animals brought into the laboratory must be inspected closely for signs of the disease. Because the disease is highly infectious, it is better to destroy any animals showing symptoms of red leg. Transfer other frogs to a weak solution of potassium permanganate (a light pink color), with only enough solution in the container so that the frogs are able to hold their heads above the surface. Empty the infected containers, and soak them for several days with a purple solution of potassium permanganate. Rinse the containers thoroughly with clean water before returning frogs to them. Reportedly, the infection may be prevented by keeping several copper pennies in the frog container. As well, peat moss in a frog container may provide sufficient iodine to prevent the infection. Gibbs (1963) describes a method of treating animals with 5 mg tetracycline HCl in 0.2 ml distilled water for each 30 g of frog weight. The dosage is administered twice daily for 6 days by means of a polyethylene stomach tube inserted through the mouth and esophagus.

Induction of ovulation. When fertilized eggs cannot be collected in the field, ovulation and fertilization may be produced within the laboratory. Alternatively, fertile eggs can be purchased from biology suppliers. Special kits containing male and female frogs, with instructions for inducing ovulation, are available from commercial suppliers. Ovulation occurs when the gonadotropic hormones of the pituitary gland are carried by the blood to the ovaries. Ovulation is induced, then, by injecting pituitary glands into the female abdomen. Mature males and females are required. Rugh (1962) states that the *Rana pipiens* female should be at least 74 mm and the male at least 70 mm in body length from snout to anus. Pituitary glands of the female are about twice as potent as those of the male, and the required number of glands for injection varies with the season. Rugh suggests the following numbers:[1]

---

[1]From Roberts Rugh, 1962, *Experimental Embryology*, 3d ed., p. 93, Burgess Publishing Company, Minneapolis. By permission.

	MALE PITUITARIES	FEMALE PITUITARIES
September to January	10	5
January to February	8	4
March	5	3
April	4	2

Before beginning the procedures described below, prepare Holtfreter's stock solution by adding 3.5 g NaCl, 0.05 g KCl, 0.1 g $CaCl_2$, and 0.02 g $NaHCO_3$ to 1 l of distilled water. For best results, all dishes used here and below must be scrupulously clean. Wash dishes thoroughly in hot, soapy water, and rinse with at least 10 changes of distilled water.

Techniques for exposing the pituitary gland are shown in Fig. 4-14. The pinkish, oval pituitary gland will be seen attached at the anteroventral region of the brain, or pulled loose and adhering to the bony flap as shown in Fig. 4-14. Use forceps to carefully remove the gland and pick away any attached white tissue. Put the gland in 1 or 2 ml distilled water (or 10% Holtfreter's solution). Draw the whole gland into a hypodermic syringe barrel before attaching needle. Avoid puncturing the glands. Attach a no. 18 needle, and adjust the volume of fluid in

Figure 4-14 Techniques for removing frog pituitary gland. *(Biological Sciences Curriculum Study.)*

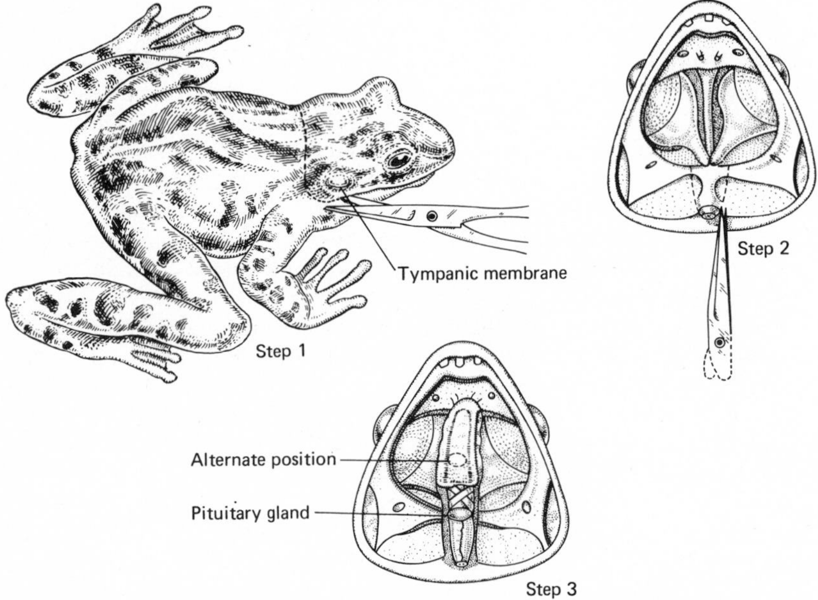

A

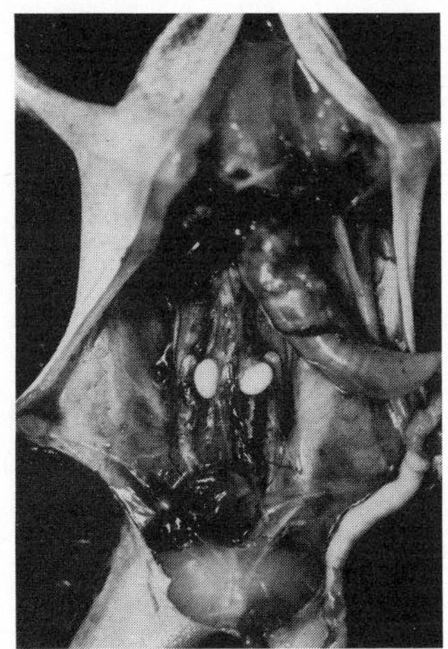

C

B

Figure 4-15 Artificial fertilization of frog eggs. **A** Stripping eggs from female. **B** Male, ventral view with body wall opened to expose internal organs. **C** Same as **B** with most abdominal organs removed to expose the pair of white testes lying at anterior edge of kidneys. The enlarged stomach, filled with food, and a portion of the intestine have been pulled to the extreme right.

the syringe to no more than 2 ml. Hold the syringe and needle parallel to the lower abdominal wall, with the needle pointing toward the head. With a forward movement insert the needle through the skin and muscle at a point just anterior to the base of a hind leg. Do not point the needle downward into the internal organs. Hold the needle in position under the muscle until the glands in the syringe settle at the base of the needle. Then quickly inject the glands and fluid into the abdomen. Leave the needle in place for about 30 sec; then pinch the skin tightly together at the puncture and slowly withdraw the needle. Again draw 1 to 2 ml of fluid into the syringe, and if any glands remain in the barrel, repeat the injection. Transfer the injected females to a covered jar containing about ½ in of water. Maintain at 18 to 20°C.

At the end of 48 h, check for ovulation by carefully holding the frog's hind legs in the left hand (if right-handed), laying the frog on its back in the palm of the right hand, and gently closing the right fingers and thumb around the abdomen while pressing in the direction of the cloaca (see Fig. 4-15). If only a fluid is emitted from the cloacal opening, replace the frog in the jar and wait 24 h before testing again. If eggs are emitted, follow the instructions below for artificial fertilization.

Artificial fertilization. Prepare a 10% Holtfreter's solution by adding 1 part of the stock solution (above) to 9 parts of distilled water. Remove a pair of testes from a male frog and place them in a small dish containing 10 ml of the 10% Holtfreter's solution. Use a glass rod to macerate and crush the testes to release the sperm. Add 10 more milliliters of the 10% Holtfreter's solution and allow the suspension to set for 10 min, during which time the sperm become active and begin to swim.

At the end of 10 min, strip a few eggs from a previously injected female as described above. Use a paper towel to remove and discard the first eggs. Then strip about 100 eggs into the sperm suspension, moving the female around to distribute the eggs evenly throughout the sperm suspension. Use a *clean* medicine dropper to bathe the eggs several times with the sperm suspension. Allow eggs to remain in the suspension for 20 min; pour off the sperm suspension and cover the eggs with pond water (or with 20% Holtfreter's solution if pond water is not available). As water is absorbed, the layers of jelly surrounding an egg become swelled. At the end of about an hour, fertilized eggs, now loosened from the jelly, will rotate and become oriented with the dark area, the animal pole, turned upward and the heavier cream-colored yolk area turned downward.

In place of removing and injecting with pituitary glands, the simplest method, and often the most successful, is to purchase three or four males with several females that have been injected with pituitrine at a biological supply house shortly before shipment. The males and females must be separated immediately when received in the laboratory. With the shipment, most companies supply information about the day and hour of injection. Attempt to strip eggs at 48 h after the time of injection and continue these attempts every 12 to 24 h. Otherwise, females may ovulate spontaneously, a condition to be avoided if possible, as eggs must be fertilized immediately after their release. Occasionally, a shipment may be delayed in transit, with the result that ovulation occurs very shortly after the frogs are received. Therefore, all preparations for artificial fertilization should be completed soon after an order is placed.

Descriptions of the early cleavage and later development of the frog embryo will be found in many manuals and books, including Rugh (1962—excellent for a detailed study), Moog (1963), and Glass (1965).

REPTILES

Reptiles are characterized by a dry skin, covered with leathery or horny scales. Most reptiles are adapted to land living; a few are aquatic but lay eggs in muddy banks or on sandy beaches. All have internal fertilization and are oviparous or ovoviviparous. There is no larval stage, and the young resemble the adult. The animals are lung-breathers throughout life. Except for turtles, reptiles move rapidly. They are predominantly carnivorous and use teeth, horny jaws, and tongue for capturing live food. Three orders are found in North America: Chelonia (turtles); Crocodilia (crocodiles and alligators); and Squamata (lizards and snakes).

Turtles

USES: For studying (1) anatomy and physiology, (2) reproduction and development, (3) nutrition, and (4) locomotion. See Carr (1952), and other references to reptiles.

DESCRIPTION: Terrestrial, fresh water, and salt water. A broad, shortened body, encased in an upper shell, the carapace, and a lower shell, the plastron. Shells composed of bony plates, each covered with a horny layer. Head, tail, and four legs can be drawn completely or partially into the shells. No teeth; jaws covered with a horny layer shaped into a sharp beak.

Oviparous, and soft- or hard-shelled eggs are laid in sand or mud. No care is given to eggs or young. In males, generally the anus is posterior to the edge of carapace. In females, anus is even with the posterior edge of carapace. Males may have a concave plastron and females a flat or convex plastron. With many species, the male has horny patches on inner side of each hind leg, longer claws on front feet, and a more conspicuous horny claw at tip of tail. As a rule, the base of the tail is heavier and longer in the male. Often the male is smaller in body size, and the plastron is shorter than the carapace. In the United States, the terms "tortoise" and "terrapin" are used loosely, with the first term generally referring to land turtles and the second referring to freshwater turtles. Small types or the young of large types of land and freshwater turtles can be maintained without difficulty, including those of the following groups (see Fig. 4-16).

A *Courtesy of Carolina Biological Supply Company*

B

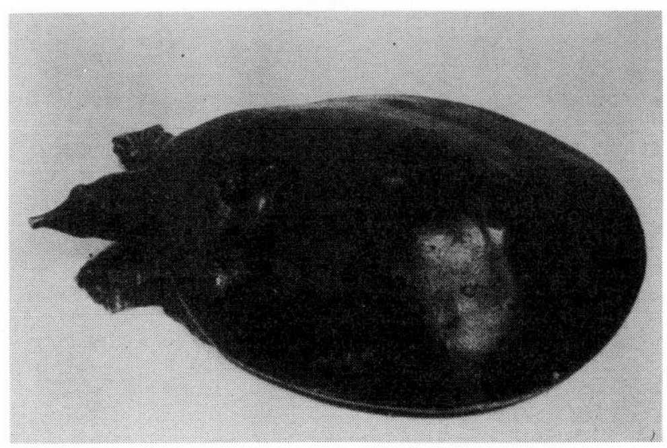

C

D *Courtesy of Carolina Biological Supply Company*

Figure 4-16 Common turtles. **A** *Chrysemys picta*, painted turtle. **B** Pseudemys sp. **C** *Trionyx spinifer*, a softshell turtle. **D** *Graptemys* sp., a map turtle. *(B and C Courtesy of CCM: General Biological, Inc., Chicago.)*

Chrysemys (painted turtle). Relatively abundant in shallow ponds and marshes over most of United States. Conspicuous patterns of red, yellow, orange, and olive-green. Sometimes observed in groups resting in the sun on logs, rocks, or stumps. Length of carapace to 6 or 7 in.

Pseudemys (cooters and sliders). Large, basking turtles of freshwater streams and ponds in southeast, southern, and central United States. Reach 10 to 12 in in carapace length. Generally, posterior edge of carapace is saw-toothed. Brown or olive color. Head stripes and spots are usually yellow or red.

Trionyx (softshell turtles). Aquatic. Soft, leathery, and flexible shell with no scales. Flattened body. Sharp claws, strong beak, and a long neck and snout. Attain carapace length of 14 to 16 in; gray or brown color; often spotted, but markings vary. In muddy bottoms or at the banks of ponds and lakes. Often vicious and irritable, and require careful handling. Unlike other freshwater turtles, *Trionyx* swims rapidly and at times will strike with great speed at other water animals. The spiny softshell turtle, *T. ferox,* and its subspecies are fairly common over most of eastern and central United States.

Graptemys (map turtle). In rivers and lakes over the United States; most abundant in eastern United States. Generally olive-green with yellow or orange markings. Timid and wary, and difficult to catch. Carapace length to 8 to 10 in.

Terrapene (box turtle). Land turtles. Brown to black. High arched carapace. Upper jaw turned down into a beak over lower jaw. Distributed over much of the United States. Omnivorous. Most easily captured and most docile of turtles. Carapace length to 6 to 7 in.

SOURCES: To collect water turtles stretch a strong net in water along one side of a log or other basking place. When turtles have climbed onto the log, approach the log from the side opposite to the net, and turtles will jump into the net. Lift and carry turtles by tails or by holding each side of the carapace. Avoid the sharp beaks and claws. Land turtles are collected by hand or with a strong net. See Appendix B for commercial sources.

CULTURE: Put freshwater turtles in a container of pond or conditioned tap water. Provide rocks or sticks for resting areas above water surface; or put water at one end of container with sand and rocks slanted above water surface at other end. Water line must be well below the top of walls to prevent animal escape. Place land turtles on soil in boxes of sufficient depth to prevent animal escape. Provide a dish of water for the turtle's occasional bathing. Turtles bask in sunlight and should receive several hours of sunlight or incandescent light each day.

As a rule, turtles eat a wide variety of food, both plant and animal. Feed aquatic turtles with live earthworms, mealworms, flies, crickets, and other live insects. Try canned or frozen seafoods, chopped liver and beef, and dried turtle food from an aquarium store. Give lettuce leaves several times weekly. Remove uneaten food within several hours. If water is not running through the container, wash out container several times weekly, and add fresh water. Do not allow water to become stagnant. Land turtles will accept a wide range of food including live insects, worms, chopped meat, leafy vegetables, celery, carrots, apples, and other fruits.

Diseases. Generally, turtles are not susceptible to disease when adequate food and clean housing are provided. On rare occasions, eye infections may appear, which can be treated by washing the eyes with borax solution. An occasional respiratory infection may be eliminated by placing the animal under a lamp for heat, or by raising the water temperature. An inadequate diet may result in a softened shell and a lethargy. Young turtles should have ample minerals, principally calcium, for shell growth. Furnish calcium in powdered form (from aquarium houses) or in a small beef bone on which the animal chews.

Lizards

USES: Especially interesting for studying behavioral patterns of dominance, territoriality, and courtship. See Follansbee (1965).

DESCRIPTION: Elongated and generally narrow, tubular body. Usually two pairs of limbs. If held by tail, lizards may shed it by breaking the tail vertebrae with a muscular movement. Ordinarily, a smaller tail is regenerated. All lizards have teeth and when restrained will attempt to bite, although only the larger animals are able to inflict a painful wound.

The male has a stout tail that is much heavier at the base than that of female. Males, more than females, display a vivid coloration of blue, yellow, or red, particularly on the ventral surface. Often the males have a thick gular fold, the dewlap, across the ventral neck region, and more prominent femoral pores (rows of tiny pits on scales under the thighs). Most species are oviparous, laying from 1 to as many as 30 or 40 eggs in loose soil. A few North American species (for example, some of the night lizards, *Xantusia*) have developed a primitive placenta and give live birth. In temperate climates, lizards mate during spring and lay eggs in late spring and early summer. Eggs hatch from midsummer into autumn.

Apparently, all North American lizards can be maintained without difficulty in the laboratory, although the venomous Gila monster of southwest United States and the closely related, venomous Mexican beaded lizard of western Mexico cannot be considered appropriate for the teaching laboratory. Among the more common and more easily maintained lizards are those described below (see Fig. 4-17).

Anolis carolinensis (American "chameleon"). Abundant throughout southern United States, on fences, in trees and shrubs, or on ground; often in yards of homes and outbuildings. Length, including tail, up to 6 or 7 in. A well-developed ability to change skin color from green to gray or reddish brown, apparently related to environment and emotional changes. Adult male larger than adult female and with a much larger dewlap of a more brilliant pink or red color. Move rapidly; toe pads used to scale glass walls of a container.

Eumeces (skinks). A large group, abundant over entire United States. Shy and evasive, and difficult to capture. Mostly day feeders, taking shelter under rocks and logs at night or when endangered. Smooth shiny coat of scales; generally brown or black.

Suitable for the laboratory is the five-lined blue-tailed skink *(Eumeces fasciatus)*, which at birth has five white or yellow stripes on the black body and a bright blue tail. The stripes become less distinct in adults. Length including the tail may reach 5 or 6 in. Tails are extremely fragile and are readily snapped loose from the body.

A *Courtesy of Carolina Biological Supply Company*

B *Courtesy of Carolina Biological Supply Company*

C

D *Courtesy of Carolina Biological Supply Company*

Figure 4-17 Common lizards. **A** *Anolis carolinensis,* American "chameleon." **B** *Eumeces fasciatus,* blue-tailed skink. **C** *Sceloporus magister,* desert spiny lizard. **D** *Phrynosoma cornutum,* horned lizard. *(C Permisson of Ward's Natural Science Establishment, Inc., Rochester, N.Y.)*

Sceloporus (Spiny lizards). Distributed over all of United States; particularly abundant in dry regions of southwest United States. Sometimes called swifts or fence lizards. Bold in disposition, not quickly frightened, and thus are captured more easily than skinks. Rough-pointed dorsal scales, which account for the name of spiny lizards. Mostly on ground and not in trees. Light brown to gray body. Males of most species with bright blue patch on each side of belly. Females with faintly colored blue patch or no marking. Adult body length, including tail, 4 to 6 in. Length of spiny lizards of lower Rio Grande Valley, 12 or 13 in.

Phrynosoma (horned lizards). Sometimes called horned "toads." Range over all of United States but most abundant in dry regions of Southwest. A bizarre flattened and ovoid body, sometimes almost round. Usually with prominent, sharp-pointed spines or "horns" on head, and large pointed scales on body. Relatively short tail that broadens against the posterior body. Adult length, including tail, 3 to 5 in. Light gray or tan color. Ground-dwelling.

For protection, *Phrynosoma* moves into mammal burrows, under rocks, or quickly buries itself in sand. It does not spontaneously snap off its tail, but the tail is regenerated when accidentally severed. Although we have not observed the phenomenon, Pope (1955), Stebbins (1954), and others describe the horned lizard's extraordinary defense mechanism of ejecting blood from the eyes for a distance of several feet.

SOURCES: Use a long-handled net or a looped snare to capture lizards. Quietly approach an animal from behind and quickly lay the net over it. If using a snare, slowly place the snare loop around the animal's head, pull the loop backwards over the head and tighten it around the neck or upper body. If approached quietly, lizards generally are very patient about the process, although they may snap at the loop and sometimes dislodge it.[1] Transport animals in any type of container, covered tightly and punctured with air holes. Do not crowd animals and do not store container in direct sunlight. Observe, and separate animals if fighting occurs. Separate spiny lizards from others, as they generally attack and kill other types. At times, they kill the smaller animals of their own species. See Appendix B for commercial sources.

CULTURE: Lizards require a warm temperature of 22 to 25°C and should be placed in a glass-walled container about 18 in deep (to avoid air drafts), and with a fine-mesh screen or glass cover. If necessary, provide warmth with a 100-W lamp above the cage. Ample space is necessary, as many lizards establish individual territories from which they operate. Each lizard should be given at least 1 ft^2 of bottom surface. Cover the bottom of container with loose soil for woodland lizards, and with sand for desert lizards. Add stones for hiding, and dry branches and leafy twigs for climbing and resting. Provide a constant supply of drinking water on a moistened sponge in a dish, and by daily sprinkling of water over the leafy twigs.

Live food is required and should be given several times weekly. A simple method is to keep opened bottles of developing *Drosophila* in the cage. The large *D. virilis* is especially good, and lizards become filled out and plump on the diet. The cage must be covered with several layers of muslin cloth to prevent escape of the flies. Although most lizards eat mealworms, a constant diet can be deleterious, as the mealworm's exoskeleton may produce an intestinal stoppage. Live moths, ants, roaches, and spiders can be given also. For a long period of maintenance, animals should be exposed to direct sunlight during several hours daily or to 10-min exposures twice weekly of ultraviolet radiation from a short wavelength ultraviolet lamp (2537 Å wavelength). To lift and carry lizards, slowly extend a hand from the rear and tightly grasp the animal behind the head. Do not lift or carry the animal by the tail.

Snakes

USES: Studies of (1) behavior, (2) morphological adaptations for locomotion, (3) reproduction and development, and (4) feeding habits.

---

[1]On several occasions, the author has observed high school and college students in Puerto Rico who during several hours' time patiently captured a great many anoles and spiny lizards by means of a loop tied at the end of a 7- to 8-in blade of grass.

DESCRIPTION: Extremely elongated and tubular body. No appendages and no vestigial pectoral girdle. Vestigial pelvic girdle in pythons and boas. Many vertebrae and ribs. No external ear openings and no middle ear. Probably do not perceive airborne sounds but are sensitive to ground vibrations. A forked extensible tongue for tactile and olfactory perception, but not for hearing or stinging. No movable eyelid, accounting for the fixed stare of snakes. Transparent cover over eye that is shed with the skin. Most snakes with backward-slanting teeth in both jaws. Most venomous snakes with certain teeth that are grooved or hollow for conducting venom. Movable skull bones and jaw halves become dislocated to allow a considerable expansion of mouth opening for ingesting and swallowing large foods. Body covering of scales, usually overlapping.

Most are terrestrial; some are principally arboreal; and many are both aquatic and terrestrial; a few are solely aquatic. Outer skin is shed periodically by loosening the skin about the mouth and by turning the skin outward and backward from anterior to posterior, thus turning the skin inside out. All snakes may be irritable, particularly before and during skin-shedding, first indicated by a dullness of skin color and a milky opaqueness of skin over eyes. During winter months, most snakes become sluggish and hibernate. To induce activity at this time, suspend a lamp over cage. Ground snakes generally move rapidly. Some arboreal snakes, such as boas and pythons, are heavy bodied and slow-moving.

Mature males generally have a broader tail base because of the two copulatory organs (hemipenes) that extend posteriorly, one from each side of vent, into a long tube-like pocket. Ordinarily, mating occurs in the spring, and eggs are laid, or young are born, during summer months. Most snakes are oviparous and, depending on species, lay from 1 to 100 white or cream-colored eggs, covered with a tough membrane. A few species guard eggs; no care is given to young. Snakes with live birth include garter snakes *(Thamnophis)*, water snakes *(Natrix)*, rattle snakes *(Crotalus)*, and boas *(Lichanura)*. Embryo develops in a transparent sac from which it emerges during birth or shortly thereafter. Occasionally, a female garter snake carrying embryos may be collected in the field. The condition is recognized by the swollen and heavy body. The developing young can be studied by overetherizing the female and slitting open the body to expose the developing embryos, each enclosed in a transparent sac.

Under no circumstances can venomous snakes be considered appropriate for maintenance in the teaching laboratory. Probably all other snakes native to continental United States can be kept with little difficulty, although some types are more nervous than others and may strike with great force. Several types that adapt readily to captivity are described below (Fig. 4-18).

Thamnophis (garter and ribbon snakes). Moderately slender with head distinct from neck; adult body length, 18 to 24 in. Many with a pronounced dorsal median, white or beige stripe and a less pronounced lateral stripe on each side. All with live birth. Average litter of 20 young. Mostly diurnal, living in or near water. When handled, they may discharge fecal matter and emit a foul-smelling fluid from anal glands. As a rule, all garter snakes become gentle, although all may bite if handled roughly or abruptly. Probably these are the most satisfactory snakes for the laboratory; found over all of continental United States.

Heterodon (hognose snakes). A burrowing animal that needs several inches of loose soil in cage. Heavy-bodied and generally no constricted neck region. A large upturned scale forms a snout used for burrowing, and accounting for the name. Adult body length, 18 to 24 in. Slender, tapered tail. Variable color and markings. Oviparous with 5 to 10 or more eggs in late spring. Incubation period of 2 to 3 months. Found in woody areas and sandy prairie regions over most of United States. Particularly interesting because of behavior in feigning death, and bluffing with violent hissing, puffed-up anterior body, and flattened head. Sometimes called "puff adders" or "spreading adders." Actually the snakes are harmless; they rarely bite and may become gentle in captivity.

Lampropeltis (kingsnakes). Generally medium sized; common kingsnake, *L. getulus*, occasionally 4 to 5 ft long. Color and markings variable; most often black or brown with white or yellow marks. Oviparous with 6 to 12 eggs, or more, during late spring. Incubation period, 2 to 3 months. Young, 6 to 8 in at hatching.

Kingsnakes are distributed over most of United States. Generally they are mild-mannered, but if handled roughly they may hiss, vibrate the tail, and strike vigorously. As a rule, kingsnakes move on the ground of woody areas but may climb trees and shrubs in search of lizards and bird nests with eggs or young. Caged animals should be given branches for climbing. A mature kingsnake should be kept in an individual cage, as it eats other snakes including those of its own kind. At times it may swallow a snake longer than itself. Many species kill their prey by constriction. In nature, kingsnakes eat venomous snakes, including the rattlesnake, but probably they do not pursue or prefer rattlers as described in folk tales. Kingsnakes are relatively resistant to the rattler's venom, as are most other snakes.

The scarlet kingsnake *(L. doliata)* is a notorious example of mimicry in its resemblance to the venomous coral snake. The differences are that the scarlet kingsnake has yellow rings separated from red rings by a black band, whereas the coral snake has yellow rings that adjoin the red rings.

Other snakes that can be maintained successfully but which generally remain nervous and irritable are the following types (Fig. 4-18).

Pituophis (bull snakes or gopher snakes). Land snakes. Good climbers. Heavy-bodied and up to 6 ft long. Oviparous. Remain nervous in captivity and may strike viciously.

Coluber (racers). Slender body, to 6 ft long. Fast-moving. Generally stay on ground but are good climbers. Oviparous. Nervous in the laboratory but may become mild-mannered.

Masticophis (whip snakes and coachwhips). Slender. Fast-moving. Generally found on ground but are good climbers. Adults 5 ft or more in length. Oviparous. Generally remain nervous and aggressive in the laboratory. Can inflict a painful wound and flesh laceration with their sharp teeth.

Natrix (water snakes). Heavy-bodied. Adult to about 4 ft long. Always near or in water. Most active at dusk or during the night. Most species ovoviviparous. Remain irritable and pugnacious in the laboratory and must be handled carefully. Although the garter snake is not as aquatic in nature, it serves as a better example of a water snake for laboratory studies. The *Natrix* may be easily confused with the venomous cottonmouth *(Agkistrodon)* and should not be collected by amateurs.

A *Courtesy of Carolina Biological Supply Company*

B

C *Courtesy of Carolina Biological Supply Company*

D *Courtesy of Carolina Biological Supply Company*

Figure 4-18 Common snakes. **A** *Thamnophis ordinoides,* garter snake. **B** *Heterodon* sp., hognose snake. **C** *Lampropeltis getulus,* kingsnake. **D** *Lampropeltis doliata,* a harmless kingsnake that mimics the venomous coral snake shown in **E**. In the kingsnake the yellow bands do not adjoin the red bands. **E** *Micrurus fulvius,* coral snake in which the yellow bands adjoin the red bands. **F** *Pituophis melanoleucus,* bullsnake. **G** *Coluber* sp., a racer. **H** *Masticophis bilineatus,* Sonora whipsnake. **I** *Natrix* sp., water snake. *(B, G, and I Courtesy of CCM: General Biological, Inc., Chicago. H Permission of Ward's Natural Science Establishment, Inc., Rochester, N.Y.)*

SOURCES: In many states snakes are protected by law. A collector must obtain a permit from the state wildlife or game bureau. Collectors must be able to recognize poisonous snakes and must be equipped with a snakebite kit. Poisonous snakes should be collected only by trained persons who have a valid reason for obtaining the animals.

Snakes emerge into the open on warm sunny days but remain under cover during extreme temperatures. A snare, a forked stick, or a net are effective tools for capturing the animals. When using a snare, approach the animal quietly from behind and place the loop quickly around the animal's head. Pull the loop back over the head and tighten it around the neck or upper body. If a net is used, press the hoop firmly against the snake and at the same time cautiously move the hoop so that the entire body is caught in the net. Generally, snakes attempt to escape, and the collector must be quick in looping or in pinning the animal to the ground.

E *Courtesy of Carolina Biological Supply Company*

F *Courtesy of Carolina Biological Supply Company*

G

H

I

All snakes bite, and the beginner should wear thick gloves when transferring animals to a bag for transporting to the laboratory. Be certain all seams of the bag are tight, as snakes often manage to escape through small openings. Do not stack one bag on top of another, and allow ample ventilation with car windows open. Take care not to store bags of snakes in open sunlight. See Appendix B for commercial sources.

CULTURE: Lift and carry snakes by extending the hand slowly from the rear and grasping the animal tightly behind the head. Occasionally these animals will become quite gentle when handled frequently, although they may be very irritable immediately before and after feeding, and when shedding the skin. Put snakes in a tight cage of appropriate size, with screen or glass walls. The animals escape easily through open cracks, and the side or top opening must fit tightly against the cage walls. Cover the cage bottom with several layers of newspapers that are removed as necessary, and replaced with fresh paper. If one wishes, the paper may be covered with a layer of dead leaves, although the leaves make cage cleaning more difficult and may allow an accumulation of excrement as well as animal parasites. Burrowing snakes require a bottom layer of several inches of sand or loose soil. Add rocks, sticks, or an inverted box for hiding and sleeping. Provide a water dish, large enough for the snake to submerge its complete body, and heavy enough so that the snake cannot upset the dish. Wash the water dish and give fresh water daily. Place cage at 21 to 25°C. Avoid extreme temperatures.

Most snakes feed on whole living animals, although some take dead animals and others eat raw eggs. The diet varies among species and even among individuals of the same species, thus requiring that a variety of foods be provided until acceptable kinds are identified. For the snakes described above, the following foods generally are suitable. Garter snakes: live fish, frogs, toads, salamanders, tadpoles, and earthworms. Try chopped raw fish or beef, live mealworms, and small, live mice. Hognose snakes: live toads, frogs, tadpoles. Try chopped raw fish and live mealworms. Kingsnakes: snakes, lizards, mice, rats, birds, bird eggs. Try frogs, toads, and chicken eggs. Water snakes: live fish, frogs, tadpoles. Try chopped raw fish or beef, and live mealworms. Many snakes fast for several weeks or longer, particularly during hibernation in winter months. Do not leave live mice or rats in a cage overnight, as they may bite and kill a snake.

When fed and housed properly, snakes are generally resistant to disease. Wash skin wounds and apply Mercurochrome or other similar medicine. Treat eye infections by washing with borax solution or by applying argyrol. A snake mite, *Ophionyssus natricis*, is fairly common and appears as small white patches on the animal's skin. Use a magnifying lens to determine if white specks are mites. The mites may increase greatly in number, particularly if the cage is not cleaned regularly. To control and prevent an infestation, wash cages periodically and dust pyrethrum powder (or other insect powder) over cage and into crevices. A layer of powder under the bottom newspaper gives added protection. Put heavily infested snakes into a bath of potassium permanganate water (a light pink color) for 10 to 15 min. Probably lightly diseased snakes fare better if they are returned to their natural outdoor habitat.

BIRDS

Young chicks are used regularly in the teaching laboratory. Live, adult birds are not so commonly studied, although the smaller types, for example, the pigeon, the bantam chicken, and the Japanese quail, may provide valuable studies. The tiny Japanese quail is an especially interesting laboratory bird. Ordinarily, the collecting of wild birds for a laboratory course cannot be considered desirable or practical. Although individual research may be very appropriate, good conservation practices preclude collecting and confining wild birds in sufficient numbers for class studies. Most wild birds are protected by law, thus requiring a permit for collecting. Furthermore, wild birds do not adapt readily to confinement and laboratory maintenance.

Domestic Chicken *(Gallus domesticus)*

USES: Studies of (1) behavioral patterns, (2) genetics, (3) embryology, and (4) hormone control in embryos and young chicks. See Hollander (1966), Moog (1963), Rugh (1962), and Schlesinger (1966).

DESCRIPTION: When young chicks are studied for only several weeks, most any strain is suitable, although the lighter-colored strains can be marked and identified more easily than dark-colored chicks. The smaller types, for example, bantams, are suitable for longer periods of maintenance. With young chicks, a hand lens is used to detect the penis tip during the first 24 h after hatching. Occasionally, sexed chicks can be purchased at local hatcheries.

SOURCES: Obtain eggs or chicks from local hatcheries and poultry farms. Commercial sources for bantams are listed in Appendix B.

CULTURE: *Young chicks.* The young chicks can be satisfactorily maintained for several weeks in a brooder made from a large cardboard carton, to which is added a bottom cover of several layers of newspaper, changed once or twice daily. A carton 24 in long, 18 in wide, and 18 in deep is sufficient for six chicks up to 5 or 6 weeks of age. The depth must be increased if chicks begin to fly. Suspend a 100-W lamp over the carton to maintain a temperature of 25 to 26°C within the carton. At all times, provide water and starter mash in heavy-bottom bowls. Chick brooders, holding 50 or 100 young chicks, can be purchased from local suppliers, including Sears, Roebuck and Co. Also commercial food and water dispensers can be purchased locally. The chicks' constant cheeping makes it necessary to locate the animals away from the laboratory. A cloth, placed loosely over the carton, helps to quiet the chicks, although care must be taken to provide adequate ventilation and to be certain the lamp is placed at a safe distance.

Incubation of eggs. Commercial incubators of 50- or 100-egg capacity are available from biological suppliers and from local stores (Fig. 4-19). Incubation time is 20 to 21 days, depending on the stage of development at the time when eggs are

Figure 4-19 Laboratory incubator. *(Courtesy of Lab-Line Instruments, Inc.)*

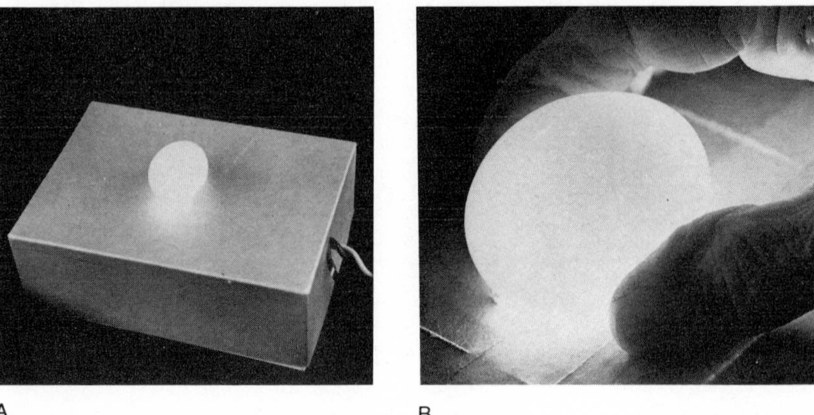

A B

Figure 4-20 Equipment for candling eggs. **A** The box is lined with aluminum foil and equipped with a 100-W lamp. The egg is placed over an opening cut out of top wall. **B** Close-up view of candling technique.

placed in the incubator. If necessary, the eggs may be stored at 15°C for as long as 7 to 10 days before incubation, although hatchability may decrease. Select clean, well-shaped, strong-shelled eggs, and lay them on their sides in the incubator. Incubate at 100°F (37.9°C) and with a 50 to 60% humidity. Generally, the humidity is obtained by keeping a pan of water in the bottom of the incubator. Eggs must be turned at least every 12 h (preferably every 8 h) to prevent adhesion of the embryo to the shell. Beginning at about 48 h incubation, the eggs can be candled to learn if embryos are living. A simple candler can be made by inserting a light socket through one side of a small box and cutting a hole in the opposite wall of the box. A 100-W lamp is placed in the socket. Line the inside wall of the box with aluminum foil or asbestos. Darken the room, and hold the egg outside the box opening so that a narrow beam of light shines through the egg (Fig. 4-20). Infertile eggs are clear. The blastoderm of live embryos will be seen at about 48 h development, and the yolk-sac circulation will be seen at about 60 to 72 h. Dead embryos show a dark outer ring of blood that has separated from the embryo. After the thirteenth day, the embryo appears more and more opaque in the candled egg, with the air space accentuated at the blunt end. A dead embryo at this stage is a cloudy mass that floats around when the egg is rotated. About 80 to 85 percent of the eggs should hatch. Allow the newly hatched chicks to dry in the incubator for 1 or 2 h, and then transfer them to a brooder chamber or to a heated cardboard carton.

Incubation in plastic bags. Hollander (1966) and others have described simple techniques for incubating eggs in clear plastic bags. The egg is removed from the shell and transferred to a bag approximately 15 cm wide and 19 cm long. The bag is then closed with flexible wire or cord that is used as a suspension hook (Fig. 4-21). Hollander recommends adding the albumen of a second egg to give more support and to hold the folds of the bag away from the yolk. He reports that development through the tenth day is easily obtained by this method, and that various treatments can be administered through the top opening when the bag is sup-

Figure 4-21 Incubation of egg in a plastic bag that is hung in an enviromental chamber or incubator. The albumen of a second egg was added to support the yolk.

ported in a bowl of warm water. Other persons have reported successful incubation when many bags of eggs are suspended in a culture chamber (or a closet or room), equipped to maintain a temperature of 35 to 37°C and a humidity level of 60 to 70%.

Diseases. Ordinarily, diseases are prevented by following strict routines for cleanliness and proper feeding. Unless the chicks are expensive or have unusual traits, generally it is better to dispose of diseased animals. Symptoms of fungal, viral, or protozoan infestations are apparent when animals show a stunted growth, a loss of appetite, or when loose droppings appear. Lice and mites often occur and must be constantly guarded against. These are best prevented by dusting the bird and cage with 5% Malathion or other dust or spray from a veterinarian or commercial poultry supplier.

Japanese Quail *(Coturnix coturnix japonica)*

USES: For studies of (1) behavior, (2) genetics, (3) embryology, and (4) physiology of the developing embryo and young. See Calhoun (1968), Daniel (1966), Farris (1967), Fitzgerald (1970), Howes (1964), Mills (1966), and Padgett and Ivey (1959).

DESCRIPTION: *Coturnix* is a native of eastern Asia and has been raised as a domestic bird in Japan for many years. The bird and its eggs are considered a food delicacy in Japan, and the pickled eggs often are imported into the United States as an hors d'oeuvre. The Japanese quail was introduced into the United States during the 1950s and recently has become an important research animal. Because the bird is relatively unknown in the teaching laboratory, a considerable description of the bird and its maintenance are given here.

The quail is hardy and easy to raise. The short life cycle is an advantage in genetic studies, and the high metabolic rate makes the bird very useful in testing for drugs and metabolites. The adult bird is small and stub-tailed (Fig. 4-22). Adult body length is 10 to 11 cm. Weight of adult male is 95 to 105 g; of adult female, 110 to 120 g. The bird triples in size and weight during the first week after hatching; flies at about 2 weeks; and matures at 6 to 8 weeks. Sex differences in plumage can be recognized at 3 weeks or earlier. Females display a gray- and black-speckled breast; males have a more evenly colored brown breast with only a few speckles.

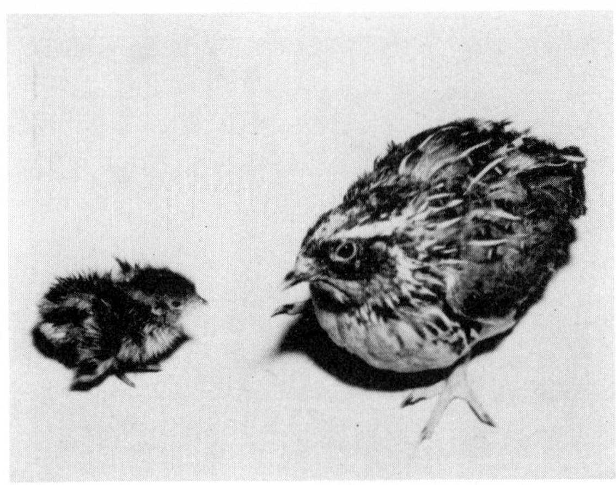

Figure 4-22 Japanese quail, *Coturnix coturnix japonica*, adult quail and day-old bird. *(Courtesy of Truslow Farms, Inc.)*

Homma et al. (1965) describe a method for identifying the sex on the first day by using a hand lens to locate a pointed protuberance in the vent of the male and a deep furrow at the dorsal side of the protuberance. These two structures are not fully developed in the female. The furrow can be seen better if a dye solution is applied and excess moisture is blotted with a tissue.

Most researchers agree that greatest hatchability is obtained from eggs laid during the first 100 to 220 days of growth, and that fertility and hatchability decline at 7 to 8 months. Eggs are about 1 in long and may be white or cream-colored with blue specks or a mottled brown color (Fig. 4-23). Number of eggs in one hatch may range from 6 to as many as 16 or 17. Padgett and Ivey (1959) state that breeding normally begins in April and extends through December. A minimum light-day of

Figure 4-23 Eggs of Japanese quail. **A** Chicken egg at left in contrast to Japanese quail egg at right. **B** Close-up of eggs showing variations in markings. See text for color description. Note small egg carton useful for egg incubation. Cartons are available from suppliers listed in Appendix B. *(A Courtesy of Truslow Farms, Inc.)*

A

B

13½ h is required for egg laying. With increased light, egg laying may occur during 10 or 11 months of the year.

SOURCES: Obtain eggs and birds from breeding farms (see Appendix B). Occasionally the quail may be obtained from state conservation and wildlife agencies. Not all suppliers will ship the eggs because of their extreme fragility. Birds are shipped at about 4 weeks old when they are full-feathered. Eggs and birds cannot be shipped during severely cold weather. A start of eggs or birds is relatively expensive, and if many are required on a regular basis, the most desirable procedure is to maintain breeding colonies.

CULTURE: *Incubation of eggs.* Incubation time is 16 or 17 days at 100°F (37.8°C). Normal egg hatchability is 60 to 70 percent. If necessary, the eggs can be stored for a week before incubation without loss of hatchability if placed at 55 to 60°F (12 to 15°C) with a 50 to 60% humidity. When laying-hens are the source of eggs in the laboratory, put the hens in a cage with a sloped bottom so that the eggs roll under the side wall into a trough that is lined with soft paper to prevent breakage. Handle the eggs with extreme care, as much handling causes fine cracks that cannot be seen without magnification. Place the eggs on their sides in an egg flat (from a commercial supplier) or on soft tissue within the incubator. Maintain a high humidity with a pan of water in the incubator. Temperatures above 103°F are fatal to embryos. Turn the eggs at least every 8 h (Howes, 1964, says every 3 h). Allow the hatched birds to dry before transferring them to a brooder chamber.

Care of hatchlings. During the first 3 days, the young birds are unable to maintain a normal body temperature, and chilling at an early age is the greatest cause of mortality. Place the hatchlings in a brooder chamber or in a heated cardboard carton as described earlier for chicks. During the first 3 days, maintain a temperature of 95°F within the brooder chamber and reduce to 75°F by the end of the first 2 weeks. Feed chicken starter mash or Game Bird Startena (from Ralston Purina Company). For the first 3 days, put food on a paper, spread over the bottom of the brooder, after which place the food in a small trough. Give water at all times in a small bowl that is filled with small pebbles to prevent the young birds from drowning.

Care of young and adult birds. At 16 to 18 days, transfer birds to rearing cages. Twenty-four birds can be housed in a cage that is 2 ft wide, 2¼ ft long, and 6 to 12 in high. If breeding animals are to be maintained, the birds are placed in colonies of one male for each two or three females. Cage walls and floor may be ½-in mesh hardware cloth, with a dropping pan or with a solid floor covered with shavings or paper. Remove droppings at least three times a week and preferably daily. For food, Padgett and Ivey (1959) suggest that Game Bird Growena be added to Game Bird Startena with increasing amounts until at 4 weeks old only Growena is fed. At 8 weeks, begin Game Bird Layena to which is added a small portion of crushed oyster shell to increase the hardness of the egg shell. During winter months, provide extra carbohydrate with scratch food such as cracked corn.[1] Many pet stores sell the crushed shell. Howes (1964) recommends a 12 h light and 12 h dark

---

[1]All named foods are Ralston Purina products, obtained from a feedstore or from Ralston Purina Co., Checkerboard Square, St. Louis, Mo. 63102.

photoperiod until 4 weeks of age and then, for breeding stock, that light be increased to 14 h and then gradually increased to 18 h until 6 months of age, after which the useful breeding life is completed. We have maintained birds constantly at 14 h light and 10 h dark with satisfactory laying and growth.

We have not attempted to raise the birds outdoors, although Padgett and Ivey (1959) describe satisfactory results when birds at 2 weeks old are put in outdoor cages with walls of ½-in mesh hardware cloth and dimensions of 3 by 6 by 1.5 ft. This size cage holds 20 birds. For breeding purposes, they use 5 males and 15 females. More males produce fighting, and fewer males may not give adequate fertility of eggs. Because the birds may become excitable when flying within a cage of this size, and may injure their heads against the top of the cage, the top cover should be of burlap or other heavy cloth. The animals must be protected from predatory animals, including dogs and cats. The birds withstand temperatures to freezing, if one end of the cage is boxed with wood walls and a wood cover. At lower temperatures, the cage must be put indoors or given heat.

The Pigeon *(Columba livia)*

USES: For studying (1) learning and other behavioral patterns, (2) genetics, (3) nutrition, and (4) embryology and development. See Clarkson et al. (1963), Hollander (1954), Hughes et al. (1957), Levi (1957), Naether (1958), and Silvan (1966).

DESCRIPTION: Pigeons are docile and easily tamed. Weight of adult male is 24 to 26 oz; of adult female, 22 to 24 oz. Life span is probably 12 yr or longer. Pigeons mature at 6 months and mate in pairs, usually for life, although they can be separated and remated, generally without difficulty. Breeding occurs throughout the year, except during molting. Eggs are usually laid in clutches of two; second egg is laid about 48 h after the first. A pair may produce 18 to 20 young per year.

Incubation time is 17 to 18 days. Both parents incubate eggs and feed young. During the first week, nestlings are fed "pigeon milk," a fat, rich material formed in the crop of each parent and resembling the curd of cow's milk. By the end of the first week, the young take small grain with pigeon milk. During third or fourth week, young begin to feed entirely on grain. Feathers appear at end of first week. Young leave nest at 4 or 5 weeks. First molting is at 6 weeks; thereafter molting occurs once each year, usually in August or September.

Hughes (1957) says that sexing is possible with 3-week-old squabs, at which time the male has a wider space between the eyes, and the top of the head is flatter. In adults, the male is somewhat larger, and feathers are coarser. The adult male struts and coos and may drag the tail feathers. The female holds the body in a more horizontal position and seldom coos. Sexes can be distinguished at mating time, when the male chases the female and performs an elaborate courtship pattern. Clarkson et al. (1963) describe a technique for sexing adult pigeons by spreading the cloacal opening and observing in the male a sperm-duct opening at the tip of a papilla on each side of the cloaca, and in the female, only an oviduct opening on the left side.

For breeding programs, the young squabs are separated by sex at weaning time, the third or fourth week. Birds are mated at 7 or 8 months old by confining a pair

A

B

Figure 4-24 Strains of pigeons used for teaching and research. **A** Pair of White Carneau pigeons; male at left and female at right. **B** Pen of Silver King youngsters ready to mate. *(A and B Courtesy of Palmetto Pigeon Plant, Sumter, S.C.)*

within a cage for several days, after which the two birds will generally remain together when released. See Fig. 4-24.

SOURCES: Pigeons are distributed over much of the United States, sometimes in great numbers within urban areas. Although local, wild pigeons can be used for much laboratory work, the birds must be carefully examined for disease and parasites. As a rule, animals from commercial breeders or local bird fanciers are less likely to carry disease. The White Carneaux strain is used commonly in research programs. See Appendix B for commercial sources.

CULTURE: During short-term studies, individual pigeons may be kept in metal cages, 18 by 15 in and 10 in high, with a ½-in wire-mesh floor and a dropping pan, or with a solid floor covered with pine shavings. The construction and care of cages is discussed in Chap. 1.

Breeding animals are best housed in a pigeon room which opens outdoors into a wire-enclosed exercise cage, preferably on the south side to allow maximum sunlight (Fig. 4-25). A portion of a building can be walled off for the pigeon room, with about 3 ft^2 of floor space for each pair of birds. In mild climates, the outdoor cage may be unnecessary, as the birds once accustomed to the house by confinement for several days will return after flights. However, this procedure allows exposure to diseases of wild pigeons.

Inside the pigeon room, hang a double nest box on the wall for each pair of birds, which often maintain, at the same time, incubating eggs in one nest and young squabs in the other. A suitable nest box is shown in Fig. 4-26. The solid-bottom box, 14 by 28 in and 14 in high, is divided into two nests by a 6-in-high board partition. A landing board is added. A one-way trapdoor may be desired for penning birds inside the box.

In each nest, place a shallow nesting pan of clean sand. Cover the floors of the room and the outdoor exercise cage with a 2- or 3-in layer of dry sand. Provide drinking water in a commercial drinking device or in a jar inverted into a bowl. Place the water container on top of a box or table with ample surrounding space

Figure 4-25 Housing pigeons. A front view of a 17-pen loft showing a drinking cup protruding in front of each pen, the water control unit in box at left, and a bath pan, turned up, at front of each pen. *(Courtesy of Palmetto Pigeon Plant, Sumter, S.C.)*

for birds to alight. During several hours each day, put a large open pan of bathing water in the room or outdoor cage. Otherwise, the birds will attempt to bathe in the drinking water.

Feed grain in shallow trays on a shelf or table with sufficient surrounding area for birds to alight. Hand-feeding helps to tame the birds. Pigeon food is available at most feedstores and generally consists of a mixture of corn, wheat, peas, and milo. At all times provide a commercial pigeon grit, usually consisting of oyster shell, sand, common table salt, and charcoal. Supplement with lettuce or cabbage leaves several times weekly. Probably a pelleted, complete food, such as Purina Pigeon Chow, is the simplest and most satisfactory type of diet. No supplementary foods are required with the pellets, although the manufacturer recommends that medium-sized oyster shells (from feedstores) be supplied at all times. As a rule, breeding stock are given a constant supply of food. Some laboratory workers prefer to give food for a 20-min period two times daily.

Several times weekly, remove droppings from floors by raking the top surface of the sand and adding a fresh layer of dry sand. Each day, clean the feeding table,

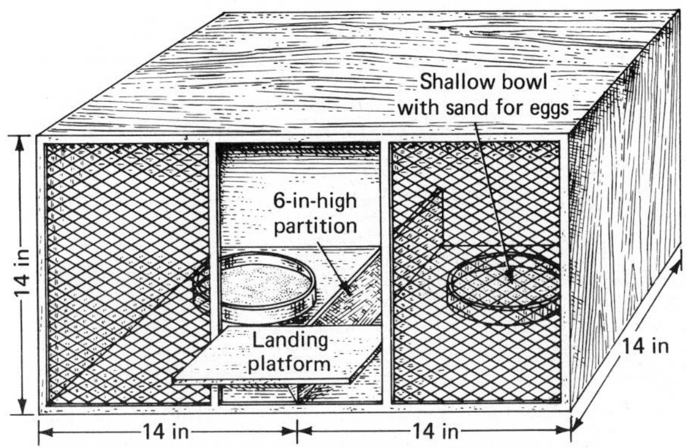

Figure 4-26 Diagram of a double nest box for a pair of pigeons. Each nest contains a bowl of sand in which an egg is laid. A breeding pair may have, at the same time, a young squab in one nest and an incubating egg in the other nest. *(From Raising Laboratory Animals, James Silvan, 1966, The Natural History Press.)*

drinking bottle, and nest boxes, with as little handling as possible of eggs and young birds. At least twice a year, scrub nest boxes with hot, soapy water, and air-dry in open sunshine if possible, and replace all sand on floor and in outside pen. Other general practices for cleaning cages and animal room are described in Chap. 1.

A proper diet and adequate cleanliness lessen the danger of disease. Nevertheless, birds should be observed daily and those removed that show symptoms of diarrhea, eye discoloration, labored breathing, general emaciation and lethargy, and unusual behavior such as shaking the head and twisting the neck. Unless the animals are costly or unique, probably the best course is to dispose of a diseased bird by overanesthesia or by dislocating the neck when the head is quickly twisted and pulled forward.

Lice may become a considerable problem, particularly if nest boxes and the room are not cleaned regularly. These infestations can be controlled with regular applications of insecticide dust and sprays in the nest boxes and crevices of the pigeon room.

MAMMALS

Mammals used most commonly in the biology laboratory are the mouse, rat, hamster, guinea pig, and rabbit. A source list of mammals is given in Appendix B. Occasionally, breeding animals can be obtained from local research laboratories of government agencies and universities. Local pet shops may be another source if animals are selected carefully in relation to breeding age and health. Housing of animals is described in detail in Chap. 1. Requirements described here and in Chap. 1 are based on the guide of the Institute of Laboratory Animal Resources (1968).

Possibly mammalian reproduction receives far less attention in the laboratory than does reproduction of lower animals, despite the fact that the evolutionary significance of embryonic membranes and the placenta are best understood when studied in the development and birth of *living* mammals. For this reason, and also because the information generally is not included in laboratory manuals, a considerable description of reproductive cycles, growth of young, and breeding habits is included in the discussion of each animal in this section.

DISEASES: Proper diet and cleanliness are by far the best measures for preventing and controlling diseases. Other general practices include the following: (1) Newly acquired animals should be isolated and observed for at least 1 week before they are put with other animals. (2) Air drafts and abrupt changes in temperature and humidity must be avoided. (3) Food and litter materials should be stored in such a way that they are not contaminated by insects, wild mice, and rats. In spite of strenuous practices, mammals do become infected and parasitized. Unless they are irreplaceable because of cost or unusual traits, the best procedure is to immediately remove diseased animals and destroy them by humane killing. Most infectious diseases are difficult and tedious to treat. For some diseases, no satisfactory treatment is known. More importantly, retaining an infected animal for treatment involves the risk of spreading the infection to other or all animals. Particular diseases are described in later discussions of individual animal types. General symptoms of common diseases are given below.

Respiratory diseases. Infectious catarrh (bacterial); bronchiectasis (viral); pneumonia (bacterial, perhaps initially viral) and apparently indigenous to mice, rats, guinea pigs, and rabbits (only occasionally in hamsters). So far as known, the diseases are not transmitted to other mammals, including man. All are highly communicable within the species, with slow onset, long duration, and rare recovery. Animals may "chatter," a noise due to bronchial tube stoppage; often they have a nasal discharge and a middle ear infection, indicated when an animal moves in a circle or carries head in a bent position. Destroy animals and sterilize cages and equipment by autoclave or chemicals (for example, Lysol in water solution).

Intestinal diseases. Typhoid (Salmonellosis—bacterial), infantile diarrhea (viral), coccidiosis (protozoan). The first two are common in mice, rats, guinea pigs, and rabbits. Coccidiosis is common in rats, guinea pigs, and rabbits. Symptoms of these diseases are an acute diarrhea, loss of weight, and emaciation. The diseases are highly infectious and extremely difficult to eradicate. Sacrifice diseased animals and sterilize equipment. With a severe outbreak, destroy all animals, including those without disease symptoms, and sterilize all equipment and entire room.

Skin diseases. Ringworm (fungal), occasionally in mice and rabbits. Symptoms are loss of hair in patches, and a scaling and encrusting of skin. Generally, treatment is extremely tedious and not worthwhile. Destroy animals, and sterilize cages and equipment. Mange (mite): symptoms are loss of hair and inflammation of skin; occurs most often in mice and rats. Remove animals and treat with insecticide powder, or spray or dip animals in Malathion solution. Sterilize cages. It is almost impossible to completely eradicate the mite once it is in a room. Use extreme care in examining new animals for ringworm and mange before bringing them into the animal room. On one occasion, with valuable animals, we dipped all animals (including undiseased ones – a total of 250 animals) once weekly for 4 weeks in a water solution of mercuric chloride (one tablet, from drugstore, in 2 gal of water).

Mercuric chloride is *poison*, and the animals' eyes, nose, and mouth were protected, and we wore rubber gloves. The entire room and all equipment were washed with the solution and opened to outside air or sunned for drying. Animals were dried in the sun. The fungus was eradicated but only with a tremendous effort and much valuable time. Prevention or disposal is highly recommended.

The Mouse *(Mus musculus)*

DESCRIPTION: The albino mouse is used most often for laboratory studies. Numerous strains have been developed, particularly for genetic and cancer research. Depending on strain, the weight of adult male is 20 to 35 g; of adult female, 25 to 40 g. Sexual maturity is usually at 4 to 5 weeks. Except during gestation and lactation, estrus occurs for a 10- to 12-h period every 4 or 5 days. A single estrus appears at 20 to 24 h after birth of young, when mating may take place. Gestation is for 19 to 21 days, unless mating and fertilization occur during the estrus immediately after the birth of young, when gestation is approximately 26 to 28 days. Average litter is 10 to 12 young. With a larger litter, reduce the number to 12 by humane killing on the second or third day after birth. Otherwise, unless necessary, do not handle young or bedding material for at least 1 week after birth of young, as the female may cannibalize the "contaminated" litter.

Newly born have no hair, and eyes are closed. Hair appears on second or third day; eyes open at about day 12 or 13. Teeth appear at about day 11, when young begin to nibble on solid food. Weaning is on about the twenty-first day, at which time the young should be separated by sex. Sex is determined by noting the greater distance between anal and genital openings in the male, or by applying a slight pressure immediately anterior to the genital region, which in the male causes the penis tip to protrude.

CARE: When many mice are maintained, small plastic cages are particularly suitable (see Chap. 1). A pair of mice with 12 young may be maintained in a cage 12 in long, 6 in wide, and 6 in deep. With growing mice, a minimum space of 18 by 12 by 6 in is required for 10 to 15 animals. Provide drinking water and food pellets at all times. A special pellet for mice is most desirable (see sources in Appendix B), although dog biscuits or rabbit pellets are satisfactory when supplemented several times weekly with raw carrots, apples, lettuce, or cabbage. Remove uneaten fresh foods after several hours. Maintain at 21 to 24°C and 45 to 50% humidity. Chloroform and ether vapors are detrimental to mice and should not be used in the animal room.

Gentle handling is reflected in the animal's disposition and even in its well-being. Capture a mouse by cupping a hand under its body so that it rests on the palm. Gently close the hand around the body with the head protruding between the thumb and forefinger (Fig. 4-27). Grasping a mouse from the top may produce fright and biting. To observe or treat the animal, move fingers to hold mouse by the loose skin at the back of the neck. Turn the hand over so that the animal rests on its back in the palm with its tail held between two fingers (Fig. 4-28). Also the mouse may be picked up and held with fingers around the base of tail. If held by the tip of the tail, the animal may twist around and bite.

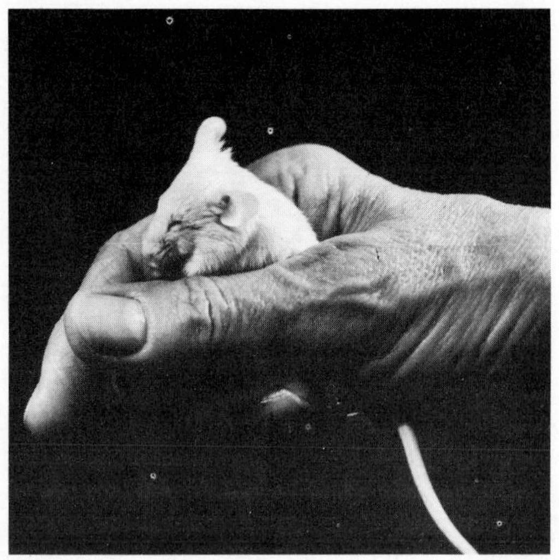

A

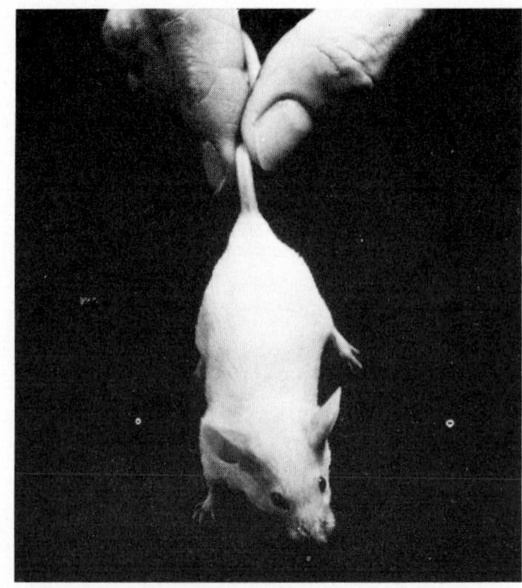

B

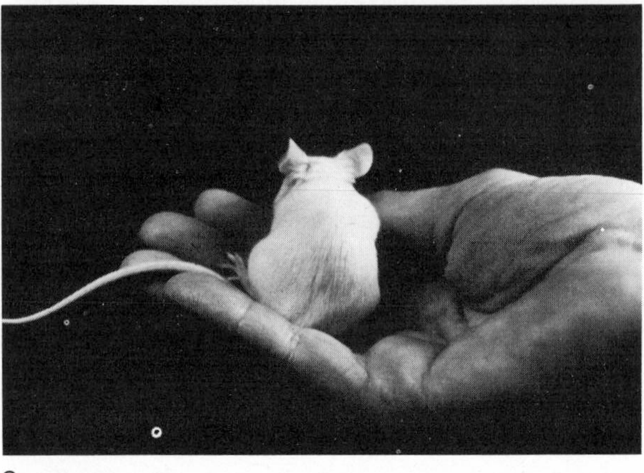

C

Figure 4-27 Picking up and holding a mouse. **A** Place hand gently over back of mouse and encircle body. **B** Method for picking up mouse by holding at the base of the tail. **C** If a mouse is handled gently, often it will move into a hand placed flat on the surface before the mouse.

The Rat *(Rattus norvegicus)*

DESCRIPTION: The laboratory-bred Norway rat is docile and often becomes a pet to students. The familiar albino has pink eyes and poor eyesight. Many strains have been developed, including Sprague-Dawley, Wistar, Charles River, and Carworth. Probably all laboratory strains are suitable for class studies, although highly inbred strains may be more expensive and less vigorous. Wild rats are not suitable for routine studies, as they are generally fierce animals and may carry parasites and infectious diseases.

Weight of adult male is 300 to 400 g; of adult female, 250 to 300 g. Most animals mature at 80 to 90 days old, but should not be mated before 100 days old. At matu-

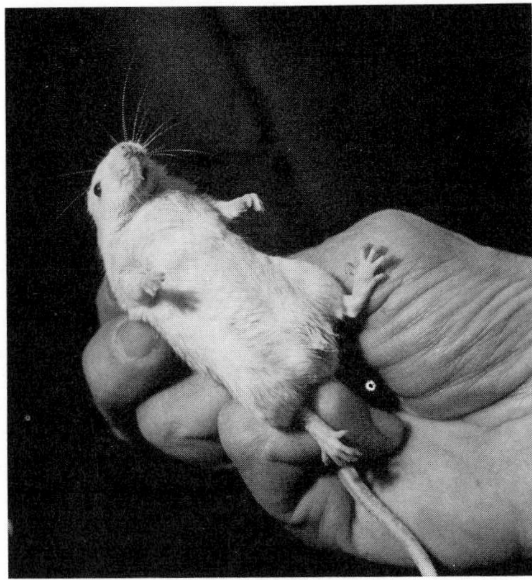

Figure 4-28 Holding a mouse for treatment.

rity, a female is in estrus during a 10- to 20-h period every 5 days, except when pregnant and when nursing young. An estrus appears shortly after birth of a litter, when mating may occur. Gestation requires about 20 to 22 days, unless female is also nursing young, when gestation is 28 to 30 days. Young are born without hair and with eyes closed. Hair appears at 3 to 5 days. Eyes open at 10 to 12 days. Teeth appear at about tenth day, and young begin to eat solid food at about twelfth day. Young are weaned at 21 days, when they should be separated by sex. At this age, sex is easily determined by the male's distinct scrotum in the lower abdomen and by the greater distance between the anal and genital openings of the male.

CARE: Maintain animals at 21 to 24°C and 45 to 50% humidity. Metal cages with solid floors are suitable for breeding animals. Individual animals are often put in plastic cages. The Institute of Laboratory Animal Resources (1968) recommends a minimum floor area of 100 in^2 for a female rat with a litter of any size, and at least 35 in^2 for each adult that is caged individually or in a group. Rats are mated in pairs, or in colonies of one or two males with three or four females. Pregnant females are generally moved from a colony to a single cage before young are born, and returned to the colony when the young are weaned.

Generally, laboratory rats become gentle when handled regularly. To lift the animal, place a hand from the rear over the animal's forebody, with the thumb under the lower jaw (Fig. 4-29). If the rat attempts to bite, press against the jaw taking care not to press against the windpipe. A frightened animal may inflict a painful bite, and the inexperienced person should wear gloves, particularly with newly acquired animals.

Except when experiments require restricted feedings, food is supplied at all times. If many animals are maintained, a pelleted complete food, such as Purina Lab Chow (for rats, mice, and hamsters) or Purina Rat Chow, may be the least expensive and the simplest type to dispense to feeders. With only a few animals, satisfactory foods are dry dog biscuits and rabbit pellets (from grocery and pet

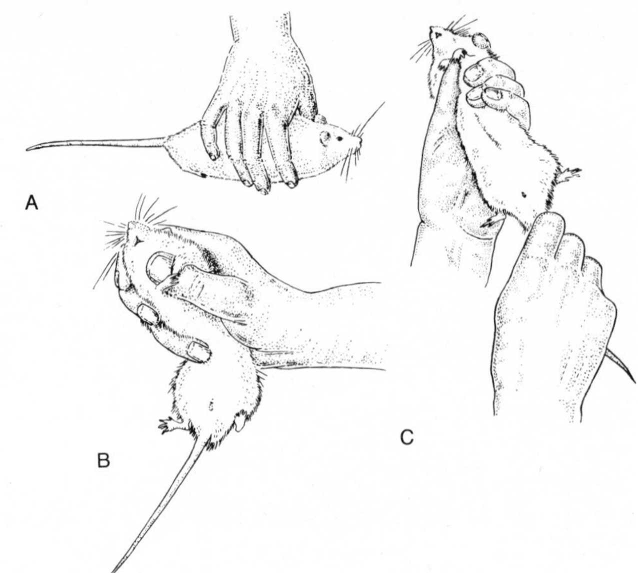

Figure 4-29 Lifting and holding a rat. **A** Picking up a rat. **B** and **C** Holding a rat. Slip thumb under the lower jaw to prevent biting during treatment. Take care not to press thumb against windpipe. *(A–C From Short and Woodnott, 1969. The I.A.T. Manual of Laboratory Animal Practice and Techniques. Crosby Lockwood/Charles C Thomas.)*

stores), supplemented with lettuce, carrots, and apples. Uneaten vegetables and fruits should be removed within 24 h. Clean drinking water must be provided continuously.

The Syrian or Golden Hamster *(Mesocricetus auratus)*

DESCRIPTION: The golden hamster of Syria and surrounding countries was first brought into the United States in 1938, where it has since become an important laboratory animal, particularly for cancer and drug research. The hamster's behavior, its remarkable resistance to disease, and its appealing appearance make it a useful and attractive animal for the teaching laboratory.

As a rule, hamsters become docile when handled regularly, although an occasional animal may retain a fierce temperament. Most hamsters will bite (with sharp teeth) when abruptly disturbed. A female may become fierce if handled when pregnant or while nursing young. Generally, the male is more easily tamed. Nevertheless, the hamster, more so than other mammals, requires frequent handling if it is to survive. Strangely, the lack of regular attention may produce a general debilitation and even sickness.

Color of upper coat is a golden brown; underbody is white or light gray. Mutant color strains have been developed, including albino, rust, cream, and piebald. Animals are nocturnal and may be very active in cages during the night. They mature at 4 to 6 weeks but should not be bred until 8 weeks old or older. A 10- to 12-h estrus and possible mating occur during the first night of a 4-day estrus cycle. No estrus and no mating are present after birth of young until the litter is weaned. Gestation period is 16 days. Average litter size is six to eight. Young are born without hair and with eyes and ears closed. Hair appears on third or fourth day.

Ears open on fifth day, and eyes open on fourteenth or fifteenth day. Young begin to eat solid food at about 1 week, and are weaned at about the twenty-first day. Adult female weight is about 110 g; adult male weight, about 100 g. Sexes are distinguished by the greater distance between the anal and genital openings of the male. Little or no breeding occurs after 12 to 16 months old. Hamsters from local pet stores may have passed the breeding age. Average life span is approximately 2 yr.

The large cheek pouches are of unusual interest. Often an enormous amount of food is stored in each pouch, producing at times an extremely droll appearance in the little animal. See Fig. 4-30 A and B. The inner tissue of cheek pouches is uniquely lacking in immunological reactions. Thus, the site is especially useful for foreign transplants, in particular tumor grafts, where one cheek receives the implant and the other serves as a control.

A

B

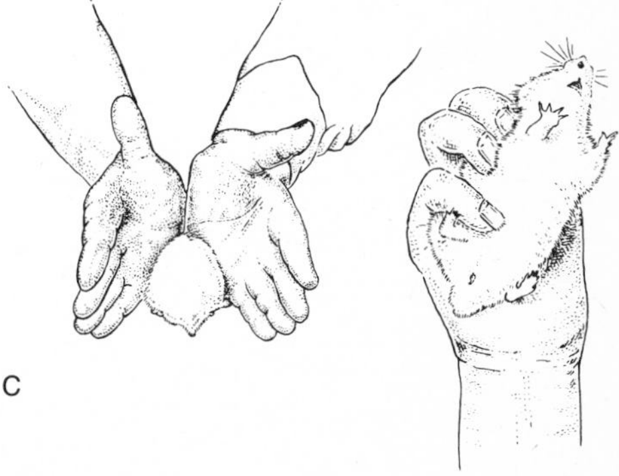

Figure 4-30 Golden hamster, *Mesocricetus auratus*. **A** An animal eating and beginning to store food in pouches. **B** The same animal, later, with filled pouches. **C** Lifting a hamster. **D** Holding a hamster for treatment. *(C and D From Short and Woodnott, 1969. The I.A.T. Manual of Laboratory Animal Practice and Techniques. Crosby Lockwood/Charles C Thomas.)*

At room temperatures below 9 or 10°C the hamster curls up tightly and goes into hibernation. The body becomes cold and rigid, and the heartbeat and respiratory rates become greatly decreased. This behavior allows interesting class studies in hypothermia. A hibernating animal must not be awakened quickly, as the reaction may produce shock and on rare occasion may be fatal. In any event, when quickly aroused, the animal wakes up "mad" and vicious. Therefore, it should be placed in a warm room for several hours and allowed to awaken slowly without handling.

CARE: Lift a hamster by gently cupping both hands around its body, or by taking hold of the loose skin at the back of the neck (Fig. 4-30). Gloves should be worn when first handling newly acquired animals. Put hamsters in metal or plastic cages with solid floors covered with pine shavings or other litter. The Institute of Laboratory Animal Resources (1968) recommends that a female with any size litter be provided a minimum floor space of 150 in^2 and a depth of no less than 6 in; and that each mature animal be given a floor space of at least 15 in^2 and preferably more. See Chap. 1 for a general discussion on housing. Room temperature should be 20 to 23°C and humidity approximately 50%. The animal is *extremely* sensitive to abrupt changes, including those of food, bedding, and sudden loud noises, although they can be housed within a laboratory where they become accustomed to the ordinary movement and noise of classes.

Provide pelleted food and clean drinking water at all times. When many animals are maintained, a general laboratory-food suitable for small mammals is probably least expensive and easiest to dispense. The diet described for mice is suitable for the hamster, although the hamster appears to require a greater supplement of lettuce, carrots, and apples. Any uneaten vegetables and fruit should be removed within 24 h. Silvan (1966) recommends that pregnant and nursing females be given about 10 ml of milk daily in a water bottle.

Examine regularly the incisor teeth and toenails of animals, since both may grow excessively and will need to be clipped. Use strong, sharp scissors or clippers so that the nails and teeth are clipped evenly without splitting. Remove only the excess growth, and take care to avoid clipping into underlying tissue. Apparently, the excessive incisor growth is inherent in some animals. Eventually, the long curved teeth will keep the animal from closing its mouth, and it will starve.

Animals are most easily mated in pairs, although colonies of one male and three or four females can be successfully established if they are watched carefully when first put together. The prime difficulty is that a female may fight and kill a male. To avoid this, place a male alone in a cage for several days before adding the female. Always place the female in the male's cage (never the reverse). Watch the animals for at least 15 to 20 min, and examine them periodically during the next 10 to 12 h. If fighting begins, remove the female. A very pugnacious female can be enclosed in a small cage which is then placed in the male's large cage for 24 h or longer before the female is released. Females may be bred at about 2 months old. The male should be 3 or 4 months old to help reduce male fatalities.

Mating occurs during the late evening or night of the first day of the 4-day estrus cycle. The indication of estrus, described by Short and Woodnott (1969), is a vaginal discharge on the second day of the cycle. The regularity of estrus and its indication on the second day have made the hamster a valuable animal in research programs. Note that only during an emergency should the young and the nesting material be handled during the first week after birth, as the female may cannibalize the "contaminated" young.

Guinea Pig *(Cavia porcellus)*

DESCRIPTION: The guinea pig is native to South America, and wild populations still live in the Andes Mountains. Although a rodent, the animal is more closely related to the chinchilla and porcupine than to mice and rats. Guinea pigs are sometimes called "Cavies," a derivative of the generic name. These are unusually attractive and gentle animals for the laboratory classroom. They are useful in nutrition studies, particularly for a vitamin-C deficiency, which in the guinea pig quickly produces scurvy. Apparently little study has been made of their behavior, which may offer interesting explorations for class work.

Three principal types are recognized: the English, with hair about 1½ in long; Abyssinian, with short, rough hair in rosettes; and Peruvian, a less common type with silky hair about 6 in long (Fig. 4-31). Strains developed from the English type are used most commonly in the laboratory, including those named English shorthair and Dunkin-Hartley. Colors may be albino, black, agouti, red, brown, or

A

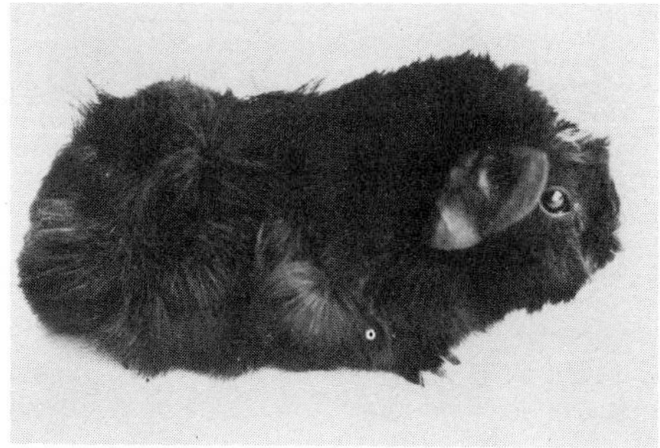

B

C

Figure 4-31 Guinea pig, *Cavia porcellus.* **A** English guinea pig, characterized by straight hair about 1 in long. **B** Abyssinian guinea pig with short, rough hair in rosettes. **C** Peruvian or angora guinea pig with long hair. *(A and C Osceola Cavies and Small Animal Farm, Inc., St. Cloud, Fla.)*

mixtures of two or three of these colors. Most commercial stock is albino (see Appendix B).

Guinea pigs have no tails. Body form is robust, and ears and legs are short. Body length is about 225 to 355 mm. Adult male weight is 1,000 to 1,200 g. Adult female weight is 850 to 900 g. Although gregarious and peaceable in colonies, they are sensitive and shy, but thrive with attention. When disturbed, the animal emits a high-pitched squeak. Abrupt environmental changes may be detrimental. A sudden fright may cause a colony to "circus" by running wildly in a circle around the cage, sometimes producing broken bones and other injuries. Nevertheless, we have kept colonies within the laboratory-classroom, where they soon adjusted to noise and lost their timidity because of student attention and handling.

Females may mature as early as 4 or 5 weeks old. Males mature at about 8 weeks. Most breeders recommend that mating be delayed for both sexes until age 3 months. Breeding life is about 3 yr. Life span is 5 or 6 yr. Estrus cycle is variable among strains. Apparently for most, an 8- to 10-h estrus appears during a 16- to 18-day estrus cycle. Average gestation is 68 days. An estrus and possible mating appear within 6 to 8 h after birth of young.

Litter size is one to six young; average number is three or four. Young are well developed at birth, with eyes open, with hair, and a full set of teeth. Young walk within an hour after birth and, while still nursing, may begin to eat solid food on the second or third day. Within a colony, young will nurse any lactating female and should be removed at 10 days old if newborn pigs are deprived of nursing. Otherwise, young are weaned and separated by sex at 21 days old. Sex is difficult to identify and is best done by applying pressure around the genital area, which in the male causes the penis tip to protrude.

CARE: See Chap. 1 for a detailed discussion of housing. Animals without young may be put on ½-in wire-mesh floors. Those with litters require a solid floor covered with pine shavings or other litter. For a single female with young, the Institute of Laboratory Animal Resources (1968) recommends a minimum floor space of 225 in^2, with a cage depth of no less than 10 in, and for each adult within a colony, a minimum floor space of 180 in^2. More space is required when litters are present in a colony. As a rule, a colony contains one male and as many as eight or ten females.

Pelleted food and drinking water are given in the same manner as for mice and rats. More than other small mammals, guinea pigs require a daily food source of vitamin C (ascorbic acid), supplied either in a complete pelleted food for guinea pigs or in *daily* feedings of green leafy vegetables. Vitamin C in pelleted food deteriorates within 9 weeks of manufacture. Therefore, only food with the date of manufacture should be purchased. Deterioration of vitamin C is greatly increased when pellets become moist or when they remain in the feeding dish for more than 2 or 3 days. A deficiency of vitamin C produces scurvy in guinea pigs within 7 to 10 days. Animals exhibit difficulty in walking, a steady weight loss, swollen joints, bleeding gums, and eventual death in another 10 to 14 days. Although a complete pelleted food, manufactured especially for guinea pigs, generally contains sufficient vitamin C and minerals, many breeders prefer to supplement the minerals, principally calcium, with daily feedings of clean, dry hay. If other types

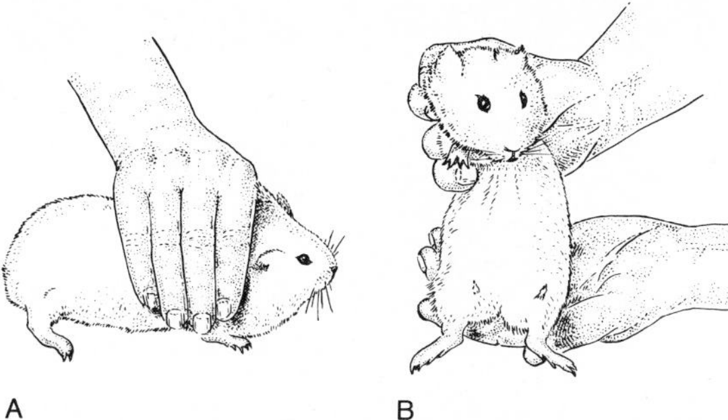

Figure 4-32 Lifting and holding a guinea pig. **A** Picking up a guinea pig, **B** Holding a guinea pig for treatment. *(From Short and Woodnott, 1969. The I.A.T. Manual of Laboratory Animal Practice and Techniques, Crosby Lockwood/Charles C Thomas.)*

of food are used, the hay is a definite requirement. Ordinarily, animals fed on pelleted food have no difficulty with an overgrowth of incisor teeth, as teeth are ground away when chewing on pellets. If excessive growth appears, the teeth must be clipped with sharp scissors or clippers and with care taken to avoid cutting the gum tissue. Maintain a temperature of 21 to 23°C and a humidity of approximately 50%.

When picking up the guinea pig, if possible approach from the rear because of the animal's extreme shyness. Pass a hand gently over the top of the body to around the neck, taking care not to press against the windpipe. When carrying the animal, support its lower body in the other hand or against the folded arm (Fig. 4-32).

Domestic Rabbit *(Oryctolagus cuniculus)*

DESCRIPTION: Probably because of size, rabbits are not used as commonly as smaller mammals in a teaching laboratory. At times, however, the rabbit is a unique requirement for physiological studies. Field rabbits may be as appropriate as domestic strains, although they should be isolated when first brought into the laboratory and observed for disease and ectoparasites. The infectious disease, coccidiosis, common in wild rabbits (and in laboratory rabbits), may be controlled by adding 3.5 g sulfaquinoxaline to each gallon of drinking water for a 3-week period (Thomsen and Evans, 1964).

Oryctolagus cuniculus, sometimes called Old World rabbits, were first found in southwestern Europe and northwestern Africa and are now distributed over most of the world. They are characterized by long hind legs, long ears, and large eyes. Unlike the fast-running rabbits and hares that build nests on the ground surface, these animals are slow runners and, in nature, live in burrows they dig in the ground. They are gregarious, and sometimes a "warren" of many burrows may be found at one site, with the burrows scattered about in close proximity to one another. The rabbit is not a rodent, as sometimes believed, but instead belongs to the closely related order Lagomorpha.

The American Rabbit Breeders Association recognizes at least 66 breeds and varieties of domestic rabbits. Of these, probably three types are used most commonly in the laboratory: the New Zealand White, a heavy animal averaging 10 to 11 lb adult weight; the Dutch, 4½ to 6 lb average adult weight, and with various

Figure 4-33 Laboratory rabbits; Dutch Belt, left; New Zealand White, right. *(Courtesy of The Charles River Breeding Laboratory, Inc., Wilmington, Mass.)*

fur colors including black, brown, blue, gray, and tortoise; and the Polish, a smaller rabbit (2½ to 3 lb), extremely hardy and gentle, although litters are small and breeding is irregular. See Fig. 4-33 for photographs, and Appendix B for commercial sources.

Sexual maturity varies with type of animal; smaller types may be first mated at 5 to 6 months old, and heavier strains at 6 to 7 months old. No definite breeding season is present, and estrus is prolonged in the absence of a male. Ovulation occurs only after mating and may result in a pseudopregnancy of 14 to 16 days if no fertilization takes place. Pseudopregnancy results from hormone secretions of corpora lutea, and is characterized externally by enlargement of mammary glands and internally by enlargement of uterus. Gestation is about 30 days. Estrus appears during third week following birth of young, when female may be placed with male for a short mating period and then returned to the nursing young. To ensure fertilization and avoid pseudopregnancy, many breeders return the female to the male's cage about 5 h later for a second short mating period.

Litter size is five to ten. Young are born without hair and with eyes closed. They begin to eat solid food when they first emerge from the nest box at about 3 weeks old but continue to nurse until about 6 to 8 weeks old. Sex can be most easily distinguished at about 6 weeks old by pressing gently around the genital opening, which, in the male, causes the rounded tip of the penis to protrude and, in the female, exposes the slanted slit of the vaginal orifice.

CARE: General housing and equipment are described in Chap. 1. Pregnant females are placed in a solid-floor cage, which contains a removable nest box at least 12 in long, 12 in wide, and 15 in high. A nest box of greater dimensions is required for large rabbits. The box is totally enclosed except for the top half of one side, which provides a passageway for the female as required. Generally the top cover is hinged to facilitate cleaning the box and to allow examination of young. The floors of the nest box and cage are covered with pine shavings or other litter. Single, mature animals may be kept on ½-in wire-mesh floors, a mesh large enough to allow droppings to fall into a waste pan but small enough to prevent animals from trapping their feet in the wire. A clean, flat resting board (about 12 by 8 in) is desirable, although cleaning chores are thus increased. When on wire mesh, an occasional animal may develop sore feet which require treatment with a disinfectant and transfer of the animal to a solid floor.

The Institute of Laboratory Animal Resources (1968) recommends a minimum floor space, including space for the nest box, of 1,080 in^2 (2½ by 3 ft) for a nursing female and litter; and a minimum floor space of 360 in^2 for a single mature animal. Cages should have a depth of at least 18 in. Maintain animals at 20 to 25°C and 50% humidity. Avoid exposures to extreme temperatures. Probably the commercial rabbit pellets are the most satisfactory food. Several times weekly, the pellets should be supplemented with lettuce, carrots, and apples. Pellets and clean drinking water are supplied continuously except when restrictive experiments are in progress.

Rabbits should not be lifted or carried by their ears. To lift the animals, place one hand over the head and gently slide the other hand under the animal's belly to lift its body in the lower outspread hand. Carry the animal in an upright position, pressed against one's chest, with the animal's weight supported in the lower cupped hand (Fig. 4-34).

Rabbits have the unique trait of forming two kinds of feces, a dry pellet that drops from the animal, and a moist, soft pellet which the animal eats from the anus, generally at night. From his studies, Blount (1957) concluded that the trait (sometimes called coprophagy) is not indicative of a diet deficiency but is a normal

Figure 4-34 Carrying and holding a rabbit. **A** Manner of supporting a rabbit against the chest. **B** Holding a rabbit for treatment. *(From Short and Woodnott, 1969. The I.A.T. Manual of Laboratory Animal Practice and Techniques. Crosby Lockwood/Charles C Thomas.)*

part of nutrition, perhaps an evolutionary feeding device for animals that remained in burrows during long periods of avoiding enemies.

Rabbits are subject to several types of infectious respiratory diseases, including the fairly common and highly contagious "snuffles," so named because of the characteristic nasal discharge and sneezing. In the acute form, the disease is rapidly fatal. Infected animals should be removed from the room and sacrificed. The bodies should be buried or incinerated, and cages and other equipment should be sterilized.

The rabbit's incisor teeth grow continuously, as is true of the other mammals described in this section. Normally, an excessive length is prevented when the upper and lower incisors grind together when eating pelleted food. In an occasional animal, the lower incisors do not occlude with the upper teeth, allowing the lower incisors to curve inward and sometimes to protrude from the mouth as tusks. The excessive length interferes with the animal's eating and requires that the teeth be cut back with bone forceps.

REFERENCES

Axelrod, H.R., C.W. Emmens, D. Schulthorpe, W. Vorderwinkler, and N. Pronek: 1967, *Exotic Tropical Fishes*, T.F.H. Publications, Jersey City, N.J.

Axelrod, H.R., and L.P. Schultz: 1955, *Handbook of Tropical Aquarium Fishes*, McGraw-Hill Book Company, New York.

Bishop, S.C.: 1943, *Handbook of Salamanders*, Comstock Publishing Associates, Cornell University Press, Ithaca, N.Y.

Blair, W.F.: 1958, "Mating Call in the Speciation of Anuran Amphibians," *Am. Nat.*, **92:** 27.

Blair, W.F., A.P. Blair, P. Brodkorb, F.R. Cagle, and G.A. Moore: 1968, *Vertebrates of the United States*, 2d ed., McGraw-Hill Book Company, New York, 616 pp.

Blount, W.P.: 1957, *Rabbit Ailments*, Fur and Feather Publishers, Bradford, England.

Briggs, J.C., and N. Egami: 1959, "The Medaka *(Oryzias latipes)*. A Commentary and a Bibliography," *J. Fish Research Board Canada*, **16**: 363.

Calhoun, W.H.: 1968, "Courtship and Mating in Japanese Quail," Exercise 29 in *Animal Behavior in Laboratory and Field*, A.W. Stokes (ed.), pp. 127–129, W.H. Freeman and Company, San Francisco.

Carlander, K.D.: 1969, *Handbook of Freshwater Fishery Biology*, vol. I, Iowa State University Press, Ames, 752 pp.

Carr, A.: 1952, *Handbook of Turtles*, Comstock Publishing Associates, Cornell University, Press, Ithaca, N.Y.

Clarkson, T.B., R.W. Prichard, H.B. Lofland, and H.O. Goodman: 1963, "The Pigeon as a Laboratory Animal," *Lab. Animal Care*, **13** (6): 767–780.

Conant, Roger: 1958, *A Field Guide to Reptiles and Amphibians of Eastern North America*, Houghton Mifflin Company, Boston, 366 pp.

Daniel, J.C.: 1966, "Some Easily Maintained Vertebrates as Sources of Embryos," *Am. Biol. Teacher*, **28** (4): 297–304.

Daugs, D.R.: 1967, "*Betta splendens*—the Siamese Fighting Fish," *Am. Biol. Teacher*, **29** (7): 528–530.

Farris, H.E.: 1967, "Classical Conditioning of Courting Behavior in the Japanese Quail, *Coturnix coturnix japonica*," *J. Exptl. Anal. Behaviour*, **10:** 213–217.

Fitzgerald, T.C.: 1970, *The Coturnix Quail: Anatomy and Histology*, Iowa State University Press, Ames, 325 pp.

Follansbee, H.: 1965, *Animal Behavior*, BSCS Laboratory Block, D.C. Heath and Company, Boston. Student Manual, 70 pp.; Teacher's Supplement, 51 pp.

Gibbs, E.L.: 1963, "An Effective Treatment for Red-Leg Disease in *Rana pipiens*," *Lab. Animal Care*, **13** (6): 781–783.

Gibbs, E.L., T.J. Gibbs, and Peter Van Dyck: 1966, "*Rana pipiens*: Health and Disease," *Lab. Animal Care*, **16** (2): 142–153.

Glass, Bentley: 1965, *Genetic Continuity*, BSCS Laboratory Block, D.C. Heath and Company, Boston. Student Manual, 154 pp.; Teacher's Supplement, 60 pp.

Goin, C.J., and O.B. Goin: 1971, *Introduction to Herpetology*, 2d ed., W.H. Freeman and Company, San Francisco, 353 pp.

Hollander, W.F.: 1954, "Care and Use of Pigeons in Research," *Proc. Animal Care Panel*, **5**: 71–80.

Hollander, W.F.: 1966, "Back to Chickens?" *Am. Biol. Teacher*, **28** (8): 631–632.

Homma, K., T.D. Siopes, W.O. Wilson, and L.Z. McFarland: 1965, "Sex Determination in Quail Chicks by Cloacal Examination," *Quail Quart.*, **2** (3).

Howes, J.R.: 1964, "Managing *Coturnix* Quail for Research," *Quail Quart.*, **1** (3 and 4). (Order *Quail Quart.* from J.R. Howes, Editor, P.O. Box 46, College Station, Texas 77840.)

Hughes, D.L., K.J. Sparrow, and D.P. Eeles: 1957, "*The Pigeon*," chap. 59 in *The UFAW Handbook on the Care and Management of Laboratory Animals*, 2d ed., A.N. Worden and W. Lane-Petter (eds.), Universities Federation for Animal Welfare, London.

Institute of Laboratory Animal Resources: 1968, *Guide for Laboratory Animal Facilities and Care*, 3d ed., National Academy of Sciences–National Research Council, Public Health Service Bull. 1024, U.S. Govt. Printing Office, Washington, D.C., 57 pp.

Kirchen, R.V., and W.R. West: 1969, *The Japanese Medaka: Its Care and Development*, Carolina Biological Supply Co., Burlington, N.C., 36 pp.

Levi, W.M.: 1957, *The Pigeon*, 2d ed., Levi Publishing Co., Sumter, S.C., 667 pp.

Marler, Peter: 1968, "Response of Male Fighting Fish to Visual Stimuli: A Study of Habituation," Exercise 37 in *Animal Behavior in Laboratory and Field*, A.W. Stokes (ed.), pp. 171–173, W.H. Freeman and Company, San Francisco.

Mills, D.A.F.: 1966, "The Japanese Quail (*Coturnix coturnix japonica*)," *J. Inst. Animal Technicians*, **17**: 74–79.

Moog, Florence: 1963, *Animal Growth and Development*, BSCS Laboratory Block, D.C. Heath and Company, Boston. Student Manual, 83 pp.; Teacher's Supplement, 60 pp.

Müller-Schwartz, D.: 1968, "Reproductive Behavior of the Three-spined Stickleback," Exercise 21 in *Animal Behavior in Laboratory and Field*, W. Stokes (ed.), pp. 93–97, W.H. Freeman and Company, San Francisco.

Naether, C.A.: 1958, *The Book of the Pigeon*, David McKay Company Inc., New York.

Padgett, C.N., and W.D. Ivey: 1959, "*Coturnix* Quail as a Laboratory Research Animal," *Science*, **129** (334).

Pope, C.H.: 1955, *The Reptile World*, Alfred A. Knopf, Inc., New York.

Regal, P.J.: 1966, "Feeding Specializations and the Classification of Terrestrial Salamanders," *Evolution*, **20**: 392.

Rugh, Roberts: 1962, *Experimental Embryology*, 3d ed., Burgess Publishing Company, Minneapolis, 501 pp.

Schlesinger, A.B.: 1966, "Plastic Bag Culture Method for Chick Embryos," *CUEBS News*, **2** (3): 10–11.

Schmidt, R.S., and W.R. Hudson: 1969, "Maintenance of Adult Anurans," *Lab. Animal Care*, **19** (5): 617–620.

Short, D.J., and D.P. Woodnott (eds.): 1969, "Breeding of Common Laboratory Animals," chap. 22 in *The I.A.T. Manual of Laboratory Animal Practice and Techniques*, 2d ed., Charles C Thomas, Publisher, Springfield, Ill., 462 pp.

Silvan, James: 1966, *Raising Laboratory Animals*, Natural History Press, Garden City, N.Y., 225 pp.

Southwick, C.H.: 1968a, "Aggressive Display in Paradise Fish, *Macropodus opercularis*," *Turtox News*, **46** (2): 57–60.

Southwick, C.H.: 1968b, "Display Patterns in Anabantid Fish," in *Animal Behavior in Laboratory and Field*, A.W. Stokes (ed.), pp. 99–101, W.H. Freeman and Company, San Francisco.

Stebbins, R.C.: 1954, *Amphibians and Reptiles of Western North America*, McGraw-Hill Book Company, New York, 528 pp.

Thomsen, J.J., and C.A. Evans: 1964, "The Feral San Juan Rabbit," *Lab. Animal Care*, **14** (2): 155–160.

Tinbergen, N.: 1953, *Social Behaviour in Animals*, Methuen & Co., Ltd., London.

Other Literature

Bailey, R.M., et al.: 1960, *A List of Common and Scientific Names of Fishes from the United States and Canada*, pp. 1–102, American Fisheries Society Special Publication 2.

Barker, W.: 1964, *Familiar Reptiles and Amphibians of America*, Harper & Row, Publishers, Incorporated, New York.

Barnett, S.A.: 1963, *The Rat, A Study in Behavior*, Aldine Publishing Company, Chicago.

Bishop, S.C.: 1943, *Handbook of Salamanders*, Comstock Publishing Associates, Cornell University Press, Ithaca, N.Y., 555 pp.

Bond, Charlotte, R.: 1945, "The Golden Hamster: Care, Breeding, and Growth," *Physiol. Zool.*, **18**: 52–59.

Cockrum, E.L.: 1955, *Laboratory Manual of Mammalogy*, Burgess Publishing Company, Minneapolis.

Green, E.L. (ed).: 1966, *Biology of the Laboratory Mouse*, 2d ed., McGraw-Hill Book Company, New York.

Peterson, R.T.: 1947, *A Field Guide to the Birds*, 2d ed., Houghton Mifflin Company, Boston.

Peterson, R.T.: 1961, *A Field Guide to Western Birds*, 2d ed., Houghton Mifflin Company, Boston.

Pettingill, O.S., Jr.: 1956, *A Laboratory and Field Manual of Ornithology*, 3d ed., Burgess Publishing Company, Minneapolis.

Roberts, M.F.: 1957, *Guinea Pigs*, T.F.H. Publications, Jersey City, N.J., 26 pp.

Roberts, M.F.: 1962, *Pigeons*, T.F.H. Publications, Jersey City, N.J., 64 pp.

Skinner, B.F., and C.B. Ferster: 1957, *Schedules of Reinforcement*, Appleton-Century-Crofts, Inc., New York.

Skinner, B.F.: 1961, *Cumulative Records*, Appleton-Century-Crofts, Inc., New York.

Walker, E.P.: 1968, *Mammals of the World*, 2d ed., The Johns Hopkins Press, Baltimore.

Wright, A.H., and A.A. Wright: 1949, *Handbook of Frogs and Toads of the United States and Canada*, 3d ed., Comstock Publishing Associates, Cornell University Press, Ithaca, N.Y., 640 pp.

Wright, A.H., and A.A. Wright: 1957, *Handbook of Snakes of the United States and Canada*, Comstock Publishing Associates, Cornell University Press, Ithaca, N.Y., 2 vols.

5

SPECIAL TECHNIQUES FOR HANDLING LABORATORY ANIMALS

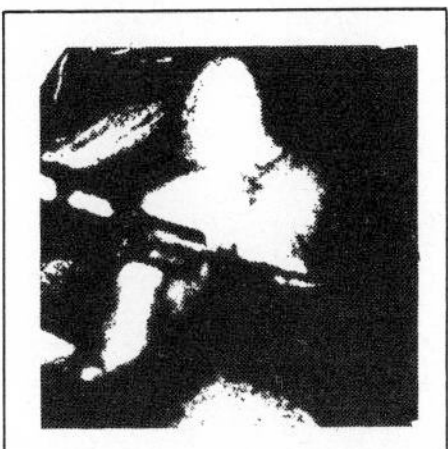

The reliability of data and thus the success of animal studies depend on the manner in which animals are handled. In this regard, both vertebrates and invertebrates require gentle, unhurried motions. A proper hold provides a sense of security and may prevent struggling, with possible injury to the animal or handler. Many animals become docile when handled frequently and carefully.[1] Gravid females are not handled unless necessary, and then only with great care. Similarly, mammals with very young offspring are not disturbed if this can be avoided.

This chapter describes techniques for handling animals when marking, measuring, injecting, anesthetizing, and preserving. Although a person acquires proficiency only with experience, the following descriptions may assist a beginner. In reality, a person soon adapts techniques to his own methods of ease and efficiency.

---

[1]Techniques for lifting and carrying animals are described in Chaps. 2 to 4, with the discussions of individual groups.

MARKING ANIMALS FOR IDENTIFICATION

When only a few animals are maintained, no elaborate method of identification is required. Placing single animals in separate containers may be sufficient, although individual markings are required in situations where more than one animal may escape at the same time. Animals are marked with color stains, ear tags, ear tattoos, leg bands, identification tape, and by various other methods peculiar to a type of animal. Sources for obtaining supplies are listed in Appendix B.

Marking Arthropods and Other Invertebrates

Segal and Gross (1967) suggest the system shown in Fig. 5-1 for marking land isopods. Comparable systems may be devised for insects, spiders, and other invertebrates. Fingernail polish and waterproof paints are suitable for marking.

Marking Fish

Place groups of fish in separate containers according to type of test or treatment. If individual identification is required within a group of fish, clip or hole-punch the dorsal or tail fin, if the change in fin activity does not interfere with a particular study (see Fig. 5-2). A disadvantage of this method is that fins may be damaged and the mark lost as a result of nibbling or fighting among fish. A more permanent label is obtained when metal or plastic tags are fastened at the corner of the mouth, providing, of course, that the tag is fitted properly. The method is used more often in field work but is equally satisfactory in the laboratory, particularly for fish maintained in large tanks where colored tags will allow identification from a distance. Do not use metal tags in salt water or tags coated with metal paint.

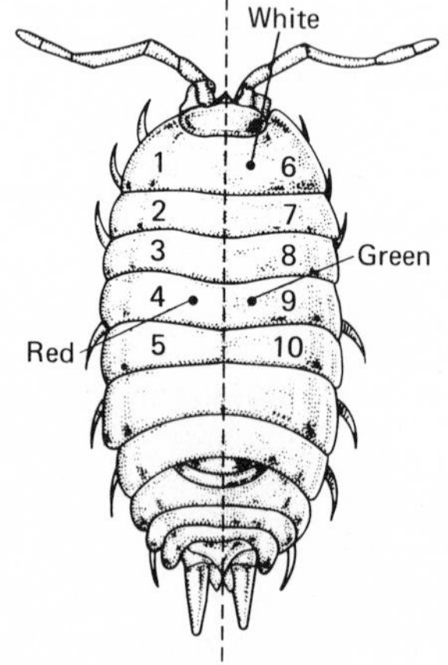

Figure 5-1 Marking arthropods. Identification code on the first five visible thoracic segments of an isopod. Designate the left side as numbers 1-5 and the right side as numbers 6-10. Mark the segments with a pen dipped in different colors of lacquer. In the illustration, white = units, red = tens, and green = hundreds. Above animal number is 946. *(Biological Sciences Curriculum Study.)*

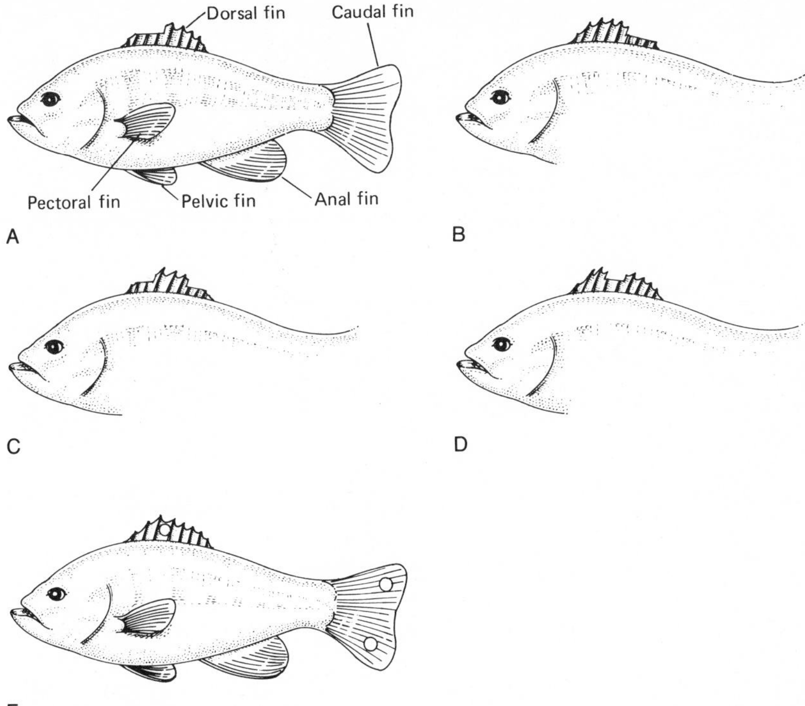

Figure 5-2 Marking fish. Drawings illustrate several types of marking by either clipping or hole-punching one or more fins. **A** Anterior clipping of dorsal fin. **B** Posterior clipping of dorsal fin. **C** Anterior and posterior clipping of dorsal fin. **D** Clipping middle portion of dorsal fin. **E** Hole-punch of dorsal fin and/or tail fin. A variety of systems can be used when one or more fins are marked and when the clipping and punching are combined.

Marking Frogs, Salamanders, and Lizards

Clipping off the toes is a common procedure for marking amphibians, a method that allows a variety of patterns for number of toes and combination of the four feet. Mark lizards with water-resistant paint; colored fingernail polish is satisfactory. Again, various patterns can be obtained by marking different body sites and combining several colors. Observe animals for loss of mark when the skin is shed.

Marking Snakes and Turtles

Ordinarily in a teaching laboratory, only a few snakes and turtles are maintained at one time. Because of differences in size or type, generally no individual markings are required. If necessary, however, water-resistant paint may be used for various marking systems. Observe animals regularly for necessary re-marking, particularly after a snake sheds its skin. As well, snakes are identified by clipping

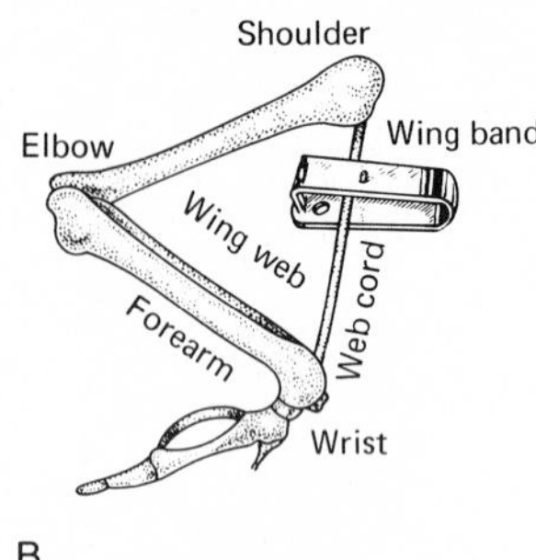

B

A

Figure 5-3 Wing tags for birds. **A** Wing tags on young chicks. **B** Diagram of wing bones and web, showing how band is doubled over to be clamped on wing web of skin and muscle. *(National Band and Tag Company, Newport, Ky.)*

their ventral scales in various patterns of number and location, although the scales are also lost when the skin is shed.

Marking Birds

Wing and tail feathers may be clipped in various patterns. Also stains may be applied in various combinations of color and body site. Water-resistant paints are desirable for long-term studies; food coloring may be used for short-term observations. For a better penetration of the stain, wash the feathers or the down of the marking site with detergent water, wipe with alcohol, and allow to dry before applying the stain. Isolate the animal until the stain is dry. The preparation of appropriate stains is described below. When a more permanent marking is desired, metal or plastic leg bands and wing tags are used (see Fig. 5-3).

Marking Mammals

Color stains are commonly employed if only a few animals are to be marked. Preparation of stains is given below. Before applying the stain, thoroughly clean the fur of the marking site with detergent water and alcohol. After staining, isolate the animal until the stain is dry.

Other methods are more desirable when many animals are to be studied. Common devices are ear tags, ear hole-punches, ear tattoos, and plastic tape for leg or collar bands (see Fig. 5-4). Although the electric tattoo instrument is initially more costly than the hand forceps, the electric machine is easier to operate and allows

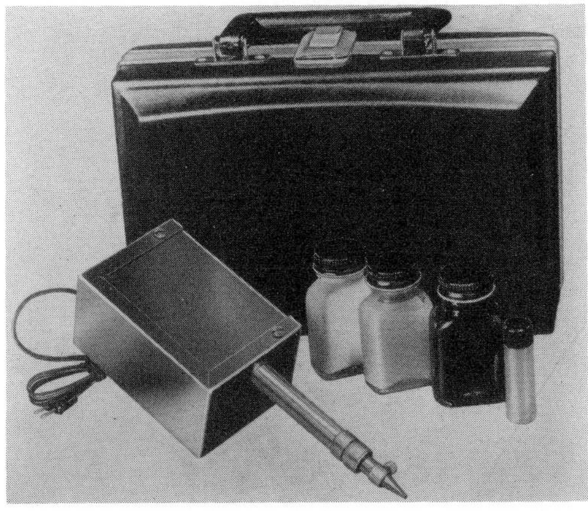

Figure 5-4 Electric tattoo outfit. *(Courtesy of Ancare Corporation, New York)*

the use of most any type of tattoo. When operated properly, the electric needle barely penetrates the skin and inflicts less pain than do the hand forceps. As well, the chance of infection is reduced. Ear tags are satisfactory if properly fitted, although they may be torn loose during animal fights. Plastic bands for the neck or leg have been introduced during the last several years and are gaining use because of ease in marking, their permanence, and the elimination of pain to the animal if bands are properly fitted (Fig. 5-5).

Figure 5-5 Plastic bands for tagging animals. **A** Rabbit with plastic collar band. **B** Guinea pig with plastic collar band. Instructions are included with an order. *(Courtesy of Hollister, Inc.)*

A

B

Preparation of Stains

Add the following dry dyes to 70% ethyl alcohol to prepare 3% to 5% solutions: *Red color,* fuchsin (acid, basic, or carbol); *blue color,* trypan blue; *green color,* brilliant green, ethyl green, or malachite green; *violet color,* methyl violet (gentian violet). For a *yellow color,* prepare a saturated solution of either picric acid or chrysoidin in 70% ethyl alcohol. The stains may be prepared in quantity and stored. Staining efficiency improves with aging.

MEASURING ANIMALS

Weighing Animals

If handled gently, most small laboratory mammals become accustomed to weighing procedures and remain quiet on a balance pan. Otherwise, place the animal in a container, and subtract the weight of the empty container from the total weight of the container and animal. The container may be a cloth or paper bag, a small cage, a jar, or a box. Sometimes small mammals will remain quiet when wrapped in a cloth towel. Aquatic animals are weighed by obtaining the weight of a container of water with and without the animal.

The ordinary triple beam balance and the Harvard trip balance, each with a 0.1-g sensitivity, are suitable for small animals. Use a spring balance for a larger animal with weight exceeding 1 kg. The spring balance, calibrated in metric or avoirdupois units, may be a dial-face type or a cylinder type with a hook and ring.

Balance pans and accessory weights must be kept clean. If animals are put directly on a pan, wash and rinse the pan after each use, taking care not to allow water to enter the weighing mechanism. Cover balances when not in use to protect from dust. Place the balance in a level position at a permanent location. Before each measurement, adjust the weight pointer to zero.

Measuring Body Temperature

The rectal temperature is easier to measure, and the measurement may be more accurate than that of other body areas. The ordinary laboratory thermometer can be employed with cold-blooded animals that have a large cloacal chamber, such as frogs and the larger salamanders. A small-bulb thermometer, with a scale of 0 to 50°C, is appropriate for measuring cloacal temperature of small, cold-blooded vertebrates. Most types of small-bulb thermometers are extremely fragile and must be handled accordingly. Therefore, it is well to purchase one that is equipped with a metal case to avoid breakage in storage.[1] The common laboratory thermometer is satisfactory for measuring rectal temperatures of adult warm-blooded vertebrates, although more accurate readings can be obtained with a clinical thermometer (range, 94 to 110°F), or with a veterinary thermometer (range, 90 to 108°F), obtained from local drug stores and most biological supply companies.

[1] A small-bulb thermometer with a scale of 0 to 50°C may be purchased from Schultheis Corp., 812–814 Wyckoff Ave., Brooklyn, N.Y.

Measuring Body Length

Measuring the body length of land animals presents no particular problem so long as a uniformity is established concerning the anterior and posterior points of measurement. A rule, string, or tape can be used.

Measuring aquatic animals may be more difficult. The following system is satisfactory for small swimming animals, such as fish, tadpoles, and worms. Place the animal in a transparent glass or plastic bowl of water, and set the container on a sheet of millimeter graph paper. Read the measurement when the animal is moving slowly or is at rest, or when an invertebrate is in the desired state of relaxation or contraction. If necessary, an animal can be anesthetized lightly. (See Anesthetics, later in chapter.) Alternatively, read the measurement from a small portion of a plastic rule that is laid flat on the inside bottom surface of a container. Some animals can be removed from water for measurement, although most vertebrates will increase their activity and many invertebrates will contract.

INJECTING ANIMALS

Injection Equipment

Purchase syringes and needles from biological or medical supply houses and from local drug stores. Syringes range in size from the ¼-ml tuberculin syringe to the 50-ml veterinary syringe. The 1-ml to 5-ml syringes are adequate for most biology work. If funds are limited, syringes of 2- or 2½-ml volume, graduated in 0.1 ml, should be sufficient for most routine work, although 5-ml syringes are desirable if large animals are to be injected with preservative fluid. Often the medical syringes are graduated in two scales: a milliliter scale, generally labeled as cubic centimeters (cc); and a minim scale (mn, one minim = 1/60 dram). The minim scale is rarely used in the biology laboratory; the double-scaled syringes are more expensive, and beginning biology students are prone to confuse the two scales.

Needles are sorted by length and diameter (gauge). Length is measured in inches and may range from ⅜ to 5 in. Diameter is expressed in gauge numbers, ranging from the larger diameters of 8 or 9 gauge to the smaller diameters of 27 or 28 gauge. The required size of needle varies with the type of injection, the animal size, and the kind of material to be injected. In general, needles of ⅜- to 1-in length and 20 to 23 gauge are appropriate for most work with small animals. A larger gauge produces a larger skin puncture, of course, and may result in a backflow of injected material. Thus, the material should contain particles of the smallest possible size to allow use of a small gauge. A sharp point is required for a quick, painless puncture. The point becomes dull with use and should be sharpened on fine carborundum or fine emery cloth. Generally, a bent point cannot be seen except with a magnifying lens, although it may be detected by drawing the point across the back of the hand or across a sheet of lens paper, where a bent point will pick up fibers. Retain the thin wire furnished with a package of needles for cleaning the bore of a needle.

Equipment may be sterilized by several methods: (1) dry heat of 125°C for 2½ h;

(2) autoclaving for 15 min at 15 lb pressure; (3) boiling in water for 10 min; or (4) soaking in 70% alcohol for 20 min. If the equipment is contaminated with spore-bearing organisms, use one of the first three methods with the time doubled. For best protection against spore contaminants, repeat the sterilization after 24 h, thus providing greater assurance that resistant spores, induced to germinate with the first sterilization, will be destroyed with the second treatment. Manufacturers recommend that glass syringes be sterilized by boiling, and that tap water be used in place of distilled water, since distilled water leaches silicates from glass. Nevertheless, because deposits from tap water may necessitate a tedious cleaning of syringes, many persons prefer to use distilled water.

When only occasional injections are required, probably the sterile-packaged and disposable plastic syringes with attached needles are the most practical and least expensive. For short-term studies where bacterial contamination is not a critical factor in a brief demonstration, the equipment can be sterilized adequately with a thorough wash and rinse, followed by storage in 70% alcohol for at least 20 min. In this way, plastic syringes may be used repeatedly. If the equipment is cleaned well and if the animal is to be sacrificed, it seems doubtful that alcohol sterilization is required for short-term studies, particularly with cold-blooded animals. On the other hand, much care must be taken to assure that sterile equipment and techniques are employed with all studies in which data are to be collected during an extended time with controlled and repeated tests.

A troublesome problem arises when glass syringes expand or contract so that the barrel and plunger no longer fit together. The problem can be avoided to some extent by placing the plunger inside the barrel during sterilization and by purchasing the more expensive syringes that are heat resistant and equipped with frosted plungers. Additionally, it is advisable to purchase glass syringes with interchangeable barrels and plungers, so that unbroken parts can be matched.

Injection Techniques

Prepare the injection site by removing any fur or feathers and by swabbing the site with alcohol. Take care to handle the syringe and needle with as little contamination as possible. Load the syringe with slightly more than the required dosage, and invert the syringe after withdrawing the needle from the vial. Allow any air bubbles to rise to the top surface, and discharge fluid through the needle until an even flow is obtained.

For a *subcutaneous injection* deposit the fluid in the tissue immediately below the skin. To accomplish this, pinch a portion of loose skin into a fold and pull it away from the body. Place the needle (3/8- to 1/2-in length) and syringe closely parallel to the body, and insert the needle at the base of the fold (Fig. 5-6). Take care not to run the needle through both layers of the folded skin, thus discharging the fluid on the opposite outer surface of the skin. After discharging the material under the skin, leave the needle in position for several seconds; then pinch the skin together around the needle, and withdraw the needle in one quick motion.

An *intramuscular injection* is made into a large muscle, for example, in the upper leg, with a needle of proper length for the animal's size (about 3/8 in for rats and mice). Hold the syringe and needle perpendicular to the animal's body, and

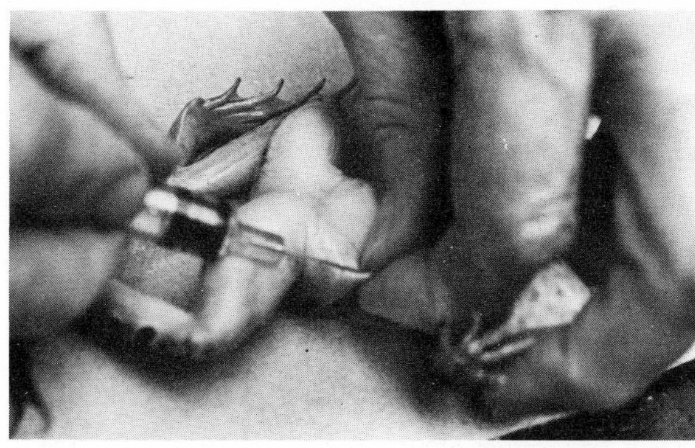

Figure 5-6 Subcutaneous injection. Skin is pulled into a fold away from the body and needle is inserted at base of fold. Illustration shows withdrawal of needle after discharge.

plunge the needle into the muscle along a straight, perpendicular line (Fig. 5-7).

Intraperitoneal injections are made near the outer edge of the ventral abdominal wall, with a ⅜- to ½-in needle. Avoid puncturing internal organs by inserting the needle at a 45° angle near the extreme outer edge of the abdomen (Fig. 5-8).

An *intravenous injection* is made directly into a large vein. This type of injection is not a usual technique for general biology studies, although it may be required in individual student problems. With rats and mice, inject the material into a lateral vein of the tail, after dilating the vein by applying dry or moist heat to the tail. (Take care not to burn the skin.) With guinea pigs and rabbits, inject into the saphenous vein at the front medial edge of the hind leg. The rabbit may also be injected in a lateral ear vein, after the ear is warmed to dilate the vein (Fig. 5-9). A ⅜-in, 27- or 28-gauge needle is used.

ANESTHETIZING AND HUMANE KILLING

In the general biology laboratory, anesthesia of animals is used (1) to quiet or slow animal activity, (2) to anesthetize before opening the animal for internal studies, (3) to prepare highly contractile invertebrates for fixing and preserving, and (4) for

Figure 5-7 Intramuscular injection.

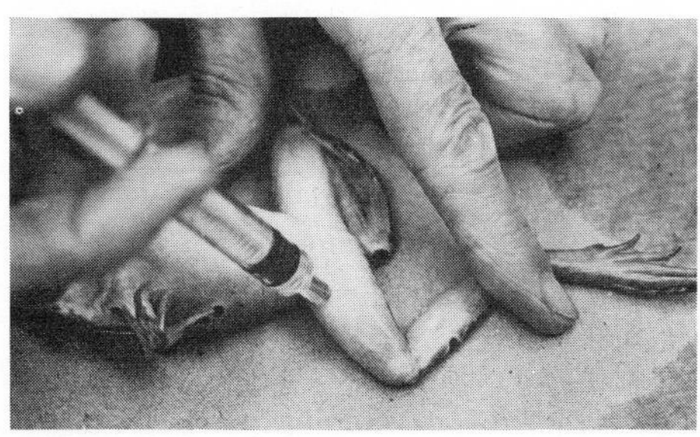

Figure 5-8 Intraperitoneal injection.

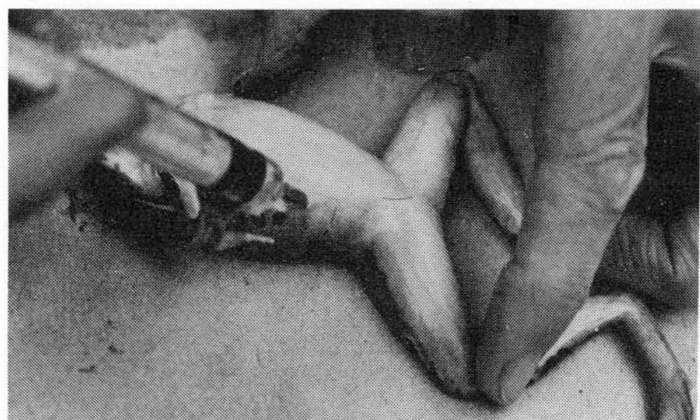

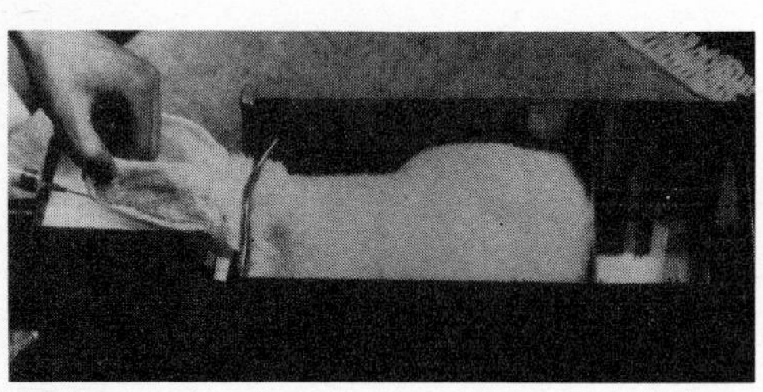

A

B

Figure 5-9 Intravenous injection. **A** Injection in ear vein of rabbit. **B** Closeup view of injection. *(Courtesy of Ancare Corporation, N.Y.)*

humane killing. Any experimentation that includes surgery and suturing of animals cannot be considered appropriate for beginning biology students. The techniques described in this section cannot be used for that purpose. It must be emphasized, therefore, that when anesthetized animals are opened for internal studies, they must be sacrificed by overanesthesia or by other humane killing upon completion of the work. Furthermore, extreme care must be taken to ensure that animals are completely anesthetized before they are opened, and that sufficient anesthesia is maintained during the studies.

Because of great variation in animal reactions to anesthetics, and because of variation in different lots of a same agent, only general techniques can be given here. In all instances, administer the anesthetic slowly while observing the animal closely. If sufficient animals are available, make preliminary tests to determine the required dosage and time for achieving a proper level of anesthesia. If proper techniques are followed closely, generally an intact animal will revive from anesthesia.

The description of agents and techniques, given below, is followed by Table 5-1 which summarizes the information according to animal groups. Obtain anesthetizing agents from biological, biochemical, or medical supply houses, and from local drug stores. Agents for which a legal permit is required, such as morphine and barbiturates, are not included here.

Alcohol (70% ethyl, 10% methyl, or 70% isopropyl). For quieting highly contractile, aquatic animals before fixing and preserving. If animal is to be revived, other agents are more desirable. Put animals in a small quantity of water; add alcohol, drop by drop, during a period of 1 to 1½ h or longer. *Note:* Methyl alcohol (wood alcohol) is highly toxic.

Butyn (butacaine sulfate). For slowing protozoa and small aquatic invertebrates. Add 0.1% aqueous solution, drop by drop, to concentrated culture of animals. Relatively expensive.

TABLE 5–1 SUMMARY OF ANESTHETICS BY ANIMAL GROUPS*

ANIMALS	AGENTS
Protozoa	Butyn, carbon dioxide, menthol crystals, methyl cellulose, tobacco smoke
Cnidaria	Butyn, carbon dioxide, chloral hydrate, clove oil, magnesium sulfate, menthol crystals, tobacco smoke, M.S. 222 Sandoz
Flatworms	Chloral hydrate, magnesium sulfate, menthol crystals, M.S. 222 Sandoz, urethane
Roundworms	Chloral hydrate, magnesium chloride, magnesium sulfate, M.S. 222 Sandoz
Segmented worms	Alcohol, magnesium sulfate, M.S. 222 Sandoz
Arthropods, aquatic	Alcohol, carbon dioxide, chloral hydrate, magnesium sulfate, M.S. 222 Sandoz
Arthropods, land	Potassium cyanide vapor fumes
Molluscs	Chloral hydrate, chloretone, magnesium sulfate, menthol crystals
Echinoderms	Carbon dioxide, magnesium sulfate, M.S. 222 Sandoz
Fish	Chloretone, chloroform, M.S. 222 Sandoz, urethane
Amphibians	M.S. 222 Sandoz, urethane
Reptiles, birds, and mammals	Chloroform, M.S. 222 Sandoz, urethane

**Note:* Some agents are more toxic than others. See the preceding discussion for techniques and effects of anesthetics.

Carbon dioxide (carbonic acid gas). For slowing protozoa, small aquatic invertebrates, and insects. *Method 1.* Slowly add charged water (ginger ale, club soda) to concentrated culture of aquatic animals, or add a drop at a time to animals in temporary mount on microscope slide. *Method 2.* Add small pieces of Dry Ice to concentrated culture of aquatic animals. *Method 3.* Bubble gas from a carbon dioxide generator (Fig. 5-10) through a tube into animal culture.

Chloral hydrate (trichlorohydrate acetaldehyde). Good results are obtained with aquatic invertebrates, and recovery is generally good. A special permit (described in suppliers' catalogs) or a medical prescription may be required. Relatively inexpensive. *Poison.* Avoid contact on skin, or in mouth and eyes. Add crystals to water surface of animal container, or slowly add 2% aqueous solution to cultures of small animals. Put large invertebrates in 5 to 10% aqueous solution. With delicate animals, such as anemones or jellyfish, add anesthetic slowly during a period of several hours. When narcotizing contractile animals prior to fixing and preserving, add agent slowly and allow animal to remain in solution for 3 to 4 h or longer.

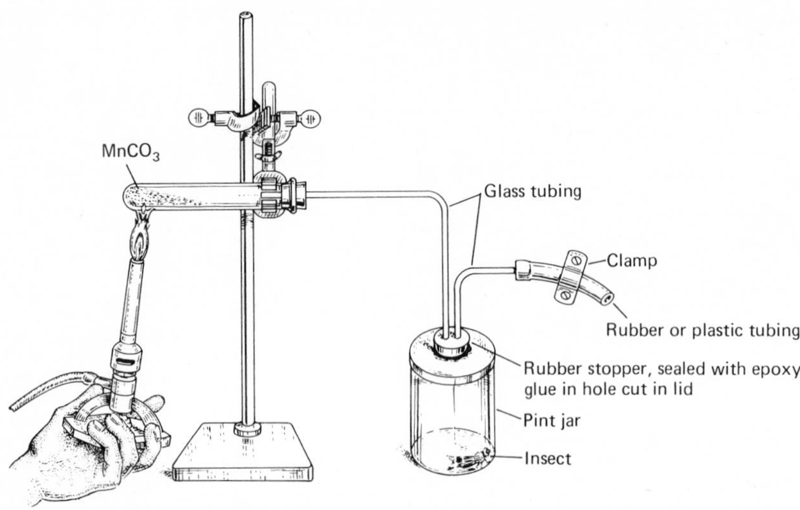

Figure 5-10 Carbon dioxide generator for anesthetizing small land invertebrates. Construct apparatus as shown in diagram. Add $MnCO_3$ to test tube. With clamp open on tubing at right, strongly heat $MnCO_3$ by passing hot flame back and forth under tube. CO_2 evolved from the $MnCO_3$ drives air out of jar through open tube at right. After several minutes, close clamp and continue heating for several more minutes. If possible, preliminary tests should be made to determine the time animal must remain in jar to reach desired state of anesthesia.

Chloretone (1,1,1-trichloro-2-methyl-2-propanol). A general anesthetic for invertebrates and lower vertebrates. Toxic, and may kill animals. Administer slowly and in small proportions. *Protozoa:* 0.1 solution, drop by drop, until animals are quiet. *Highly contractile invertebrates:* One crystal on water surface at a distance from the animal. Add second crystal if animal is not narcotized when first is dissolved. *Other small invertebrates and larvae of larger animals:* Several drops of 0.1% to 1.0% aqueous solution for each 100 ml of animal water. Allow 15 to 20 min for effect. *Larger animals (starfish, crayfish, goldfish, frogs, salamanders):* One volume of 0.5% chloretone solution for each four volumes of animal water. Allow 15 to 20 min for effect.

Chloroform. Highly toxic to animals. Use for humane killing, or for anesthesia before fixing and preserving. Do not use for anesthesia alone as satisfactory recovery of animal is doubtful. *Aquatic animals:* Place tip of dropper with chloroform below water surface. Add drops slowly until animal is anesthetized. Alternatively, spray chloroform over water and immediately cover animal container. *Land animals:* Soak cotton pad with chloroform. Place pad under a screen in bottom of a glass anesthetizing chamber (1- to 2-gal glass jar) or attach pad to the inside surface of the chamber lid. Put animal in chamber at once, and seal chamber tightly with lid. Ordinarily, the animal will go to sleep quietly if the chamber has not become saturated with chloroform vapor before adding the animal, and if the animal is not handled roughly. Do not open the chamber or remove the animal until 15 to 20 min after all sign of breathing has disappeared. Although snakes become anesthetized quickly, they should be left in the chamber for at least an hour for respiration to cease.

Clove oil. A general agent for quieting small aquatic animals. Concentrate animals in small quantity of water. Place tip of pipette with clove oil below water surface, and add 1 or 2 drops at a time.

Ether (ethyl; diethyl ether). Useful for *Drosophila* studies because of good animal recovery. Must be purchased and used in small quantity. *Not recommended for other studies,* particularly when used by beginning biology students. Extremely volatile, flammable, and explosive. All flames in vicinity must be extinguished. Sources for the generation of static electricity must be removed (a danger often overlooked). If held in large, partially empty containers, an extremely dangerous volatility accumulates, which can result in a violent explosion when the container is opened. The results of one such explosion have been observed in a research laboratory where the researcher was left completely deaf, partially blind, and badly scarred. Purchase ether in 1-pt cans. Store and handle in an area removed from the teaching laboratory. Partially used cans should not be stored for an extended time, but should be emptied outdoors or into the sink drain, and then discarded.

Magnesium chloride and magnesium sulfate (epsom salts). Both are effective agents for most aquatic invertebrates. Especially useful for quieting highly contractile invertebrates before fixing and preserving. Allow animals to expand in only enough water to cover them. Slowly add crystals or a 20 to 30% aqueous solution during a period of several hours. If animals begin to contract, reduce dosage and lengthen the time between additions of the agent. Leave large invertebrates in the solution for 12 to 14 h.

Menthol crystals. Effective for most aquatic invertebrates, particularly highly contractile animals. Sprinkle crystals on water surface and place container in dark for 10 to 12 h. Menthol may produce fragmentation of very delicate organisms, and magnesium sulfate is recommended for these animals. Quiet protozoa and hydra by adding one crystal at a time to animals in a test tube of water or on a microscope slide.

Methyl cellulose (10%). For slowing activity of protozoa and small invertebrates in a temporary slide mount. See Appendix A for preparation.

M.S. 222 Sandoz (tricaine methane sulfonate or m-amino ethyl benzoate). A highly desirable agent for a wide range of aquatic animals. Relatively expensive. Effective with most aquatic invertebrates and small aquatic vertebrates. Low toxicity, quick effect, and prolonged exposure is not harmful. Generally a quick and complete recovery when animal is transferred to clean water. Solution deteriorates and must be prepared fresh each time. *For larger animals,* use 1 g/gal of water, and smaller dosages with larval forms and small invertebrates. Muscular activity ceases in 1 to 5 min. *Larval forms* can be held safely in solution for 2 to 15 min; *mature animals* can be held in weak solutions for several days, thus allowing a means for safe and easy transport.

Potassium cyanide. For killing insects. See Chap. 1 for preparation of insect jar.

Tobacco smoke. For quieting protozoa and small aquatic invertebrates. Bubble tobacco smoke through a tube into a small quantity of animal culture. Alternatively, invert a temporary slide mount over the top of a smoke-filled test tube.

Urethane (ethyl carbamate). Highly toxic. Less desirable than other agents because of toxicity and carcinogenic property. Agent should not come in contact with handler's skin. Use with large aquatic invertebrates and small aquatic and land vertebrates. Animal excretes urethane slowly, and recovery may require 10 to 12 h. *Aquatic animals:* Submerge mature animals in 5% solution. Upon immobilization, immediately transfer animal to clean water if purpose is anesthesia and not killing. *Land vertebrates:* Inject 10% aqueous solution subcutaneously in the amount of 0.1 ml for each 20 g of animal weight.

Vapor fumes of benzene, ammonia, or carbon tetrachloride. For quieting or killing land arthropods. Put agent on cotton pad in bottom of bottle. Cover pad with a cardboard circle and on top of cardboard place a circle of blotting paper. *Carbon tetrachloride is toxic; do not inhale fumes.* Keep bottle tightly closed except when adding animals. Replenish agent occasionally by raising the covering layers and adding drops to cotton. Alternatively, soak rubber bands or rubber tubing in agent, a method that provides longer-lasting fumes.

PRESERVING ANIMALS

The following discussion pertains to fixing and preserving animals for later dissections or for museum displays. Histological preparations are not included here and may be found in Humason (1967), Conn et al. (1960), and other literature. As a rule, better specimens are obtained if animals are anesthetized before killing.

Formalin and ethyl alcohol are used most commonly for fixing and preserving animals. Commercial formalin is a 40% solution of formaldehyde gas in water. Unless specified otherwise, formalin dilutions are prepared from the 40% commercial product. For example, 5% formalin is prepared in the proportions of 5 ml of the 40% formaldehyde solution with 95 ml of water. Formalin hardens tissue and produces brittleness to a considerably greater degree than alcohol. Nevertheless, it is very successful for short periods of preservation, particularly with vertebrates. All formalin specimens eventually become hard and brittle if not transferred to alcohol. When transferring specimens from formalin to alcohol, washing is not necessary unless one wishes to remove the formalin odor. Formalin is not suitable for animals with lime spicules or shells, as the commercial solution is generally acid and softens the calcareous portions. Neutralizing the formalin will help to eliminate the problem (see Formalin, Buffered in Appendix A). Formalin preserves the color of specimens for a longer time than does alcohol, although probably no known preservative retains color indefinitely.

Delicate animals should be fixed and preserved in alcohol. Soft or gelatinous organisms require a gradual fixation, depending on density and size, of 2 to 6 h for each of 30, 50, and 60% alcohol solution, following which they are stored in 70% alcohol. Although ethyl alcohol is most often prescribed, isopropyl alcohol and denatured ethyl alcohol (ordinary medicinal, rubbing alcohols) serve equally well as preservative fluids and often are more available, for example, during trips when medicinal alcohol may be purchased at a drug store.

PORIFERA: Wash with fresh water and put in 70% alcohol for 24 h. Dry in open air, in sunlight if possible. Store calcareous types in boxes with soft packing tissue. Alternatively, fix in 70% alcohol for 24 h. Transfer to fresh 70% alcohol for storage or display. Large specimens, or many specimens in one jar, require 85 to 95% alcohol.

HYDRA: Slowly anesthetize with menthol crystals, magnesium sulfate, or chloral hydrate (described earlier in chapter). When animals no longer respond to touch, remove fluid and fix by flooding with warm (45 to 50°C) Bouin's fluid (see Appendix A). To inhibit contraction of mouth and tentacles, first drop the fluid at the basal disc and then along the body to the oral end. After 5 min, pour off the first Bouin's and cover with fresh Bouin's for 30 min to 1 h. Wash in several changes of 30% alcohol. Transfer to 50% alcohol for 1 h. Transfer to 70% alcohol for storage.

JELLYFISH AND OTHER MEDUSAE: Place in flat dish with only enough seawater to cover animals. Narcotize with menthol, magnesium sulfate, or chloral hydrate (described earlier in chapter) until animals are insensitive to probing. Do not overanesthetize before preserving. Transfer to 30% alcohol and 50% alcohol, each for 30-min periods (1 or 2 h for large specimens). Store in 70% alcohol, with large specimens turned upside-down or suspended in the jar to allow exposure of tentacles.

ANEMONES: Cover animals with a small amount of seawater and allow them to fully expand. Slowly add several crystals of menthol or magnesium sulfate (see Anesthesia, earlier in chapter). Place in dark for 10 to 12 h. Remove excess water, leaving animals covered with water at all times, and add 10% formalin. Transfer to fresh 10% formalin or to 70% alcohol for storage.

HORNY CORALS, SEA PENS, SEA WHIPS, AND OTHER FORMS: Anesthetize as for anemones. Transfer to 70% alcohol for storage. For dry preservation, put animals in 5% buffered formalin (see Appendix A) for 15 min; remove from solution and place in open air for thorough drying. Store in covered boxes with paradichlorobenzene crystals.

STONY OR "TRUE" CORALS: To preserve in solution, treat as for horny corals. To dry, soak the specimens in seawater for 3 to 4 days. Wash thoroughly and scrub to remove dead tissue. If desired, soak in 10% solution of household bleach for 24 h; rinse thoroughly in fresh water, and air dry.

PLANARIA AND OTHER TURBELLARIA: Do not feed for 2 to 3 days to empty the gut. Narcotize slowly with menthol or chloral hydrate crystals (described earlier in chapter) for 2 to 3 h until response to touch is barely perceptible. Pour off solution. Cover with 35% alcohol for 3 to 4 h. Transfer to 50% alcohol for 3 to 4 h. Transfer to 70% alcohol for storage. Alternatively, transfer from narcotizing solution to 5% formalin for storage. If flattened specimens are desired for microscopic studies, place an animal between two glass slides, weighted together with a heavy glass bowl, during the first 2 h of fixation in alcohol or formalin.

TREMATODES (PARASITIC "FLUKES"): Put specimens in 1% sodium chloride solution and shake container to clean animals. Pour or pipette off solution and add a second portion. Gently shake container to wash animals and to cause them to relax. Fix by adding to the salt solution an equal portion of hot (55 to 60°C), saturated aqueous solution of mercuric chloride (sometimes called corrosive sublimate). *Note:* Mercuric chloride is toxic. Do not inhale vapors or place hands in the solution. Do not use metal instruments or metal containers. Gently shake the tube to mix the two solutions, and leave animals in the tube for 24 to 48 h. Wash in running water for 4 h. Store in 5% formalin, or transfer to 50% alcohol for 4 h, followed by a transfer to 70% alcohol for storage. Alternatively, fix by transferring animals from second wash of sodium chloride solution to FAA or Gilson's fixative (see Appendix A) for 24 h, followed by storage in 5% formalin or in alcohol as described above. If flattened specimens are desired for microscopic study, place animals between two glass slides or plates, weighted together with a heavy glass bowl, during the first several hours of fixation.

CESTODES (TAPEWORMS): Wash by shaking animals in a 1% sodium chloride solution. Wrap long worms around a glass bottle and secure with a string that is wrapped several times around animal and bottle. Place in 1% chromic acid solution (1 g chromic acid in 99 ml distilled water) until animal is dead. Fix in FAA (see Appendix A) for 24 h. Transfer to 50% alcohol for 4 h. Store in 70% alcohol.

NEMATODES: Put small nematodes in 1% sodium chloride solution and wash by vigorously shaking the container. Allow worms to settle to bottom. Pour off top water and add fresh salt solution for one or more additional washings. Pour off last wash solution and place in 70% alcohol solution for several minutes. Transfer to fresh 70% alcohol for storage. With large nematodes (for example, *Ascaris*), straighten out animals on a square of thin cloth. Carefully roll up the animals and cloth; grasp the roll at each end and dip it into near-boiling water for several seconds. Transfer animals to 5% formalin for 24 h. Store in 10% formalin. Alternatively, transfer animals to 50% alcohol for 24 h, and store in 70% alcohol.

EARTHWORMS: Dip animals in fresh water for cleaning. Place in shallow bowl and cover with fresh water. Anesthetize by slowly adding enough 95% alcohol, drop by drop, to make an approximate 10% solution in the bowl. When the animal no longer responds to touch (2 or 3 h may be required), transfer to 5% formalin for storage. If earthworms are intended for later dissections, use the following procedure. Inject the anesthetized worm, at a point about 1 in posterior to the clitellum, with sufficient 1% chromic acid (1 g chromic acid to 99 ml distilled water) to fill body and render it turgid. Immerse the straightened worm in 1% chromic acid solution for 4 h. Wash in running tap water for 12 to 16 h. Hold for 24 h in 30% alcohol and again for 24 h in 60% alcohol. Transfer to 85% alcohol for storage.

LEECHES: Cover animals with a small amount of water and allow them to expand. Slowly narcotize with alcohol as described above for earthworms, or with magnesium sulfate (see Anesthesia, earlier in chapter). When animals no longer respond to touch, straighten them in a flat bowl and flood with warm 10% formalin. Inject large animals with 10% formalin. Allow animals to remain in the 10% formalin for 24 h. During the first 2 h in formalin solution, specimens may be flattened between 2 glass plates weighted down with a heavy glass bowl or jar. For storage, transfer to 8% formalin, or put in 50% alcohol for 2 h and transfer to 70% alcohol.

AQUATIC MOLLUSCS: *Gastropods and pelecypods.* Cover with their natural habitat water (fresh or seawater), and allow them to expand. Narcotize by adding enough alcohol, drop by drop, to make a 10% solution (see Anesthesia, earlier in chapter). When bivalved specimens are completely relaxed, place a small wood peg, about ¼ in thick, between the two shells. After several hours, test animals for response to touch. When animals are no longer responsive, slowly add 40% formalin during a 1- to 2-h period to kill the animals. Remove from formalin solution and wash in fresh water. Place in 30% alcohol and then 60% alcohol, 2 to 4 h each, depending on size. Transfer to 70% alcohol for storage. *Cephalopods.* Cover small squids or octopods with a small amount of seawater and place in a darkened area. During a 2- to 3-h period, slowly anesthetize by adding small portions of alcohol to make a final 10% solution. Alternatively, narcotize with magnesium sulfate or chloretone (see Anesthesia, earlier in chapter). Probe tentacles to test for a reaction. Allow animals to remain in narcotizing solution for another 30 or 40 min after they show no response. Before transferring to formalin or alcohol, drop a small amount of the preservative on a tentacle. If contraction occurs, leave the animal in the narcotizing solution for at least another 30 min. Position the animal in a storage container and cover with 5% formalin. After 24 h, replace with fresh 5% formalin or, if animals are large, with 10% formalin. For display specimens, replace the first formalin solution with 50% alcohol and leave for 24 h. Replace with 70% alcohol for final storage.

LAND SNAILS AND SLUGS: Put animals in a jar and submerge the open jar in fresh water. When jar is filled, tightly cover it with a lid while holding jar under water. Leave animals in air-tight jar until suffocated, a period of 24 h or longer. Transfer to 30% alcohol and 60% alcohol, for 2 to 4 h each, depending on animal size. Store in 70% alcohol.

MOLLUSC SHELLS: Place specimens in a sieve or cloth and lower into a container of cold water that is gradually heated to near boiling. Allow water to cool before shells are removed. Generally the inner, cooked tissue will fall from the shell, or it may be picked loose. With large snails, wash out tissue by forcing a strong stream of water into the shell. This cannot be done with fragile shells. If tissue becomes lodged within a spiraled cone, carefully drop 10% potassium hydroxide into the shell and leave the shell in an upright position for 24 h, after which wash out the softened tissue with water. If the snail has an operculum, pack the shell with cotton and glue the operculum into place on the cotton.

SMALL CRUSTACEANS: Place animals in small vials of fresh or seawater, according to natural habitat. Add 10% formalin, drop by drop, to poison and kill animals. Animals will drop to bottom of vial. Pipette off the formalin solution and add 50% alcohol. After 10 min, replace with 70% alcohol (or 5% formalin) for storage.

LARGE CRUSTACEANS: *Barnacles.* Cover with a small amount of seawater and allow animals to expand. Slowly narcotize with menthol crystals or chloretone (see Anesthetics, earlier in chapter). Leave for 5 or 6 h and test for reaction to touch. When animal is no longer responsive, add sufficient formalin to make a 5% solution. After 2 h, wash in fresh water and put animal in 30% alcohol and 60% alcohol, each for 2 h. Transfer to 70% alcohol for storage; or transfer directly from first formalin solution to fresh 5% formalin for storage.

Crayfish, shrimp, lobsters, crab. To prevent animals from automatically shed-

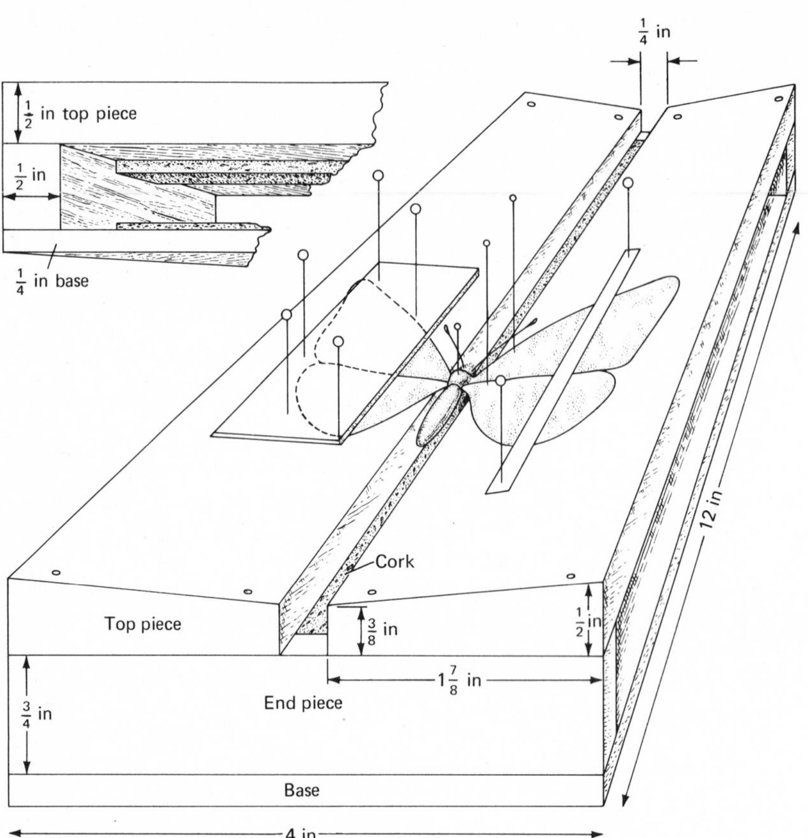

Figure 5-11 Insect spreading board. The insect is pinned with its body in the groove and with legs, wings, and antennae placed as shown. *(Courtesy of Illinois State Natural History Survey.)*

ding their appendages, kill by injecting chloroform into the body cavity through the soft tissue at the base of a walking leg. (Four or five millimeters may be required for large crabs and lobsters.) For this procedure, place animals in individual containers, since fighting may occur, with an increased shedding of appendages. Marine animals will die in 6 or 8 h, without chloroform, when put in one volume of seawater to which is added three volumes of fresh water. Alternatively, narcotize by leaving animals in their natural water and gradually adding chloroform. Transfer to 8 or 10% formalin for storage, or pass through 30 and 60% alcohols, each for 3 or 4 h. Store in alcohol-glycerol solution (95 ml of 70% alcohol with 5 ml glycerol).

ARACHNIDS: Put in insect-killing jar (see Chap. 1). Transfer to 70% alcohol for preserving and storing. Alternatively, drop directly into vials of 70% alcohol for killing and preserving. If alcohol becomes cloudy, transfer to a fresh portion at the end of 24 h.

CENTIPEDES AND MILLIPEDES: Kill in insect-killing jar (see Chap. 1) or in 70% alcohol. Remove animals and straighten them in a flat bowl; cover with fresh 70% alcohol for 2 or 3 days. Transfer to vials of 70% alcohol for storage.

INSECTS: Put in an insect-killing jar (see Chap. 1). Larval and pupal stages, as well as many adult forms, may be preserved in liquid by transferring them from the killing-bottle to vials of alcohol-glycerol solution (95 ml of 70% alcohol and 5 ml of glycerol). More often, adults are dry-mounted on pins, according to the techniques described below.

After killing, put the specimens on a spreading board to dry, with wings, antennae, and legs positioned, as shown in Fig. 5-11. The boards may be purchased, or they may be constructed from fine textured fiberboard. The dried insects are mounted on black-enamelled insect pins, 1½- to 2$^1/_{16}$-in length and numbered by diameter size from the smallest to largest as 000, 00, 0, and 1 to 7. Pins of numbers 1, 2, and 3 are used most commonly. A uniform method has been established for pinning the various insect orders, as shown in Fig. 5-12. A small or

Figure 5-12 Pinning insects. Hard-shelled and medium-hard-shelled insects are pinned in the manner shown in **A.** The black spots indicate the location of the pin for bees, flies, and wasps (**B**); stink bugs (**C**); grasshoppers (**D**); and beetles (**E**). *(Courtesy of Illinois State Natural History Survey.)*

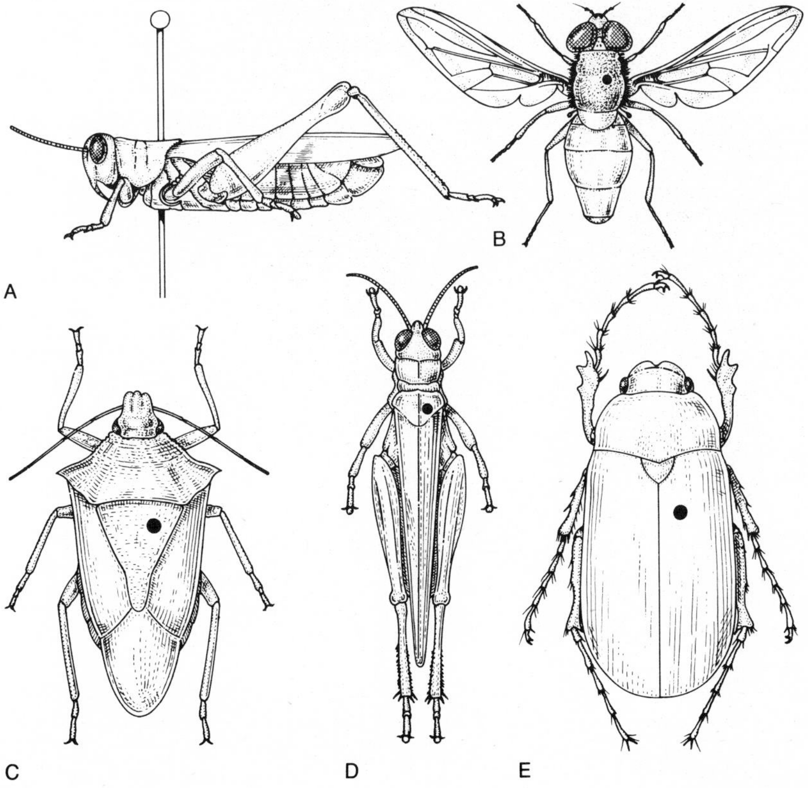

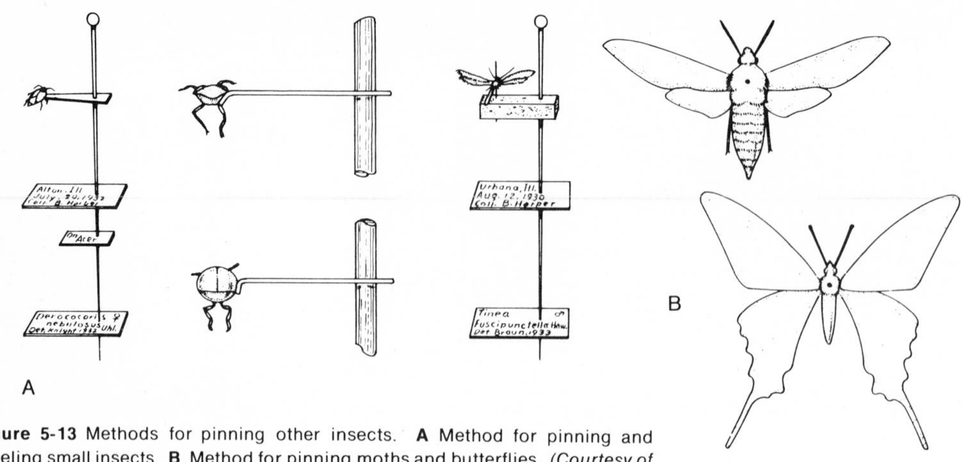

Figure 5-13 Methods for pinning other insects. **A** Method for pinning and labeling small insects. **B** Method for pinning moths and butterflies. *(Courtesy of the Illinois State Natural History Survey.)*

fragile insect is cemented on its right side to the tip of a small paper triangle cut from lightweight stiff paper (Fig. 5-13). Clear fingernail polish may be used as a cement. Before the insect is cemented to the point, a no. 2 or 3 pin is put through the base of the triangle. Labels containing the name of insect and collecting data are placed on the pin below the insect. The pins are then inserted into the cork or balsa wood floor of insect boxes. Many persons store their collections in cigar boxes or in clear plastic boxes, into which they fit a bottom lining of cork, balsa wood, or corrugated cardboard. To protect the collection from insect infestations, small chunks of paradichlorobenzene are pinned (with a heated pin) into several corners of the tightly closed box (Fig. 5-14). More specific details regarding preparation of insects will be found in Ross (1966).

ECHINODERMS: *Starfishes, serpent stars, brittle stars.* To prepare dried specimens, tape the arms of an animal to a flat surface and dry in air or put in an oven at

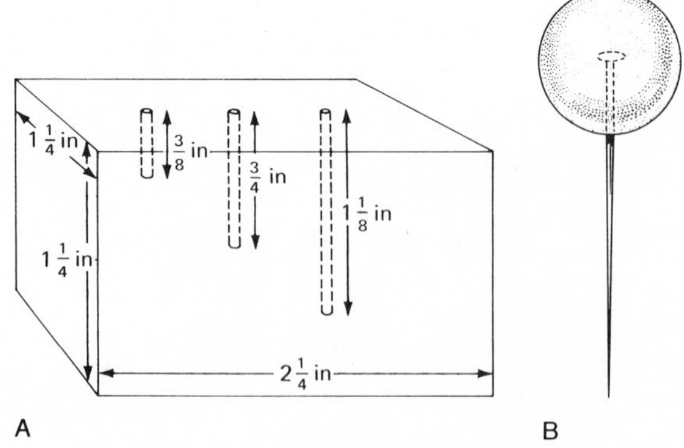

Figure 5-14 Other equipment. **A** Wood pinning block, 1¼ × 1¼ × 2¼ in. Holes are drilled to the depths shown, with diameter of holes slightly greater than the largest pin that will be used. A specimen is pinned, and the pin is inserted into one of the holes until the pin touches bottom. Thus, insects may be pinned uniformly at a desired height. **B** A naphthalene mothball pinned inside an insect box serves to repel invading insects that might cause damage. Prepare by sticking the pin point in a cork. Hold by the cork while heating the pin head and pushing the heated head into the mothball. *(Courtesy of the Illinois State Natural History Survey.)*

37°C. To prepare specimens in solution, cover animals with seawater and place container in the dark. Sprinkle about 0.5 g of magnesium sulfate over water surface each hour during 6 or 7 h. Let stand overnight. Test animal for reaction to probing. When no response is present, test with several drops of formalin on an arm. When animal is insensitive to formalin, transfer to a flat dish with the ventral surface turned upward to prevent flattening the tube feet, and with the arms straightened to a natural position. Cover with 5% formalin for 24 h. Wash in fresh water and put in 50% alcohol for 1 or 2 h. Transfer to 70% alcohol for storage. *Sea urchins and sand dollars.* Dry in open air or in an oven at 37°C. To preserve in fluid, treat as described above for starfishes. Alternatively, the animals may be put directly in 70% alcohol or 5% neutral formalin (see Appendix A). The color of urchins is retained longer in neutral formalin, although the specimen eventually becomes brittle and spines drop off. *Sea cucumbers.* Handle these animals gently and no more than necessary, as a slight disturbance may cause an animal to disgorge the viscera through the mouth opening. Cover animal with seawater and narcotize by sprinkling increasing amounts of magnesium sulfate crystals over the water each hour for 6 or 8 h. Leave in water for an additional 10 to 12 h until the animal is completely unresponsive to gentle probing. Transfer to 10% formalin for 2 h. Wash in water and inject 70% alcohol through the body wall at several points. Store in 70% alcohol.

FISH AND AMPHIBIANS: Cover with small amount of water and anesthetize with M.S. 222 Sandoz, chloretone, or urethane (see Anesthesia, earlier in chapter). Alternatively, kill by adding a small quantity of formalin to the water. Inject 8% neutral formalin into body cavity, and also into leg muscles of large animals. Position animals into a flat bowl and fix tissue by covering with 8% formalin for 48 h, or for as long as 1 week with larger animals. Store temporarily in 8% formalin. For display specimens, remove formalin by soaking animal in water for 2 days or more. When the formalin odor has disappeared, and before tissue has become soft, transfer to 70% alcohol. At the end of 24 h, inject 70% alcohol into the body cavity and also into the leg muscles of large animals.

REPTILES: These animals are difficult to kill. Knudson (1966) describes a freezing technique that may be the simplest and most satisfactory method. The animals are put into a cloth bag that is placed in the freezing compartment of the refrigerator for 12 or 14 h. The thawed animals are then injected with 10% neutral formalin until the body cavity is filled. Lizards and turtles must also be injected in the leg muscles. Snakes, unless very small, are injected along one side at each inch of the entire body length. Store in neutral 10% formalin (see Appendix A).

BIRDS AND MAMMALS: Ordinarily, whole specimens are not preserved in the teaching laboratory. Taxidermic techniques are described in a number of publications, including Knudsen (1966). Instructions for preparing dry skeletons will be found in Egerton (1968), Mahoney (1966), Taylor (1967), Schmitt (1966), and in the Turtox Service Leaflet No. 9.

CLEARING AND STAINING SKELETONS: This is a highly desirable method for displaying the stained and preserved skeletons of small vertebrates and their embryos (see Fig. 5-15). Various techniques have been described, most of them based on the method of Hollister (1934).

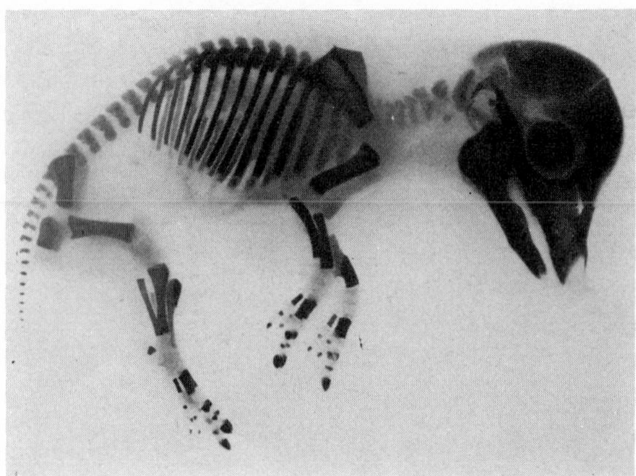

Figure 5-15 Pig embryo, cleared and stained by Method 2 in text.

Courtesy of Carolina Biological Supply Company

Method 1. The author has successfully preserved embryos and small vertebrates by the following method modified from Avis (1957):

1. Completely skin the animal. Eviscerate by making a small incision in the midventral abdominal region and carefully pulling out soft organs. In mammals, remove diaphragm and carefully pull out lungs and heart.

2. Rinse thoroughly in cold water.

3. Gently tease away the submaxillary glands, thymus glands, and fat pads in neck region of mammals.

4. Place adult animals in 2% aqueous potassium hydroxide for 24 h or longer, depending on size. Put small or embryo animals in 1% potassium hydroxide for 4 to 12 h.

5. Rinse in cold *distilled* water and place in 1% KOH until femur is clearly visible through thigh muscles.

6. Rinse in cold *distilled* water and place in 1% KOH to which a few drops of 0.5% aqueous alizarine red S have been added.

7. Remove when femur is well stained, and rinse in distilled water.

8. Blot on paper toweling and place in glycerol.

9. Within 48 h specimen should clear so that muscles are transparent, with the skeleton, stained red, showing through the muscles.

10. Specimen may be stored indefinitely in glycerol.

Method 2 (modified from Humason, 1967).

1. Skin and eviscerate animal.

2. Fix in 70% alcohol for 1 to 2 weeks, depending on size.

3. Rinse in *distilled* water. Place animal in 2% KOH until skeleton shows through musculature: 2 to 4 h for small embryos, 48 h or longer for larger animals.

4. Transfer to alizarine working solution (given below) until skeleton is a deep red, 6 to 12 h or longer.

5. Transfer to 2% KOH (0.5% to 1% for small specimens) until stain is removed from soft tissue.

6. Clear in clearing solution no. 1 (see below) for 1 or 2 days.

7. Transfer to clearing solution no. 2 (see below) for 1 day.

8. Store in glycerol, with several crystals of thymol added as a preservative.

Alizarine stock solution (Hollister, 1934):	
Alizarine red S, C. I. 58005, saturated solution in 50% acetic acid	5.0 ml
Glycerol	10.0 ml
Chloral hydrate, 1% aqueous	60.0 ml
Alizarine working solution:	
Alizarine stock solution	1.0 ml
1 to 2% KOH in distilled water	1,000.0 ml
Clearing solution no. 1 (Hood and Neill, 1948):	
2% KOH	150.0 ml
0.2% formalin	150.0 ml
Glycerol	150.0 ml
Clearing solution no. 2 (Hood and Neill, 1948):	
2% KOH	100.0 ml
Glycerol	400.0 ml

REFERENCES

Avis, F.R.: 1957, *About Mice and Man, An Introduction to Mammalian Biology*, Weston Walch Publisher, Portland, Me.

Conn, H.J., M.A. Darrow, and V.M. Emmel: 1960, *Staining Procedures*, 2d ed., The Williams & Wilkins Company, Baltimore, 289 pp.

Egerton, C.P.: 1968, "Method for the Preparation and Preservation of Articulated Skeletons," *Turtox News*, **46** (5): 156–157.

Hollister, Gloria: 1934, "Clearing and Dyeing Fish for Bone Study," *Zoologica*, **12:** 89–101.

Hood, R.C., and W.M. Neill: 1948, "A Modificationof Alizarine Red S Technic for Demonstrating Bone Formation," *Stain Technol.*, **23**: 209–218.

Humason, G.L.: 1967, *Animal Tissue Techniques*, 2d ed., W.H. Freeman and Company, San Francisco, 569 pp.

Knudsen, Jens W.: 1966, *Biological Techniques*, Harper & Row, Publishers, Incorporated, New York, 525 pp.

Mahoney, Roy: 1966, *Laboratory Techniques in Zoology,* Butterworth & Co., Ltd., London, 404 pp.

Ross, H. H.: 1966, *How to Collect and Preserve Insects,* Circular 39, Illinois Natural History Survey Division, Urbana, Ill., 71 pp.

Schmitt, D.M.: 1966, *How to Prepare Skeletons,* Ward's Natural Science Establishment, Rochester, N.Y.

Segal, Earl, and W.J. Gross: 1967, *Physiological Adaptation,* BSCS Laboratory Block, D.C. Heath and Company, Boston. Student Manual, 109 pp.; Teacher's Supplement, 60 pp.

Taylor, W.R.: 1967, "An Enzyme Method of Clearing and Staining Small Vertebrates," *Proc. U.S. Nat. Museum,* **122** (3596): 1–17.

Other Literature

Davis, D.D., and U.R. Gore: 1947, "Clearing and Staining Skeletons of Small Vertebrates," *Fieldiana:* Technique no. 4: 3–16.

Fremling, C.R., and D.L. Hemming: 1965, "A New Method of Taxidermy Using Polyethylene Glycol as an Impregnation Medium," *Am. Biol. Teacher,* **27** (9): 697–701.

Gantert, R.L.: 1967, "The Preparation of Skeletal Mounts," *Am. Biol. Teacher,* **29** (7): 531–534.

McFarland, W.N.: 1960, "The Use of Anesthetics for the Handling and the Transport of Fishes," *California Fish and Game,* **46** (4): 407–431.

Merryman, H.T.: 1960, "The Preparation of Biological Specimens by Freeze-Drying," *Curator,* **3** (1): 5–19.

Merryman, H.T.: 1961, "The Preparation of Biological Museum Specimens by Freeze-Drying: II. Instrumentation," *Curator,* **4** (2): 153–174.

Richmond, G.W., and Leslie Bennett: 1938, "Clearing and Staining of Embryos for Demonstration of Ossification," *Stain Technol.,* **13:** 77–79.

PART

two

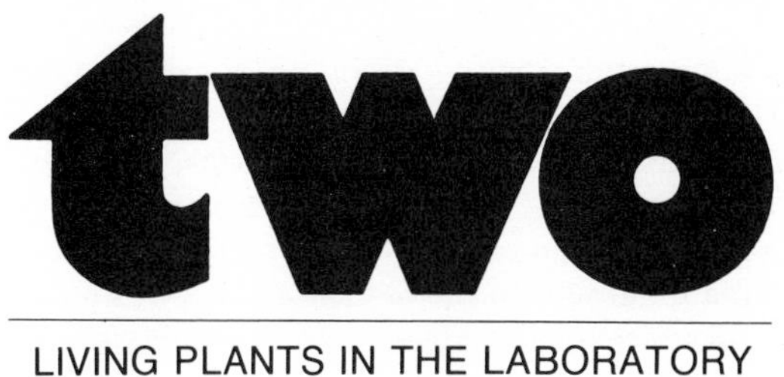

LIVING PLANTS IN THE LABORATORY

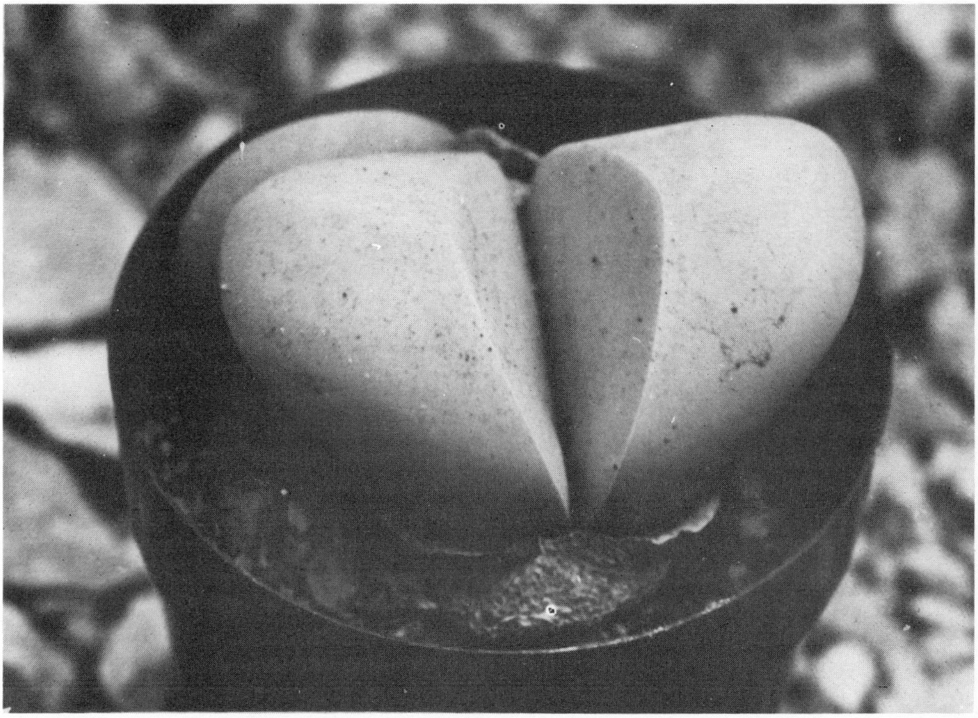

Although much literature is available regarding the culture of animals in the laboratory, little can be found about the care of plants for laboratory use. It is true that a few plant types are common to many teaching laboratories, including *Coleus*, geranium, *Tradescantia*, and *Anacharis* (elodea), as well as beans, corn, and onion sprouts. As useful as these plants are, neither the teacher nor the student needs to be confined to the monotony of a few types. A great many other plants are as hardy and as useful, or even more useful, for particular studies.

Indeed, a growing plant within the laboratory requires no greater reason for its presence than its own elegance and beauty. A group of well-nurtured plants, consisting of a wide variety of types, colors, and textures, can be as fascinating to students as laboratory animals. Botanists claim that plants are more exciting and many biologists agree. We know that plants do not escape from cages, and they do not require daily feedings nor regular removal of wastes. Furthermore, a preserved plant can in no way replace the living one. Although herbarium collections are always necessary for particular botanical studies, most often a dead, preserved plant is of little value in stimulating students within the biology teaching laboratory. The plant's magic and many of its secrets in growth, behavior, physiology, and reproduction are stopped.

Possibly, then, the chapters of this section can help teachers and students to become better acquainted with plants and to perceive the very real pleasures that come from a close association with living plants. As a practical aspect, horticulture as a career and gardening as a hobby may well have their beginnings in the biology laboratory. For all these reasons, the living plants described in these chapters are not intended solely for scientific inquiry and dissection. Most certainly, the growing plant with all its wonder and beauty needs no superimposed purpose for its glorification.

GENERAL CARE OF PLANTS

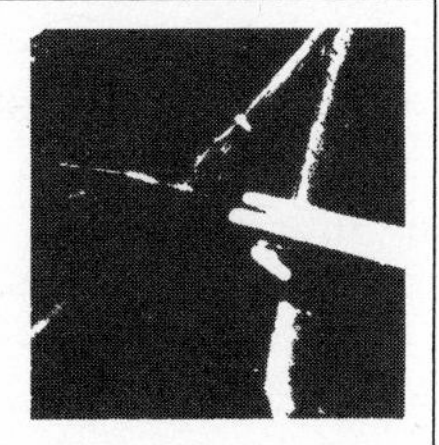

As with caged animals, indoor plants are confined to a restricted area, under conditions unlike those of their natural surroundings. Although plants are generally more easily maintained than animals, many plants have a narrow optimum range of light, temperature, and humidity beyond which they do not flourish. Generally little is said in botanical literature about the day-to-day chores that must be performed if one is to achieve a successful plant-growth program. Most methods discussed here are applicable to both the cultivated and wild plants and their seeds. Only general methods are given in this chapter; specific procedures are included in the next chapter with the description of individual plant types.

Although much satisfaction is derived simply from growing plants successfully, the procedures described here may also generate many questions that often can be answered with student-devised experiments. In reality, the number and kind of questions become the exciting part of working with plants, and the nature of questions depends on the interest and enthusiasm of students and teachers. It follows, however, that neither interest nor enthusiasm can be expected until the simple techniques have been learned for managing plants. Therefore, the following discussions purport to describe techniques and, at the same time, to suggest a great many avenues for experimentation.

HOUSING PLANTS

A first requirement of a teaching laboratory is the availability of plants to classes, teachers, and individual students. In many institutions, one greenhouse or a complex of houses serves all biological sciences. Often the greenhouse becomes a growing and storage place for plants that are removed for teaching or research and then returned to the plant caretaker. Occasionally, because of financial restrictions, the greenhouses are reserved for the research of professors and their graduate students, with perhaps a small area set aside for a few standard plants that may, or may not, find their way into a teaching laboratory. Midway situations exist, of course, where research and teaching personnel are able to satisfactorily share the greenhouse facilities. In general, however, it seems very evident that one facility, alone, cannot adequately serve the two purposes.

A solution to this difficult situation may be to construct one or more of the less costly lean-to greenhouses (described below), which then can adjoin or be located near the laboratory, with ready access by both students and teachers. In regions of mild climates, an outdoor plant area, for example, an enclosed yard, a patio, or a screened room, may often supplement the less accessible campus greenhouses. As well, the laboratory, itself, may provide growing space for plants that are put in window light or under artificial light. Indeed, many types of plants, particularly the foliage plants, can be successfully maintained within the laboratory. Thus, a very desirable—even necessary—living laboratory can be established for learning about living plants.

The Greenhouse or Plant Room

Planning and building a greenhouse are best accomplished under the direction of a greenhouse specialist. Requirements vary from school to school, and only general suggestions can be given here. The names of several manufacturers are listed in Appendix B.

Types and construction. Greenhouses are of two main types: (1) a free-standing house, set apart from other buildings, and (2) an attached room that may be a lean-to room, attached to one side of an existing building, or an even-span room connected at one end to an existing building (Fig. 6-1). The selection of type depends primarily on the most desirable location and the size of budget.

Each type has advantages and disadvantages. The *free-standing greenhouse* can be located at most any site where water, electricity, and heating can be provided. Orientation for maximum winter sunlight becomes a minor problem; ventilation and temperature often are more easily controlled. *Attached rooms* are less expensive to construct, but they require careful planning for location and attachment to another building. The *even-span type*, at the end of a building, provides more growing space than a *lean-to* and costs less per square foot. Both the free-standing and even-span types may be remodeled or enlarged more readily than a lean-to room. Heating an attached room costs less than that for a separate house.

Prefabricated aluminum and glass greenhouses of various sizes and types are available for "do-it-yourself" installations. Parts are labeled with instructions for assembly, thus allowing a skillful person to put together an attached room or a separate house with a minimum of tools. Because of mass production, the cost for a prefabricated unit is much less than that for building a single greenhouse of similar materials. Additionally, the valuable, and generally necessary, consultant service is available from the manufacturer.

The catalog price of a greenhouse, however, is actually about one-half of the total cost. Except in mild climates where freezing seldom occurs, any type of greenhouse requires a masonry foundation on a footing that extends below the frost line in the soil. Additionally, the planning must include provisions for ventilation, temperature, lighting, and humidity. A work table, the plant benches, a sink, and electrical and water outlets are other costs to be considered. Most of these installations are done by local craftsmen, according to specifications supplied by the greenhouse company. An aspect, sometimes overlooked, is that of adding a glass or aluminum wall-partition to give diversity of temperature and light with one area for warm plants and the other for cool plants. Furthermore, an attached greenhouse needs two entrances: a doorway that leads from the adjoining building and an outside door for bringing in soil and other supplies. The cost of flooring materials is also an added expense. Many persons consider a concrete floor to be the most desirable type, although others prefer a dirt floor covered with gravel or flagstones. Generally, a concrete floor is more expensive than others.

Unless trees or vines shade the house during the summer, outside shades of wood, metal, or plastic are usually necessary (Fig. 6-2). Less expensive shading materials are available and serve equally well. McDonald (1971) describes a green, vinyl plastic film that is sold in rolls. A sheet is cut to fit a glass panel and is then pressed flat against the inside, wetted surface of the glass. According to McDonald, the water does not dry out, and the plastic remains in place until removed the following autumn. The plastic sheet is then rolled onto a stick and labeled for its location on the wall or roof. The sheets may be used for many summers. Another inexpensive shading material is a paste-paint, obtained from greenhouse suppliers. Two coats of paint are applied annually to the outside surface of glass, one coat in the middle of spring and a second coat in midsummer. If the paint has not flaked off by fall, it must be washed from the glass at that time.

Plastic greenhouses. Although not as durable as a glass house, a plastic greenhouse can be built for about one-third of the initial cost of a glass house. Many commercial growers, in regions with very mild climates, find that a plastic house is practical and satisfactory. The wooden frame (redwood is preferable) is designed in the pattern of the conventional attached or free-standing greenhouse. The frame may be covered with *polyethylene sheeting*, of about 6- to 8-mils thickness (a mil is equal to 0.001 in), or with rigid, sheet plastic. Polyethylene sheeting must be replaced after about three summer months or nine winter months. Ultraviolet rays of sunlight destroy most plastics, although an ultraviolet-resistant polyester film *(Mylar)* is now available and lasts for at least several years. Commercial growers generally use two layers of sheeting; the air space between the layers gives a more even distribution of temperature. The inner layer may be 4-mils polyethylene sheeting. Polyethylene can be obtained in wide sheets, often as wide as the rooftop; thus the sheeting can be installed rapidly and inexpensively. *Polyvinyl sheeting* is more durable than polyethylene sheeting but about

C

D

Figure 6-1 Types of greenhouses. **A** A small free-standing house. **B** A larger free-standing house at St. Olaf College, Minnesota. **C** A lean-to type attached to a building. **D** An even-span plant room, attached at one end to a building. **E** Rooftop glass room at Newark State College, New Jersey, for plants and animals. **F** Plant room attached at second floor level, Westfield High School, New Jersey. **G** Dome plant room (center) at end of attached plant room (right of center). **H** Interior of greenhouse, Wheaton High School, Maryland. *(A–H Courtesy of Lord and Burnham, New York.)*

twice as expensive. It is manufactured in 6-ft widths and thus requires a longer time to install.

Fiber glass, a rigid plastic that is not nearly as clear as glass, is made of glass fibers embedded in plastic resin. It is easily installed; the plastic erodes, however, and dust and dirt catch on the exposed glass fibers, further reducing the light. Hence, periodic applications of acrylin resin are needed as the glass fibers become exposed. Also, some color change occurs in fiber glass after several months' use. Rigid, *polyvinyl chloride panels* have been introduced more recently. The material is less expensive than fiber glass, although it is also lacking in clearness.

Orientation of the greenhouse. In addition to accessibility, the location of the greenhouse is related to its orientation for capturing the most November to Febru-

E

F

G

H

ary sunlight. A free-standing house can be oriented in almost any direction as long as it receives at least 3 h daily of winter sunlight. As a rule, commercial people prefer that the house run from east to west, and ideally that it be placed slightly northeast to southwest. For protection from strong summer sunlight, it is well to select a site shaded by trees that lose their leaves in the winter. When possible, attached rooms should have a south, southeast, east, or southwest exposure, in that order of preference. An east exposure is preferable to a west exposure, since the latter requires heavy shading from summer sun. Although a north exposure is satisfactory for foliage plants, more inside heat is required, and most flowering plants will need artificial lighting to produce blooms.

Furnishings. The tables, generally called plant benches, are the most important furnishings. These should be made from rot-resistant materials, such as cedar, cypress, or redwood, raised to waist height. With prefabricated units, greenhouse

A

B

Figure 6-2 Shades for a greenhouse. **A** Wood roll shades. **B** Plastic roll shades. *(A and B Courtesy of Lord and Burnham, New York.)*

companies often include bench legs to which table tops must be added. Most companies also sell aluminum or steel-pipe frames with asbestos-cement board tops, a material that is practically indestructible and resistant to moisture and soil bacteria. The asbestos-cement board is corrugated to allow drainage toward the front and back sides (Fig. 6-3). Drainage from wood tables is obtained by leaving 1-in spaces between the boards. The table top is enclosed by side and end pieces, 5 or 6 in high, that form a top tray. Plant containers are placed directly in the tray or on a layer of gravel spread over the tray. As well, plants may be grown directly in soil within the top tray. Many persons prefer to place the soil on a tray lining of

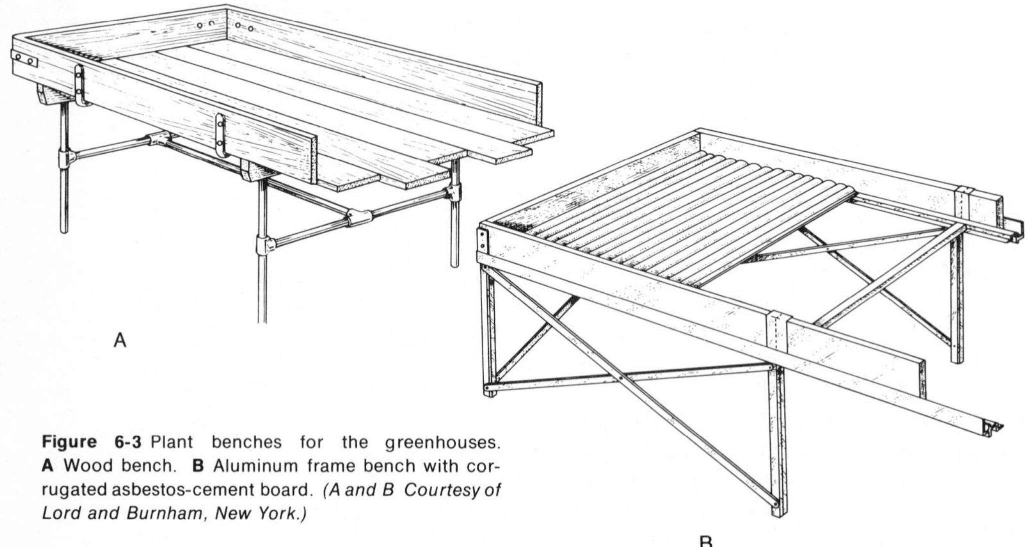

A

B

Figure 6-3 Plant benches for the greenhouses. **A** Wood bench. **B** Aluminum frame bench with corrugated asbestos-cement board. *(A and B Courtesy of Lord and Burnham, New York.)*

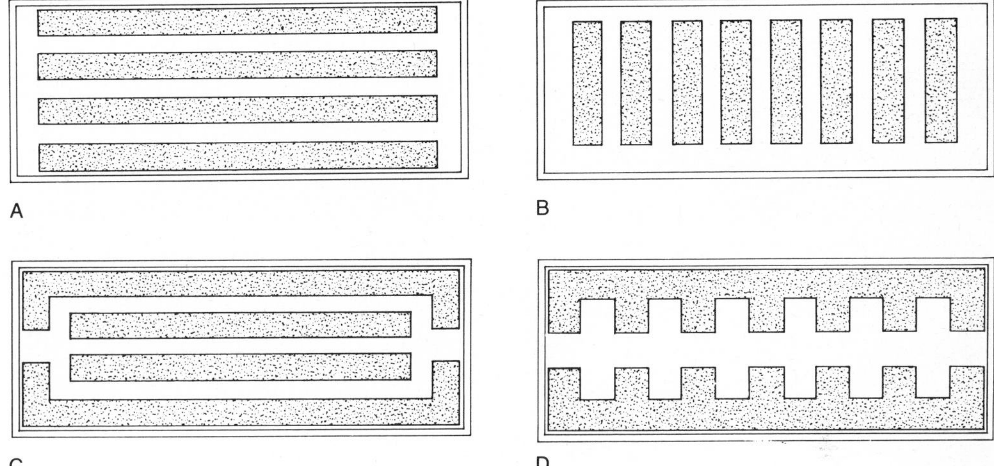

Figure 6-4 Schemes for arrangement of benches. **A** Longitudinal arrangement; **B** Island arrangement; **C** Combination arrangement; **D** Peninsula arrangement. *(Courtesy of Lord and Burnham, New York.)*

polyethylene sheeting that is punctured for drainage. If only pot plants are to be grown, no top soil tray is necessary, and the table top may consist of heavy, meshed wire. The room dimensions determine the number and placement of plant benches (Fig. 6-4).

Space must be provided for a work table with a sink, and for storage shelves and bins above and below the table (Fig. 6-5). The growing area of a greenhouse can be

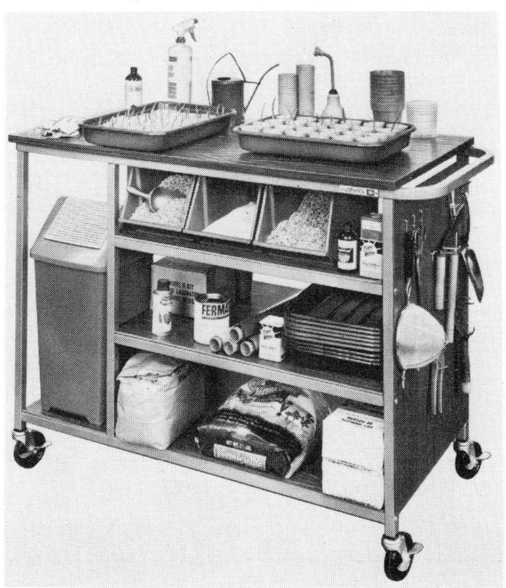

Figure 6-5 One type of arrangement for work table and storage when space is limited. This type of table does not include a sink, which must be located at a wall space. *(Courtesy of Jewel Aquarium Division.)*

A

B

Figure 6-6 Growth chambers and controlled environment rooms. **A** Growth chamber with tomatoes grown from seed in the chamber. **B** Multiple installation of growth chambers at Michigan State University. **C** Controlled environment room for plants and animals, University of Nebraska, Omaha. *(A–C Courtesy of Sherer-Dual Jet.)*

C

enlarged by transferring the work table and supplies to a preparation room or to an adjoining laboratory. Glass shelves with beveled edges will provide extra growing space on walls above the benches. Manufacturers offer a great variety of accessories for furnishing a greenhouse and for the care of plants. The catalogs should be studied carefully before placing orders.

Growth Chambers

Growth chambers range in type and cost from the large floor-chambers to the smaller table-models (Fig. 6-6). An instructor must understand how to operate the chamber properly, and must supervise the students in their use of the chamber. Operation and maintenance vary with different types and should be learned from a representative of the manufacturing company and from manuals that accompany the equipment. The names of several manufacturers are listed in Appendix B. Miller and Fagan (1970) describe their method for converting a used home-freezer into a controlled-environment chamber that can serve for both plant and animal studies (Fig. 6-7).

Plant Tables

A growth table affords versatility and often it is constructed easily. Figure 6-8 shows a plant table with a lamp fixture suspended about 12 in above the plant foliage. The fixture contains one fluorescent growth lamp (Sylvania Gro-Lux, Westinghouse Plant Gro, or GE Plant Light), and one cool-white fluorescent lamp, with 25 to 30 W of light for each square foot of growing area. Growth lamps may be obtained at electric or hardware stores, or at most aquarium stores. In place of suspending the fixture on a frame, as in Fig. 6-8, it may be hung from the ceiling on chains that allow the fixture to be lowered to about 8 in above seed germination boxes and raised to 10 or 12 in above growing plants. A sheet of polyethylene film is placed under the plant containers to protect the table surface.

A plant table may be easily converted to a plant growth chamber, similar to the one shown in Fig. 6-9. Add three or four 1-in strips of lumber, metal, or other building materials around the table edges to make a tray that is then lined with black polyethylene sheeting. Build an aluminum-wire frame from the table to above and around the light fixture. Tack double layers of black plastic sheeting over the frame and above the light fixture. Be certain the plastic is placed at a proper distance from the lamp to avoid heat damage. If desired, use double layers of black polyethylene as a middle wall to divide the enclosure into two or more compartments. The light intensity can be varied among compartments by the number and wattage of lamps; the photoperiod can be controlled by timer clocks, obtained at a relatively low cost from hardware stores or Sears, Roebuck and Co. Heat control may be provided with a thermostat connected to heating tape (from a hardware store) that is laid over hardware cloth on the table and covered with an inch of sand, on top of which the polyethylene film is placed. The sand, which protects the plastic sheeting from the heated tape, should be kept moist to give an even spread of heat.

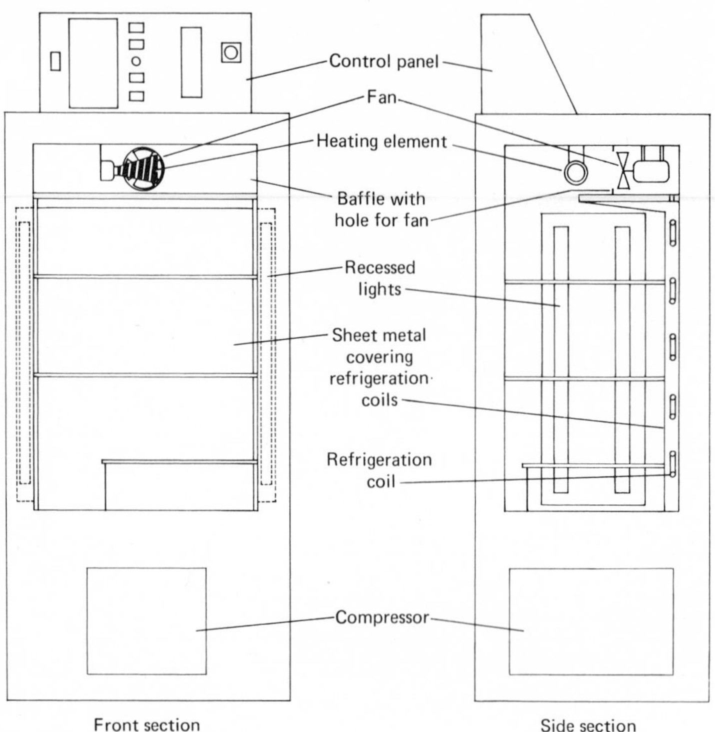

A

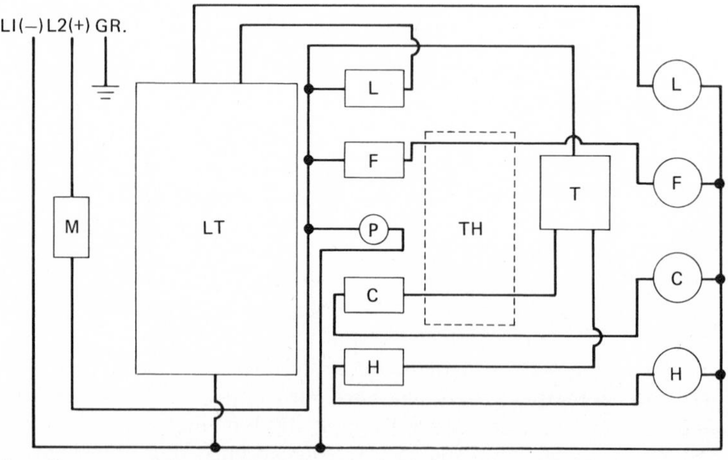

Key: M, master switch; LT, light timer; TH, thermometer; T, thermostat; P, pilot light; L (rectangle), light switch; F, fan switch; C, compressor switch; H, heater switch; L (circle) lights; F, fan; C, compressor; H, heater.

B

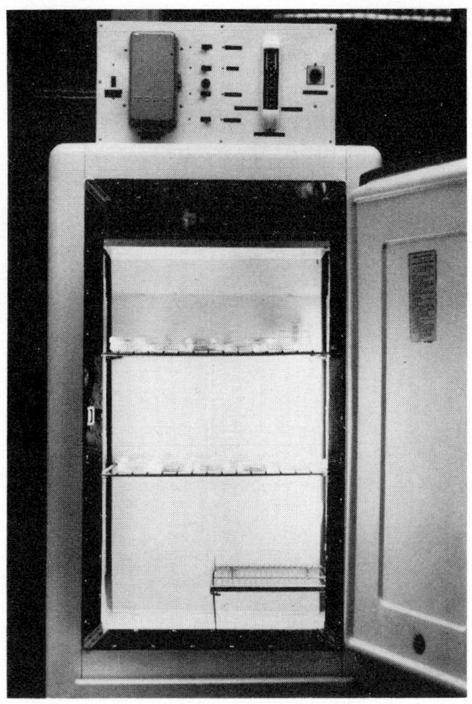

C

Figure 6-7 A growth chamber constructed from a home freezer. **A** Diagrams of front and side sections. **B** Diagram of control panel. **C** Completed unit in operation. *(A and B After Miller and Fagan, 1970; courtesy of CCM: General Biological, Inc., Chicago. C Courtesy of Miller and Fagan.)*

A Window Greenhouse

When a laboratory has ample window space, the simplest method for growing many plants is to place containers on window ledges or shelves. Depending on the amount of direct sunlight exposure and the type of plant, it may be necessary to provide additional light or, conversely, to shade plants during a portion of the day. During winter months, the plants must be protected or removed from cold windows; also, plants must not be placed above or near heat radiators.

Figure 6-8 Plant table with light fixture containing two 40-W fluorescent lamps.

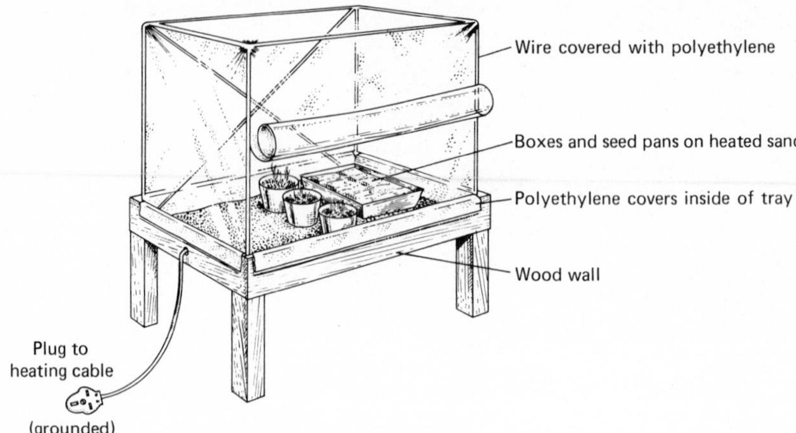

Figure 6-9 Conversion of a plant table to an enclosed plant chamber. *(After C. D. Bingham.)*

Bingham (1968) describes an enclosed window box, called a Wardian case. Figures 6-10 and 6-11 show the box on the outside wall, with access from inside the room through the vertical, sliding window. The temperature is regulated with a controlled-heat tape that is placed in several inches of sand or gravel in the manner described above for a plant table.

When a window case is sealed tightly against the outer wall, the sliding windows may be removed, thus providing simply an extension of the room. Other variations are possible; for example, an old showcase from a store can be equipped

Figure 6-10 An enclosed window box. *(After C. D. Bingham.)*

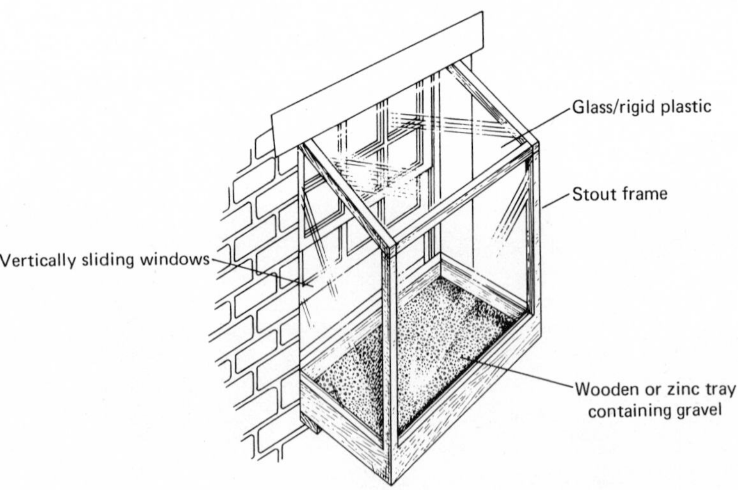

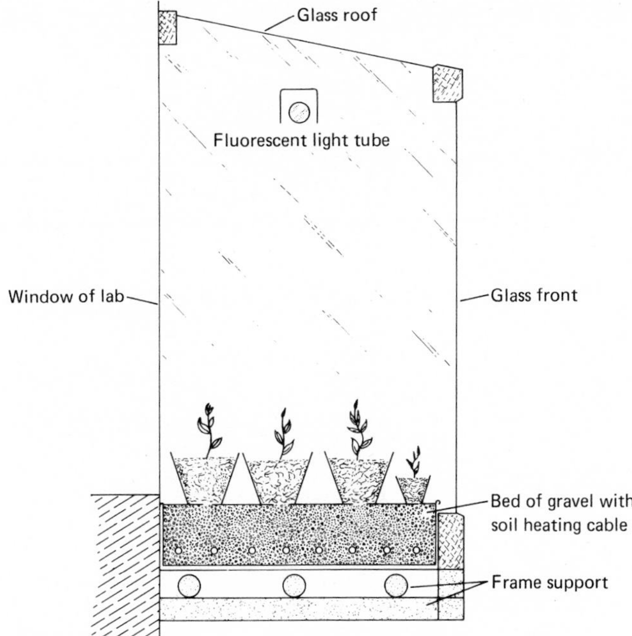

Figure 6-11 Side view of window box. *(After C. D. Bingham.)*

with a bottom tray, then supported with an outer frame and sealed against a window case.

Prefabricated window greenhouses may be purchased from greenhouse suppliers (Appendix B). Possibly these are the simplest to construct and are the most satisfactory, since all necessary climate controls are available with the unit. Additionally, the instructions for assembly and the suggestions from the manufacturer will help to ensure making a tight seal against the house for preventing water and air leakage.

Plant Carts

A plant cart, such as the one shown in Fig. 6-12, affords the advantage of easily locating plants against the walls or in a corner, and of moving plants about in the laboratory as desired. A skillful person may be able to mount wheels on a set of shelves, with a light fixture attached over one or more shelves, or he may be able to have the work done at the school shop. Otherwise, the cost of a commercial cart is well justified for growing plants in a laboratory of limited space. A unit with three shelves, each 18 in deep and 52 in long, provides about 18 ft^2 of planting area, a space probably greater than that provided by the window ledges of many laboratories. Blackout covers are furnished with some types of plant carts, or they may be made from double layers of black polyethylene sheeting, with the edges heat-sealed with a clothes iron.

Figure 6-12 Plant cart, useful when space is limited. *(Courtesy of Jewel Aquarium Division.)*

Other Methods for Extending the Plant Area

Glass terrarium. A terrarium provides a simple method, requiring only a little attention, for maintaining small plants. Appropriate animals may be added to make a desert, woodland, or semiaquatic situation, as described in the discussion of *The Terrestrial Vivarium* in Chap. 1. The covered glass terrarium is particularly advantageous for growing plants that require a high humidity.

Most any glass container of appropriate size is suitable, including an old aquarium tank, a widemouthed gallon jar, a battery jar, or a commercial glass terrarium with a metal frame and a slanting glass front (Fig. 6-13). Generally, a container of at least 6- to 10-gal volume is the most desirable size for culturing a variety of plants of both low-growing and erect types.

Figure 6-13 Glass terrarium with small woodland plants. *(Courtesy of Jewel Aquarium Division.)*

To prepare a terrarium, follow these general procedures:

1. Use a window-glass cleaner to wipe and polish the inner and outer surfaces, including the glass cover. Use a lintless cloth for the final polishing. A plastic film cover is as satisfactory as a glass cover, if the plastic is sealed tightly and replaced when necessary with fresh film.

2. Allow the container to air for 24 h to dispel any toxic fumes from the cleaning solution.

3. Add a bottom layer (1 in deep for a 6- to 10-gal tank) of equal parts of coarse gravel and coarse charcoal.

4. *For a woodland terrarium*, add a 2- to 4-in layer of moistened and loosely textured soil mixture, for example, a mixture of equal parts of peat moss, sand, and garden loam. Arrange plants in the soil or on the soil surface. Suitable plants are mosses, ferns, lichens on twigs or rocks, mushrooms from biological supply companies, and not the mushrooms from grocery stores (add humus soil), and low-growing woodland flowers and vines. See Chaps. 7 and 8.

When planting is finished, clean debris from the inside walls, sprinkle water over plants, and put the cover in position. Place in indirect light at a cool temperature (18 to 21°C). If water condenses on the inside walls, remove the cover for several hours at a time, or insert a wooden match between the glass cover and container, or make several small holes in the plastic cover. When the soil begins to feel dry, sprinkle water over plants (about every 3 or 4 weeks). Avoid overwatering and the accumulation of water in the bottom of the container. Trim back excessive plant growth. If mold develops, remove the molded portions and lightly dust the plants and soil with powdered sulfur; reduce the moisture.

5. *For a bog terrarium* of insectivorous plants, add to the bottom layer of gravel and charcoal a 2- to 3-in layer of water-soaked, acid soil from a supplier (or use a mixture of one part garden soil and two parts peat moss). Cover with a top layer of soaked *Sphagnum* moss. The moss and acid soil may be ordered with a set of insectivorous plants from a biological supply company (see Appendix B). Before placing the plants in the soil, wrap *Sphagnum* loosely around the plant roots. Set pitcher plants deep enough for the roots to grow into the water that accumulates in the bottom gravel. Set the venus flytrap about 1 in higher, with root tips in the soil but above the lower water, Place the roots of the sundew at the top of the soil and in the layer of *Sphagnum*. Note: The plants should be handled gently and as little as possible; avoid touching the leaves. Do not allow the plants to become dry when transferring them from the shipping or collecting bag to the terrarium. Maintain the plants in indirect light and at a temperature of 20 to 24°C. Several weeks may be required before the plants begin to grow. One to two hours of daily sunlight are required for the plants to produce flowers. Take care, however, not to leave the plants in direct sunlight for a prolonged period, as the temperature will rise inside the glass container and the plant may be burned or killed. Add water to keep the soil moist but not soggy, although a layer of accumulated water is desirable in the bottom gravel for the roots of pitcher plants. Ordinarily, mold does not

A

B

Figure 6-14 Constructing a planted plank. **A** Wire has been pulled under the board, brought up and over the front edge, and then pulled back on the top side to a distance of about 1 in from front edge. The edge of the wire has been stapled along the length of the board. (Staples and wire are only faintly visible at front top of board.) **B** The completed plank, with an inner core of peat and vermiculite and an outer layer of *Sphagnum* covered with wire.

grow on the acid soil and moss. See Chap. 7 for a greater description of insectivorous plants.

A planted plank. The idea of a planted plank is not new to the home gardener and might well be introduced into the laboratory as a space-saver and as an unusual method for displaying the growth of succulent plants. The plank may vary in dimensions; an appropriate size might be 18 in long, 4 in wide, and 1 in thick. For these dimensions, cut a rectangle of chicken wire that is 30 in long and 16 in wide. Staple a long edge of the wire under one side of the board (Fig. 6-14). Line the wire with a 1-in layer of moist *Sphagnum* moss; over the *Sphagnum* spread a moistened mixture of two parts peat moss and one part vermiculite or perlite. Carefully pull the wire into the position as shown in Fig. 6-14B; then miter the corners of the piece of wire, and staple the wire into position as shown. If the wire cover is not completely filled, push more of the peat mixture through one end of the wire before closing it.

Plants that are easily grown in this manner include the rosette-type succulents, for example, *Escheveria, Sedum, Sempervivum* (all described in Chap. 7), and many others, including the trailing plants. To plant the board, punch through the wire to make a hole in the moss. Lightly wrap moist *Sphagnum* around the plant roots. Gently press the roots into position in the peat mixture. Stem cuttings may be started by planting them in the same way. Select small, lightweight plants that will remain anchored within the wire. Hang the board at a window or on a wall. Protect the window or wall with a sheet of polyethylene under the board. Divide or trim back the plants to prevent excessive growth and weight (Fig. 6-15).

The planting mixture dries quickly and must be watered regularly, perhaps daily. Give plant food in water solution about every 2 to 3 weeks. At that time, first remove the plank from the wall and place it in a flat position on a table. After adding the liquid food, leave the plank on the table for several hours to prevent an accumulation of nutrients at the lower end of the plank.

Plastic-bag growth chamber. A simple and temporary growth chamber may be devised by inserting a plant tray or pot into a plastic bag (e.g., a plastic clothes

Figure 6-15 Planted plank hanging on wall; a portion showing *Scindapsus* (devil's ivy) at top sides, one large and four small rosettes of *Sempervivum* (hen and chickens), and branches of *Sedum* (stonecrop) scattered over the structure.

bag), with the bag supported over the top of the plant by wooden stakes or wire arches (Fig. 6-16). The method is particularly useful for obtaining a high humidity when seeds are germinated or when cuttings are rooted. Also, the technique allows an excellent means for conserving moisture over a weekend and during vacation periods. If placed in the appropriate lighting situation, most potted plants can be held in this manner for 3 or 4 weeks without additional water. Generally, the best location is one with no more than 2 or 3 h daily of direct sunlight. Because the temperature increases in the plastic bag, too much light and heat may burn or kill a plant.

A bottle garden. Long used as a novelty among home gardeners, the bottle garden often appeals to students, who find an added fascination in certain plants displayed in this manner. Once the bottle garden is established, it requires as little care as its larger counterpart, the glass terrarium.

To prepare the bottle, follow the procedures outlined above for the glass terrarium. Because of the small bottle opening, use long wires or sticks for cleaning the inside walls and for putting the plants in place. The results of planting a gar-

Figure 6-16 Plastic-bag plant chamber used in this instance for stem cuttings of *Sansevieria*. Fluorescent lamp is above.

Figure 6-17 Bottle garden.

den are illustrated in Fig. 6-17. Add water through a glass or paper straw. Remove excessive growth with a razor blade fastened to one end of a wire or stick. Close the top with a stopper, a lid, or with plastic film. When moisture accumulates on the inside walls, remove the cover for several hours. Place the bottle garden in diffuse light; flowering plants should have 2 or 3 h daily of sun or artificial light. See Chaps. 7, 8, and 10 for descriptions of plants that are especially suitable for bottle gardens.

Plant Containers

Many types of containers are satisfactory, including those of clay, plastic, glazed ceramics, copper, brass, and wood. All types have their advantages and disadvantages. The container must have rigid walls so that roots are not damaged by the pressure of shifting movements when the container is handled. A container must provide ample space for root growth and ample depth for support of the plant. Water drainage from a container requires special attention and is described later in this section.

Unglazed clay pots, plastic pots, and plastic trays are used most commonly. Because unglazed clay pots allow a better drainage than plastic containers, they are used almost exclusively in many plant nurseries and greenhouses. Cleaning the pots is a difficult chore that may require scrubbing with steel wool. Both the inner and outer surfaces of a clay pot must be cleaned and completely dried before use. Plastic containers are generally less expensive than clay pots; they do not break as easily; and cleaning is a relatively easy chore. Many other kinds of containers have been converted to satisfactory use, including glass jars and bowls, plastic fruit boxes, the discarded plastic containers of kitchen products, and various types of rigid, waxed-paper containers, such as milk cartons and paper cups.

The size of commercial pots is based on the top, inner diameter. "Standard" pots, as they are called, have equal dimensions of height and top diameter. The so-called azalea pots (suitable also for many kinds of low-growing plants) are three-fourths the height of a standard pot. Pot sizes range in top, inner diameter from 2 to 7 in in half-inch increments; from 7 to 12 in in 1-in increments, and from 12 to 18 in in 2-in increments. The required size of a pot depends on the size of plant, of course; ordinarily, seedlings are put into 2½- or 3-in pots, from which they are transferred to larger containers as the plant grows.

Other Equipment

Florists and garden stores display a large assortment of tools and accessories. Generally, simple garden tools are adequate; the selection of hand spades, trowels, scoops, and watering equipment is mostly a matter of individual preference. Labels are often required for individual plants, and these are inserted in the soil or attached to a plant. Labels may be purchased from florist shops, or they may be fashioned from discarded aluminum containers, for example, frozen-dinner trays, or from discarded plastic containers. Use a nail or other pointed object to write on an aluminum strip; probably a lead pencil is best for writing on a plastic strip. Bamboo or other types of stakes are required for supporting tall or climbing plants. A water-mist sprayer is a very desirable accessory for washing and moistening individual plants.

SOIL FOR INDOOR PLANTS

A soil mixture must be porous to allow air circulation and water drainage; it must contain roughage materials by which the roots become anchored; and it must have water-holding capacity. A tightly-packed soil, that is pasty when wet, will become waterlogged and deficient in oxygen. Horticulturists describe soil as desirably friable when it crumbles loosely but retains a spongy resilience. A good garden loam may be used alone for potted plants. Most soils, however, need other materials for drainage and absorbency. Several media are described below.

Standard Soil Mixture

The old, standard mixture consists of 1 part each of garden loam, sharp sand (or perlite), and peat moss. The mixture is satisfactory for most plants; the materials are relatively inexpensive and may be obtained at a garden store. The peat moss can be made more readily absorbent by first soaking it in boiling water. Allow the moss to cool and remove excess water before adding the moss to the other materials. A coarse grade of builder's sand from quarries is used and not the soft, riverbed sand. Perlite, an expanded volcanic ash, may be substituted for sand. The chemically inert perlite absorbs water but remains rigid, thus producing a coarse texture with good drainage.

Peat-lite Mixture

A person may wish to experiment with other formulas, of which an abundance are described in horticultural and gardening publications. The following formula, used rather commonly in large commercial greenhouses and in research situations, was developed at Cornell University.

Vermiculite (Terralite)	4 qt
Shredded peat moss	4 qt
20% super phosphate (powdered)	1 teaspoon
Ground dolomitic limestone	1 tablespoon
Plus either of the following (not both):	
33% ammonium nitrate	1 tablespoon
5-10-5 commercial fertilizer	4 tablespoons

Moisten the mixture well before using it. Begin feeding liquid plant food at 5 or 6 weeks after potting the plants. Follow instructions on the container for diluting the plant food.

Terralite is a brand name for vermiculite, a medium often used alone for germinating seeds or for growing plants in a nutrient water solution. Vermiculite is made by heat-expanding mica particles at 1,400°F. The material helps to prevent plant damage from overfertilizing because of its capacity to absorb and hold nutrients in reserve, gradually releasing them to the plants. As well, vermiculite provides a buffering action that tends to resist a rapid pH change in the soil.

PLANT FOODS

Fertilizers may be obtained at plant nurseries and garden stores. General directions for their use are given on the packages. Numbers expressing the percentages of nitrogen, phosphorus, and potassium are printed, in that order, on the label of the container. A common fertilizer is 5-10-5, the numbers indicating 5% nitrogen, 10% phosphorus, and 5% potassium. Except for minute quantities of trace elements, the other approximately 80% of the contents is a filler material. The 5-10-5 fertilizer is often used in water solution (not dry) as a starter food for freshly potted plants, in the proportions of 1 cup of fertilizer dissolved in 3 gal of water. Apply by pouring 1 cup of the solution on the soil medium when the plant is first potted. Five or six weeks after potting, begin feeding houseplant fertilizer from a garden store. In general, plant growers agree that frequent feedings in small amounts are more beneficial than large feedings several times yearly. The liquid fertilizers, diluted in water, are probably the easiest and most effective for use with indoor plants. Several aspects must be emphasized: (1) always water a pot plant before adding the liquid fertilizer to the soil surface; (2) do not apply a dry fertilizer, and do not add a liquid fertilizer to dry soil; (3) feed plants only when they are in an active growth stage; (4) do not feed dormant or resting plants.

Foliar Feeding

Many horticulturists feed plants by spraying fertilizers on the foliage. Tests have shown that plants absorb the nutrients directly through the leaves, a process that offers possibilities for laboratory studies and tests. Commercial preparations for foliar feeding are available at garden stores; concentrations must be adjusted according to instructions on the container. Nutrients supplied in this manner are absorbed and used quickly, and concentrations must be reduced considerably to avoid "burning" the leaves. Applications are made every 7 to 10 days, in contrast to soil nutrients that are applied every 3 or 4 weeks.

LIGHTING FOR LABORATORY PLANTS

As with all aspects of plant culture, many conflicting opinions have been expressed concerning appropriate lighting for indoor plants. Although much can be learned by consulting the literature, and this is necessary in plant research, the essentials of lighting for culturing laboratory plants can be reduced to a short list. The requirements for individual plant types are described in Chaps. 7 and 8.

1. Evergreen plants need daily light the year round. Laboratory plants in this category include the succulents, cacti, and many of the foliage plants, such as the philodendrons. Plants with tubers, bulbs, or corms require light only during the growth period, after which they are stored in low light or darkness during resting periods.

2. Many plants produce good foliage in diffused light; thus they will grow well in almost any location within the laboratory, except perhaps in full sunlight.

3. A number of plants will thrive and flower under a combination of fluorescent lamps, for example, cool-white fluorescent lamps combined with daylight fluorescent lamps (see Fig. 6-8).

4. As a rule, flowering plants bud and flower most satisfactorily in complete light. Complete or nearly complete light may be furnished by:

a. Window light.

b. Special horticultural lamps, for example, Sylvania Gro-Lux, Westinghouse Plant-Gro, or GE Plant Light, purchased at electrical, garden, or aquarium stores.

c. A combination of incandescent lamps and cool-white fluorescent lamps, in a wattage ratio of 1:5 (incandescent watts: fluorescent watts). Klein (1970) recommends a greater proportion of incandescent light and states that a ratio of 1:2 (incandescent: fluorescent) is not too great. A diagram for spacing the lamps is shown in Fig. 6-18. The bank of lamps is placed 12 to 15 in above mature, potted plants, as shown. If leaves begin to curl, the lamps are too close; if stems become long and spindly, the lamps are too far away.

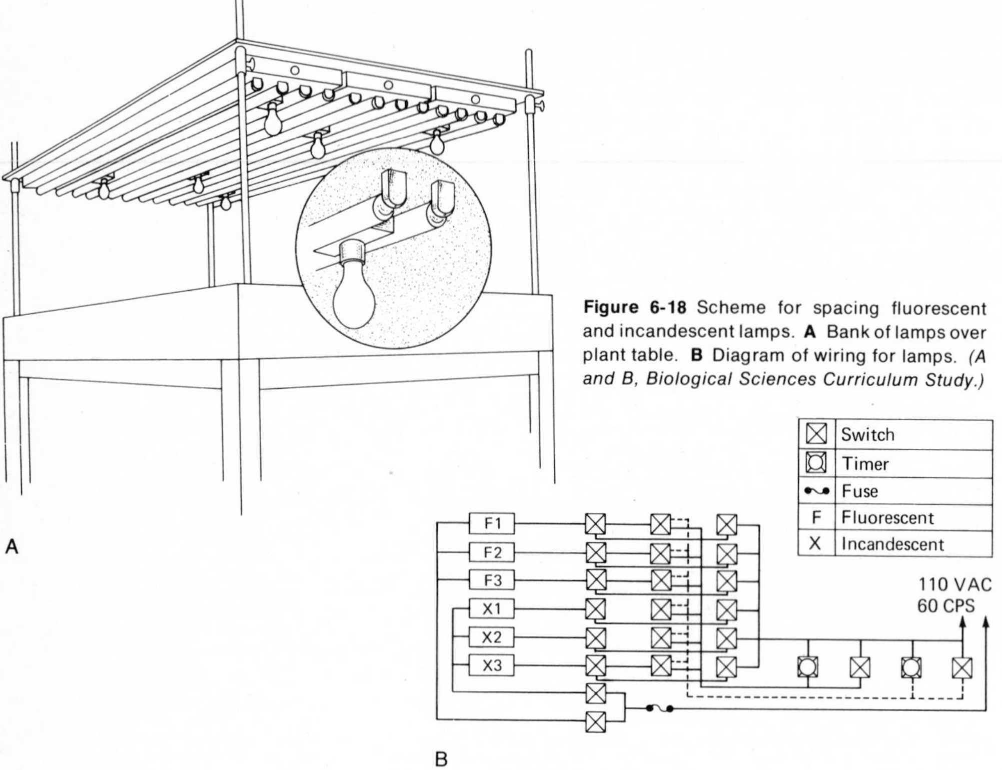

Figure 6-18 Scheme for spacing fluorescent and incandescent lamps. **A** Bank of lamps over plant table. **B** Diagram of wiring for lamps. *(A and B, Biological Sciences Curriculum Study.)*

5. North windows furnish daylight but no direct sunlight. Many laboratory plants grow well, and a few produce flowers, in this location. Plants that require sunlight should be put in a south, southeast, or southwest window, or under artificial light. A west location may be too warm.

6. Most indoor plants described in Chaps. 7 and 8 grow well with 14 to 16 h of light and 8 to 10 h of dark each day. They require the dark period as well as the light. A timer clock is a good investment for regulating photoperiod.

7. Plants must be guarded from excessive heat of incandescent lamps by allowing air spaces around and above the lamps. If necessary, a fan can be installed to blow air horizontally across the lamps and away from the plants.

Light Quality (Wavelength) Requirements for Plant Growth

Optimum plant growth requires radiation energy of both the blue band (4,000 to 5,000 Å) and the red band (6,500 to 8,000 Å). In red light alone, plants mature more rapidly and tend to have a tall, spindly growth pattern. In blue light alone, plants develop short, thickened stems and exhibit reduced flowering. Direct or in-

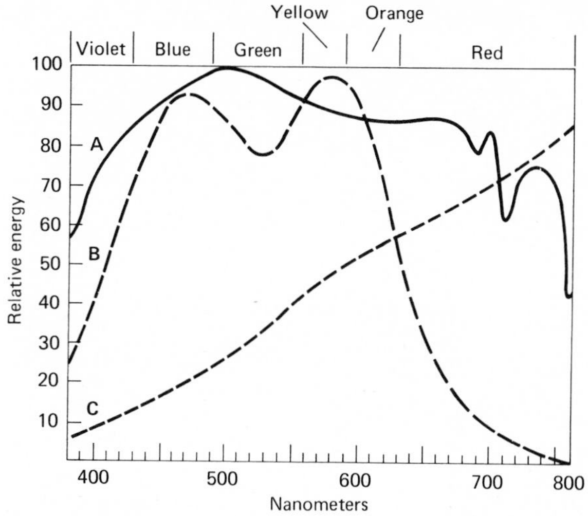

Figure 6-19 Comparison of spectral energy distribution curves of: **A** Solar radiation reaching the earth's surface; **B** Daylight fluorescent lamp; and **C** Incandescent lamp. Compare this figure with Fig. 6-20. Nanometer (nm) has become the accepted term for millimicron (1 nm = 10Å). *(Modified from Biological Sciences Curriculum Study.)*

direct sunlight gives a complete light spectrum that is difficult to achieve with artificial light. By comparing in Fig. 6-19 the spectrum of sunlight and the spectra of the three types of lamps mentioned in the preceding section, one can see more clearly the reasons for the particular lamp combinations that are used to gain an approximate radiation of complete light. Also see Fig. 6-20.

Light Intensity Requirements for Plant Growth

Most commonly, the foot-candle (ft-c) is used as a unit of measurement for light intensity. The foot-candle is defined as the amount of light falling on a surface 1 foot square from an "international" candle (a candle of special wax at a special temperature). A photographic light meter may be used to measure light intensity at plant level. With most meters, the readings must be converted into foot-candle units according to the manufacturer's instructions. General Electric Company now manufactures a light meter (GE 213) which reads directly in foot-candles with a range of

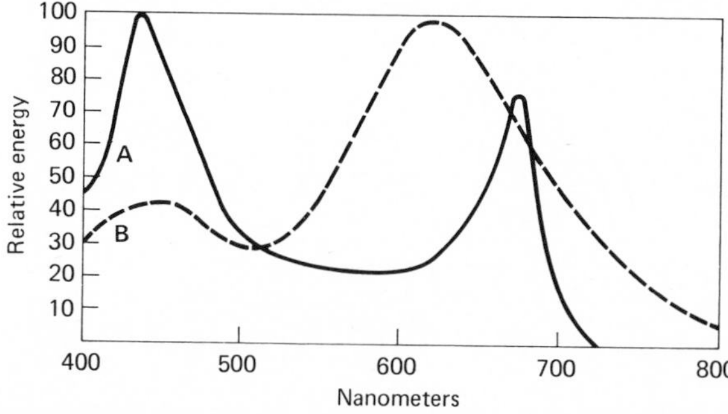

Figure 6-20 Comparison of the action spectra of photosynthesis with the Gro-Lux, wide-spectrum lamp of Sylvania Electric. **A** Photosynthesis curve. **B** Curve of the Gro-lux/WS fluorescent lamp. *(Modified from GTE Sylvania Incorporated.)*

0 to 5,000 ft-c. Included with the GE meter is a calculator for photographic purposes.

Intensity requirements for common plants vary from 100 to 2,000 ft-c. Although sunlight intensity often exceeds 10,000 ft-c, plants requiring full sunlight may grow and flower with only 1,000 to 2,000 ft-c. Research scientists may use levels as high as 8,500 ft-c in growth chambers.

Although a measurement of light intensity in foot-candles is a vital part of much plant research, the amount of light for laboratory plants can be adequately managed by regulating the lamp wattage and the distance between lamps and plants. Seedlings require a greater amount of light than mature plants, and thus the lamps are placed about 6 in above the tops of seedlings. As the plant matures, the lamps are raised to a final distance of 12 to 15 in from the plant tops. Much less specific rules, used satisfactorily by many plant hobbyists, are those suggested by Potter (1967):

1. *For seeds and cuttings:* 10 W for each square foot of growing area. Place lamps 6 to 8 in above the soil.

2. *For low-energy plants:* 15 W per square foot of growing area. Put light source 12 to 15 in above plant tops. Many household plants belong in this classification.

3. *For high-energy plants:* 20 W per square foot of growing area. Place light source 12 to 15 in above plant tops. Tomatoes, beans, cucumbers and most other vegetable plants are in this group.

Photoperiod Requirements for Plant Growth

It is well known that many plants display a rhythmic night and day growth rate of the shoot, with a greater growth at night than during the day, providing the night temperature does not drop too low. Furthermore, the flowering of many plants is determined by the length of day, or more correctly, by the length of night. For example, spinach begins to flower after exposure for a 2-week period to days that are 13 to 14 h long, but it will not bloom if the days are less than 13 h long. Such plants are called long-day plants. On the other hand, *Poinsettia* produces flowers in the winter when the days are short. These plants are called short-day plants. A large group of other plants apparently are unaffected by the length of day and night, and these plants are designated as day-neutral or indeterminate plants.

The determining factor is the length of the dark period for each 24 h and not the duration of illumination. In fact, plant scientists now know that by simply interrupting the dark period, sometimes for only a few minutes, they can effectively control and modify the flowering pattern of long-day and short-day plants. Night length affects other processes also, including seed germination, tuber and bulb formation, and the color and enlargement of leaves. Although a number of plants have been categorized into the three light-dark systems, this manner of categorization represents an over-simplification. Much variation exists among and within plant groups, and factors other than day and night length contribute to growth and flowering patterns. Because the photoperiod has not been identified for many indoor plants, specific requirements cannot be indicated for most plants described in Chaps. 7 and 8. The day length for the normal blooming season of a flower (winter flowers or summer flowers, etc.) is the most appropriate gauge for determining the light and dark periods of indoor plants.

TEMPERATURE AND HUMIDITY FOR INDOOR PLANTS

Most plants tolerate a wide temperature range, providing the change is gradual and extreme temperatures are not sustained longer than several hours. Temperatures on the cool side are less damaging than those above the optimum range, primarily because humidity drops rapidly as temperature rises. Of greater damage than temperature fluctuations is a decreased humidity, which in classrooms and homes may often go as low as 10 to 12%. Under such circumstances, several techniques will help to increase the atmospheric moisture around plants. A simple method is to maintain potted plants in a tray or bowl containing a layer of small stones that are nearly covered with water. The bottom of the pots must rest above the water level to prevent waterlogging of plant roots. As the water evaporates, the humidity increases in the air around the plants. If the room becomes extremely dry, both plants and persons in the laboratory will benefit from an increase in the air moisture when a steaming (not boiling) pan of water is kept on an electric plate. Boiling water may put too much moisture into the air for personal comfort. The most desirable technique, of course, is to install a portable, room humidifier that can be controlled at 40 to 50% humidity.

Thermoperiod in Plants

Plant research indicates that the growth of flowering plants is influenced by a daily rhythm of temperature change. Went (1957) developed this concept of thermoperiod in plants, an idea that offers many possibilities for exploration by biology students. Many experienced greenhouse and indoor gardeners, for example, Abraham (1967) and McDonald (1966), recognize that flowering plants require a night-temperature drop of 3 to 5°C (5 to 10°F) for a sustained budding and blooming, and that foliage plants show improved growth with a night-temperature drop of at least 2 to 3°C. In many laboratories and plant rooms, the indoor night temperature may change when the outdoor night temperature drops. In situations where the night temperature does not drop several degrees, or where the night temperature cannot be controlled (in some situations the temperature may rise), the choice of plants may be limited to foliage types, most of which will tolerate a wide range of temperature changes.

Controlling Temperature for Plant Tests

Plant tests often require a controlled situation, with temperatures above or below ambient temperatures. Much equipment is available or can be devised. Long-term, precise studies require a growth chamber, equipped with adequate controls for temperature, light, and humidity. With short-term and less precise studies, a water bath, a refrigerator, or an oven may suffice, providing that light is supplied to green plants.

Controlling temperatures at below ambient. Generally at least three levels of temperature can be established in a refrigerator: (1) a below-freezing temperature in the freezing compartment; (2) 8 to 10°C on the top shelf of the refrigerator compartment; and (3) 4 to 6°C on the bottom shelf. When required, a low light intensity can be maintained in a refrigerator by dismantling the light switch on the door frame, and then pulling the switch into the chamber. Thus, the refrigerator light is not disconnected when the door is closed. Alternatively, a 75-W or 100-W incandescent lamp can be hung inside the box from an electric cord that extends through the door. The soft gasket that lines the door frame allows the door to be closed against the electric cord.

Controlling temperatures at above ambient. If a controlled growth chamber is unavailable, probably small polyethylene chambers or glass terrariums, placed under lamps for heat, offer the most satisfactory means of providing a raised temperature with sufficient humidity for short-term studies. However, establishing a steady temperature under these conditions is probably impossible. Furnishing light to a green plant in an oven or an incubator becomes a more difficult task, as moisture must be added to compensate for the drop in humidity. Lighting may be supplied by threading the electric cord through the door or through a small opening drilled into one wall. The drop in humidity may be offset by keeping a pan of water in the chamber.

HANDLING SEEDS

As a rule, seeds of the best quality are obtained from seed stores or seed growers. Those from grocery or variety stores may be several years old, and storage conditions may have been inadequate. When quality seeds are not available locally, they may be ordered by catalog from established seed companies which guarantee that the seeds are fresh, and properly processed by specialists. Also, seed-testing companies provide germination data with their shipments, when requested, and asssist with research problems through their contract services. Biology supply companies furnish a wide variety of seeds for germination tests and for genetic and radiation studies. Generally biological supply houses are the best sources for unique seeds, for example, those of the carob tree, *Ceratonia siliqua;* the primitive cycad, *Zamia floridana;* and the coconut palm, *Cocos nucifera.* The names of seed suppliers are listed in Appendix B.

Drying and Storing Seeds

Most seeds, wild or cultivated, collected in the temperate zone should be dried before storage. Probably all seeds from supply companies are dried before they are packaged and sold. However, if seeds are purchased in bags that are not moisture-proof, the safest procedure is to dry the seeds again before storing them. The drying method is determined by weather conditions. In all cases, the seeds are spread into a thin layer within a tray or, preferably, over a fine-mesh screen that is placed on a rack to allow better circulation of air. On warm dry days, the seeds may be dried outdoors in direct sunlight for 5 or 6 h. When the indoor air is warm and relatively dry, the seeds may be processed by exposure to indoor air for several days. During cool or wet seasons, a better method is to heat-dry the seeds in an oven for about 4 h at 35 to 38°C. Higher temperatures will damage most seeds.

After drying, the seeds should be stored in a dry and, if possible, a cool or cold (near freezing) location. Many seeds, however, will retain good viability for several years in warm or even hot temperatures if they have been properly dried before storage and are kept in moisture-proof containers. Occasionally, special moisture-proof bags are available in seed stores; often the seeds are purchased already packaged in moisture-proof bags or in hermetically sealed cans. Otherwise, probably the simplest procedure is to store seeds in glass jars with tight screw-on lids and to keep the jars in a refrigerator at 6 to 10°C. Ordinary plastic bags are not satisfactory except for short storage periods of no longer than several months. All containers should be labeled with the name of the seeds and the date of their receipt.

Not all seeds tolerate a low internal-moisture content or a long storage period. Seeds of many tropical and subtropical plants will die if the internal-moisture content drops below a certain level. Citrus fruit seeds are not easily stored and should be planted soon after their removal from ripe fruit, thus illustrating how seeds of many tropical plants germinate almost at once in nature when released from the fruit to proper growing conditions. On the other hand, wild rice seeds must be stored in water at near-freezing temperature to retain a maximum germination percentage. These few examples illustrate how the natural environment of plants is reflected in the viability patterns of seeds. This subject, alone, of the evolutionary aspect of complementarity of environment and seed growth-patterns can represent fascinating topics for student explorations.

Seed Germination

All seeds begin germination by taking up water, a process called imbibition. Each seed type must absorb a fairly definite amount of water before germination will begin; the amount of water depends on the structure and composition of the seed. For most seeds, however, excessive moisture restricts respiratory activities, and germination may stop. As a general practice, most persons hasten imbibition by soaking seeds in water for 12 to 24 h and then transferring the seeds to a slightly moistened germination medium.

For those plants commonly grown in a greenhouse or laboratory, the optimum temperature for seed germination is in the range of 25 to 30°C. At room temperatures, which most often are between 20 and 23°C, satisfactory germination occurs in most seeds, although germination time will be increased by one or more days. Light has no influence on the germination of most kinds of seeds that are used for routine laboratory studies. Such seeds germinate equally well in the light or in the dark, although they are commonly placed in the dark or planted below the soil level to avoid loss of moisture due to light-generated heat. Other kinds of seeds have a definite germination requirement of light or darkness. Some species of garden cress (*Lepidium* L.), lettuce (*Lactuca* L.), and the sage (*Salvia* L.) show a definite requirement for light during germination. Grand Rapids lettuce seeds are often used to demonstrate a germination-light requirement in contrast to seeds of the Great Lakes lettuce which are not influenced by the light or the dark. The seeds of certain other plants, including species of *Phacelia*, *Nigella*, and *Delphinium*, are known to require the dark for germination. Light-sensitive and light-insensitive seeds may be obtained from biological or seed supply companies. (See Appendix B.)

It is important to recognize the interdependence and interaction of light, temperature, moisture, genetic traits, and other factors. Thus, a variation in one factor may change the effect of another in such a way that germination processes become enhanced or inhibited. Special requirements of plant types, as well as seed germination time, are indicated in Chap. 7 with the discussions of individual plants.

Procedures for germinating seeds. Seeds may be germinated in almost any type of container that provides the required conditions described above. To hasten germination, the seeds are soaked in water for 12 to 24 h before they are put into a germination chamber. A number of methods have been suggested for germinating seeds. A simple technique is to space the seeds over moist filter paper, paper toweling, or vermiculite in a covered container. When only a small number of seeds is involved (25 to 50), adequate containers are petri plates, bowls or pans covered with plastic film, or cardboard boxes lined with aluminum foil or polyethylene and covered with plastic film.

When a greater number of seeds are germinated, sets of rolled paper towels may be more efficient. Also, this method allows a person to obtain seedlings that are growing in straight and unmatted positions as shown in Fig. 6-21. To prepare a paper towel germinator, put two or three damp towels together in layers on a sheet of waxed paper or plastic film. Place the seeds about ½ to 1 in apart in a row across the layers of towels, with the embryo shoot of each seed pointed toward the top edge of the paper. Carefully cover the seeds with two or three more moist towels and roll the towels and waxed paper together as shown. Use string or rubber bands to fasten the roll, lightly but firmly, around each end and around the center. Put the roll in an upright position in a container with the lower end of the roll (seedling root end) in 1 or 2 in of water.

HANDLING SEED PLANTS

Starting plants from seeds is a more arduous task than buying the young annuals or the potted perennials at a nursery. In certain instances, however, to grow plants from seeds is a desirable part of a study or is necessary when unusual plants are wanted. The procedures for planting seeds depend on the future uses, which generally are: (1) The plants will be used in great numbers as seedlings or young plants for short-time laboratory studies; or (2) they are to become mature plants for long-time studies during which time they may flower and produce seeds. When only seedlings are needed, any of the techniques are suitable of those described earlier in this chapter for the germination of seeds.

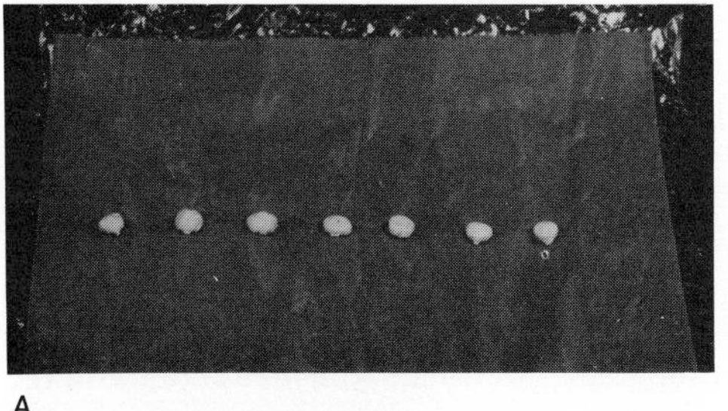
A

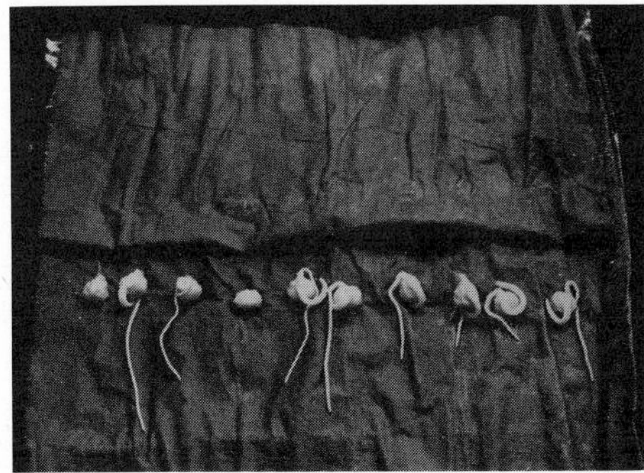
D

B

C

Figure 6-21 Germinating seeds in a roll of paper towels. **A** Soaked seeds on damp paper towels with a sheet of plastic film under paper towels. **B** Several damp towels are placed over seeds, and towels and plastic sheet are made into a roll. **C** The roll is loosely fastened with a rubber band at each end and placed in jar with water. **D** The opened roll of germinating seeds several days later.

Starting Plants from Seeds

When many young plants are required, the most efficient method is to plant the seeds in flats or trays and to thin out the young plants after they have sprouted. The trays should be 4 to 5 in deep and of a rigid material, such as heavy plastic or wood, so that the containers can be moved about without flexing and disturbing the young roots. Also, the weight of soil may cause a flexible container, such as a cardboard carton, to collapse when moved about. The trays must have bottom openings for drainage and should be sterilized by soaking them for 20 min in a solution of 1 part of household bleach and 9 parts of water. In the tray, place a ½- to 1-in layer of gravel and horticultural charcoal, mixed in the proportion of 2 parts of gravel and 1 part of charcoal. To this add a 2½- to 3-in layer of sterile soil mix.

Although sterile soil and pots are not a strict necessity for mature potted plants, sterilizing the tray and soil mixture is essential when growing seedlings. A fungal disease, commonly called "damping-off" and produced by the fungus, *Rhizoc-*

tonia, is always a threat to the fragile seedlings, causing the stems to break at the soil level and the plants to topple over and die. Apparently, the fungal spores are present in all soils, and the disease is practically impossible to control once it is started in a seed tray. Premixed, sterile soil may be purchased at garden stores, and this may be the simplest and least expensive method when only a few trays are required. Otherwise, a mixture can be prepared of 1 part, each, of garden loam, sand, and peat moss, which is then sterilized in an oven at 180°F for 30 min. Alternatively, the soil may be soaked with a fungicide, for example, a water solution of Captan, Ferbam, or Terrachlor, all prepared according to the instructions on the container.

Before planting seeds, set the tray in shallow water until the moisture is drawn through the bottom of the tray to the top of the soil, thus firming-down the soil. Avoid excessive wetness of the soil. Then add enough finely sieved, dry soil mixture to fill the tray to ½ in below the top of tray walls. With a trowel, a pencil, or the edge of a ruler, mark off planting rows spaced 2 in apart. Place seeds in the rows, two to five seeds to each inch depending on seed size, and cover with soil to a depth of about twice the thickness of the seed. Very fine seeds of grass, petunias, begonias, gloxinias, and others are not sowed in rows, but are sprinkled thinly over the soil and left uncovered. The side of a flat board is then used to gently press the fine seeds slightly into the soil. The large seeds of corn and beans, and the hard-coated seeds germinate more quickly if soaked in water 24 h before planting. Many persons presoak all types of seeds for at least 12 h. Label the trays or the seed rows, using a lead pencil or waterproof ink.

If the trays are in a humid atmosphere of a greenhouse, no cover is necessary. Otherwise, cover the trays with a glass plate or with plastic film. A simple method is to insert a tray into a plastic clothes bag with the folded end of the bag tucked under the tray. Place the trays in diffused light or in the dark, and inspect daily. The soil should remain slightly damp; if necessary, add water by gently spraying the surface with a hand bulb-sprayer or a mist bottle. If mold appears, immediately remove the moldy soil and any moldy seeds, and sprinkle sulphur or another dry fungicide over the area. Excessive moisture is conducive to mold growth; if water condenses inside the plastic cover, open the bag for several hours.

When at least half of the seed sprouts are above the soil, remove the cover and put the trays in indirect light. As the seedlings grow and toughen, gradually transfer them to a more intense light, avoiding, however, the intense heat of direct sunlight or artificial light. Add water less often and allow the surface to dry between watering, thus forcing the roots to grow downward into the soil. Remove weak plants and those of crowded areas to allow about 1-in space between each plant.

When many seeds are planted on a regular basis, an electric seed-starter provides an easy and almost foolproof method. Electric seed-starter kits may be purchased from the larger garden and seed stores, and from electrical companies, including Easy-Heat Company, 555 North Michigan Street, Lakeville, Indiana 46536. Ordinarily, a kit contains a plastic planting tray, soil cables, a thermostat, and a polyethylene cover. The cable is used to heat the soil, producing faster growth. There is reason to believe that fast-germinating seeds are not strongly affected by the damping-off fungus.

Transplanting Seedlings

Young plants grown in a tray do not need to be transplanted if they are used only for short-time studies. When mature plants are required, however, the young plants must be transferred from the starting trays to boxes or pots, or the seeds may be started in pots. Young plants should be transferred when the first foliage leaves (not the seed leaves, or cotyledons) open fully. This is generally 1 or 2 weeks after sprouts appear above the soil and when most plants are 2 or 3 in high. The transfers should not be made later than when the second set of foliage leaves appears, since the rapidly growing roots become matted within the tray and are damaged when taken from the soil.

Transfer one plant to a size 2 or 2½ pot, or several plants to a size 4 pot. With a pot of size 4 (or larger), the lower layer should consist of about 1 in of large-grained gravel or broken pieces of pottery (shard) which provide drainage and prevent the soil from running out the bottom openings. Many persons mix horticultural charcoal with the gravel to prevent a build-up of acid in the lower soil. For smaller pots of sizes 2 and 3, no drainage material is required, but a curved piece of shard should be placed over the drainage hole to prevent the loss of soil.

Add soil to the pot, building up the sides and leaving a hollowed-out portion for the plant. Before removing plants, water the tray well so that the soil clings to the roots during the transfer. Then insert a flat blade or, for small plants, a pointed stick at one side and well below the plant to pick up the plant roots and the surrounding soil. Gently support the seedling within the pot in such a way that the plant will be covered with soil to the same depth that it had in the tray, and with the soil filled-in to about ¾ to 1 in below the top of the pot rim. Water the pots thoroughly, but gently, to settle the soil around the roots. Cover the plants or put them in diffused light during the first week, after which place them in appropriate lighting for their particular type, as described in Chap. 7. At the end of the first week, feed with liquid plant food and continue to feed regularly every 3 to 4 weeks, according to directions on the plant-food container.

In place of trays, small peat pots made of compressed peat may be used for starting seeds. Two or three seeds are planted in each pot; at transplanting time the entire peat pot is buried in the soil of a larger pot. The roots grow through the walls of peat; thus damage to roots during transplanting is eliminated (Fig. 6-22).

Pruning and Shaping Plants

Most potted plants require trimming to produce a fuller growth and a proportioned shape. When a plant has started growing well, pinch off the main stem bud to promote a bushier growth of side branches and to produce larger flowers. For foliage that has small, inconspicious blooms, it may be desirable to remove all flower buds or flowers to gain large, showy leaves. *Coleus*, in particular, should have all flower buds and blooms pinched off, if large colorful leaves are wanted.

Pruning the stems is a more drastic treatment than removing buds. Nevertheless, if the pruning is a gradual process, many herbaceous and softwood plants can be cut back with no ill effect. Use sharp clippers and do the pruning in steps, several branches each week until the desired size and shape are gained.

A

B

C

Figure 6-22 Peat pots. **A** Small pots molded from peat moss. For planting, the pots are filled with a mixture of peat and vermiculite. **B** Compressed peat pellets that expand when soaked in water from a flat pellet to a 1½- to 2-in-high mass of peat encircled with a plastic netting. Several seeds are embedded in top surface and entire mass is placed in pot of soil after seeds sprout. Plastic netting is removed when peat and seedling are transferred to a pot of soil. **C** A group of expanded peat pellets with young pepper plants at stage to be inserted into pots of soil.

Asexual Propagation

Stem cuttings. This is a particularly desirable method for plants that become unsightly as they age, forming tough, bare-leafed stems, for example, *Coleus* and geranium. Take cuttings from side branches that are not blooming; select from well-developed branches and not from the old brown stems or the young light-green shoots. Use a sharp, clean blade, to remove a stem tip, 3 to 5 in long, which contains at least two leaf nodes. Remove the lower leaves and cut off any remaining portion of the stem below the lower node. New roots sprout from a node region; a fragment of internode tissue eventually dies and rots, and may slow the growth of new roots. Cuttings must not become dry and should be put at once in water or in the growing medium.

Stem cuttings of most herbaceous plants grow roots within 1 or 2 weeks in water alone. The rooted stems are then transferred from water to soil in the manner described for transplanting earlier in this chapter. Stem cuttings of softwood plants are rooted in a 3- to 4-in-deep tray of rooting medium which may be sand, vermiculite, perlite, peat moss, *Sphagnum* moss, or any combination of these materials. Before planting, set the tray in shallow water until the moisture is drawn through the soil to the top surface. Remove the tray before the soil becomes soggy.

With a pencil or sharp stick make planting holes in the medium to a depth of 1 or 1½ in, spaced far enough apart so that leaves of adjoining cuttings do not touch. Root growth can be hastened by dipping the cut end of the stem in a root hormone powder (from a garden store). Tap the stem with a pencil to remove excess powder, leaving only a thin coat on the stem.

Two rules must be observed for successful root growth on stem cuttings: (1) The cut end of the stem must rest firmly against the bottom surface of the planting hole. The necessary callus-growth forms only in response to pressure against the cut surface. (2) Once the cutting is in position, firmly pack the medium into the planting hole and around the stem. New roots do not grow if the stem remains loose and if it is occasionally rocked back and forth when the tray is touched or handled.

Cuttings require a high atmospheric humidity. If this cannot be provided in a greenhouse, construct a tent over the tray by bending two strips of wire into arches and placing one at each end of the tray. Drape a single layer of cheesecloth over the arches and down around the tray sides. Slip the tray and cheesecloth cover into a plastic bag. Close the bag opening with a rubber band, or fold the end under the tray. Put under artificial light with the lamps about 2 in above the tent; or put in a warm window that receives sunlight part of the day. Avoid a temperature above 24°C inside the tent. The cheesecloth diffuses the light, and the plastic film holds the moisture which condenses on the inner surface of the film. Generally, no additional water is necessary during the time required for the roots to grow. If the condensation disappears and the plastic film becomes dry, the bag must be opened to gently spray water over the tray.

At the end of 2 or 3 weeks, remove a cutting and examine it for the growth of a swollen callus at the cut end. If a callus has formed, probably the roots will develop in another week or two. Otherwise the callus and roots may require 2 or 3 months to develop. Continue to inspect a cutting at weekly intervals, each time returning the cutting to its position in the tray. When roots appear, open the plastic bag for 1 h the first day, and then for increasingly longer periods each day for a week. Finally, let the tray remain in the opened bag for a continuous period of 2 days, and then remove the tray from the bag and tent. Water the cuttings with a weak solution of plant food of about one-half the concentration given on the container for regular feedings. Put the tray under bright light for another week to adapt plants to a lower humidity and a greater light intensity. Then transfer cuttings to pots as described earlier in this chapter for transplanting.

Leaf cuttings. Plants with fleshy leaves and petioles, including practically all succulents and semisucculents, may be propagated easily by leaf cuttings. With semisucculent plants, select leaves that are fully developed, and use a sharp, clean blade to cut through the petiole about 1 in below the leaf. Dip the cut edge of the petiole in rooting hormone powder, and insert the petiole and the base of the leaf into a tray or pot of rooting medium of the same type as described above for stem cuttings. Put the container in a plastic bag and in indirect light. If only a few leaf cuttings are prepared, each may be put into a small pot and covered with a glass jar. Small new plants appear at the base of the leaves within several weeks. Do not remove the parent leaf until a cluster of plantlets has grown to about one-half the size of the original leaf. Then separate the small plants and put them in potted soil according to instructions for transplanting, given earlier in this chapter.

Propagating succulent plants by leaf cuttings is an easy chore and almost always successful. Remove a leaf with a portion of the petiole. Allow the cutting to air dry for 24 h and put into rooting medium according to the instructions given above for semisucculent leaves. Cuttings of fleshy cactus stems are propagated in the same manner except for one additional and important treatment. Before plant-

A

B

Figure 6-23 Asexual propagation in *Begonia*. **A** and **B** Method of slitting large veins on under surface of leaves. **C** After lightly dusting the slits with a root hormone powder, the leaf is held by stones against a soil mixture.

C

ing, allow the cut surface of the cactus stem to dry in air for at least 1 week. If the cut surface is not completely dry, most likely the cutting will rot when planted. Other treatments have been suggested and might be tried. Abraham (1967) recommends searing the cut end by passing it back and forth through a flame until dry. Johnston and Carriere (1964) suggest that the cut surface may be rubbed with charcoal until the "bleeding" stops.

Leaves with large, prominent veins may be propagated in a different manner. Make several cuts along the main veins on the ventral side of the leaf. Sprinkle a very fine layer of root hormone powder over the leaf's ventral surface. Tap the leaf to remove excess powder. Flatten the leaf on the rooting medium with the cuts against the surface of the medium. Pin down the leaf with toothpicks or small pebbles as shown in Fig. 6-23. Place in a plastic bag and treat in the manner described above for a leaf cutting.

Leaf cuttings of the *Sansevieria*, commonly called snake plant or mother-in-law tongue, are particularly interesting to study and may stimulate provocative questions and experimentation by students. (See the description of *Sansevieria* in Chap. 7.) The long, leathery blades are cut crosswise into 1½- to 2-in sections that are dipped into a root hormone powder and set at a slant in moist sand or other rooting medium (Fig. 6-24). Since all sections except the one from the leaf tip have two cut edges, it is natural to immediately ask, "Will roots grow from either end of a section that has two cut surfaces, or must the sections be oriented in their original up-and-down position? And if so, why? If not, will similar sections from other leaves respond in the same manner?" The tests are simple; the answers

Figure 6-24 Propagation of *Sansevieria* by stem cuttings. Sections of a long stem have been placed in a tray of peat, sand, and vermiculite. The tray was then transferred to a plastic-bag chamber, shown in Fig. 6-16.

may become very complex, indeed, particularly if a student becomes interested in the meristematic tissue and its role in the growth of various parts of a plant. In fact, studies with *Sansevieria* can lead to much speculation. For example, it might occur to a student to test for a propagation gradient from the base to the tip of the *Sansevieria* leaf, a test comparable to those made for the regeneration gradient from posterior to anterior ends of a planarian.

Dividing plants. By far the quickest and easiest method of propagation is to divide plants that have many shoots or that form small, new plants by offsets. Plants that form only one stem cannot be divided, of course. If the plant is very large, probably it must be knocked out of the pot by tapping the turned-over pot against the edge of a table. Pull apart or cut the roots and soil into sections, taking a portion of the shoots with each section. Use a clean, sharp blade, and dust the cut surfaces with sulphur or another fungicide before repotting, to prevent mold growth and perhaps loss of the plant. Plants with thickened underground roots and stems are separated in the same way.

Offsets from young plants may be cut from the parent plant and transferred to an individual pot. Both *Chlorophytum cosmosum* (spider plant) and *Sempervivum pectorum* (hen and chickens) are excellent examples of plants that produce offsets. Both are easily grown in the laboratory and will live under adverse conditions of light and humidity, although their growth may be slowed.

Air layering. Asexual propagation by air layering is not a common technique, but is useful at times with certain plants that develop long, bare stems. Students may become interested in the unique process, by which roots develop at the midlength of a stem on an intact plant.

Certain foliage plants, including *Dieffenbachia* and *Ficus* (rubber plant) tend to drop their lower leaves as the erect stem lengthens. Eventually, the plant consists of an unattractive tall, bare stem with a top cluster of leaves. Because the stems are thick and old, they cannot be easily rooted as stem cuttings. Therefore, the intact stem is induced to grow roots before the top portion with leaves is removed and potted.

Figure 6-25 shows the steps of the technique. To perform this, select a location on the stem at a distance below the leaves which will form an attractive plant

Figure 6-25 Propagation by air layering in *Ficus*. **A** A transverse cut is made halfway across the stem. **B** A toothpick has been inserted in the cut; **C** The stem has been wrapped with a layer of *Sphagnum*. **D** Finally, plastic film has been wrapped around the stem.

when removed. With a clean, sharp knife make a transverse cut halfway across the stem at that point. Now make a 2-in longitudinal cut, upward from the transverse cut and as deep as the transverse cut. At the top of the longitudinal cut, insert a small, clean piece of wood or a pebble to hold the cut open. Lightly cover the cut surface with a thin coating of a rooting hormone. Pack moist *Sphagnum* moss into the opening, and wrap more of the moist *Sphagnum* around the stem to cover the stem about 2 in above and below the cuts. Wrap a sheet of transparent plastic film around and over the moss. Tightly fasten the moss and plastic sheet at each end with twine or tape.

Within several weeks roots will grow from the stem and through the moss.

Eventually the roots will be visible through the plastic film. When roots have filled the *Sphagnum* cover, cut through the stem at a point below the roots. Transfer the removed top portion to a pot of fresh soil. Occasionally, when the unremoved stem is cut off at a point 2 to 3 in above the soil, the remaining lower portion will sprout and put on leaves again.

Watering Plants

Emphasis must be given to the importance of proper watering. Probably plants are lost, more than for any other reason, because of too much or too little water. Although plants differ in individual requirements, a general rule is to give enough water to soak through the potted soil and then to withhold water until the soil feels dry to a depth of one-half inch below the surface, at which time the pot is thoroughly watered again. Frequent, shallow applications may be detrimental to the plant, since roots will remain near the top surface. Many factors control the frequency of watering: room temperature, relative humidity, size of plant and its rate of growth, size and type of container, and the kind of potting medium. Growing and budding plants need more water; plants in bloom need less. More frequent watering is necessary if plants are in small pots or in clay pots. Plants in large containers, and those in plastic pots, require less frequent watering. Hanging plants in baskets need more moisture because of the warmer and drier air near the ceiling of the room. Use water that is at room temperature or slightly warmer. Add water at the top surface of the soil. Do not allow a potted plant to stand in a saucer of water.

Treatment during Plant Dormancy

Except for a few tropical species, most plants pass through periods of suspended growth. Apparently, dormancy in wild plants is most often related to survival during adverse seasonal changes, but it seems that this is not always the case with cultivated plants and those of tropical zones.

Annual plants, whether indoors or outdoors, give clear evidence of seasonal growth and change, and of survival through seeds. Indoor plants with corms, bulbs, or tubers also clearly indicate a dormancy period by their loss of foliage. However, dormancy is not always so easily recognized in perennials. Often a person may believe that a dormant plant is diseased and dying, and he may discard the plant. Since the requirements of a dormant plant are generally quite different from those of a growing plant, a person must recognize resting periods and give proper treatment at this time. Treatments vary considerably among plant species. In general, food, water, and light are reduced, or perhaps completely withheld, until the time of renewed growth. The requirements for individual plant types are described in Chap. 7.

Cleaning Foliage Plants

The growth and appearance of foliage plants are improved when their leaves are given a regular cleaning. Gas fumes and dust can form a thick surface film that is greatly detrimental to plant growth. Large, tough leaves may be cleaned with a cloth dipped in dilute detergent water. When the leaves are extremely dirty, place aluminum foil or plastic film tightly over the top of the pot and around the plant stems. Then invert the plant and hold the leaves under the detergent water while scrubbing them gently.

Foliage plants benefit from complete immersion of both the pot and the plant in water about once a month. Place the plant upright or on one side in a deep container of water, and leave it immersed for about 30 min. The treatment not only cleans away dust but helps to remove insects and other pests.

Many persons like to improve the appearance of foliage plants by polishing the leaves about once each month. Florists follow this same practice by wiping the top surface of leaves with a few drops of glycerin on a cloth. Although a number of products are marketed for this purpose, probably none is better than glycerin, and some products may be harmful. Do not wipe glycerin or the commercial products on the under surface of leaves, as the film may close stomata.

Care of Plants during Vacations

When cooling or heating systems are in operation, a greenhouse cannot be left unattended for even a day. During mild seasons when temperature-control equipment is not needed, possibly the greenhouse can go without attention for several days if vents are adjusted for air circulation. Otherwise, the fans, heaters, thermostats, electricity, water system, and practically all equipment and all facilities require a daily inspection. The regular, daily care of a greenhouse, then, becomes an important consideration, and one that generally involves the additional expense of wages for a caretaker.

When plants are kept in the laboratory or in an adjoining plant room, arrangements can often be made for a vacation period of at least several weeks. If the usual system of light and temperature control is continued within the building during a short vacation, the primary chore is to furnish moisture. Several arrangements are possible, the simplest being to put one or more potted plants into a plastic bag that is closed with a string or a rubberband. If a plant or its blooms are fragile, put several stakes in the pot to hold the bag away from the plant. Water the plant well before placing it in the bag, and leave the enclosed plant under its normal lighting situation. The bag holds the moisture for at least 3 weeks and perhaps longer. For large plants or vines, place the container in a plastic bag, and after thoroughly watering the plant, draw the top of the bag up over the soil and closely around the base of the stems. Tie the bag as tightly closed as possible without injuring the plants. Plants can be sustained for at least 2 weeks in this manner and perhaps longer.

An arrangement often used by home gardeners is to put the plant pots on clay bricks or on inverted clay pots in a sink or large tub with enough water to reach the tops of the bricks or the inverted pots. The water must not reach the base of the plant pots. The water soaks through the clay and into the bottom of the plant pots. The system is more efficient if the plants are also in clay pots and if a sheet of plastic film is fastened as tightly as possible over the sink or tub. If sufficient window light does not reach the plants, light must be made available from lamps on a timer. This system should provide moisture to the plants for at least 2 weeks. Another device used by both greenhouse and home gardeners is that of inserting a fiber-glass wick from the lower portion of soil in the pot, through the drainage opening, and into a reservoir of water below. Another method is to sink clay pots of plants into a box of damp peat moss, sand, vermiculite, or perlite.

Summer vacation periods of perhaps several months duration may present difficult problems, indeed, if regular school personnel are not available to care for plants. Some plants, including geraniums and *Coleus*, unless of special named types, are not worth carrying over indoors and probably should be discarded or transferred to outdoor beds. These can be replaced early the next autumn from local nurseries or transferred indoors by cuttings. Bulbs, corms, and tubers are dormant at this time and need little or no care. The annuals that flower during the regular school year have finished or nearly finished their blooming. They can be discarded or given away. Succulents, cacti, and other foliage plants, however, are often expensive, and probably a person will wish to save all of them. These should be carried through a summer vacation. The maintenance of these perennials becomes an individual matter that most often depends on whatever arrangements can be made. A desirable solution is for the school or department to furnish a part-time worker who can be trained to care for the plants. Otherwise, portable plants can be loaned during summer months to business offices and to friends, although plants, the same as animals, don't "like" to be moved about. As well, the plants may receive questionable care. Occasionally, arrangements can be made to take the most highly prized plants to a local nursery, where they will receive professional attention.

PESTS AND DISEASES OF INDOOR PLANTS

In large commercial greenhouses, where the same soil is used year after year, plant disease is reduced by sterilizing the soil, perhaps once each year. Ordinarily, this practice is not required for small greenhouses or for laboratory plants, where it is simpler to dispose of pots of soil with diseased plants, or to replenish the soil each year. When necessary, however, sterilize soil by moistening it thoroughly and spreading it in shallow pans that are put in an oven at a temperature of 180°F for 30 min. If pots and tools are heat-resistant they may be sterilized in the same manner, or they may be soaked in household bleach solution (1 part bleach and 9 parts water) for 20 min.

Infestations can be controlled or prevented by (1) proper conditions for plant growth; (2) constant watchfulness for pests and disease; (3) isolation and observation of new plants for at least a week; and (4) a regular biweekly or monthly cleaning of plant foliage (described earlier in this chapter). When only a few plants are lightly infested, a person may be able to pick off the pests with tweezers or to cut off the infested plant portion. If practical, isolate a plant as soon as its infestation or disease is discovered. Discard a heavily infested plant unless there is a particular reason to try to save it.

If diseased plants cannot be incinerated, treat them in another manner that will prevent later transmission of the pests or disease. Probably the simplest method is to heat the plant and soil in a dry oven at 180°F for 1 h. Other methods are to soak the materials in a 10% formaldehyde solution or in a solution of one of the chemicals mentioned below as a treatment for the pest or disease. Before reuse, all containers must be sterilized in the oven or, if the container is not heat resistant, by soaking it for several hours in a solution of 1 part household bleach and 9 parts of water.

As a rule, only a few diseases appear on indoor plants, including mainly the mildews and the "damping-off" fungus of seedlings (described earlier in this chapter). Both diseases are caused primarily by excessive dampness and are treated by sprinkling the area with dry sulphur powder or by spraying with a safe fungicide from a garden store. Infestations of insects and related pests are more common to indoor plants than are diseases. Treatments consist of dry powders, sprays, and dips. The sprays and dips are preferred because the concentrations may be better controlled, and because the wet solution adheres better to plant parts. Many types of commercial push-button sprays are available, some containing a combination of two or more pesticides. The instructions given on a container must be carefully read and followed.

Abraham (1967) suggests that the plant be laid on one side in a large carton (about 4 ft^3), and that the box interior and plant be sprayed from a distance of 18 in for 4 sec, or 1 sec/ft^3. After spraying, close the box for 2 min before removing the plant. Do not inhale the fumes when removing the plant.

All pesticides are toxic; when improperly used they may damage animals and man. The following precautions *must be observed.* Use pesticides only when necessary and handle them with care. Do not inhale the powders or vapors; do not contaminate the hands or other body areas. Wear gloves if much spraying is to be done, and wear a face mask in a confined area. If skin contamination occurs, immediately wash the area thoroughly with soap and water. If the chemical gets in the eyes, immediately flush with water for 5 min and get medical attention at once. If clothes become contaminated, remove them at once, and launder the clothes before wearing them again.

An excellent reference for a description of common insects and pests of indoor plants is the USDA Bulletin 67 (1969).

REFERENCES

Abraham, G.: 1967, *The Green Thumb Book of Indoor Gardening*, Prentice-Hall, Inc., Englewood Cliffs, N.J., 304 pp.

Barthelemy, R.E., J.R. Dawson, Jr., and A.E. Lee: 1964, *Innovations in Equipment and Techniques for the Biology Teaching Laboratory*, D.C. Heath and Company, Boston, 116 pp.

Bingham, C. D.: 1968, "The Culture and Use of Plants in School," *J. Biol. Educ.*, **2** (4): 353–363.

Johnston, V., and W. Carriere: 1964, *An Easy Guide to Artificial Light – Gardening for Pleasure and Profit*, Hearthside Press, N.Y., 192 pp.

Klein, R.M., and D.T. Klein: 1970, *Research Methods in Plant Science*, Natural History Press, Garden City, N.Y., 756 pp.

McDonald, Elvin, 1966: *The Flowering Greenhouse Day by Day*, D. Van Nostrand Company, Inc., Princeton, N.J., 158 pp.

McDonald, Elvin (ed.): 1971, *Handbook for Greenhouse Gardeners*, published by Lord and Burnham Greenhouse Specialists, Irvington-On-Hudson, N.Y. 10533, 107 pp. Approx. $1.00.

Miller, K.I., and A.S. Fagan: 1970, "Construction of a 'Growth Chamber'," *Turtox News*, **48** (2): 82–84.

Potter, Charles H.: 1967, *Greenhouse, Place of Magic*, E. P. Dutton & Co., Inc., New York, 255 pp.

Went, F.W.: 1957, *Environmental Control of Plant Growth*, Chronica Botanica, Waltham, Mass.

Other Literature

General

Butler, W. L., and R. J. Downs: 1960, "Light and Plant Development," *Sci. Am.*, **203** (6) 56–63.

Cherry, E.C.: 1965, *Fluorescent Light Gardening*, D. Van Nostrand Company, Inc., Princeton, N.J.

Cruso, Thalassa: 1969, *Making Things Grow*, Alfred A. Knopf, Inc., New York.

Gunther, P.P., and R.H. Wagner: 1972, "An Inexpensive Growth Chamber for Research and Teaching in Plant Science," *BioScience*, **22** (1): 32.

Hendricks, S.B.: 1957, "The Clocks of Life," *Atlantic Monthly*, **200**: 111–115.

Koller, Dov: 1959, "Germination," *Sci. Am.*, **200**: 75–84.

Mahlstede, J.P., and E.S. Haber: 1957, *Plant Propagation*, John Wiley & Sons, Inc., New York.

Martin, A.C., and W.D. Barkley: 1961, *Seed Identification Manual*, University of Calif. Press, Berkeley.

Mayer, A.M., and A. Poljakoff-Mayber: 1963, *The Germination of Seeds*, The Macmillan Company, N.Y., 236 pp.

McClure, D. S.: 1957, "Seed Characteristics of Selected Plant Families," *Science*, **31**: 649–682.

Nelson, K.S.: 1966, *Flower and Plant Production in the Greenhouse*, Interstate Printers & Publishers, Inc., Danville, Ill., 335 pp.

Pirone, P.P.: 1970, *Diseases and Pests of Ornamental Plants*, 4th ed., The Ronald Press Company, New York.

Reed, E.W.: 1966, "Instant Greenhouse," *Turtox News*, **44** (9): 230–231.

Reed, E.W., and E.P. Hill: 1969, "Using a Greenhouse Efficiently," *Am. Biol. Teacher*, **31** (6): 390–391.

U.S. Dept. of Agriculture, 1969, *Insects and Related Pests of House Plants*, Home and Garden Bulletin 67, U.S. Govt. Printing Office, Washington, D.C. 20402, 16 pp. 10 cents.

Walker, J.C.: 1969, *Plant Pathology*, 3d ed., McGraw-Hill Book Company, New York, 768 pp.

Collecting and Preserving Plants

Baker, G.E.: 1949, "Freezing Laboratory Materials for Plant Science," *Science*, **109** (2838): 525.

DeWolf, G.P., Jr.: 1968, "Notes on Making an Herbarium," *Arnoldia*, Jamaica Plain, Mass., **28** (8–9): 69–111. The article contains a bibliography of 217 listings. A copy of this very useful publication may be ordered from The Arnold Arboretum of Harvard University, Jamaica Plain, Mass. 02130. Price for single copy is 60 cents.

Lonert, A.C.: 1962, "Turtox Silica-gel Flower Preservative," *Turtox News*, **40** (2): 54.

MacDougal, T.: 1947, "A Method for Pressing Cactus Flowers," *Cactus and Succulent J.*, **19**: 188.

Pike, R.B.: 1964, "Plant Pressing with Plastic Sponges," *Rhodora*, **66** (766): 172–176.

"Preserving Botanical Specimens," *Turtox Service Leaflet No. 3*, General Biological, Inc., 8200 S. Hoyne Ave., Chicago, Ill. 60620.

Ruth, F.S.: 1965, "Laminating Method for Preservation of Herbarium Specimens," *Turtox News*, **43** (8): 212–214.

Steyermark, J.A.: 1968, "Notes on the Use of Formaldehyde for the Preparation of Herbarium Specimens," *Taxon*, **17** (1): 61–64.

Tehon, L. R.: 1966, *Pleasure with Plants*, Illinois Natural History Survey Circular 32, 32 pp. Order from Illinois Natural History Survey, Natural Resources Bldg., Urbana, Ill. 61801.

Uhe, G.: 1967, "A Portable and Expansible Drier for Herbarium Specimens," *Turtox News*, **45** (4): 98–101.

Journals

Horticulture, monthly, The Boston Horticultural Society, 300 Massachusetts Ave., Boston, Mass. 02115.

J. Cactus and Succulent Soc. Am., bimonthly, 132 West Union Ave., Pasadena, Calif.

Under Glass, bimonthly, for home greenhouse gardeners, P.O. Box 114, Irvington-On-Hudson, N.Y.

Workbench, bimonthly, Dept. 729, 4251 Pennsylvania Ave., Kansas City, Mo. 64111.

Correspondence Courses

Pennsylvania State University, College of Agriculture Extension Service, University Park, Pennsylvania. Write for bulletin that describes courses. Recommended courses are: Correspondence Course No. 139, *The Home Greenhouse* ($2.65 for 6 lessons); and Correspondence Course No. 144, *House Plants for Your Home* ($4.65 for 11 lessons).

7

SEED PLANTS FOR THE LABORATORY

Seeds and potted plants may be obtained from local sources or ordered from the catalogs of plant suppliers. The names and addresses of suppliers are listed in Appendix B.

Because many hybrids and cultivars are constantly in development by plant breeders, nomenclature for cultivated plants is often a difficult and confusing matter. Moreover, the many common names add to the confusion, as local gardeners and plant suppliers tend to identify by colloquial names only. The botanical names of this chapter are based primarily on Bailey's *Manual of Cultivated Plants* (1949), perhaps the most generally recognized reference for cultivated plants. All generic categories were cross-checked with those of Willis (1966). With only a few exceptions, the two sources agree in nomenclature; in all cases of disagreement, the nomenclature of Willis is used here. Descriptions of plants are given by the alphabetized generic names in three later sections of this chapter: (1) Gymnospermae; (2) Angiospermae: Dicotyledonae; and (3) Angiospermae: Monocotyledonae.

Categories of general uses for laboratory plants are given below. The lists are not all-inclusive, and many experienced biology teachers will have knowledge of other plants and other uses that might well be included. A number of unusual plants, which may have a more limited use, are not included in the following lists, although they are described in the later sections of the chapter.

GENERAL USES FOR PLANTS CITED IN THIS CHAPTER

(G = Gymnosperm; D = Dicot; M = Monocot)

Aromatic Plants: Fragrant Flowers or Scented Leaves

Boussingaultia (madeira vine)—D
Calonyction (moonflower)—D
Cestrum (night jessamine)—D
Citrus—D
Hyacinthus (hyacinth)—M
Lobularia (sweet alyssum)—D
Narcissus (jonquil)—M
Pelargonium (geranium)—D
Tagetes (marigold)—D

Evolutionary Convergence

Cacti—D
Euphorbia—D
Stapelia (starfish flower)—D

Fast-Growing Plants for Study of All Life Stages

Arabidopsis—D
Cucumis (cucumber)—D
Capsella (shepherd's purse)—D
Ipomoea (morning glory)—D
Lobularia (sweet alyssum)—D
Lycopersicon (tomato)—D
Nigella (fennel flower)—D
Petunia—D
Phaseolus (garden bean)—D
Pisum (garden pea)—D
Raphanus (radish)—D
Sorghum—M
Spinacea (spinach)—D
Tagetes (marigold)—D

Plants Grown for Flowers

Aglaonema (Chinese evergreen)—M
Amaryllis—M
Anthurium—M
Begonia—D
Bougainvillea—D
Bromeliads—M
Calonyction—D
Centaurea (cornflower)—D
Ipomoea (morning glory)—D
Lobularia (sweet alyssum)—D
Narcissus—M
Nigella (fennel flower)—D
Passiflora (passion flower)—D
Pelargonium (geranium)—D
Petunia—D
Rhoeo (moses-in-a-boat)—M

Cestrum (night jessamine)—D

Citrus—D

Colchicum—M

Cucumis (cucumber)—D

Euphorbia (poinsettia)—D

Hyacinthus (hyacinth)—M

Impatiens (garden balsam)—D

Salvia (sage)—D

Sinningia (gloxinia)—D

Stapelia (starfish flower)—D

Tagetes (marigold)—D

Tradescantia (wandering Jew)—M

Zebrina (zebra plant)—M

Foliage Plants

Aglaonema (Chinese evergreen)—M

Araucaria (Norfolk Is. pine)—G

Asparagus (asparagus "fern")—M

Bromeliads—M

Capsicum (pepper)—D

Chlorophytum (spider plant)—M

Citrus—D

Codiaeum (croton)—D

Coleus—D

Dieffenbachia (dumb cane)—M

Dracaena—M

Ficus (fig)—D

Maranta (prayer plant)—M

Monstera (split-leaf philodendron)—M

Philodendron (ivy)—M

Podocarpus (yew)—G

Sansevieria (snake plant)—M

Scindapsus (devil's ivy)—M

Genetic or Chromosomal Studies

Arabidopsis (mouse earcress)—D

Capsella (shepherd's purse)—D

Nigella (fennel flower)—D

Pisum (garden pea)—D

Rhoeo (moses-in-a-boat)—M

Sorghum—M

Tagetes (marigold)—D

Tradescantia (wandering Jew)—M

Vicia (broad bean)—D

Zea (corn)—M

Zebrina (zebra plant)—M

Plants for Hanging Baskets

Boussingaultia (madeira vine)—D

Bryophyllum (air plant)—D

Chlorophytum (spider plant)—M

Hedera (English ivy)—D

Philodendron (ivy)—M

Tradescantia (wandering Jew)—M

Zebrina (zebra plant)—M

Insectivorous Plants

Dionaea (venus flytrap)—D

Drosera (sundew)—D

Pinguicula (butterwort)—D

Sarracenia (pitcher plant)—D

Utricularia (bladderwort)—D

Mimicry Plants

Ariocarpus (living rock)—D

Lithops (stoneface)—D

Plants for Physiological Studies

Arabidopsis (mouse earcress)—D

Avena (oat)—M

Begonia—D

Bryophyllum (air plant)—D

Dieffenbachia—M

Impatiens (garden balsam)—D

Lemna (duckweed)—M

Lycopersicon (tomato)—D

Nigella (fennel flower)—D

Pelargonium (geranium)—D

Phaseolus (garden bean)—D

Pisum (garden pea)—D

Poa (grass)—M

Raphanus (radish)—D

Rhoeo (moses-in-a-boat)—M

Scindapsis (devil's ivy)—M

Sorghum—M

Spinacea (spinach)—D

Tradescantia (wandering Jew)—M

Zea (corn)—M

Zebrina (zebra plant)—M

Pollen, Pollination, and Fertilization Studies

Amaryllis—M

Cestrum (night jessamine)—D

Cucumis (cucumber)—D

Impatiens (garden balsam)—D

Lycopersicon (tomato)—D

Narcissus (jonquil; daffodil)—M

Phaseolus (garden bean)—D

Pisum (garden pea)—D

Rhoeo (moses-in-a-boat)—M

Sinningia (gloxinia)—D

Tradescantia (wandering Jew)—M

Zebrina (zebra plant)—M

Seed and Seedling Studies

Avena (oat)—M

Ceratonia (carob tree)—D

Cocos (coconut palm)—M

Raphanus (radish)—D

Ricinus (castor bean)—D

Sorghum—M

Glycine (soybean)—D

Phaseolus (garden bean)—D

Pisum (garden pea)—D

Poa (grass)—M

Vicia (broad bean)—D

Zamia—G

Zea (corn)—M

Sensitive Plants

Dionaea (venus flytrap)—D

Drosera (sundew)—D

Ipomoea (morning glory)—D

Maranta (prayer plant)—M

Mimosa (sensitive plant)—D

Pinguicula (butterwort)—D

Utricularia (bladderwort)—D

Succulent Plants

Aloe (medicine plant)—M

Bryophyllum (air plant)—D

Cacti—D

Crassula—D

Echeveria—D

Euphorbia—D

Lithops (stoneface)—D

Sedum (stonecrop)—D

Sempervivum (hen and chickens)—D

Stapelia (starfish flower)—D

Plants with Thickened Underground Parts

Amaryllis (bulb)—M

Chlorophytum (rhizome)—M

Colchicum (corm)—M

Hyacinthus (bulb)—M

Narcissus (bulb)—M

Sansevieria (rhizome)—M

Sinningia (tuber)—D

Twining and Climbing Plants

Boussingaultia (madeira vine)—D

Calonyction (moonflower)—D

Cucumis (cucumber)—D

Hedera (English ivy)—D

Ipomoea (morning glory)—D

Monstera (split-leaf philodendron)—M

Passiflora (passion flower)—D

Phaseolus (garden bean)—D

Philodendron (ivy)—M

Pisum (garden pea)—D

Scindapsus (devil's ivy)—M

Plants for Demonstrating Asexual Propagation

Begonia—D

Bryophyllum (air plant)—D

Bromeliad: Ananas (pineapple)—M

Cacti—D

Chlorophytum (spider plant)—M

Coleus—D

Crassula—D

Dieffenbachia (dumb cane)—M

Dracaena—M

Echeveria—D

Lemna (duckweed)—M

Pelargonium (geranium)—D

Philodendron (ivy)—M

Poa (grass)—M

Sansevieria (snake plant)—M

Scindapsus (devil's ivy)—M

Sedum—D

Sempervivum (hen and chickens)—D

Tradescantia (wandering Jew)—M

Zebrina (zebra plant)—M

GENERAL METHODS FOR CULTURING SEED PLANTS

Many plants described in this chapter may be cultured by the same method, designated here as the "standard culture method," and described below. Each item is discussed in greater detail in Chap. 6.

Standard Culture

SOIL MIX: 1 part each of garden loam, sharp sand (or perlite), and peat moss.

WATER: Water thoroughly when top ½ in of soil becomes dry. Reduce water during dormancy or rest periods.

FERTILIZER: Give liquid plant food to growing plants every 3 or 4 weeks. Reduce food when plant begins to bloom. Give no food during dormancy or rest periods.

HUMIDITY: 40 to 60%.

LIGHT: The following are the various levels of optimum intensities for growth or flowering in particular plant types. Most plants can be *maintained* at a lower intensity, although growth may be slow and flowering may not occur. See discussion of photoperiod in Chap. 6. The appropriate level is stated with the discussions of individual types later in the chapter.

Bright light. In east, south, or west window; protect from excessive heat in west window. Alternatively put under lamps of 1,000 ft-c or more for 14 to 16 h daily. Most plants require bright light to flower.

Medium light. North window, or several feet from east or south window, or under lamps of 500 to 1,000 ft-c for 14 to 16 h daily.

Low light. Any location within room except in sunny window, or under lamps with 50 to 500 ft-c for 12 to 14 h daily.

TEMPERATURE: These are optimum temperatures. Most plants tolerate a wider range. The appropriate temperatures for a particular plant type are included in the later discussions of individual genera.

	Day	Night
Cool	15 to 18°C	10 to 15°C
Moderate	18 to 21°C	15 to 18°C
Warm	21 to 25°C	18 to 21°C

Culture of Xerophytic Succulents

For desert cacti, desert euphorbs, and other xerophytic succulents, the standard culture method is modified as outlined below.

SOIL MIX: 2 parts of sharp sand (quarried sand), 1 part each of garden loam and peat moss.

WATER: Water thoroughly when top ½ in of soil becomes dry. Occasionally, allow soil to dry completely between waterings.

FERTILIZER: Same as for standard culture. Dormancy or rest periods generally occur during winter months when no food is given.

HUMIDITY: 30 to 50%. Many xerophytes tolerate a lower humidity.

LIGHT: Low to medium light for *maintenance;* medium to bright light for flowering, as described under standard culture.

TEMPERATURE: Day temperature of 20 to 25°C; night temperature of 15 to 20°C.

APPROPRIATE SEED PLANTS FOR THE LABORATORY

The descriptions of seed plants include technical terms that may be familiar only to a plant scientist. Therefore, a person may find it necessary to refer to a plant morphology text when reading the descriptions.

GYMNOSPERMAE

Several tropical or semitropical gymnosperms are useful as indoor plants. *Podocarpus* and *Araucaria* may be grown readily in the greenhouse or laboratory. Because of the endospermous, dicotyledonous seed, *Zamia* provides an interesting supplement to the studies of other dicotyledonous seeds.

Araucaria (Norfolk Island Pine; *A. excelsa* R. Br.)

USES: (1) For studying the foliage of a tropical gymnosperm. (2) As an ornamental, foliage plant.

DESCRIPTION: An attractive evergreen tree that is grown indoors in the juve-

Figure 7-1 *Araucaria excelsa*, Norfolk island pine; a tropical gymnosperm.

nile stage. Native to islands of South Pacific, where the pyramid-shaped tree may grow to 200 ft high. Graceful, whorled, plumelike, and flattened branches to 18 in long, extending from a central stem. Stiff, needle leaves, to ½ in long, spirally arranged on branches, and giving the appearance of fern fronds (Fig. 7-1). Mature leaves, a dark green; young leaves curved and sharp-pointed, and forming a light, green tip on branch. Outdoor trees mature to form staminate catkins, to 2 in long, and terminal ovulate heads that mature into large, woody, fruiting cones, 3- to 6-in diameter, with many two-winged scales, each scale with one seed.

CULTURE: Standard method (p. 250), with moderate temperature and medium light. A hardy plant that tolerates less than optimum conditions, except that regular watering is required. Grows 2 to 3 ft high in 12-in pot, or higher in a tub. May grow 6 to 8 in in 1 yr. When necessary to limit size, trim back side branches and remove top of main stem.

PROPAGATION: By a cutting from the top 4 to 6 in of the main stem. Cuttings from side branches will root also, but apparently do not make a nicely shaped plant. (See Asexual Propagation in Chap. 6.)

Podocarpus (Yew; *P. macrophylla* D. Don)

USES: (1) As an ornamental gymnosperm. (2) To study reproductive structures which form on the outdoor tree in southern and southwestern United States. Preserved reproductive structures can be obtained from biological supply companies.

DESCRIPTION: An evergreen tree of Japan and warm regions of southern hemisphere. Grows 40 to 50 ft high outdoors. Small horticultural types are used as ornamental foliage plants that require only low to medium light (Fig. 7-2). Bright green leaves, ½ in wide and 2 to 3 in long. Dioecious (male and female reproductive structures are on separate plants). In mature outdoor trees, the receptacle ripens into a small, fleshy, green or purple fruit with a naked seed on top of fruit, accounting for the name: Podo (foot), carpus (fruit).

CULTURE: Standard method (p. 250), with low, medium, or bright light and moderate or warm temperature. Indoor plants grow slowly, to 6 to 8 ft in a 10- or 12-in pot. Trim back plant as necessary to restrict size. Hardy indoors and easily cultured as a pot plant.

A

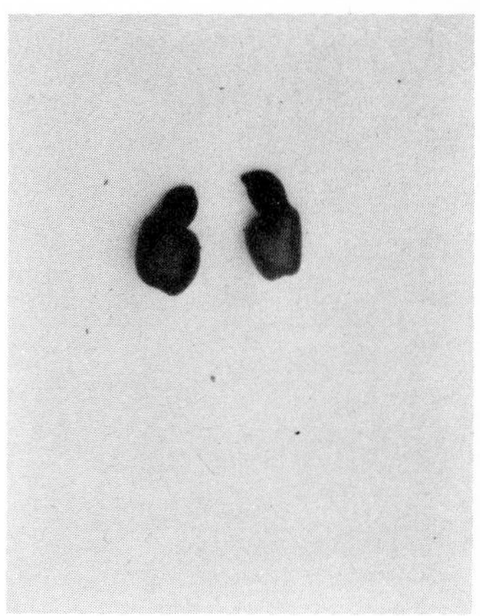

B

Figure 7-2 *Podocarpus macrophylla*, yew, a tropical gymnosperm. Juveniles are grown indoors as ornamental foliage plants. **A** Mature outdoor tree. Leaves, 6 to 12 in long on outdoor plants; male reproductive structure at top center. **B** Two seeds with fruit; naked seed at top and fruit at bottom of each, accounting for the generic name, podo = foot, and carpus = fruit.

PROPAGATION: Difficult to propagate indoors. Might try to root stem cuttings, using a root hormone.

Zamia (*Z. floridana* DC.)

USES: (1) For study of seed structure and its germination. (2) If ovules are available, they can be studied for the large female gametophyte and egg, that are readily visible when seed coat is removed and a median longitudinal cut is made through ovule (see Fig. 7-3). Preserved ovules may be obtained from biological suppliers. (3) Possibly as an ornamental gymnosperm.

DESCRIPTION: *Zamia* represents one of ten remaining genera of cycads that 200 million years ago constituted a large portion of the earth's flora. Today about 100 species grow in widely separated areas, mostly in the tropics. *Zamia floridana*, found in southern Florida, is the only species native to the continental United States.

Low, fleshy, trunklike stem, 6 to 12 in above soil. Pinnately compound leaves, about 4 ft long and curved downward similar to palm. (Palms are monocotyledonous angiosperms, however.) Dioecious, and pollen (male) and seed (female) cones are borne on separate plants. Mature pollen cone, 3 to 4 in long; mature seed cone, 4 to 6 in long, borne on a short stem at the base of leaves. Many megasporophylls, each with two ovules. Wind pollination; about 5 months interval between pollination and fertilization. Mature seed about 1 in long. Seed coat of three layers; bright orange and leathery outer layer; tough, stony, middle layer; and thin, papery, inner layer. Embryo with two cotyledons and a large mass of nutrient materials from the female gametophyte (Figs. 7-3 and 7-4).

Figure 7-3 Seed of *Zamia* sp.; entire seeds showing leathery, orange outer coat. Each seed is about 1 in long.

CULTURE: Purchase a plant or start from seed. Hooft (1970) reports that seeds germinate in about 48 h if the seed coat is removed and the seed is half buried in a horizontal position on vermiculite. Cover the container with a glass plate or plastic film to increase humidity. The germination process can be observed through the transparent cover. Hooft describes the methods for germination. When the first leaf has developed and opened, in 20 to 25 days, transfer the seedling to soil. Culture by standard method (p. 250), in medium light and moderate or warm tem-

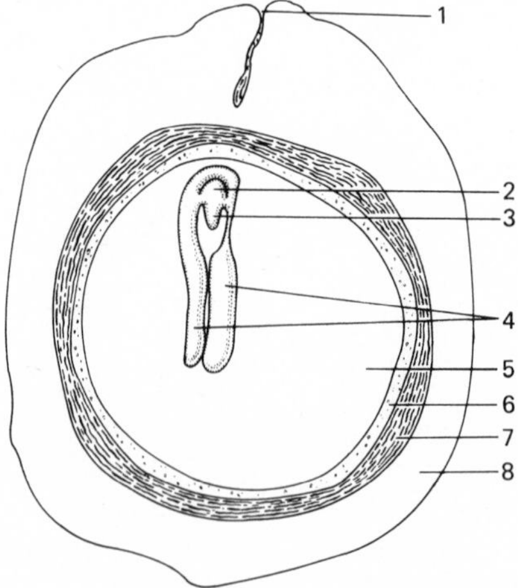

Figure 7-4 Median section of *Zamia* seed; 1, micropyle; 2, radicle or root of embryo; 3, plumule or shoot of embryo; 4, cotyledons; 5, female gametophyte; 6, inside fleshy layer of coat; 7, stony layer of coat; 8, outer, leathery layer of coat.

perature. Reduce water during winter months when the plant is resting. Very slow growth; cones are formed during ninth or tenth year.

PROPAGATION: From seeds.

ANGIOSPERMAE: DICOTYLEDONAE

Arabidopsis (Mouse-earcress; *A. thaliana* Heynh.)

USES: (1) A fast-growing plant for study of complete life cycle. (2) Much used in genetic studies; sometimes called the "botanical *Drosophila*"; $n = 5$. (3) Callus can be cultured in an artificial medium.

DESCRIPTION: A small desert annual, much used in plant research. Grows in dry fields, along roadsides, and in waste areas from Massachusetts to Michigan, and from Illinois to the Southern states. Life cycle from seed to seed is 30 to 35 days. Mature plant is 2 to 20 in high. Narrow, oblong leaves form a rosette. Inflorescence, a raceme. Small, perfect, inconspicuous flowers; white petals, 3 to 4 mm long. Superior ovary. Self-fertilizing. Very small seeds, 0.1 to 0.5 mm long. Fruit, a silique; 1,500 to 1,600 seeds per plant.

CULTURE: Standard method (p. 250), with bright light and moderate temperature. Acid soil, pH 5 to 6.

PROPAGATION: Start seeds at any time of year. Seeds germinate in 2 to 3 days.

Begonia

USES: (1) For ornamental foliage and flowers. (2) To demonstrate asexual propagation by leaf and stem cuttings, and particularly by the technique of slitting the large leaf veins. (3) Study of stomatal patterns. (4) The trailing begonias are excellent for hanging baskets.

DESCRIPTION: Over 900 tropical and subtropical species. Many horticultural types, with highly descriptive common names and comprising some of our most handsome, ornamental plants.

Perennial herbs with watery stems that may grow 1 to 3 ft high. Alternate leaves, waxy or glossy, and asymmetrical, with one side larger than other, giving a lopsided appearance. Inflorescence, a cyme, with staminate and pistillate flowers. Generally, staminate flowers bloom first. Staminate perianth with two petals and two petallike sepals. Pistillate perianth with two to five petals and sepals, collectively. A winged, inferior ovary. Fruit, a winged capsule, with numerous minute seeds.

Three general root types: fibrous, rhizomatous, and tuberous. The first two types are easily grown and produce flowers or showy leaves the year round.

Fibrous-rooted types. Common house plants that are easily grown. Bloom year round. *Begonia semperflorens* Link & Otto, the "Wax Begonia," consists of a large group with single or double, pink, white, or red flowers. Other fibrous types that are easy to grow include those with the common names of (1) "angel wing," with 2- to 4-in glossy, green or red, lopsided leaves that account for the name; and (2) "Indian maid," with 2- to 4-in, bronze-colored leaves (Fig. 7-5).

A

B

C

Figure 7-5 *Begonia*, fibrous-rooted types. **B** *B. semperflorens* "Cinderella," a wax begonia. **B** Close-up of flower from plant shown in **A.** **C** Angel-wing begonia ("Corallina de Lucerna"). (*A and C Merry Gardens, Camden, Me.*)

Rhizomatous types. Commonly grown is the *B. rex-cultorum* Bailey, often called "Rex Begonia." Most have large, showy leaves of rich colors, and varied textures and patterns. Many with a metallic sheen in leaves, and some with hairs on stems and leaf undersurface. Bloom in winter and spring. An unusual type is *B. sanguinea* Raddi, or "Beefsteak Begonia," with leaves that are a blood-red coloring on the undersurface and a smooth, shiny green on top surface. See Fig. 7-6.

CULTURE: Standard method (p. 250), with medium light and moderate temperature. When plant reaches desirable height, pinch off stem tip for bushier growth and larger flowers. Double-flowered varieties give longer lasting blooms.

PROPAGATION: By stem and leaf cuttings; from seeds. Germination time is 14 to 21 days.

Bougainvillea (*B. glabra* Choisy)

USES: For study of floral parts, particularly the colored bracts that enclose the flowers.

DESCRIPTION: Tropical and subtropical vine that can be grown over a window frame. Ovate, petioled, alternate leaves. Small, inconspicuous flowers. General-

Figure 7-6 *Begonia*, rhizomatous types. **A** *B. sanguinea* "Bessie Buxton," beefsteak begonia. **B** *B. sunderbruchi*, finger leaf begonia. (*A and B Merry Gardens, Camden, Me.*)

A

B

Figure 7-7 *Bougainvillea* sp. A vine with colored floral bracts (leaves) enclosing small flowers. (*Courtesy of CCM: General Biological, Inc., Chicago.*)

ly, three flowers enclosed by three large purple, red, pink, or white bracts (leaves), that are produced year round. Flowers in cymes; five petals; five to ten stamens; superior ovary, one long style. Fruit, an achene. See Fig. 7-7.

CULTURE: Standard method (p. 250), with bright light and warm temperature. Usually shows vigorous growth and requires regular pruning for desired shape and size. At times, covered with dense growth of long-lasting and colorful bracts. In early spring, trim vines back to about 12 in long. If possible, put outdoors in partial shade during summer. Periodically throughout the year, the plant becomes semidormant for several weeks. Give no plant food during these periods and reduce water. Start food and normal watering again when growth of stems and leaves reappears.

PROPAGATION: Take stem-tip cuttings during March or April. Cuttings may require 3 or 4 months to grow roots.

Boussingaultia (Madeira or Mignonette Vine; *B. gracilis* Miers)

USES: (1) Fragrant vine for ornament and for regular supply of plant parts. Excellent for hanging baskets. Hardy and easily grown in situations where time is limited and conditions are less than optimum for other plants. (2) To study twining action of stem.

DESCRIPTION: A hardy, vigorous vine, that generally needs regular trimming to limit the size. Blooms during summer and into late fall, with 1-ft long racemes of small, fragrant, white flowers. Perennial, herbaceous, twining vine that grows 10 to 20 ft in one season. Tuberous root. Little tubercles form in leaf axils. Ovate leaves, 1 to 3 in long. Plant is not known to set fruit.

CULTURE: Standard method (p. 250), with bright light and warm temperature. Obtain tubers locally or by catalog. Put one tuber in no. 6- or 8-size pot. Cover tuber with 1-in layer of soil. Train the vine on stakes, cord, or a wire frame. Cut off stem tip to limit growth and to increase flowering. If roots become potbound, transfer to larger pot of fresh soil.

PROPAGATION: Use a sharp blade to divide tuberous root into several portions. Transfer each root portion to a pot of fresh soil.

Bryophyllum (Air Plant; *B. calycinum* Salisb.)

USES: (1) To demonstrate the unusual type of propagation by young plantlets formed in notches of leaves. (2) To study xerophytic traits, including the sunken stomata in waxy cuticle.

DESCRIPTION: Perennial, branched xerophytes (Figs. 7-9). Commonly in laboratories, where it is used to demonstrate the propagation by producing young plantlets in the notches of leaf margins. Leaf comprised of three to five fleshy leaflets, each 3 to 4 in long. Generally a detached leaf will continue to produce plantlets, thus accounting for the common name "air plant." May produce flowers if kept in sun or bright light (800 to 1,200 ft-c). Clusters of yellow, red, pink, or purple flowers. Four sepals; four petals fused into a tubular corolla; generally eight stamens and four pistils. Fruit a follicle with very small seeds.

CULTURE: See *Culture of Xerophytic Succulents* (p. 251).

PROPAGATION: By leaf cuttings or from plantlets at any time of year. Plant seeds in March or April. Germination time is 7 to 14 days. Very fine seeds; sow on soil surface.

Figure 7-8 *Bryophyllum* sp. Entire plant.

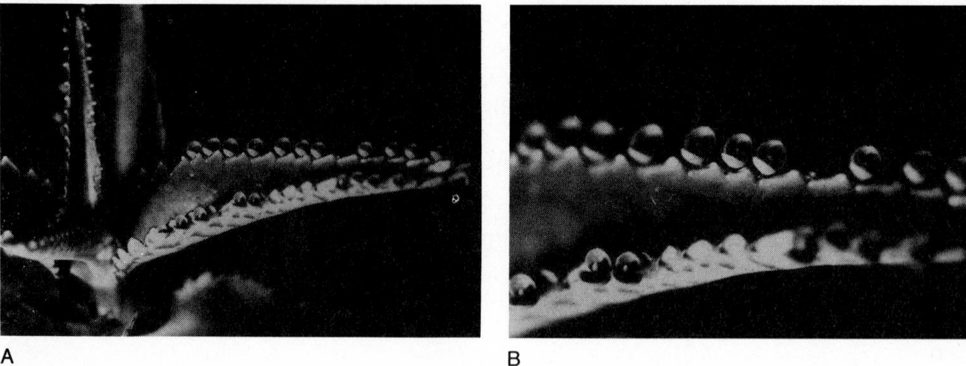

Figure 7-9 *Bryophyllum* sp. **A** Close-up of leaf with plantlets on margin. **B** A close-up of plantlets.

The Cacti

USES: (1) To study (a) adaptation to desert life (xerophytes); (b) to study jungle and mountain life (epiphytes). (2) To illustrate by grafts the close relationship of cacti genera. (3) As hardy ornamental plants that require little care. (4) Use rock cactus *(Ariocarpus)* as an example of plant mimicry.

DESCRIPTION: General traits of the cactus family are given here. The identifying characteristics of 12 genera are outlined in a following section. Many other types are suitable for indoor culture and are described in plant catalogs. The hybrids are particularly desirable because of their large, very attractive flowers.

Cacti are succulent plants of the western hemisphere, distinguished from other succulents by the large, fleshy, and generally leafless stems, and by clusters of spines (on most types) that develop within areoles (small, round, or oval pits) arranged spirally on the stem. Some with small, barbed bristles, glochides (pronounced glō kĭds), for example, the prickly pears *(Opuntia)*. Most genera with prominent ribs, or with tubercles on stems that may be tubular, globular, or flattened. Most cacti have shallow root systems of long, fleshy roots. Stems are covered with a thick cuticle; stomata are sunken. Inner tissue is mostly parenchymatous for water storage.

Large, showy flowers, usually perfect, with inferior ovary. A short or long floral tube covered with scale leaves. Many sepals and petals that intermix and may be the same color. Many stamens. One style with a lobed stigma. Many ovules in the one-chambered ovary. Fruit, a fleshy or dry berry that is edible but not always palatable. See Fig. 7-10.

CULTURE: General requirements of most types are given here. Specific requirements are given later with the descriptions of individual genera. All types included here are hardy indoors and at times may be maintained under less exact conditions than those described.

Desert cacti. Soil mix: 2 parts sharp sand, 1 part each of garden loam and peat moss. Clay pots are recommended to ensure no buildup of water around roots. If plastic or glazed pots are used, add 1 part of perlite to soil mix, or double the

A

Figure 7-10 Large, showy cactus flower. **A** Side view showing long, floral tube covered with scale leaves (*Opuntia* sp.). **B** Open flower (*Opuntia* sp.)

B

amount of sand. Medium to bright light. Most types begin active growth in spring, and bloom in summer. During this time, add water when top ½-in soil layer becomes dry (probably daily in midsummer), and give liquid plant food every 3 or 4 weeks. Occasionally, allow soil to dry completely between watering periods. Day temperature: 21 to 30°C; night temperature: 15 to 20°C. During summer vacations, plants can be put outdoors in semishade where, if necessary, they can be left with only little care. Under these conditions, they probably will not bloom, and they will die if not watered occasionally. Decrease water and food during resting periods of the winter months.

Jungle forest cacti. These cacti are found in humid-moist, tropical forests or in cool mountains. Several genera are epiphytes, growing on forest trees along with bromeliads, orchids, and ferns.

Soil mix of 1 part each of sandy loam, coarse leaf mold, and sand. Add water when top soil surface becomes dry. Day temperature: 20 to 22°C; night temperature: 13 to 19°C. Medium light indoors, or semishade outdoors. Most types grow and bloom during winter and spring. During this time, they require liquid plant food every 3 to 4 weeks. Reduce food during summer and fall months.

PROPAGATION: All may be propagated by cuttings, by grafting, and by seeds. Because of slow growth, generally plants are not started from seeds except for special studies. Most types grow easily and quickly from cuttings. Cuttings are made during the growing season when the plants are active.

Grafting cacti. Grafting together two or more types of cacti may produce unusual and interesting plants. The simple techniques are illustrated in Fig. 7-11. Two types of grafts are possible: a *flat graft* and a *cleft graft.* For a flat graft, cut horizontally across the rootstock and the scion (portion that is transferred) to provide flat, cut surfaces, that are then bound tightly together with rubber bands or cord. Be certain to make a clean cut with a sharp blade. Do not use a sawing motion. Keep the cut surfaces clean, and put the two surfaces together immediately to avoid drying. Even a slight drying may result in graft failure. Sprinkle or brush sulfur over any exposed cut surfaces after the two are put together. Cover the graft with a paper bag to protect from dirt and dryness.

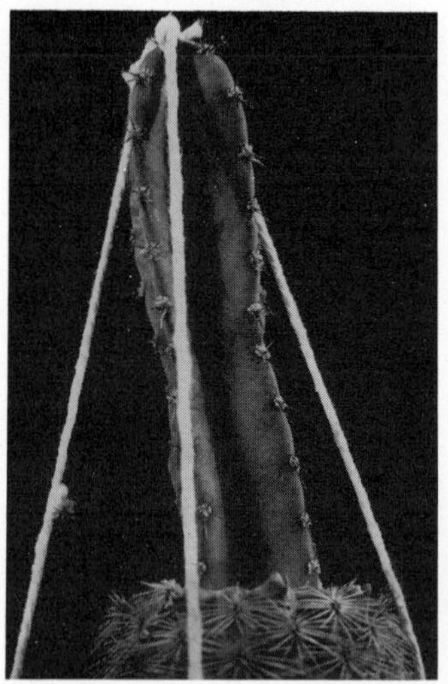

A

B

Figure 7-11 Techniques for grafting cactus. **A** A flat graft showing top scion placed on lower stock. Powdered sulfur has been dusted on exposed areas of stock, and the two portions have been tied together. **B** A cleft graft showing the top scion fitted into the cleft of the lower stock. A cord has been tied around the two portions to hold them together.

Figure 7-12 *Aporocactus* sp., rattail cactus. A portion that shows resemblance to rat tails.

For a cleft graft, make a 1- to 1½-in perpendicular cut (cleft) in the rootstock, as shown in Fig. 7-11. Trim the scion to fit the cleft, and immediately put the cut surfaces together. Bind the stock and scion together; sprinkle sulfur over any exposed cut surfaces, and cover the graft with a paper bag. With either type of graft do not move or shake the pot for at least 1 week. The cut surfaces will begin to unite within 5 to 7 days. Take care not to handle the plant roughly and not to splash water into the joined surface until the two portions are well fused.

Aporocactus (*Rattail cactus; A. flagelliformis* Lem.). Vinelike cactus. Slender and creeping or long and pendant stems, with 10 to 12 close ribs (Fig. 7-12). Covered with small red to brown spines that account for the name. Aerial roots develop from stems. Pink or red flowers, 1 to 3 in long. Small, red-brown fruit with brown seeds. Stems are weak and break easily. An old favorite as an indoor plant because of curious taillike stems. Hardy and easy to grow. Culture as for forest cacti, described earlier in this section.

Astrophytum (*sea urchin cactus; A. asterias* Lem.). Low and dome-shaped; about 1 in high and 2 to 3 in in diameter; starlike depression (Fig. 7-13). No spines. Eight prominent ribs, each with a row of woolly areoles. Greenish, coppery color and covered with white scales. Flowers 1 in long, and yellow with red throat. Small, round berry-type fruit, covered with scales and wool; dark-brown seeds. Hardy and attractive plants. Culture as for desert cacti, described earlier.

Astrophytum (*bishop's cap; A. myriostigma* Lem.). Globular as a juvenile,

Figure 7-13 *Astrophytum* sp.

becoming cylindrical at maturity. Usually five prominent ribs. Covered with white scales. Spineless. Grows slowly but eventually reaches about 4- to 6-in diameter and 12- to 20-in height. Bright, yellow flowers, 1 to 2 in across. Fruit similar to *A. asterias*. Hardy and easily grown. Culture as for desert plants, described earlier.

Ariocarpus (*living rock; A. fissuratus* Schum.). Plants resemble rocks and are exceedingly difficult to locate on rocky ground except when blooming. Much of stem is underground. Above ground are circles 4 to 6 in across, of overlapping triangular-shaped tubercles, that appear to be pointed leaves (Fig. 7-14). Surface of tubercles is warty and with a tan, leathery appearance. Thick, leathery wool at top, where white to pink flowers, 1 to 2 in across are borne. Small green fruit and black seeds. Plants need very porous soil, obtained by adding sand. Let soil throughout pot become moderately dry, and then water thoroughly. Otherwise, culture as for desert cacti, described earlier.

Cephalocereus (*old man cactus; C. senilis* Pfeiffer). Erect, slender, columnar stem that may become branched and 20 to 30 ft high in nature. Small plants, 4 to 6

Figure 7-14 *Ariocarpus fissuratis*, ing rock that mimics rocky surroundings and is difficult to locate on the ground.

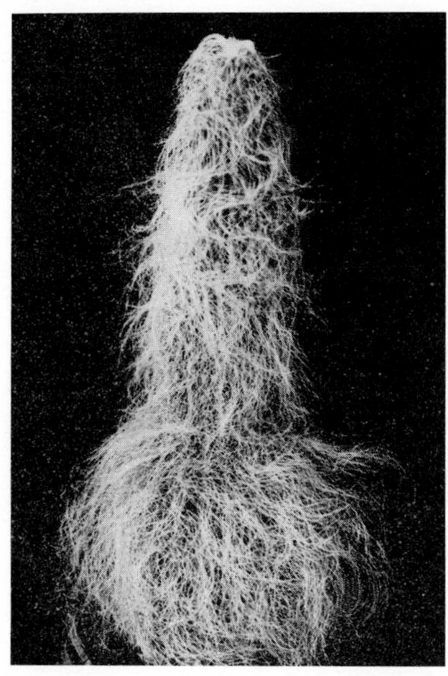

Figure 7-15 *Cephalocereus senilis*, old man cactus.

in high, are used indoors (Fig. 7-15). Long, white hairs grow from areoles at top, or hairs may grow over entire stem of small, indoor plants. Yellow spines. Mature plants form rose-colored flowers that open at night. Slow growing. Let soil throughout pot become moderately dry and then water thoroughly. Use very porous soil. Otherwise, culture as for desert cacti, discussed earlier.

Chamaecereus (*peanut cactus; C. silvestrii* Britt. and Rose). A small cactus, native to Argentina. Plant forms a cluster of short, cylindrical branches, 1 to 2 in high, somewhat peanut-shaped, and covered with soft, white spines. Stem shows six to nine ribs. Large pink to scarlet flowers, 2 to 3 in long, arise from sides of stem. Dry, woolly, small globular fruit. Black seeds. Culture as for desert cacti, described earlier.

Echinocereus (*hedgehog or cob cactus; E. reichenbachii* Haage Jr.). One of the easiest cacti to culture and to bring to flowering. Large, attractive flowers begin at early stage of growth. See suppliers' catalogs for other species and wide range of horticultural types, for example, the rainbow cactus (*E. melanocentrus*), with large pink to rose flowers. Cylindrical stems, 1 to 8 in high, and 1 to 3 in in diameter, sometimes in clusters (Fig. 7-16). Covered with short spines. Ribs, 12 to 18. Fragrant, purple flowers, 2 to 3 in long. Green fruit, ½ in long. Black seeds. Culture as for desert cacti, described earlier.

Epiphyllum (*night-blooming Cereus or orchid cactus; E. phyllanthus* Haw.). Many hybrids have been developed from several epiphytic genera, including *Epiphyllum* and others. A person must study the suppliers' catalogs to learn about the wide range of flower colors that is available. Often plant hobbyists call the orchid cactus simply "epiphyllum cactus." The stem of hybrids is similar to *E. phyllanthus*, described here, although flower color, size, and blooming time

Figure 7-16 *Echinocereus* sp.; hedgehog or cob cactus. (*Courtesy of CCM: General Biological, Inc., Chicago.*)

may differ. As well, many orchid cacti have spines on long pendant stems, which, unless confined, may fasten to the clothing of persons working around them. Long, flattened, upright stems (*E. phyllanthus*), 2 to 3 ft long and 3 in across, with toothed margins (Figs. 7-17, Fig. 7-18). Large, fragrant, white flowers, 10 to 12 in long, and night-blooming. Red fruit, 2 to 3 in long. Many black seeds. Culture as for forest or jungle cacti, described earlier.

Hylocereus (*night-blooming Cereus; H. undatus* Britt. and Rose). One of the largest night-blooming epiphytes. Climber with aerial roots. Three-sided stems, 3 to 8 in long, with wavy margins and small spines. Beautiful, night-blooming, white flowers, 10 to 12 in long. Fleshy and edible red fruit, 3 to 4 in long. Perhaps the most popular of all night-blooming cereus plants. Fasten stems to stakes or trellis. Culture as for forest or jungle cacti, described earlier.

Mammillaria (*pincushion cactus*). Many species are available. Graf (1970) describes more than 150 species, all suitable for indoor culture. Most types are at-

Figure 7-17 *Epiphyllum* sp., orchid cactus. Entire plant approximately 4 ft across.

A

B

Figure 7-18 *Epiphyllum* sp. **A** Close-up of stem with flower. **B** Close-up of flower.

tractive and hardy. Globular or cylindrical stems, generally small, 1 to 4 in in diameter, and covered with hairs, bristles, and spines that are often white (Fig. 7-19). Surface of stem is broken into small mammary-like tubercles (swellings) which account for the name. Flowers 1 to 2 in across, yellow, red, orange, white, or pink, borne on upper sides of stems. Clusters of these small globules are unusually attractive, as indicated by their common names, for example, snowball pincushion, scarlet ball, pink powder puff, and golf ball. Culture as for desert cacti, described earlier.

Figure 7-19 *Mammillaria* sp., pincushion cactus.

Figure 7-20 *Opuntia* sp., prickly pear.

Opuntia (*prickly pear; O. ficus-indica* Mill.). Many species and horticultural types are available. Most types produce large, showy flowers and edible fruit. All are hardy and easily grown as small indoor plants. In nature, *O. ficus-indica* is bushy, or may become a tree, 10 to 15 ft high, sometimes cultivated in the tropics for its fruit. Stem is jointed into thick, flat, fleshy pads up to 12 in long (Fig. 7-20). Spineless, or with small, white or yellow spines. Young plants have small, fleshy leaves that fall off as plant matures. The deciduous leaves indicate that *Opuntia* is primitive to leafless cacti. Yellow flowers, 3 to 4 in across, borne toward upper part of stem pads. Purple or red fleshy fruit, 2 to 3 in long, and edible. White, flat seeds. Culture as for desert cacti, described earlier.

Pereskia (*lemon vine or Barbados gooseberry; P. aculeata* Mill.). A primitive cactus with leaves on woody stems. Resembles a woody shrub or vine. Young plant is an erect shrub that eventually forms vinelike stems, up to 30 ft long in nature (Fig. 7-21). Spines on lower part of stems. Oblong leaves, 2 to 3 in long. Clusters of white, yellow, or pink flowers, 1 to 2 in across, and lemon scented. Planted in tropics for small yellow fruit about ¾ in in diameter. Useful in labora-

Figure 7-21 *Pereskia* sp., lemon vine or Barbados gooseberry. A primitive cactus that bears leaves on a woody stem and short spines on proximal end of stem.

A

B

Figure 7-22 Other cacti easily maintained indoors. **A** *Notocactus* sp., yellow ball. Note diagonal line of yellow blossoms. **B** *Homalocephala* sp., horse crippler. The long, stout spines (2 to 3 in long) are avoided by horsemen.

tory as an example of a primitive cactus, and as an ornamental plant. Culture as for forest or jungle cacti, discussed earlier. See other cacti in Fig. 7-22.

Calonyction (Moonflower; *C. aculeatum* House)

USES: (1) To study twining behavior of stem. (2) To maintain as a hardy indoor plant that produces large, fragrant flowers.

DESCRIPTION: Twining, perennial herb, of interest because of its large night-blooming and very fragrant flowers. Stems may grow to 10 to 20 ft long, unless stem tips are removed. Related to morning-glory (see *Ipomoea*). Leaves simple, alternate, and 3 to 8 in long. Large, fragrant flowers, most often white, but purple in some varieties; 5 to 6 in across. In some horticultural types, flowers stay open the next morning until noon. Sepals five; corolla with long tube and broad, open tips; five stamens; two-lobed stigma. Fruit a dehiscent capsule, ½ to ¾ in long.

CULTURE: Standard method (p. 250), with medium to bright light and moderate or warm temperature. Hardy and very fast-growing. Place only one plant in a pot. Provide stakes or trellis for twining, or train vine on cord around window. Pinch stem tip and trim plant to control rampant growth.

PROPAGATION: Grow from seeds, obtained locally or by catalog. Soak seeds in water for 24 h before germination, or treat as described for *Ipomoea*. Seeds germinate in 5 or 6 days after soaking.

Capsella (Shepherd's Purse; *C. bursa-pastoris* Medic.)

USES: (1) For study of complete life history of a fast-growing plant. (2) For genetic studies.

DESCRIPTION: A common weed that is used regularly in plant research. Self-pollinating annual. Leaves in a rosette, and often with edges toothed or cleft. Grows 4 in to 2 ft high. Small, inconspicuous flowers in a raceme. Fruit a small

A B

Figure 7-23 *Capsicum* sp., ornamental pepper plant. **A** Plant in bloom. **B** Pepper plant with fruit of green, yellow, and red pods.

silique, about 5 to 10 mm long, with a characteristic purselike shape that accounts for the common name. Seed 1 to 5 mm long. Germinates in 6 days; flowers in 30 days; makes new seeds in 15 additional days.

CULTURE: Standard method (p. 250), with medium to bright light and moderate to warm temperature. Unaffected by photoperiod.

PROPAGATION: Plant seeds year round in greenhouse or laboratory.

Capsicum (Pepper Plant; *C. frutescens* L.)

USES: An interesting plant because of long-lasting, bright red fruit. If desired, pepper pods may be dried and ground to make red cayenne pepper. An attractive ornamental plant.

DESCRIPTION: Grows to about 16 in high (Fig. 7-23). Dark-green, ovate leaves, 1 to 2 in long. Small, white, perfect flowers, less than 1 in in diameter; short calyx with very short points; five-lobed corolla; five stamens; two compartments in inferior ovary; two-lobed stigma. Fruit a many-seeded, podlike berry with thick skin. Blooms in fall. Fruit develops quickly and holds color into late fall or to end of year. Sold by florists as the Christmas pepper plant. Bright red, purple, or cream-colored fruit stands out from branches as an erect cone, about 1 in long. Fruit is edible but very hot.

CULTURE: Standard method (p. 250), with bright light and warm temperature. The potted plants are "tender" perennials, in that they may not grow vigorously once they produce fruit. Many persons treat them as annuals and discard the plant after it blooms and fruits. The dried seeds may be retained for planting in pots or outdoors. More easily, a young plant is purchased during the early fall. Otherwise, the old plant may be trimmed back to about 6 in in height in the spring and repotted in fresh soil. Add water when top of soil is barely dry.

PROPAGATION: Grow from seeds. Germination time is 6 to 14 days. Germinating seeds require a warm soil.

Centaurea (Cornflower; Bachelors Button; *C. cyanus* L.)

USES: (1) As an annual that blooms during winter and spring. (2) To demonstrate preservation of flowers by drying. (3) To study structure of a composite flower.

DESCRIPTION: Slender annual, 1 to 2 ft tall. Young foliage is covered with white, cottonlike hairs. Linear leaves, 3 to 6 in long. Floral head, 1 to 1½ in in diameter, blue, pink, white or purple, and consisting of many small florets massed together on the flattened receptacle. Marginal raylike flowers are enlarged and sterile; central disc flowers are perfect; whorls of bracts, forming the involucre (beneath the flower head), are black-fringed.

CULTURE: Obtain young plants from a local nursery, or grow plants from seeds. Plant seeds in August or September for January or February bloom, or plant in December or January for spring bloom. Transplant seedlings to planters or pots containing a soil mix of 2 parts loam, 1 part peat moss, and 1 part sand. Put in bright light or full sun. Keep soil slightly moist; allow ample ventilation around plants by thinning young plants when necessary. As soon as plants are growing actively, feed liquid plant food biweekly. When a flower is mature, cut with long stems for drying. This is best done during a hot, dry day. Hang plants in bunches upside down in a dry, warm place for several weeks.

PROPAGATION: Grow plants from seeds. Germination time is 14 to 28 days.

Ceratonia (Carob Tree; St. John's Bread; *C. siliqua* L.)

USES: To study the interesting fruit and the endospermous, dicotyledonous seed.

DESCRIPTION: A small evergreen tree, native to the Mediterranean. Grown widely throughout tropical and subtropical regions. Ordinarily, the plant is not grown indoors. The seeds (Fig. 7-24A) are used in the laboratory. The large brown or dark-red leguminous pod, 4 to 10 in long and 2 to 3 in wide (Fig. 7-24B), is filled with a juicy pulp that is fed to livestock.

The seeds are said to have been used by jewelers as the original carat. According to old stories, John the Baptist ate the fruit when he lived in the wilderness, thus accounting for the name, St. John's bread. However, the sweet gummy pulp

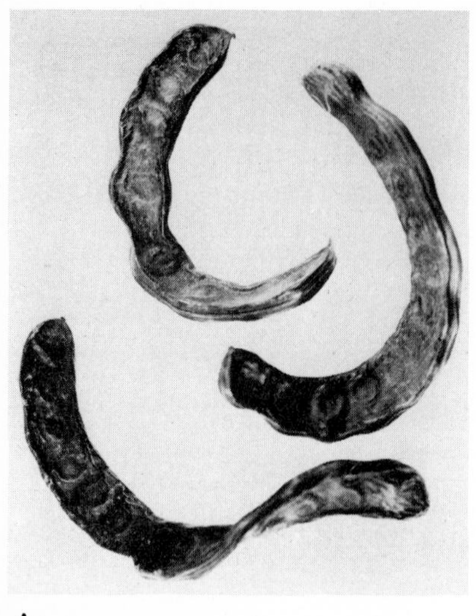

A

B

Figure 7-24 *Ceratonia siliqua*, carob tree or St. John's bread. **A** Dry pod with seeds. **B** Seed pods and leaf. (*B Courtesy of CCM: General Biological, Inc., Chicago.*)

of the fruit is nauseous to most persons and is eaten only when no other food is available.

CULTURE: Pods of *Ceratonia*, with viable seeds, may be purchased from biological companies, for example, General Biological, Inc. (Turtox). Germination is slow and irregular, but can be hastened by filing through the hard coat in several places. Take care not to damage the internal parts.

Cestrum (Night Jessamine; *C. nocturnum* L.)

USES: As an example of a plant with very fragrant night-blooming flowers.

DESCRIPTION: A hardy perennial shrub that is pruned back when grown indoors (Fig. 7-25). Night-blooming flowers are very fragrant, and scent spreads to other parts of building. Thin, glossy leaves, 4 to 8 in long and 1 to 2 in broad. Greenish-white to cream-colored flowers in axillary clusters; slender, tubular corolla about 1 in long; five-toothed calyx; five stamens attached to corolla tube. Superior ovary with two chambers; three to six ovules in each chamber. Fruit a small berry.

CULTURE: Standard method (p. 250), with medium light and warm temperature. Prune back as necessary to retain a convenient size. Pinch stem tips to obtain bushy growth. Give liquid plant food every 3 to 4 weeks. Plant is easily grown in a well-lighted laboratory or near window light.

PROPAGATION: Root 4-in stem-tip cuttings from February to June in sand or other medium.

Figure 7-25 *Cestrum nocturnum*, nightblooming jessamine. (*Merry Gardens, Camden, Me.*)

Citrus (Citrus Fruit Plants)

USES: (1) As an example of a citrus fruit-bearing plant that is attractive and interesting throughout the year. (2) To demonstrate artificial pollination. (3) For fragrant flowers intermittently throughout the year.

DESCRIPTION: Evergreen shrubs or trees that are pruned to an appropriate size for indoor growth (Figs. 7-26, 7-27). Will have fragrant flowers, green fruit, and ripe fruit practically the entire year. Leaves are glossy, dark green, simple, and alternate. Small, aromatic oil glands on leaves. Fragrant, white flowers, 1 to 1½ in at top diameter. Flowers are perfect, and solitary or in axillary clusters. Stamens,

Figure 7-26 *Citrus limonia ponderosa*, lemon tree with large, edible fruit. Each lemon is 3 to 4 in in diameter.

Figure 7-27 *Citrus mitis*, ornamental orange with small edible fruit. The oranges are 1 to 2 in in diameter.

15 or more in a few bundles. Superior ovary with 8 to 15 chambers; generally several ovules in each chamber. Fruit a berry with leathery outer rind (epicarp).

Abraham (1967) recommends the following types for indoor culture: *C. aurantifolia*, a small edible lime; *C. limonia ponderosa*, a large edible lemon; *C. mitis*, a miniature edible orange. Many other types are available for ornamentals and for fruit production.

CULTURE: Standard method (p. 250), except change soil mix to 2 parts peat moss, 1 part loam, and 1 part sand. Moderate to warm temperature and medium to bright light. Plant needs ventilation; do not crowd with other plants. Increase air humidity by placing pot on a bed of small gravel with water, and with base of pot above water. Plant tolerates a fairly wide range of temperatures, 16 to 25°C. For best fruiting, night temperature should be 15 to 18°C during winter months. Transfer to pot of fresh soil every 2 or 3 yr, or more often if roots become potbound. May be placed outdoors in a partially shaded spot during summer months. Keep soil barely moist during summer. Transfer indoors before frost.

PROPAGATION: By stem-tip cuttings in sand or other medium during May to July. Because plants grow slowly, buying a potted plant may be more satisfactory. Plants can be grown from fresh seeds of market fruit but most likely will produce only foliage and no fruit. For this purpose, take fresh seeds from mature fruit, soak them overnight, and plant them about ½ in below soil, with two or three seeds in a pot. Place 6 in below lights or in warm, sunny window. Keep soil moist until sprouts appear.

Coleus (Flame Nettle; *C. blumei* Benth.)

USES: (1) For studying many plant processes: photosynthesis, water and mineral uptake, conduction, transpiration, auxin control, leaf abscission, propagation by cuttings, and others. (2) For the showy ornamental leaves of most species.

DESCRIPTION: Perennial herb, little branched, 2 to 3 ft tall (Fig. 7-28). One of most easily grown pot plants. Many varieties available at local stores. Widely cultivated for large brightly colored, variegated leaves. Ovate leaves, generally with sharply toothed edges, and with yellow, white, red, or pink colors variegated with green. Very small flowers, blue, lavender, or white, in terminal spikes. Lower leaves fall off as new leaves form at stem tip, leaving an increasing length of naked lower stem. Pinching off the tip of stem, and removing flower buds or spikes helps to increase branching and to avoid a single, long, partially defoliated stem.

CULTURE: Standard method (p. 250), with warm temperature and bright light. Partial shade if outdoors in summer. Pinch off stem and flower buds regularly to make a bushy growth. Remove lower dead leaves. Very susceptible to mealybug infestation. (See *Plant Pests* in Chap. 6.) Probably best to discard any heavily infested plants, as new plants can be obtained easily and quickly from cuttings. If infestation is light, pick off mealybugs by hand. (Feed them to the goldfish.)

PROPAGATION: By stem cuttings in water at any time of year. Transfer to soil when well rooted, generally 12 to 14 days. Obtain seeds locally or by catalog. Germination time is 10 days. May be planted at any time of year.

Crassula

USES: (1) For study of xerophytic succulents. (2) As ornamental plants that display a wide range of interesting, sometimes fascinating, forms and colors. (3)

A

Figure 7-28 *Coleus blumei.* **A** Entire potted plant showing flowers and variegated leaves. **B** Close-up view of **A**.

B

A

B

Figure 7-29 *Crassula*, a perennial succulent. **A** *C. perforata* or necklace vine. **B** *C. multicava*, another suitable species with white or pink blooms. (*B Merry Gardens, Cambden, Me.*)

To demonstrate asexual propagation by leaf and stem cuttings, and by offsets of young plantlets.

DESCRIPTION: Perennial succulent herbs with typical xerophytic characteristics of fleshy leaves and stems, often growing in tufts or rosettes (Fig. 7-29). Leaves generally closely packed together, with waxy surface, and sunken stomata. Native mostly to South Africa. Often the leaves are colored or edged with colors of red, pink, purple, or silver. Flowers generally in showy, compact clusters and white, red, yellow, orange, pink, lavender, or purple. Flower parts, generally five of each type. Petals are separate, or joined only at base. Reproduction often by offsets of plantlets and by rhizomes.

Probably the most common indoor plants are varieties and cultivars of *C. argentea* (the jade plant), but many other types are as interesting, or more interesting, and as easily cultured. The common names listed in suppliers' catalogs give evidence of why many persons grow succulents as an avocation: jade necklace, rattlesnake crassula, pyramid crassula (an unusually striking plant), arab's turban, Morgan's pink crassula (a large cluster of pink flowers at top of a base of flattened, soft, gray leaves), crassula royal purple, scarlet coral, alabaster towers, crassula deceptrix (an erect column of thick, powdery white leaves that grow in a pleated fashion).

CULTURE: See *Culture of Xerophytic Succulents* (p. 251). These are among the plants (including cacti) that are grown most easily. Hardy indoors, and survive under neglectful conditions, although, as with all plants, they produce healthy and attractive growth and flowers only when given good care. May be placed outdoors during summer, in partial shade.

PROPAGATION: By leaf or stem cuttings at any season. Remove offset plants and transfer them to fresh pots of soil. Obtain plants locally or by catalog. (See *Asexual Propagation* in Chap. 6.)

Cucumis (Cucumber; *C. sativus* L.)

USES: (1) Excellent for studying the flower parts. (2) For tests involving the very sensitive, simple tendrils.

DESCRIPTION: Several horticultural varieties are available, with differences mainly in type of stems and size of fruit (Fig. 7-30). For home use, the choice of variety depends on whether the cucumber is to be eaten fresh or used for pickling. Those for pickling have a more compact and bushier growth with smaller fruit. The "tiny dill" types are interesting and easy to grow indoors in a pot. Each bush may bear many small cucumbers. The types with larger fruit are more useful for studying tendrils. The latter produce longer vines that can be trained to grow on stakes, or on heavy cord or wire alongside window frames. Vines grow rapidly; flowers mature in 50 to 60 days. At maturity, the plant blooms almost continuously for at least a month. Probably the fruit will remain small on all indoor plants.

Stems may be rough and hairy. Simple tendrils are very sensitive and illustrate in good fashion all the phenomena of tendril climbing. Alternate leaves, triangular-ovate, and 3 to 6 in long. Open, bell-shaped, bright yellow, single flowers, 1 to 2 in across. Flowers are unisexual, with both pistillate and staminate flowers on one plant (monoecious plant). Five-lobed calyx, five-lobed corolla, five stamens (two pairs united), inferior ovary generally with three chambers. Fleshy fruit with a rind, and a more or less spongy, seedy center (a pepo fruit).

CULTURE: Standard method (p. 250), with warm temperature and medium to bright light. Will grow and produce small fruit under normal room lighting. Pinch off stem tips to make bushier growth and more flowers.

PROPAGATION: Plant seeds at any season of year. Seeds germinate in 3 to 4 days. Flowers mature in 50 to 60 days.

Figure 7-30 *Cucumis* sp., vine and tendrils of a tiny-dill type.

Dionaea (Venus Flytrap; *D. muscipula* Ellis)

USES: For studying (1) the rapid movement of the traps; (2) the bog environment of insectivorous plants; and (3) the unusual type of nutrition, the sensitive hairs in the traps, and the enzymatic action on insects. See Di Palma et al. (1966).

DESCRIPTION: Not widely distributed. Native growth restricted to parts of North Carolina and South Carolina, and a few other regions. A perennial herb. Leaves form a basal rosette that grows from a rhizome, close to soil surface (Fig. 7-31). Leaves are 1½ to 6 in long, each consisting of a broadened two-winged petiole and a two-lobed blade that is modified into an insect trap. The two lobes are hinged at the center, edged with long bristles, and equipped with three sensitive cilia, or trigger hairs, on the top surface of each lobe. When the trigger hairs are stimulated, the lobes close together and the bristles lock, generally in 2-sec time, or less, depending on temperature and age of trap. The struggling of a trapped insect stimulates secretions of digestive enzymes that digest soft parts of the insect. Traps reopen in 4 to 20 days, depending on insect size. If traps are stimulated with a pencil or other object that is not digestible, traps reopen in about 1 day. Insects are attracted by nectar at edge of traps and by bright red colors on inner surface of trap. As leaf-traps age, they turn black and should be cut away. New leaves will replace those removed. Newly transferred plants require a week or so to adjust and become sensitive. During spring months, perfect, white flowers, about ½ in long, are produced on a 6- to 8-in erect stem that grows from center of leaf rosette.

Figure 7-31 *Dionaea muscipula*, Venus flytrap. **A** Rosette of leaves and the two-lobed insect traps. **B** Close-up view of insect trap showing the long bristles at edges of lobes.

A *Courtesy of Carolina Biological Supply Company*

B *Courtesy of Carolina Biological Supply Company*

Much research has been done, including tests that involve feeding ground beef (Schwab et al., 1969). See Mozingo et al. (1970) for a fascinating account of trigger hairs and for the electron-scanning micrographs that show the surface topography.

CULTURE: See preparation of bog terrarium in Chap. 6. Important aspects are: (1) Put plant rhizomes in an acid substratum, generally commercial peat or *Sphagnum* moss or a mixture of the two. (2) Maintain a high humidity by covering the containers with a plastic film or a sheet of glass. If the inner layer of condensed moisture disappears from the plastic film, the plant is dry. Add water and seal the cover against leaks. Open cover occasionally, perhaps once a week, for ventilation and evaporation of excess water. (3) Place roots above level of any standing water in bottom of container. (4) For best growth, put plants in medium light (about 800 ft-c) and at a temperature of 23 to 26°C.

PROPAGATION. Obtain plants with rhizomes from biological supply companies and from plant specialists. See list of suppliers in Appendix B.

Drosera (Sundew; *D. rotundifolia* L.)

USES: For studying (1) sensitive hairs on leaf pads; (2) the insectivorous nutrition; and (3) environmental requirements of a bog plant.

DESCRIPTION: Perennial, low-growing herb, related to the *Dionaea*, but much more widely distributed. Found in many parts of the world, including the bogs and swamps of eastern United States, northern Wisconsin, Michigan, Minnesota, and Canada. A basal rosette of leaves. Each leaf, ½ to 2 in long, ending in a circular, red pad covered with glandular hairs (tentacles) which secrete a sticky, odorous fluid that attracts and traps insects (Fig. 7-32). When stimulated, tentacles bend and press small or weak insects against digestive fluids, sometimes

Figure 7-32 *Drosera rotundifolia*, sundew. The photograph shows a rosette of leaves and the sticky pad at the outer end of each leaf. (*Courtesy of CCM: General Biological, Inc., Chicago.*)

covering and smothering the insect with leaf blade. The action is much slower than that of *Dionaea*. Tentacles resume original position in about 6 days, when another insect may be trapped. When stimulated by a nondigestible object, only a few tentacles bend and only a little digestive fluid is secreted. Apparently, the digested insect is metabolized by the plant, although it can live without the added food. Perfect, small white flowers, ¼ in across, on stem 2 to 12 in high.

CULTURE: See *Dionaea*.

PROPAGATION: See *Dionaea*.

Echeveria

USES: (1) To study environmental requirements of a typical xerophyte. (2) As an ornamental succulent.

DESCRIPTION: Perennial, succulent herbs with typical xerophytic characteristics. Dense rosettes of broad, flat, fleshy leaves, with waxy surface and sunken stomata (Fig. 7-33). Native to subtropical and tropical areas of North and South America. Very similar to *Crassula* in general appearance, and the two genera are difficult to separate. Differences are based primarily on arrangement and number of flower parts. Horticulturists have developed and named many types of crassulas and echeverias, which are some of our hardiest and most attractive indoor plants.

Leaves are green or gray-green, and often overlaid with purple, pink, or red. Leaves become more colorful in direct, bright light. Tubular corolla, with petals spread at tips and united toward base. Flowers borne on a stalk that rises above rosette. May bloom throughout year, but generally during winter months. Cultivars of the following species may be obtained at nurseries or by catalog: *E.*

Figure 7-33 *Echeveria pulvinata*, chenille plant that consists of a number of branches terminating in a rosette of leaves, as shown.

elegans Rose, with pink flowers and green or blue-green leaves; *E. derenbergii* Purpus, with orange or red flowers on short stems; *E. pulvinata* Rose, sometimes called the chenille plant, with velvety, blue-green leaves that are edged in bright red, and with yellow or red flowers. Ordinarily, many common types can be obtained locally. Rare and unusual *Echeveria* may be purchased by catalog (see source list in Appendix B).

CULTURE: See *Culture of Xerophytic Succulents* (p. 251). Plants survive neglectful conditions, although growth is more luxuriant with optimum care.

PROPAGATION: By stem or leaf cuttings taken from mature growth. Dry stem cuttings for about 2 weeks and leaf cuttings for about a week before they are put into rooting medium; otherwise, the cut surface will decay. A leaf cutting of most succulents dries to a very small mass as new plants grow from its base.

Many types eventually develop long, bare, and unsightly stems. The plant can be improved by cutting off the top portion, including about 1 or 2 in of the stem, and rooting the stem of the removed portion in sand. Roots will form in about 3 weeks, when the plant may be transferred to soil. Occasionally, the lower remaining portion of stem will grow new leaf rosettes.

Sometimes seeds are available from catalog suppliers, or seeds may be collected from the plants. Unless there are particular reasons for using seeds, the purchase of plants seems more desirable.

Euphorbia (Baseball Plant; *E. obesa* Hook f.)

USES: To compare the xerophytic *Euphorbia*, the cacti (Cactaceae), and *Stapelia* (Asclepiadaceae), wherein similar conditions of life have produced in three families, not closely related to one another, such great likenesses that the plants are not easily distinguished except by the specialist. *Stapelia* and the cacti are described elsewhere in this chapter. Common traits of the three types are: (1) fleshy stems, spherical, ridged, or cylindrical; (2) a green, photosynthetic outer-stem layer; a parenchymatous inner stem of storage tissue; (3) leaves reduced or aborted, or often modified to scales or thorns. Differences among the three groups are found primarily in the flowers, and in the milky juice of euphorbs that is not in the other two groups. See Webb (1968) for his laboratory investigation, "Convergent Evolution in Cactuslike Plants."

DESCRIPTION: A spineless succulent of South Africa. Generally a simple stem, almost completely round and about 3½ in in diameter (Fig. 7-34); mature plants gradually elongate to 7 to 8 in. Grayish green color, marked with purple lines, particularly in fissures between the eight ribs. Tubercles make a purple, median strip down each rib. Small, green inflorescences form at top of plant.

To observe evolutionary convergence of xerophytes, compare *E. obesa* with the cacti, *Astrophytum asterias* (sea urchin cactus), or *A. myriostigma* (bishop's cap cactus). Another interesting euphorb, *E. meloformis* (little melons), is described in suppliers' catalogs and may be used in place of *E. obesa*.

CULTURE: See *Culture of Xerophytic Succulents* (p. 251).

PROPAGATION: Difficult to propagate in laboratory. Obtain starter plants locally or by catalog. *Euphorbia meloformis*, however, produces offset plants that may be easily separated for propagation.

Figure 7-34 *Euphorbia obesa*, baseball plant, at right, and *Astrophytum* sp., at left. The euphorb and cactus illustrate evolutionary convergence.

Ficus (India Rubber Plant; *F. elastica* Roxb.)

USES: (1) As an ornamental plant. (2) For studies of stomatal pattern. (3) Excellent for studying adventitious roots. (4) Excised leaves are excellent for transpiration studies.

DESCRIPTION: Almost foolproof for indoor culture where light and humidity are at low levels, although plant growth and appearance are improved with optimum conditions. Native to southern Asia, India, and Malaya, where *Ficus* is a large forest tree and a source of India rubber. Young plants are grown as ornamentals in large pots or tubs. Thick leaves, glossy dark-green, 6 to 12 in long, with distinct parallel, lateral veins off midrib (Fig. 7-35). Petiole 1 to 3 in long. Not known to flower or fruit indoors. Long, stout, adventitious roots grow from branches to

Figure 7-35 *Ficus elastica*, India rubber plant.

the soil. The plant is closely related to the banyan tree, *F. bengalensis* L., sacred tree of India, which extends outward indefinitely by sending down large aerial roots from the branches to the ground. *Ficus elastica forma decora* is a decorative, horticultural type with red color on undersurface of leaves.

Ficus lyrata Warb. (Fiddleleaf Fig), a native tree of tropical Africa, is often grown as an indoor ornamental plant. Young plants are grown in large pots or tubs. The very large leaves are 10 to 15 in long and 10 in wide, thick and fiddleshaped with a large, rounded end and slightly wavy margins. Top surface is a dull green, and a lighter undersurface; a short petiole. Keefe (1965) reports that *F. lyrata*, when growing indoors, may produce figs that are inedible but which can be studied for the fruit type (a syconium). The common, edible fig, *F. carica*, is a member of this family.

CULTURE: Standard method (p. 250), with low to medium light and warm temperature. Put in large container, with bottom layer of pebbles. If plant becomes rootbound, transfer to larger pot. Too much water causes bottom leaves to turn yellow and drop off. Wipe dust and film from leaves as necessary, probably every 1 or 2 weeks. Generally, plant grows quickly and will reach 6- to 8-ft height, with a long heavy stem that is difficult to stake. If plant outgrows the space, or if a long, bare lower stem develops, take stem cuttings and discard old plant.

PROPAGATION: Cut off top of plant and root it in water, sand, or other rooting medium; or take leaf cuttings, with a bud in leaf axil and about 3 in of stem. Put lower end of stem in sand or other rooting medium.

Glycine (Soybean; *G. max* Merr.)

USES: (1) For seed and seedling studies. (2) Used regularly for genetic studies, and for tests with *Rhizobium*, nitrogen-fixing bacteria. Seeds and inoculum may be obtained from biological supply companies.

DESCRIPTION: Generally only the seed or seedling is used in the laboratory. An important food crop for the beans, fodder, and oil. Mature, cultivated varieties are an annual bush, 2 to 3 ft tall, with inconspicuous white or purple flowers. Fruit is a pod, 2 to 3 in long, with 2 to 4 beans. Requires about 100 days from planting seed to mature fruit. Although other beans (e.g., *Phaseolus*) are used more often for the laboratory, probably this plant is as satisfactory for most studies.

CULTURE: Standard method (p. 250), with medium to bright light and warm temperature.

PROPAGATION: Obtain seeds locally or by catalog. Germinate indoors during any season.

Hedera (English Ivy; *H. helix* L.)

USES: (1) For study of aerial roots used for climbing, and for comparison with tendril climbers. (2) Observation of positive phototropism and negative geotropism of aerial roots in contrast to the tropisms of underground roots. (3) Ornamental plant for hanging basket, or for dimly lighted areas.

DESCRIPTION: Many forms of stems and leaves. Generally, tough, triangular-

Figure 7-36 *Hedera helix,* English ivy.

shaped, broad leaves, with three to five lobes, and 1 to 5 in long (Fig. 7-36). Some types with variegated leaves. Juvenile plants climb with aerial roots; mature plants, in nature, may form nonclimbing stems without aerial roots. Inconspicuous flowers and fruit may form on mature, outdoor plants.

CULTURE: Standard method (p. 250), with low or medium light, and moderate or warm temperature. Clean dust and film from leaves as necessary. Occasionally, immerse pot and plant in water for 20 to 30 min to remove dust and infestations.

PROPAGATION: Stems will root easily in water or sand at any time of year.

Impatiens (Garden Balsam; *I. balsamina* L.)

USES: (1) Used regularly for study of complete life cycle, 65 to 75 days. (See Postlethwait and Enochs, 1967; and Creighton, 1965.) (2) Excellent for study of anatomy of plant parts. *Impatiens sultanii* Hook, with single flowers, is commonly used to study floral parts. (3) For germination of pollen of *I. sultanii* (see Stevenson, 1965). (4) To study seed-dispersal mechanism. (5) For many types of physiological studies.

DESCRIPTION: Erect, annual herb, 1 to 2 ft tall (Fig. 7-37). Alternate leaves, 1 to 6 in long, and serrated. A watery, transparent stem. Perfect flowers, 1 to 2 in across; self-fertilizing; many colors. Posterior sepal elongated into honey spur. Five each of stamens and petals. Anthers unite to form a cap over ovary. Growth of ovary eventually breaks the filaments loose at base. Ovary with five chambers. Fruit a hairy capsule, ½ to 1 in long and with many seeds. Ripened capsule explodes open, and seeds are scattered to some distance.

CULTURE: Standard method (p. 250), with medium to bright light and warm temperature. Pinch off stem tips or prune to make a bushy plant.

PROPAGATION: Buy young plants at nursery, or start from seeds during any season of year. Seeds germinate in 4 to 5 days. Put stem-tip cuttings in water, sand, or other medium.

Ipomoea (Morning Glory)

USES: To study (1) life stages from seed to seed in several of the faster growing species; (2) plant anatomy and morphology; (3) twining behavior; (4) relationship

A

B

Figure 7-37 *Impatiens balsamina*, garden balsam. **A** Entire plant. **B** Close-up of flower.

of photoperiod and morning bloom; (5) asexual propagation by root in the sweet potato.

DESCRIPTION: Herbaceous twiners, with large leaves and large colorful flowers. Generally flowers open and close in the morning, although varieties have been developed with flowers that stay open until early afternoon. Alternate leaves, entire or lobed. Perfect flowers, single or in clusters. Five sepals; funnel or bell-shaped corolla; five stamens; inferior ovary with two to four chambers; two-lobed stigma. Fruit a dehiscent capsule with four to eight seeds.

Ipomoea tricolor Cav. A tall twiner, 8 to 10 ft high. Thin leaves, 3 to 5 in long, with prominent veins. Blue flowers, 3 to 4 in across. Many showy cultivars, including the well known "Heavenly Blue" morning glory. Reed and Ihrig (1968) report that mixed varieties of *I. tricolor* completed stages from seed to seed in 90 days, and that the plant bloomed continuously indoors.

Ipomoea purpurea Lam. Common morning glory, with blue and purple flowers, 2 to 3 in long. A tall, twining annual. Many cultivars.

Ipomoea nil Roth (*Pharbitis nil* Choisy). A strong and showy plant, with purple, violet, rose, or blue flowers; some with fluted edges. Many horticultural types, including the red, pink, white, or blue giant Japanese morning glories, up to 6 in across; and the "Scarlett O'Hara," 3 to 4 in across. Postlethwait and Enochs (1967) report that the Japanese morning glory completes the life stages, from seed to seed, in 100 days.

Ipomoea batatas Lam. The edible sweet potato. Students may be interested to learn that the sweet potato is also a morning glory, and that "potato" is derived from "batata." Stems are long and trailing (not twining), and growth is from a thickened, starchy root. Light-purple flowers, 1 to 2 in long.

CULTURE: Standard method (p. 250), with medium to bright light and warm temperature. A quick growing plant that is easily cultured. Put one or two plants in pots. Can be trained around window frame. Pinch off stem tips to control growth, and to gain more flowers.

PROPAGATION: Grow all except sweet potato from seeds. Germination time is about 4 days for Japanese morning glory if seed is first scarified or soaked in sulfuric acid. Scarify the seed by gently cracking the outer coat with a weight, or by filing through the coat with a nail file. The soaking treatment is done in 70% sulfuric acid solution for 24 h, followed by washing seeds in running water. Prepare the acid solution by adding 70 ml of concentrated sulfuric acid *to* 30 ml of distilled water. Add a few drops of acid at a time, while stirring continuously. At intervals during the mixing, allow the solution and container to cool before adding more acid. (Caution: Strong Acid!)

Grow the sweet potato vine by suspending the starchy root in a jar of water with the upper one-third held above the water. The trailing vines and roots will grow for at least 5 to 6 weeks in water, alone.

Lithops (Stone Face; Living Rock; *L. pseudotruncatella* N. E. Br.)

USES: (1) To illustrate plant mimicry. Plants are almost indistinguishable from the surrounding rock. (2) To study the adaptation for receiving sunlight through "windows" on the top surface of almost buried leaves. (3) To observe method of seed dispersal. (4) As a very unique plant that may create much student interest.

DESCRIPTION: Small, succulent, stemless perennial, native to South Africa. Pairs of very succulent leaves are joined into a stonelike form with a cleavage on top from which the flower arises (Fig. 7-38). Oval, flat-topped leaves, about 1 in high and 1 to 1½ in across; gray with brown markings. Yellow flowers, 1 to 2 in across.

New pairs of leaves formed each year at right angles with the old pair; the old leaves eventually wither away. In nature, leaves may be almost completely buried in soil with only top surface exposed, an adaptation that helps prevent loss of water by transpiration and conceals the plant from animals that might feed on the juicy leaves. Windows in the top surface allow light to reach the buried chloroplast tissue. Window patterns vary among species. Windows appear opaque and, except for surface patterns, are not apparent unless the top surface is removed and held before a light source (Figs. 7-38 and 7-39).

Fruit is a capsule with five or more valves and many seeds. Valve lids open when moist; seeds are expelled when drops of water fall forcibly into a valve, thus ensuring that seeds are dispersed at a time favorable to germination. Capsules can be stored dry and used for class studies at a later time to demonstrate seed dispersal by dropping water on the valves.

Other genera of this family show similar adaptations. Several types are available from cactus suppliers, including *Pleiospilos* and *Fenestraria*. Jump (1968) presents an excellent discussion of the plants and of their uses for class studies.

CULTURE: Easy to grow in a greenhouse, a sunny window, or under lights (1,000 ft-c or more) if culture techniques are followed closely. Soil mixture of 2 parts sand, 1 part garden loam, and 1 part leaf mold. Three to five plants can be

Figure 7-38 Stone face plants. **A** Four species of *Lithops.* **B** Flowers, leaves, and capsule of *Lithops kuibisensis.* **C** Windows in the top surface of leaves; 1, leaf top and window of *Lithops bella*; 2, window of *Lithops bella* illuminated from beneath; 3, leaf top and miniature windows of *Lithops fulviceps*; 4, miniature windows of *L. fulviceps* illuminated from beneath. **D** 1, 2, and 3, progressive stages of opening of capsule of *Glotiphyllum* sp. **D** 4, side view of capsule shown in **D**, 3 (*All photos courtesy of John A. Jump, Elmhurst College.*)

put in one 5-in pot, with roots and the lower portion of leaves in the soil. Surround plants with rocks of same size to demonstrate mimicry. Water *on sunny days only* and take care *not to splash water on plants.* Add water when ½ in of top soil layer becomes dry. Give liquid plant food every 3 to 4 weeks. Plants rest during summer months, when no food should be given and when water should be added only every 2 to 3 weeks.

Figure 7-39 Other stone face plants. Left, *Argyroderma* sp.; right, *Pleiospilos* sp. (*Courtesy of John A. Jump, Elmhurst College.*)

PROPAGATION. May be grown from seeds. Seed germination time is 6 to 14 days. Plant growth is very slow, and buying the mature plant may be more desirable.

Lobularia (Sweet Alyssum; *L. maritima* Desv.)

USES: (1) For study of life stages of a fast-growing plant. Time from seed to flower is about 35 days. (2) As an ornamental, fragrant plant.

DESCRIPTION: A "tender" perennial, usually grown as an annual. The four-petaled flowers are arranged opposite each other in a square cross, typical of all crucifers and accounting for the family name Cruciferae or "cross-bearing." The varieties of sweet alyssum, listed as such in catalogs, belong to this genus and are not to be confused with species of the genus *Alyssum* L., also a crucifer.

Low-growing and spreading branches that may be 6 to 12 in high (Fig. 7-40). Short, soft hairs on stems and leaves. Alternate, entire leaves, 1 to 2 in long. Small white, purple, or pink flowers. Gold or yellow flowers on several horticultural types of perennial shrubs. Perfect flowers, self-fertilized. Four sepals, four petals,

Figure 7-40 *Lobularia maritima,* sweet alyssum.

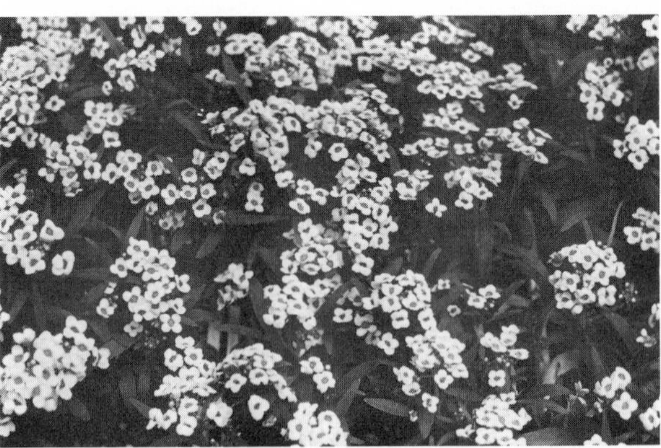

Figure 7-41 *Lycopersicon esculentum,* small cherry tomato.

six stamens, one pistil, superior ovary with two chambers. Fruit a small, round silique.

CULTURE: Standard method (p. 250), with medium to bright light and warm temperature. Recommended as an easily grown annual that gives long-lasting blooms in winter months. Requires bright light for blooming. Seeds may be planted indoors at almost any season. Put one or two plants in a 6-in pot so that roots will be bound somewhat tightly, giving more bloom.

PROPAGATION: Grow from seeds obtained locally or by catalog. Seeds germinate in 5 to 6 days.

Lycopersicon (Tomato; *L. esculentum* Mill.)

USES: For studies of (1) complete life stages of a fast-growing plant (80 to 90 days from seed to seed); (2) many aspects of physiology and structure; (3) development of chromoplasts in skin of green, yellow, and red tomato fruit; (4) root tissue culture (see Jacobs and LaMotte, 1964); and (5) radioisotope uptake and distribution.

DESCRIPTION: Erect, branching, soft, annual herb; 2 to 4 ft high (Fig. 7-41). Hairy stems and leaves; strong, pungent odor. Leaves 6 to 12 in long, made up of pinnate leaflets. Yellow, perfect flowers, each about ¾ in across; three to seven flowers in a drooping cluster. Five-lobed calyx; generally five petals united into a lobed corolla. Number of stamens same as number of petals. Superior ovary, with two or more chambers. One style, and two-lobed stigma. Fruit a yellow or red berry with many small seeds.

For indoor growth, an excellent type is "Tiny Tim," a small cherry tomato that grows 15 to 18 in high and produces many small tomatoes on a plant. Reed and Ihrig (1968) report 3 days for germination, 74 days to flowering, and 85 days to seed.

CULTURE: Standard method (p. 250), with medium to bright light and warm temperature. As plant grows, tie branches to a stake inserted at edge of pot. To be certain of pollination, which ordinarily is brought about by insects, hand pollinate by shaking the bushes when the pollen is ripe, or by gently brushing across each flower with a small, soft-haired brush.

PROPAGATION: Start plants from seeds, or buy young plants at a local nursery.

Mimosa (Sensitive Plant; *M. pudica* L.)

USES: To study (1) leaf movements and action of pulvini (see Fondeville et al., 1966; and Sibaoka, 1969); (2) relationship of photoperiod and the sleeping position of leaves.

DESCRIPTION: Perennial herb or shrub, often treated as an annual (Fig. 7-42). A common tropical and subtropical weed; grows to 1½ ft high. Hairy, spiny branches. Bipinnate, sensitive leaves. Many small perfect, purple to pink flowers in axillary clusters; four or five united petals; four to ten stamens. Superior ovary; fruit a flat pod, ½ in long and with three or four seeds. Generally blossoms in fall.

Although much study has been done of the sensitive leaves and the pulvini, the process is not yet completely understood. To even a slight touch or shake of a branch the extremely sensitive leaves react by closing against one another in the "sleeping" and drooping position, normally assumed at night. The leaflets move quickly, closing up in pairs, one after the other along the leaf. After several minutes, they reopen in a reverse action. The movement is a result of the action of pulvini. A pulvinus, one at the base of each leaflet stem, is a small, swollen bulbous structure that apparently initiates the action by turgor changes in the leaflets.

Figure 7-42 *Mimosa pudica,* sensitive plant. *(Courtesy of CCM: General Biological, Inc., Chicago.)*

One leaflet, alone, may be touched, and the transfer of stimulation from pulvinus to pulvinus can be observed.

CULTURE: Soil mix of 2 parts, each, of peat moss and loam, and 1 part of sand. Keep soil slightly moist. Give liquid plant food every 3 or 4 weeks; warm temperature and high humidity of 70% or more. Put plant in a greenhouse or in a glass chamber covered with glass or plastic film, and under bright light or in sunlight several hours daily.

PROPAGATION: May be grown from seeds that are obtained locally or from a biological supply company. Seeds may be planted indoors any season. Seed germination time is 7 to 8 days. Biological companies also furnish young plants. Wild plants may be collected from Texas to Florida, and in Mexico, Central America, and the Caribbean Islands.

Mitchella (Partridge Berry, Twin Berry, or Squaw Berry; *M. repens* L.)

USES: Ornamental plant for woodland terrarium.

DESCRIPTION: Evergreen, low-growing herb, with glossy green leaves. Bright, red, berrylike fruit is long lasting and useful in decorating a woodland terrarium. Creeping and rooting stems grow to 1 ft long. Opposite, ovate leaves, ¼ to ¾ in long, sometimes with white markings. White, fragrant twin flowers, ½ in long, with ovaries united. Four, each, of sepals, petals, and stamens. Ovary with four chambers and one ovule in each chamber. Four-lobed stigma. Fruit a scarlet, berrylike, double drupe, ⅓ in across. Native to woodland areas east of Mississippi, Florida to Texas, and north to Minnesota.

CULTURE: See description of woodland terrarium in Chap. 6.

PROPAGATION: Collect plants from outdoors, or order by catalog (e.g., from Arthur E. Allgrove, North Wilmington, Mass. 01887).

Nigella (Fennel Flower; *N. damascena* L.)

USES: (1) For cytological studies of large chromosomes. (2) To study complete life history of a fast-growing plant; time from seed to seed is about 115 days (Postlethwait and Enochs, 1967). (3) For dried floral arrangements.

DESCRIPTION: Erect, branching annual, 1 to 1½ ft high. Single, white or blue flowers, 1 to 1½ in across, located at end of branches, with many small leaves (involucre) at base of each flower. Five petallike sepals; five petals; many stamens; five pistils. Self-fertile. Superior ovaries. Fruit a capsule with many small seeds. Normally blooms during summer months, but may bloom at other times when indoors under bright lights.

CULTURE: Standard method (p. 250), with medium to bright light and warm temperature.

PROPAGATION: Obtain seeds or young plants locally or by catalog. Seed germinates in 5 to 8 days.

Passiflora (Passion Flower; *P. caerulea* L.)

USES: (1) Unusually easy to grow and can be maintained for its great beauty and its regular supply of plant parts. (2) To study the unusual flowers. (3) To test the responses of the long tendrils. Much testing has been done of mechanical stimulation of the long tendrils *in situ* and *in vitro*. See Jaffe (1970). It seems that the plant should be found more regularly in the laboratory.

DESCRIPTION: Many cultivars and hybrids have been developed. General characteristics of the basic species, *P. caerulea*, are given here. The name "Passion Flower" is derived from a story told by early missionaries, in which the white to pink sepals and petals represented the ten apostles at the crucifixion; the white and purple corona represented the crown of thorns; the five anthers were the wounds; and the three stigmas, the nails. Carrying the symbolism further, they described the coiling tendrils as cords and whips, and the five-lobed leaves as the cruel hands of the persecutors.

Slender, perennial herbaceous vine (Fig. 7-43). Prolific, vigorous growth that may require a periodic trimming-back of growing tips. Leaves deep green, divided into five lobes; the two lower lobes may be divided again. Two to four glands on leaf petioles. Perfect flowers, 3 to 4 in across. Five white or pale-pink sepals; five pale-pink petals; white crown with purple base, and shorter than petals. Five stamens united in a tube around the base of the ovary. Superior ovary with three chambers; three styles; three stigmas. Fruit, berry with many seeds.

Flowers of hybrids and horticultural types may be blue, white, blue and white, white and pink, or scarlet. Several species bear edible fruit, for example, *P. edulis*, which produces a purple fruit, 2 to 3 in long, with a hard rind. *Passiflora incarnata* is a wild passion-flower vine found along the southeast coast to Florida and the southern coast to Texas.

CULTURE: Standard method (p. 250), medium to bright light and warm temperature. Provide stakes, twine, or trellis for tendrils. If roots became potbound, transfer to larger pot of fresh soil. Will bloom most of the year. Generally best to cut back close to soil in February, and to replace as much soil as possible without disturbing root.

PROPAGATION: May be started from seed, but some types may not bloom for several years. Germination time is about 30 days. Stem cuttings can be made from March to August, or plants can be obtained locally or by catalog.

Pelargonium (Geranium)

USES: Used regularly in many laboratories for most types of plant studies. Many horticultural types are easily grown and might well give greater variety and added ornamentation to the ordinary laboratory geranium. Much variation is available in color of flowers and foliage, in leaf form, and in inflorescence of the miniatures, the ivies, the regals, and over 200 kinds of scented geraniums. Thus, we can understand why many persons have become geranium hobbyists.

DESCRIPTION: The name *Geranium* belongs to a plant genus most often found in the wild as an annual weed. A few perennial species are grown in outdoor gardens. Practically all indoor plants that are called geraniums are, in fact, species of *Pelargonium*. Many hybrids and cultivars have been developed from *Pelargonium* while the same is not true of *Geranium*. Since the greenhouse plant is commonly designated as a geranium, the name is retained here.

A

B

C

Figure 7-43 *Passiflora caerulea.* **A** Side view of flower showing: the tube formed by the united filaments of stamens; the upper portion of the filaments, each attached to an anther; the superior ovary on top of the stamens; and the three stigmas. **B** Same as **A** except that the flower is tilted forward to give a better view of the crown and the attachment of the upper portion of five filaments to anthers. The five sepals and five petals are similar in appearance and appear to be 10 petals. **C** Vine and tendrils.

Herbaceous annuals or perennials (Fig. 7-44). Often succulent. Mostly opposite leaves, palmately or pinnately veined, sometimes lobed, many times strongly scented. Perfect flowers of many colors. Calyx with nectar spur closely bound to most of the length of the pedicel. (Flowers of *Geranium* have no nectar spur.) Usually five sepals and five petals; two upper petals larger; ten stamens. Superior ovary, five-chambered, and five styles. Dry fruit with five valves that twist as they break open. Much variation, however, in numbers of floral parts among cultivars and hybrids.

Figure 7-44 *Pelargonium hortorum,* geranium.

Pelargonium peltatum Ait., an ivy geranium with red, white, or lavender flowers, is a trailing plant often used in hanging baskets. *Pelargonium hortorum* Bailey is the common plant with succulent stems. Many named cultivars for window and outdoor gardens. Flowers red, white, pink, and salmon. Foliage has a strong, fishy odor. *Pelargonium domesticum* Bailey (the regals: fancy, flower-show geraniums) are erect, soft-hairy, and stems are not succulent. Large, showy flowers, 2 in across, white, pink, dark red, or crimson. At nurseries, the scented geraniums most often will be named according to their particular fragrance, for example: lemon, apple, nutmeg, cinnamon, and rose geraniums. Miniatures are excellent for winter blooming. As a rule, they must be obtained from geranium specialists.

CULTURE: Standard method (p. 250), with medium to bright light and moderate temperature. Put one plant in 5-in pot. Keep slightly potbound, and pinch back the growing tips to make a compact, bushy growth and to induce greater flowering.

PROPAGATION: May be grown from seeds, although seeds are not available for many of the best types. Germination time is 30 days or longer. Easily started at most any time of year from stem cuttings. Obtain plants from local nurseries or by catalog from geranium specialists, listed in Appendix B.

Petunia (Common Garden Petunia; *P. hybrida* Vilm.)

USES: (1) Study of life stages from seed to flower (about 100 days). (2) Study of the perfect flower. (3) As an ornamental, flowering annual that blooms indoors most of the year.

DESCRIPTION: Herbaceous annual; or a perennial generally treated as an annual. Alternate leaves with entire margin. Flowers, 2 to 3 in long, with petals in a tube that widens into broad lobes. Much variety in petal color, from white to red or purple, striped, barred, or with a deep-colored star at base of petals. Sometimes fringed or double. Five-lobed calyx; five stamens, four in pairs and fifth smaller or rudimentary. Superior ovary with two chambers; one style, and stigma simple or lobed. Fruit a capsule with many small seeds. See Fig. 7-45.

CULTURE: Standard method (p. 250), with medium to bright light and warm temperature. Partial shade if outdoors in summer.

PROPAGATION: May be started from seeds. Seeds are very small and can be bought in pelleted form, which means they are coated with a nutrient material that increases their size and makes their handling easier. Germination time is 8 to 10 days. Reed and Ihrig (1968) report that time from seed to flower is about 105 days. Hybrid seeds sometimes have a low viability; additionally, their seedlings may require special care until started. As a rule, buying well-started hybrid plants is the most satisfactory procedure. Buy young plants locally or by catalog. Study the wide selection of types and colors described in catalogs.

Phaseolus (Mung Bean; *P. aureus* Roxb.)

USES: (1) To study gross structure of seed and seedlings. (2) To observe tropisms of seedlings. (3) For tests of viability, and germination requirements.

DESCRIPTION: Annual herb, 2 to 4 ft high; may climb. For description of flower, see *P. vulgaris*. Fruit a pod 2½ to 4 in long with external short hairs, and with small, green, yellow, or brown seeds, about ¼ in long. Seedlings are used as Chinese bean sprouts in Chinese food.

CULTURE: Standard method (p. 250), with medium to bright light and warm temperature. Provide stake or twine for climbing stems.

PROPAGATION: From seeds. Germination time is 2 or 3 days.

Phaseolus (Scarlet Runner; *P. coccineus* L.)

USES: (1) To study life stages from seed to flower (about 50 days). (2) For artificial pollination. (3) For study of flower parts.

DESCRIPTION: Perennial, grown as an annual; twining plant of rapid growth.

Figure 7-45 *Petunia hybrida,* the common petunia.

Leaflets 3 to 5 in long. Bright, scarlet flowers ¾ to 1 in long. Typical bean flower (see *P. vulgaris*). Fruit a pod, 4 to 12 in long. Broad, oblong seed ¾ to 1 in long, and black with red markings. Var. *albus* Bailey (White Dutch Runner): flowers and seeds white. Var. *albonanus* Bailey: bushy plant with white beans, sometimes grown as dwarf lima beans.

CULTURE: Standard method (p. 250), with medium to bright light and warm temperature.

PROPAGATION: From seeds. Germination time is 5 or 6 days.

Phaseolus (Kidney Bean; *P. vulgaris* L.)

USES: To study (1) gross structure of seeds; (2) seed germination, and seedlings; (3) all life stages of a fast-growing plant; and (4) general plant morphology and physiology.

DESCRIPTION: Common garden bean, sometimes called pole bean or string bean. Annual, twining plant with pinnate leaf of three leaflets, each leaflet 3 to 6 in long. White or light purple flowers, ½ to ¾ in long. Perfect flowers with two broad bracts, much like an outer calyx. Five sepals. Corolla as of a typical leguminous "butterfly" flower, with one upstanding, dorsal petal (the standard), two lateral, horizontal petals (the wings), and two ventral, united, and spiraled petals (the keel). The coiled keel is a distinguishing trait of *Phaseolus*. Ten stamens, one free and nine united into a tube that encloses the superior ovary. Slender pod, 4 to 8 in long. Seeds ½ to ¾ in long, white, blue-black, brown, or speckled. Cultivated for edible, dry bean and edible immature pods and seeds (green beans). Common horticultural types are "Pole Bean" and "Kentucky Wonder."

CULTURE: Standard method (p. 250), with medium to bright light and moderate temperature.

PROPAGATION: Grow from seeds. Germination time is 2 to 3 days. Time from seed to seed is 40 to 45 days.

Pinguicula (Butterwort; *P. vulgaris* L.)

USES: To study an insectivorous plant and its environment.

DESCRIPTION: Small perennials that form a rosette of leaves, 2 to 3 in wide, on damp earth or rocks in a cool climate (Fig. 7-46). Grows in New England states, and northern Michigan and Wisconsin, often in association with sundews, pitcher plants, and flytraps. Pale-green or yellow-green leaves, soft, somewhat fleshy, and turned upward around the leaf margin. One flower on stalk (a scape) from center of leaf rosette. Purple flower, ½ to ¾ in long.

Two kinds of glands are located on leaves: mucilage and enzyme. A small insect may become trapped in the mucilage and, perhaps as a result of the insect's movement, the turned-up edges roll toward the center of the leaf. Investigators disagree about the carnivorous nutrition, and the importance of the leaf movement in trapping insects.

CULTURE: See culture of *Dionaea* (venus flytrap).

PROPAGATION: Plant is difficult to propagate in the laboratory. Collect plants in the field, or obtain them from commercial suppliers.

Courtesy of Carolina Biological Supply Company

Figure 7-46 *Pinguicula vulgaris*, butterwort, an insectivorous plant.

Pisum (Garden Pea; *P. sativum* L.)

USES: To study (1) life stages of a fast-growing plant; (2) general plant morphology and physiology; (3) genetic and hormonal interactions (see Lee, 1963).

DESCRIPTION: A quick-growing and easily cultured herbaceous annual, climbing 3 to 6 ft high. Pinnate leaves, each with one to three pairs of leaflets, and with a branched tendril at end of leaf. Leaflets 1 to 2 in long, oval and entire. Large leaflike stipules. Perfect flowers, generally white and solitary. Flower parts similar to the common bean *(Phaseolus vulgaris)*, except keel is not coiled. Fruit, a legume pod with three to eight seeds.

CULTURE: Standard method (p. 250), with medium to bright light and warm temperature.

PROPAGATION: Grow plants from seeds. Germination time is 3 or 4 days. Time from planting seed to flowering is 25 to 30 days.

Raphanus (Radish; *R. sativus* L.)

USES: Easily grown and excellent for many types of plant studies, including (1) seed germination; (2) root hairs on seedlings; (3) tropisms; (4) morphology and anatomy; (5) effects of mineral deficiencies; (6) life stages of a fast-growing plant.

DESCRIPTION: Annual and biennial herb. Thick, fleshy, edible root. Stems 1 to 2 ft high. Pinnately lobed leaves. White to purple flowers. Typical crucifer flower of four petals arranged opposite each other in the form of a square cross, thus giving the name family Cruciferae or "cross-bearing." Perfect flower; six stamens, two shorter than other four; one pistil; superior ovary with two chambers. Fruit a silique, 1 to 3 in long, with 1 to 6 seeds.

CULTURE: Standard method (p. 250), with medium to bright light and warm temperature.

PROPAGATION: Grow plants from seeds. Germination time is 2 or 3 days. Time from seed to flower is 25 to 35 days; from seed to seed, 35 to 50 days.

Ricinus (Castor Bean; Caster Oil Plant; *R. communis* L.)

USES: To study the endospermous, dicotyledonous seed (Fig. 7-47). *Caution:* Seeds are poisonous.

DESCRIPTION: Perennial shrub or tree with herbaceous stem, 30 to 35 ft high in tropical or subtropical regions. As an annual, stems grow 5 to 15 ft high. Large leaves, 2 to 3 ft across, with 5 to 11 palmate lobes. Imperfect flowers, both staminate and pistillate flowers on same plant (monoecious plant); staminate flowers below and pistillate flowers above on terminal panicle. No petals; three- to five-lobed calyx. Many stamens. Ovary with three chambers and one ovule in each. Fruit a spiny capsule. Seed ½ to ¾ in long, with brown and white markings on seed coat. *All plant parts are toxic.*

PROPAGATION: Ordinarily, seeds are planted in outdoor beds. Germination time is 10 to 12 days.

A

B

Figure 7-47 *Ricinus* sp., castor bean. **A** Seeds, ½ to 1 in long. **B** Leaves. Plant is staked with a pole. **C** Fruit. (*C Courtesy of CCM: General Biological, Inc., Chicago.*)

C

Salvia (Sage; *S. pratensis* L.)

USES: (1) For study of flower. (2) For dried flower arrangements. (3) Plants with flowers can be placed in sealed, plastic bags and stored for several months in a freezer compartment for later study of anatomy.

DESCRIPTION: A number of dwarf, winter-blooming, horticultural types are suitable for indoor culture, including those of *S. horminum* L., *S. splendens* Sello, in addition to those of *S. pratensis*. The descriptions given in seed catalogs help in making selections of appropriate types for indoor culture.

Salvia pratensis is a perennial herb, 2 to 3 ft high, with four-sided stems typical of the Labiatae, the mint family. Leaves with edge entire, or sometimes toothed, and red-spotted. Oblong basal leaves, 3 to 6 in long. Blue, sometimes white or red, 1-in flowers, in whorls on a raceme. Two-lipped calyx and two-lipped corolla. One pair of fertile stamens; other pair rudimentary or lacking. Two anther chambers separated on a long, connecting arch that can be observed with a 5 or 10X lens. Superior ovary, with four chambers, and each chamber with one ovule. One style and two stigmas. Fruit consists of four nutlets, each with one seed.

CULTURE: Standard method (p. 250), with bright light and warm temperature.

PROPAGATION: Start plants from seeds, or from stem cuttings of perennial types. Seeds germinate in 10 to 15 days. Seed to flowering time is 85 to 90 days. Put one or two seedlings in a 6-in pot.

Sarracenia (Common Pitcher Plant; *S. purpurea* L.)

USES: To study (1) nutrition of an insectivorous plant; (2) environmental requirements of an insectivorous plant; (3) structure of the leaf trap, called the pitcher; (4) organisms living within the pitcher.

DESCRIPTION: Low, perennial herb, native to Canada, and from Minnesota to East Coast. Other species of *Sarracenia* are found from New Jersey to Florida and along the southern coast to Louisiana. Much variation in size and color of leaves and flowers.

Rosette of leaves from a horizontal rhizome. Hollow, tubular, purple or green leaves, lifting at the apex to make a pitcher or trumpet (Fig. 7-48). Single flowers, about 2 in across, on a bare stalk (scape) that rises 1 to 2 ft from center of leaf rosette. Perfect flowers with three floral bracts, five red to purple sepals, five red to purple petals; many stamens. Superior ovary with five chambers; style spreads to become a five-ribbed, inverted umbrella, with a stigma at the tip of each rib. Fruit a capsule, ½ to ¾ in across, with many light-brown seeds. See West (1965).

The hollow leaves have been much studied for their structure and processes in capturing and digesting insects. Basal portion of leaf is small and solid. Central cavity begins about an inch above the base of the leaf; the cavity expands in diameter along the midpoint of the leaf's length and decreases toward the apex, where the upper edge of the tube makes a rolled collar, called the nectar roll. The underside of the tube continues for about another inch, forming a flap around the tube opening at the end of the leaf. Insects are attracted to the nectar at the opening.

A

B

Figure 7-48 *Sarracenia purpurea*, pitcher plant. **A** A stand of pitcher plants in a bog area of Darlingtonia Park, Oregon. **B** The head of a single plant with insect on exterior.

Down-curved, stiff hairs help to trap the insect, as it slides down the slick sides of the tube into fluid at the bottom of the pitcher. (Kodachromes showing the hairs may be purchased from biological supply companies.) Enzymes in the fluid digest the soft portions of the insect, and the digested products are then available to the plant.

Many kinds of organisms may live in the fluid of the leaf, including bacteria, mosquito larvae and pupae, and maggots of flesh flies. Small spiders may be found on the inner walls of the tube above the fluid. The larvae of certain moths live within the fleshy tissue of the leaves. (See Plummer, 1966; and Plummer and Kethley, 1964.)

CULTURE: Care of the pitcher plant is described in Chap. 6 with the discussion of the bog terrarium.

PROPAGATION: Difficult to propagate in laboratory. Obtain plants from biological supply houses or from plant-specialty companies. See source list in Appendix B.

Sedum (Stonecrop)

USES: (1) To study a fleshy leafed xerophyte and the xerophytic environment. (2) To demonstrate asexual propagation with leaf and stem cuttings. (3) As an ornamental succulent that blooms most of the year.

DESCRIPTION: A large group of plants, native to the north temperate region and tropical mountain areas. Mostly low-growing, evergreen perennials, with thick, succulent leaves that are highly variable in color, size, and shape (Fig. 7-49). Clusters of small, brightly colored flowers through most of year. Species suitable as indoor plants are described below.

A

B

C

Figure 7-49 *Sedum*, stonecrop. **A** *S. spectabili*, view from top of plant showing cluster of unopened flowers. **B** S. morganianum, burro tail, consisting of long, creeping stems with plum, blue-green leaves. **C** *S. reflexum*.

Sedum acre L. Common, perennial stonecrop. Sometimes called golden moss or wall pepper. Often seen in rock gardens and on rock walls. Creeping stems form a mat, with small, bright yellow flowers on stalks 2 to 3 in high. *Sedum album* L. Common in rock gardens. Perennial, creeping, evergreen stems make a green mat. White flowers on stalks that rise 4 to 5 in. *Sedum sieboldii* Sweet. Leaves generally blue or sometimes red. Clusters of pink flowers rise at the ends of long creeping or trailing stems. *Sedum reflexum* L. Creeping, evergreen stems. Bright yellow flowers on a stalk 6 to 12 in high. *Sedum spectabile* Boreau. A showy, perennial with erect (not creeping), unbranched stems growing 12 to 20 in high from long, thick roots. Clusters of pink flowers.

Many horticultural types have been developed from the five species described above. The sedums are grown easily in the laboratory under conditions that are unfavorable for many other plants.

CULTURE: See *Culture of Xerophytic Succulents* (p. 251).

PROPAGATION: Obtain plants locally or by catalog. Plants grow quickly from leaf or stem cuttings. May be grown from seeds, although propagation is slower. Seed germination time is 4 to 6 days. May be planted any time of year.

Sempervivum (Common Houseleek or Hen and Chickens; *S. tectorum* L.)

USES: (1) To study a xerophytic succulent. (2) To observe asexual propagation by offset plants, which accounts for the name of "hen and chickens." (3) As an ornamental foliage and flowering plant. (4) May be used on a plant plank (see Chap. 6).

DESCRIPTION: Fleshy, perennial herb that requires little care (Fig. 7-50). Thick leaves, 1 to 3 in long, and 50 or more leaves in a dense rosette, 2 to 4 in across. Branched panicles of pink to red flowers on hairy, erect stalks 5 to 10 in high. Perfect flowers, ½ to 1 in across. Flower parts generally in multiples of six. Fruit, a follicle.

CULTURE: See *Culture of Xerophytic Succulents* (p. 251). Plants flower better when roots are somewhat potbound; therefore put plants in pot or tray that is only slightly bigger than the plant and 3 to 4 in deep.

PROPAGATION: May be started from seeds but growth is slow. Seeds germinate in about 15 days. Generally more satisfactory to buy a starting plant, and propagate by breaking off and potting the plantlets.

Sinningia (Gloxinia; *S. speciosa* Benth. and Hook.)

USES: (1) Excellent for study of flower, and for self-pollination and artificial cross-pollination. Time from pollination to noticeably enlarged ovules is about 14 days. (2) As ornamental flowering plants, although culture requirements are somewhat strict for continued care.

DESCRIPTION: In addition to *Sinningia*, the gesneriad family includes other interesting and widely cultivated genera, for example: African violets (*Saintpaulia* Wendl.); cape primrose (*Streptocarpus* Lindl.), and *Episcia* Mart. Gloxinias are prize greenhouse plants, and many hobbyists specialize in a variety of horticultural types. A great many cultivars are available with single, double, and ruffled flowers, and a wide range of colors and petal markings.

Low, herbaceous perennials with tuberous rhizome (Fig. 7-51). Oblong or ovate, velvetlike leaves, 3 to 6 in long. Perfect flowers. Calyx of five sepals. Five-lobed corolla, 2 to 3 in long, tubular or opening into a bell, 2 to 4 in across. Five stamens inserted at base of corolla, and arching above to unite at anthers. Half-inferior, single ovary with two chambers; single style, and one or two stigmas. Two-chambered capsule with many, very small seeds.

CULTURE: Soil mix of equal parts of garden loam, peat moss, sand, and leaf mold, or use African violet soil mix from garden store. Give little water until leaves appear on plants; then add water when top surface becomes dry. Do not splash water on foliage. Day temperature of 21 to 24°C. Biweekly feedings of liquid plant food while plant is in active growth period. Medium to bright light (600 to 800 ft-c). For a successful and prolonged blooming, a 60 to 70% humidity is required. Unless a greenhouse or plant chamber is available, a teacher may prefer to buy the budding plants from florists when they are needed for class use.

A

B

Figure 7-50 *Sempervivum*, houseleek, or hen and chickens. **A** *S. tectorum*, common houseleek, showing main plant and small plantlets. **B** *S. calcareum*, houseleek with bright-green leaves tipped with red.

PROPAGATION: May be started from seeds or tubers. The very fine seeds are lightly pressed on the surface of moist germinating medium (see Chap. 6). Germination time is about 15 days. Transfer seedlings to soil mix as described above under Culture. Time from seed to flower is about 7 months.

When starting plants from tubers, remove all but one shoot, and put tuber 1 in

A

Figure 7-51 *Sinningia speciosa*, gloxinia. **A** Entire plant with blossoms. **B** Single blossom, view from top. **C** Close-up of **B**, showing stigma, at right, and five stamens that arch and unite at center left.

B

C

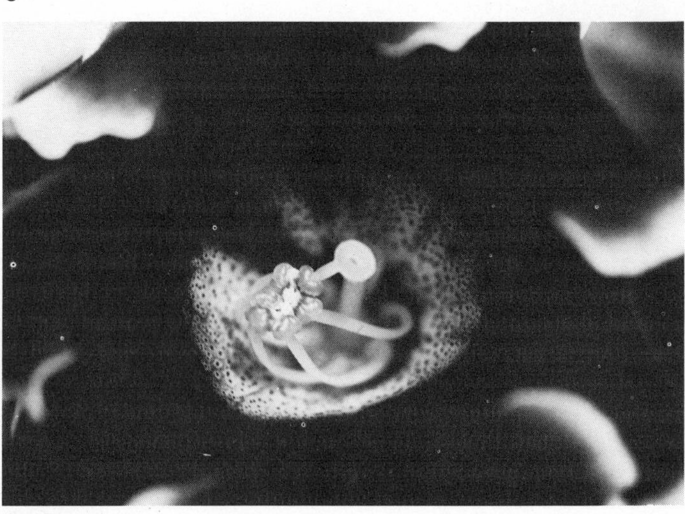

deep in soil mix. The removed shoots may be rooted in moist sand in the manner described for cuttings in Chap. 6. Start tubers in August to October for winter blooming. After bloom is finished, start leaf cuttings if additional plants are desired. Then allow soil of the original plant to dry gradually. Remove dried and yellowed leaves, and store plant in a cool, dark place, with water added every 1 or 2 weeks to keep tubers from drying out. Repot in fresh soil when growth begins again, in about 3 months. Tuberous plants may be kept for 2 or 3 years in this manner, but the plant must be transferred to larger pots as the tuber increases in size. Generally, growth is at a slower rate after the first year, and buds are not as profuse.

Spinacia (Prickly Seed Spinach; *S. oleracea* L.)

USES: (1) To study the complete life stages of a fast-growing plant. Growth from seed to seed requires 40 to 60 days, depending on plant type. (2) For study of plant parts. (3) For tests with plant hormones. (4) Tests of environmental effects on growth and morphology.

DESCRIPTION: The common, garden spinach. Annual, erect herb. Stem grows 6 to 12 in high. Dioecious plant, with small, inconspicuous light-green flowers toward end of season when growing outdoors. Staminate flowers form in leafless spikes or panicles. Pistillate flowers (on another plant) form in axillary clusters. Fruit a two- to four-spined "capsule," formed from leaf bracts and sold as the "seed."

CULTURE: Standard method (p. 250), with medium to bright light and moderate temperature.

PROPAGATION: Obtain seeds locally or by catalog. Seeds germinate in 3 to 4 days. If only the leaves are needed, it is much simpler, of course, to buy the fresh or frozen vegetable at the local market.

Stapelia (Starfish Flower; *S. variegata* L.)

USES: (1) For studying evolutionary convergence of stapeliads, cacti, and euphorbs. (Also see *Euphorbia obesa*.) For examples of convergence compare the morphology of *S. variegata* with that of *Epiphyllum oxypetalum* (orchid cactus), or *Zygocactus truncatus* (christmas cactus). (2) For studying the large and very unusual flowers.

DESCRIPTION: Most types are easily grown. The large, unusual flowers can stimulate students to learn more about living plants. Often the stapeliads are called "carrion flowers" because of the strong, putrid odor of native African flowers that attracts blow flies. Horticultural types have been developed that have large, beautiful, nonodorous flowers. These are described in the catalogs of plant specialists, listed in the source list of Appendix B.

Fleshy, cactuslike plants without leaves (Fig. 7-52). Erect stems are 3 to 8 in long, ½ to 1 in across, four-angled, and generally toothed along the angles. Large flowers, generally 4 to 5 in across, although those of *S. gigantea* may be up to 14 in across. Flowers are perfect, with five, each, of sepals, petals, and stamens. Stamen filaments are united into a tube. Two superior ovaries, each with one chamber and

A

B

Figure 7-52 *Stapelia*, starfish flower. **A** *S. gigantea*, showing 10-in flower and stems. **B** *S. variegata;* flower is 4 to 5 in across. (*A Merry Gardens, Camden, Me. B Courtesy of CCM: General Biological, Inc., Chicago.*)

many ovules. Fused styles and one disc-shaped stigma. A double crown, of five lobes each, arises from staminal tube. Anthers bend inward over stigma. Pollen is formed into one or two sticky masses (pollinia) in each anther. Fruit, a follicle.

In different species and cultivars, flowers may be purple, brown, red, yellow, yellow-green, tan, white, or pink. Many with stripes or dots on petals in contrasting colors. Some with hairy petals.

CULTURE: See *Culture of Xerophytic Succulents* (p. 251). Plants bloom from spring to fall. During summer, plants may be put outdoors in semishade. Check soil each day to determine if water is needed.

PROPAGATION: By stem cuttings. See *Asexual Propagation* in Chap. 6.

Tagetes (Dwarf Marigold; *T. tenuifolia* Cav. var. *pumila*)

USES: (1) To study complete life history of a fast-growing plant. (2) To induce polyploidy by soaking seeds in 0.1 to 2.0% colchicine solution for 6 to 8 h before planting. (3) To study a composite flower.

DESCRIPTION: Branching, annual herb, 8 to 12 in high (Fig. 7-53). Pinnately divided leaves with scented oil glands. Composite head of flowers, ¾ to 1 in across, and bright yellow or gold. Self-fertilizing. Fruit, an achene.

CULTURE: Standard method (p. 250), with bright light and warm temperature.

PROPAGATION: Start from seeds, purchased locally or by catalog. Seeds germinate in 3 to 5 days. Flowers produced in 45 to 60 days; new seeds mature in another 25 to 30 days. Plant seeds indoors at any time of year.

Figure 7-53 *Tagetes tenuifolia*, dwarf marigold.

Utricularia (Bladderworts; *U. vulgaris* L.)

USES: To study the plant's atypical morphology and its method of capturing aquatic organisms by means of the underwater bladders.

DESCRIPTION: According to Willis (1966), 120 tropical and temperate species have been described. All temperate-zone species, including *U. vulgaris*, are aquatic. The morphology of *U. vulgaris* is particularly interesting because of the absence of roots and the slight differentiation in tissues of stems and leaves. Found in slow streams and ponds over much of southern Canada and central, eastern, and southern United States. Consists of finely divided leaves growing underwater, stalks of bright yellow flowers rising above the water, and short shoots containing small leaves that grow upward to water surface (Fig. 7-54). Little difference is apparent in the tissue structure of stems and leaves, and the plant resembles an alga more than a seed plant.

Many small bladders (1 to 2 mm long) grow on the submerged leaves. Each bladder is a hollow, baglike structure with a trap door. When small aquatic crustaceans and other zooplankton brush against the bladder door, it opens inward, and the bladder suddenly enlarges causing a suction that sweeps the plankton into the bag, after which the door closes. Whether or not the plant secretes digestive enzymes is undecided. It is known that small zooplankton die inside the bladder; apparently the digested or decayed organisms are absorbed by the plant tissue. West (1966) says that the fate of ingested prey varies and that some organisms, such as *Euglena*, will live and multiply inside the bladder, whereas *Daphnia*, *Paramecium*, and many others soon die. Tissues of plant may be observed under a 10 to 100X lens. Trapped organisms can be observed inside the small bladder.

CULTURE: Obtain from local streams and ponds, or purchase from biological supply companies. Put plants in an aquarium, or in small containers of pond water (or dechlorinated tap water). We have successfully cultured the plant in Bristol's solution (Appendix A) under continuous light from a 100-W incandescent lamp, placed about 14 in above the culture bowl. Under these conditions, the plant developed many bladders within 2 to 3 days.

A

B

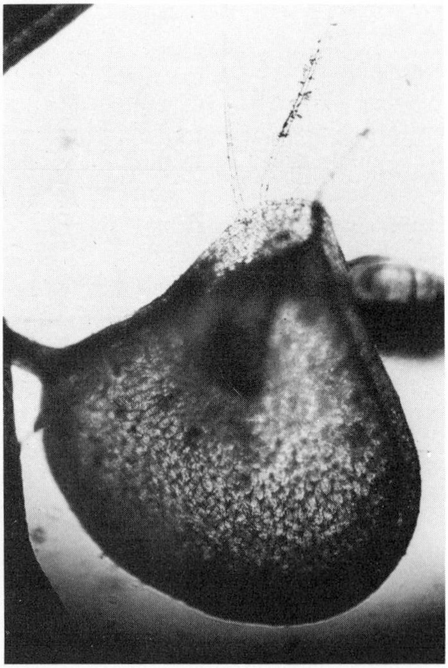

C

Figure 7-54 *Utricularia* sp. **A** Portion of a submerged plant, showing stem, leaves, and bladders. **B** Close-up of a stem, showing branches and bladders on leaves. Bladders are 1 to 2 mm long. **C** Close-up view of a bladder, showing the trap door and the surrounding hairs.

PROPAGATION: Apparently little is known about the reproduction of the plant. When strong growth is obtained, the plant may be divided and parts transferred to another bowl or aquarium.

Vicia (Broad Bean or Horse Bean; *V. faba* L.)

USES: (1) For mitotic studies of stained root-tip cells. (2) To induce polyploidy by soaking root tips in colchicine (see Glass, 1965). (3) For seed germination and seedling studies. (4) To study complete life history of a fast-growing plant.

DESCRIPTION: A climbing, annual herb; grows to 6 ft high; many leaves. White flowers, 1 to 2 in long, with large purple spot. Fruit a large, thick pod up to 1 ft long. Large, flattened seeds, tan, green, black, or purple, and to 1 in across.

CULTURE: Standard method (p. 250), with medium to bright light and moderate to warm temperature. Used most often for seed and seedling studies, although the plant is hardy and can be used readily in place of *Phaseolus*.

PROPAGATION: Start from seeds. Germination time is 3 to 6 days. Time required for growth from seed to seed is about 75 days.

ANGIOSPERMAE: MONOCOTYLEDONAE

Aglaonema (Chinese Evergreen; *A. modestum* Schott)

USES. (1) As a foliage plant that will grow at low light intensities. (2) For study of spathe and spadix floral structures.

DESCRIPTION. Herbaceous hydrophyte, native to India and Malaysia. The house plant grows to several feet in height (Fig. 7-55). Erect stems with glossy, dark-green leaves, growing 15 to 16 in long and 5 in across. Petioles, two-thirds as long as leaf blade. Inflorescence, a thickened spike called the spadix, 1 to 2 in long, with many inconspicuous, naked, imperfect flowers. Spadix is partially surrounded by a green or yellow spathe (a leaf), 2 to 3 in long, open above and shaped somewhat as a boat. Other species with variegated leaves. Red berries are more conspicuous than flowers and may remain on plant for several years.

CULTURE: Standard method (p. 250), with low light and warm temperature. Medium to bright light required for flowering. Roots must become potbound before buds will form; therefore keep mature plant in 4- or 5-in pot. Keep soil moist. Plant will grow in water alone when mineral nutrients are added. A good foliage plant and easy to grow. Tolerates neglect.

PROPAGATION: Obtain starter plant at local nursery or by catalog. Divide rootstocks, or start stem cuttings in water at any time of year.

Aloe (Medicine Plant; *A. barbadensis* Mill.)

USES: (1) As a xerophytic, succulent, foliage plant. (2) Study of sunken stomata and other xerophytic characteristics.

DESCRIPTION: The aloes are standard, indoor plants, easy to grow and attractive as foliage plants. When put in medium to bright light most plants readily produce flowers on long stalks. Other species and many horticultural types are

Figure 7-55 *Aglaonema modestum*, Chinese evergreen. (*Courtesy of CCM: General Biological, Inc., Chicago.*)

available at most local nurseries or by catalog. Generally, a very reduced stem. Rosette of succulent leaves rising from soil (Fig. 7-56). Some types with leaf rosettes on erect stems; some with stolons. Thick, succulent leaves, 1 to 1½ ft long and 2 to 3 in wide, with toothed margins, and ending in a sharp point. The drug aloes, a bitter laxative, is made from the juice of the thick leaves. Inflorescence, a raceme of yellow flowers to 1 in long, rising 1 ft or more above leaves.

CULTURE: See *Culture of Xerophytic Succulents* (p. 251). Avoid overwatering.

PROPAGATION: Separate rootstocks at any time of year. May be grown from seed, but development is very slow.

Amaryllis (Hippeastrum; A. vittata Ait.)

USES: (1) Excellent for germination of pollen tubes. (2) For study of a complete flower. (3) To observe a bulbous plant. (4) For cross-pollination studies.

DESCRIPTION: Horticultural types vary chiefly in flower color, which may be red, white, pink, or orange, and perhaps with stripes or dots on petals. Relatively large bulb, 2 to 4 in in diameter. Straplike leaves, 12 to 18 in long, that appear with or after the flowers. Several large, bell-shaped flowers, 4 to 6 in across, are borne on a long bare-leafed stalk (a scape), that rises from the bulb (Fig. 7-57). Perfect flowers; six petallike structures; six stamens, one pistil with three-lobed stigma. Inferior ovary, three-chambered. Fruit a capsule with many seeds.

CULTURE: Large bulbs produce better flowers and continue to bloom each spring for at least several years if properly maintained. Plant bulbs in November

Figure 7-56 *Aloe barbadensis*, medicine plant. Two young plantlets can be seen at lower front of parent plant.

or December. Simplest procedure may be to purchase a preplanted bulb by catalog or a plant in bloom from a local florist. Leave bulb in same pot for 2 yr or longer. Generally, blooms appear each spring.

Remove old flowers and stalks when blooming is finished. Continue water and plant food to end of summer. Return plant to moderate or low light, lay pot on one side, and give no water or food. Leaves will die in about 3 weeks. Remove leaves and give no food or water until about November 1. When growth is resumed, return plant to bright light and treat as before. Soil around bulb may be replaced, but roots must not be disturbed. Transfer bulb only when its growth has nearly filled the original pot. Make transfer during dormancy, taking care to disturb roots as little as possible.

Figure 7-57 *Amaryllis vittata*. (*Courtesy of CCM: General Biological, Inc., Chicago.*)

PROPAGATION: At time of repotting, separate offset bulbs and put them in fresh soil.

Anthurium (*A. andraeanum* Lind.)

USES: (1) A large, very beautiful "flower," useful for studying spathe and spadix floral structures. (2) For ornamental foliage.

DESCRIPTION: Short, erect stem. Broad, glossy, deep green, net-veined, oblong and sharp-pointed leaves, to 12 in long. Petioles longer than blades. A spectacular and long-lasting floral structure with heart-shaped, waxy spathe, to 6 in long, and of a coral, red, rose, white, or pink color. Yellow-white spadix, to 5 in long, rises from spathe. Small perfect flowers on spadix. Under proper conditions, plant blooms indoors during winter and spring. Other species and many horticultural types are available.

CULTURE: Probably best procedure is to purchase a potted plant during fall or winter months, and to propagate others from the starting plant. Put plant in pot of osmunda fiber or in fir bark (from local garden store). Plant needs about 70% humidity. If no greenhouse is available, lightly wrap stems in damp *Sphagnum* moss, and put pot on a tray of pebbles with water over pebbles; bottom of pot must be above water level. Or cover plant with a large plastic bag. Medium light (800 to 1,000 ft-c), or a bright north window, but no direct sunlight. Keep potting medium slightly moist, but with good drainage from bottom openings to prevent rotting of roots. Warm temperature of 25 to 30°C required for blooming, but foliage can be maintained at a lower temperature. Minimum night temperature is 18°C. Give liquid plant food every 3 to 4 weeks during growth period of winter and spring. Give no food and little water in summer during rest period, but do not let medium become completely dry. Transfer to fresh medium in the fall.

PROPAGATION: During fall months, remove any new offset plants when original plant is repotted. Can be started from seed, but growth is slow and type of plant is sometimes unpredictable. Seeds germinate in 30 days or more. Sow seeds in March or April in moist peat or milled *Sphagnum* moss. Keep trays slightly moist, at warm temperatures of 24 to 27°C and with a high humidity.

Asparagus (Asparagus "Fern"; *A. plumosus* Baker)

USES: (1) As an attractive foliage plant that can be grown in north window or under low to medium light. (2) To observe cladodes, leaflike stems that function as leaves.

DESCRIPTION: Attractive, low-growing plant, with large, flattened, lacy and wide-spreading branches (the cladodes) that arise from reduced, dry, scalelike leaves. Often used by florists as greenery in corsages and floral arrangements. Plant grows 2 to 3 ft high. A long, somewhat fleshy root. A small, inconspicuous white flower and small purple berry. Generally plant does not bloom indoors. *Asparagus plumosus* var. *nanus* is a popular dwarf plant that grows to 18 in high.

CULTURE: Standard method (p. 250), with low to medium light and moderate to warm temperature. Keep plant slightly moist, but with good drainage. Transfer to a larger pot as necessary, or divide rootstock. Cut back plant if it becomes too large.

PROPAGATION: Divide rootstock with foliage at any time of year. Will grow from seeds. Seeds germinate in about 30 days. Probably more satisfactory to buy first plant and propagate others by rootstock.

Avena (Oats; *A. sativa* L.)

USES: For physiological studies (for example, tropisms and auxin assays).

DESCRIPTION: Generally only the grain and seedlings are used in laboratory studies.

An annual cereal grass. Leaf blades about ½ in across and 10 to 12 in high. Inflorescence, a terminal panicle reaching 2 to 3 ft high. Fruit is a grain (a caryopsis fruit), typical of grasses in which the pericarp of fruit and the seed coat are closely united.

CULTURE: May germinate seeds and grow seedlings at any time of year.

PROPAGATION: From seed. Germination time is 2 to 3 days.

The Bromeliads (Pineapple Family)

USES: (1) For ornamental foliage and flowering plant. (2) Study of organisms that live in the plant's cup of water (a small ecosystem).

DESCRIPTION: A very large family of tropical and subtropical plants; mostly epiphytic and with very attractive flowers and foliage. Many horticultural types, and most are grown easily indoors. Bromeliads are as popular with plant hobbyists as are the orchids, African violets, and gloxinias. Among the hardiest of indoor bromeliads are species of *Aechmea* (urn plant), *Billbergia* (vase plant), and other stiff-leafed types (Fig. 7-58). Other popular species are those of *Cryptanthus* (starfish plant), *Neoregelia* (fingernail plant), and *Vriesia* (flaming sword). The Spanish moss (*Tillandsia usneoides* L.) and the edible pineapple (*Ananas cosmosus* Merr.) also belong to this family. See Fig. 7-59.

Most species with many long, slender leaves in a rosette; some with tightly clasping leaves which form a cup that holds water. Generally a reduced stem or

Figure 7-58 *Neoregelia carolinae*, a bromeliad. **A** Side view. **B** View from top showing cup of water, at center, in which small organisms may accumulate.

A

B

Figure 7-59 *Ananas cosmosus*, pineapple plant; a small, greenhouse plant with edible fruit.

no stem, and a reduced root system. Showy flowers, generally perfect, in clusters on spikes or panicles. Rosettes die after blooming but may form young offset plants first. Others may become unattractive because of accumulation of dead leaves at base.

CULTURE: Use growth medium of osmunda or fir bark (both from garden stores), or a mixture of 1:1:2, sand, perlite, and peat moss. Put in clay pots, and enlarge the bottom openings of the pot for better drainage. Pack medium firmly enough to hold plant in place, but not so tightly that water drainage is stopped. Roots must not become waterlogged, but leaves require relatively high humidity. Let medium become slightly dry between waterings, but keep water in the leaf cup. Increase humidity around plant by setting pot on a bowl of pebbles, with water in bowl to a level below the bottom of the pot. Occasionally, turn the plant over to empty all water from the cup and to wipe any mineral deposits from the leaves.

Medium to bright light (500 to 1,000 ft-c, or more for blooming); can be maintained under low light. If transferring from low or medium light to bright light, plant must be moved gradually during several weeks to avoid burning the leaves. Day temperature of 24 to 27°C is optimum, although plants tolerate a cooler temperature.

PROPAGATION: Purchase starting plant from local nurseries or by catalog. Separate and pot young offset plants, which require 1 or 2 years' growth to bloom.

An inexpensive and somewhat novel method of propagation may be demonstrated with the fresh pineapple fruit (*Ananas cosmosus*). Indoor gardeners have used this technique for years to grow attractive foliage plants. Start by slicing off the top of the fruit with the leaves and 1 or 2 in of the top flesh attached (see Fig. 7-60). Use a spoon to scoop from the top slice the fleshy portion encircling the central stem core, taking care not to remove or damage the tough core. Allow the slice to dry for several days. Dust the cut surface with a mold inhibitor (e.g., Mol-

A

B

C

Figure 7-60 Propagation of pineapple plant. **A** Top of fruit with leaves is removed. **B** Inside fleshy portion is taken out leaving center stem. **C** After drying, the exposed surface is sprinkled lightly with root hormone powder, and the plant is placed in germination tray. **D** The plant is then placed in a plastic chamber, under light for 10 to 12 h each day.

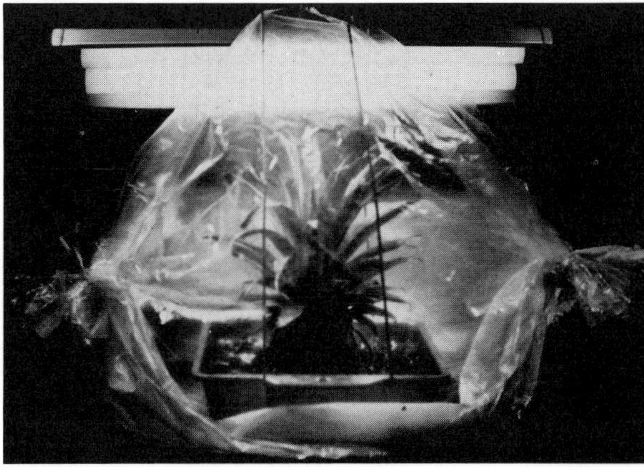

D

dex or Tegosept). Plant in a pot of moist osmunda or other medium described above for bromeliads, covering the slice of fruit to the base of leaves. Place under bright light or in a sunny window at 24 to 27°C, and with humidity as described above under culture. Keep medium slightly moist, but do not overwater. Small pineapple plants that bear fruit have been developed recently and are available at nurseries or by catalog. Culture in same way as described above for other bromeliads.

Chlorophytum (Spider Plant; *C. capense* Kuntze)

USES: (1) To study rhizome with its thickened roots. (2) For asexual propagation by offset plants that form at end of stolons. (3) As an ornamental foliage plant that may flower occasionally. (4) May be used in a hanging basket.

DESCRIPTION: Tropical, foliage plant with a horizontal rhizome from which grow many thick, white, storage roots. Slender, grasslike leaves, sometimes yellow- or white-striped, ½ to ¾ in wide, that rise from rhizome to 12 in high (Fig. 7-61). Flower stalk (scape) rises above foliage with white flowers ¾ in across in a long raceme. Scape may become a stolon by bending over to soil and producing a small terminal plant that forms roots in the soil.

CULTURE: Standard method (p. 250), with low to medium light and moderate to warm temperature. Easily cultured.

PROPAGATION: Remove and pot young offset plants.

Cocos (Coconut Palm; *C. nucifera* L.)

USES: (1) To study structure of fruit and seeds. (2) To germinate and grow young plants as a novelty.

DESCRIPTION: Tropical palm tree, to 100 ft or more tall. Single stem or trunk, generally not straight but curving slightly or leaning. A crown of large, pinnate, drooping leaves, to 15 ft long and 5 ft across. Clusters of floral stalks among leaf bases, with small pistillate and staminate flowers. Two chambers of the three-chambered ovary do not develop. A drupe fruit, 10 to 20 fruits in a cluster, each 10 to 12 in long and 6 to 8 in across. Pericarp of drupe fruit is the outer, thick, fibrous husk; endocarp is the inner, tough, coconut shell (Fig. 7-62). At base of shell are three round marks, or eyes; two are the undeveloped ovary chambers;

Figure 7-61 *Chlorophytum capense*, spider plant. **A** Plant with flowers at top and left. **B** After blooming is finished, a plantlet forms at end of stem, and stem arches downward, as shown.

A

B

Figure 7-62 *Cocos nucifera*, coconut. **A** Coconut fruit with outer dried husk (exocarp and pericarp). **B** Husk has been removed to show the inner, tough shell (endocarp) and the three eyes that mark the sites of the three ovary chambers present in the flower. **C** Opened fruit showing inner seed. Brown layer that lines the shell is the testa of seed. Testa is also seen on removed white, edible portion, the endosperm of seed. **D** Shell with endosperm removed to show small cavity at bottom right, which marks the site of the ovary chamber from which an embryo developed.

under third eye is an embryo. Seed is enclosed in the endocarp, the shell. Thin testa of seed lines the shell. White endosperm of seed ripens into an edible portion. Endosperm encloses a large cavity, filled with a somewhat sweet, watery "coconut milk" that decreases in quantity as fruit ripens. Endosperm, or kernel, is dried into copra from which coconut oil is extracted.

CULTURE: Obtain coconut in husk from a biological supply company. Keefe (1965) has described the following procedure by which he germinated the seed successfully and grew the young plant. Soak the coconut in warm water for 2 weeks. Place it in moist vermiculite, covering the half portion containing the three eyes, and leaving the upper half exposed. At least several months are required for roots and shoots to appear. Keefe records the germination time as 6 months. Transfer seedling, still attached to the coconut, to a 6-in or 8-in pot of 1:1:2 sand, perlite, and peat moss. Keep the growth medium slightly moist, with seedling under medium to bright light and temperatures of 25 to 27°C. Give liquid plant food every 3 or 4 weeks. At the time of Keefe's report, the plant had produced seven fanlike leaves.

Colchicum (Autumn Crocus or Meadow Saffron; *C. autumnale* L.)

USES: (1) To demonstrate propagation by means of a corm. (2) For unusual "dry" culture without soil or water. (3) For showy flowers during fall months.

DESCRIPTION: The plants bloom during late September and October when one to four floral tubes emerge from soil, with ovary and short flower stalk remaining underground (Fig. 7-63). Purple, white, or yellow flowers, 2 to 4 in long and 3 to 4 in across. Six petallike structures; six stamens; pistil with three-chambered ovary and three styles. Several straplike leaves, to 12 in long and 2 in wide, appear in the spring after flowering. At the same time, the underground flower stalk grows upward carrying the ovary above ground. Fruit, a capsule with many seeds. *Note: Colchicum* is poisonous. Colchicine, a poisonous alkaloid, is extracted from seeds and corm, and is used in plant research and sometimes medicinally for gout and other ailments.

CULTURE: Obtain corms in August or early September. Study catalogs and place orders well in advance. (See Appendix B.) For dry culture, lay corms in east or south window, or in a container under bright light (1,000 ft-c, or more). Corms

Figure 7-63 *Colchicum autumnale*, autumn crocus. (*Courtesy of CCM: General Biological, Inc., Chicago.*)

will produce bloom with no further care, an interesting demonstration. For soil culture, put five or six corms in a 5-in pot of soil mix, consisting of 1 part, each, of loam, peat moss, and sand. Plant corms about 1 in deep in soil. Water thoroughly and place in dark at cool temperature until flower stalk appears. Keep soil slightly moist. Transfer to medium or bright light and warm temperature. Give liquid plant food every 3 to 4 weeks. Add water when top soil surface becomes dry. Reduce water and food during summer dormancy. Plant may bloom for several years without repotting.

PROPAGATION: Use a sharp blade to separate root offsets when plant is repotted. It may be more satisfactory to buy fresh corms, which are relatively inexpensive. One source for the autumn *Colchicum* is P. de Jager & Sons, Inc., South Hamilton, Mass. 01982. Place the order before September 1.

Dieffenbachia (Dumb Cane; *D. seguine* Schott)

USES: (1) Excellent for hand-cut root sections, and for demonstrating effects of root pressure on water flow through stem (Tamhane, 1970). (2) As an ornamental, foliage plant that grows in low or medium light. (3) To study water mounts under magnification of gently macerated leaf tissue for observation of spindle structures that contain and eject needlelike calcium oxalate crystals, called raphides (see Middendorf, 1968). Note the caution given below under Description. (4) A good plant for demonstrating propagation by air layering of stem.

DESCRIPTION: Plant has been given the name "dumb cane" because of the calcium oxalate needlelike crystals, raphides, throughout the plant structures, which produce a paralysis of throat for several hours to several days if plant portions are chewed. According to historical accounts, the plants were used in the West Indies for torturing slaves. Similar effects may be produced on the external skin if the crushed plant is handled. Therefore, a person should wear gloves when working with plants, and he should take caution not to rub crystals into the eyes.

Easily cultured and very hardy. Attractive foliage with large, glossy, pinnately-ribbed, oblong leaves, to 12 in long and 4 in wide (Fig. 7-64). Leaves may be variegated. Stems 3 to 4 ft high. Plant produces typical spadix inflorescence, with a lower spathe (see description of *Aglaonema*). A great many other species and horticultural types have been developed. Primary distinctions are variations in color and pattern of leaf markings.

CULTURE: Standard method (p. 250), with low or medium light and moderate to warm temperature. A good plant for a dark corner in laboratory. If stem becomes tall and loses lower leaves, cut off top portion and transfer this top portion to another pot. Remove the lower stem, and use 4-in portions for cuttings. The last remaining few inches of stem may produce new leaves.

PROPAGATION: With stem cuttings and air layering.

Dracaena (*D. sanderiana* Sander)

USES: (1) A hardy and attractive foliage plant for growth in low to medium light. (2) For general studies of plant physiology and histology.

Figure 7-64 *Dieffenbachia picta*, dumb cane.

DESCRIPTION: Tropical plant with erect stem, 15 to 18 in high (Fig. 7-65). Drooping, long-pointed, narrow blades, 6 to 8 in long and 1 in wide, with white margins and sometimes white stripes. Small, inconspicuous clusters of green, white, or yellow bell-shaped flowers. Rarely blooms indoors. Many other horticultural types of various leaf sizes, shapes, and markings.

CULTURE: Standard method (p. 250), with low to medium light and moderate to warm temperature. Hardy indoors. May shed lower leaves.

PROPAGATION: Stem cuttings. See *Asexual Propagation* in Chap. 6.

Goodyera (Rattlesnake Plantain or Woodland Orchid; *G. pubescens* R. Br.)

USES: Ornamental plant for a woodland terrarium.

DESCRIPTION: Attractive, low-growing land-orchid, native in woodland areas of northwestern, north-central and northeastern United States, and southern Canada. Thick, tough, underground rhizome. Dark-green, oval leaves, 2 to 3 in long, and with white "rattlesnake" markings. May bloom under good terrarium conditions. Hairy flower stalk, 6 to 8 in high or higher; small, white flowers in a raceme.

CULTURE: See *Woodland Terrarium*, Chap. 6. Suppliers are listed in Appendix B.

PROPAGATION: Separate rootstock with upper leaves. A. E. Allgrove, North Wilmington, Mass. 01887, sells rhizomes in a kit for educational purposes.

Figure 7-65 *Dracaena borinquensis*, corn plant.

Hyacinthus (Hyacinth; *H. orientalis* L.)

USES: (1) To observe a bulbous plant. (2) For fragrant winter and spring flowers.

DESCRIPTION: Study the catalogs for species and many horticultural types. Globular bulbs, to 3-in diameter. Narrow leaves, 8 to 11 in long and 1 in wide. Dense cluster of flowers in a terminal raceme that rises on a stalk (scape) above the narrow leaves. Pink, purple, blue, white, yellow, or red flowers, produced indoors during winter or spring months. Each flower about 1 in long with six petallike segments, six stamens, one pistil, and three-chambered ovary with a knoblike stigma. Fruit a capsule with many seeds.

CULTURE: Obtain bulbs locally or by catalog. Medium-sized bulbs may be more desirable than large ones. "Jumbo" bulbs are more costly, and produce large, heavy flowers that must be staked. Buy precooled bulbs for planting during September to December. Blooming requires about 3 months. Depending on size, put one to three bulbs in a 6-in pot of soil mix, consisting of 1 part, each, sand, peat moss, and garden loam, with a ½-in layer of the soil mix over bulb tips. Water thoroughly; put in low light or dark at a cool temperature until shoots are several inches high. Transfer to medium light for 1 week, retaining the cool temperature; then transfer to bright light under medium temperature. Blossoms will open within several weeks. Keep soil slightly moist at all times.

Plants may also be grown easily in water. Suspend a bulb in a container of water, with the bottom of the bulb barely touching the water surface. A special hyacinth glass at a garden store is most satisfactory for this purpose. Also grown on pebbles as described for *Narcissus*. Place in a cool, dark spot and treat the same way as described for soil culture. The plant will bloom profusely if well rooted in the dark and then gradually transferred to bright light.

PROPAGATION: Buy fresh bulbs each year, as forced bulbs do not bloom well indoors during a second season. Possibly they will bloom if planted outdoors for the second season.

Lemna (Duckweed; *L. minor* L.)

USES: (1) Excellent for many types of physiological and environmental tests, for example, photoperiodism, nutrition tests, and carbon 14 cycle. (See Yurkiewicz et al., 1970, and Hodgson, 1970.) Plant is much used in research, and deserves more use in beginning courses. (2) For study of unique structure and reproduction. (3) Excellent for studying plant cells and chloroplasts, and for roots on which root cap can be seen without magnification. (4) As an aquarium plant, although it may reproduce rapidly and become a nuisance.

DESCRIPTION: This species and others are common in quiet fresh water over United States. *Lemna gibba* and *L. perpusilla* are also used in laboratories. *Lemna* is the simplest and among the smallest of flowering plants. Consists of a flat, oval, green, leaflike floating frond, 1.5 to 3 mm across, and with one rootlet (Fig. 7-66). No stem or leaves. Fronds are slightly turned up at ends, thus causing them to congregate, sometimes in large bunches, due to surface tension. Meristematic groove on each side of frond at one end, from each of which a new plant arises that may remain attached to parent plant or become separated. In autumn many small fronds, sometimes called bulblets, are formed; parent plants sink to bottom of

Figure 7-66 *Lemna minor*, duckweed. **A** Top view of floating plants. **B** Side view that shows leaflike structure on water surface and rootlets, 1 to 2 in long, hanging below in water.

A

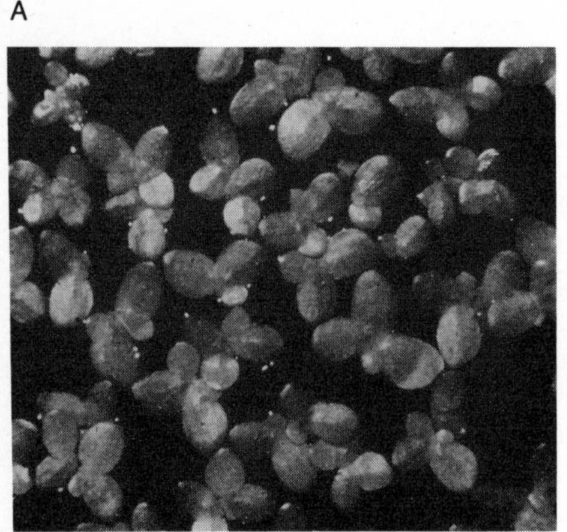

B

Figure 7-67 *Maranta leuconeura,* prayer plant. (*Merry Gardens, Camden, Me.*)

water, where they remain through winter. The following spring, they rise to the surface and begin growing again.

Very reduced flower stalks may also arise from grooves. Each stalk with two staminate flowers (each flower with one stamen only), and one pistillate flower with one pistil. Rhodes (1968) describes techniques for producing flowers in *L. perpusilla* under an 8-h day in Hoagland's (M) solution, and in *L. gibba* under a 16-h day in the same medium.

CULTURE: May be grown in freshwater aquarium or on sterile growth medium, for example, Hoagland's, modified (see Appendix A). Collect plants or obtain from biological supply companies.

PROPAGATION: Plants readily propagate by growing new fronds. Occasionally, the very small seeds may be collected from plants on growth media. Seeds germinate in 3 days (Postlethwait and Enochs, 1967).

Maranta (Prayer Plant; *M. leuconeura* Morr.)

USES: (1) To study leaf movements at night, or in dark. Movement is controlled by a swollen pulvinus at base of leaf petiole; action is similar to that of *Mimosa* but apparently leaves are not as sensitive. (2) As an ornamental, foliage plant that grows in low light or in north window.

DESCRIPTION: Perennial herb with short stem, to about 1-ft high (Fig. 7-67). Showy, oblong leaves, 7 to 8 in long, and variously marked in white, purple, or red, depending on type. Leaves fold upward at night, accounting for name. Rarely blooms indoors. Small, white, perfect flowers, several in a cluster on a flower stalk from leaf axil. *Maranta leuconeura* var. *kerchoveana* Morr. is a popular, attractive type for indoors. *Maranta arundinacea* L. (arrowroot) is grown in the tropics for its starchy rhizome, a source of arrowroot starch from which tapioca is made. Indians used the root as an antidote for poisoning from arrows.

CULTURE: Standard method (p. 250), with low to medium light and warm temperature. Keep soil slightly moist, and put pot on tray of pebbles with water over pebbles to increase humidity. Do not allow water level in tray to reach base of pot.

PROPAGATION: Make stem cuttings, or separate rootstock.

Monstera (Split Leaf Philodendron; *M. deliciosa* Liebm.)

USES: (1) To study the climbing and absorbing adventitious roots. (2) As an ornamental, foliage plant. (3) To demonstrate negative geotropism of adventitious, climbing roots and positive geotropism of adventitious, absorbing roots. (4) To study stomatal patterns.

DESCRIPTION: Tropical vine with large, leathery, pinnately and net-veined, dark-green leaves, 2 to 3 ft long (Fig. 7-68). As the young leaves grow, they become perforated, or deeply notched, due to slow growth and splitting of tissue between lateral veins. In nature, plant is rooted in soil but climbs on palms and other trees, perhaps to 100 ft high. Finally, it becomes epiphytic, as absorbing, aerial roots grow downward into soil, and as adventitious, climbing roots become attached to the support. Needlelike calcium oxalate crystals, called raphides, in stem and leaf tissue, but not as abundant or as toxic as those in *Dieffenbachia*.

Rarely blooms indoors. In nature, many perfect, sessile flowers form on erect, thick stalk (spadix) 8 or 9 in long, and encircled by a longer, white bracted leaf (the spathe). See description of floral structure of *Aglaonema*. *Monstera* often is called *Philodendron*. Foliage of both groups is similar; botanical distinction is based primarily on the flower.

CULTURE: Easily grown and tolerant of neglect, except for overwatering and poor drainage. Standard culture (p. 250), with low to medium light and moderate to warm temperature. Increase humidity by placing pot on tray of pebbles with water over pebbles. Water level must not reach bottom of pot. Water plant

Figure 7-68 *Monstera deliciosa*, split-leaf philodendron. Two aerial roots are seen at center bottom.

thoroughly at least once a week, and more often in dry atmosphere. In the soil at one side of plant, sink a gardener's "totem" pole or other water-absorbent support for climbing, for example, chicken wire fashioned into a pole and filled with *Sphagnum*. Add water to the support regularly.

Wipe dust from leaves as necessary, or wash with soap and water. Appearance is improved by wiping top surface of leaves with several drops of glycerin. Do not use on lower surfaces, as stomata may become clogged. The many long, absorbing roots become unattractive and may be pushed into soil or totem pole, or they may be cut off.

PROPAGATION: From stem or leaf cuttings that are rooted in water or soil.

Narcissus (Narcissus, Daffodil, or Jonquil)

USES: (1) Easily grown for winter blooms. (2) For cross-pollination studies; time from planting bulb to mature seed is 8 to 10 weeks.

DESCRIPTION: Technically, daffodil, narcissus, and jonquils are all of the same genus, *Narcissus*, but common usage has divided them into the three groups. Generally, daffodils are designated as those with a long trumpet or crown in the center of the flower; the narcissus as those with small flowers and a shallow cup in the flower center; and jonquils as the small, fragrant, yellow bloomers of the species *N. jonquilla* L. The designations are indefinite and overlapping.

Bulb, 1 to 2 in in diameter. Straplike leaves, 10 to 18 in long and 3/4 to 1 in wide (Fig. 7-69). Several yellow or white flowers borne on an erect unbranched stalk (a scape), 10 to 18 in high. Flowers, 1 to 2 in across (horticultural types to 5 in across). Six petallike segments, and an inner tubular crown (corona), which is a long trumpet or a shallow cup, and sometimes of different color from that of outer petals. Stamens six. Inferior, three-chambered ovary; style with a three-lobed stigma. Fruit a capsule with many seeds. The many horticultural hybrids and cultivars cannot be clearly defined botanically. Study catalogs for the wide range of types.

CULTURE: Many types are described in catalogs as suitable for indoor culture. Obtain bulbs in early fall, September to November. Select those with two or three buds, which give more stalks and flowers. Depending on size, plant four to eight bulbs in 6-in pot of soil mix of 1 part, each, loam, peat moss, and sand, with pointed tips of bulbs just above soil surface. Water thoroughly and put in the dark at a temperature of 10 to 12°C until shoots are 3 to 4 in high. Transfer to medium light and a temperature of about 15 to 16°C for 3 weeks; then put under bright light at warm temperature.

A simple and very satisfactory method is to place the bulbs among pebbles with water barely over the pebbles. Varieties often grown in water are *N. tazetta* L. var. *papyraceus* (paper-white narcissus), and *N. tazetta* L. var. *orientalis* (a robust, yellow narcissus). Put six to ten bulbs among pebbles in a bowl, twice as deep as bulbs and 3/4 full of pebbles. Bury about one half of the bulb among the pebbles. Add enough water to barely cover pebbles. Keep in cool, dark place until roots are well formed and shoots have emerged, generally about 2 to 3 weeks. Transfer to medium light and moderate temperature for 1 or 2 weeks. When shoots have

Figure 7-69 *Narcissus* sp. (*Courtesy of CCM: General Biological, Inc., Chicago.*)

turned green, put in sunny window or bright light with moderate to warm temperature. Some persons leave bulbs in the room at normal room temperature, without the first dark treatment. Lack of dark, cool treatment, however, may result in long leaves and no flowers. Both varieties bloom in 6 to 8 weeks from time of planting in water.

PROPAGATION: Bulbs that are forced to bloom early indoors cannot be successfully forced a second season. Obtain fresh bulbs each fall for indoor culture.

Philodendron (Philodendron Ivy)

USES: Same as for *Monstera*.

DESCRIPTION: Erect, shrublike or branching and climbing vines (Fig. 7-70). Many species native to tropics, and many horticultural types available. Two kinds of aerial roots: climbers that anchor to a support, and absorbers that grow downward into soil. Stems often woody and branching, with long, obvious internodes. Parallel-veined leaves, oblong, sometimes perforated or deeply notched; without net veins. Vine types need a support. Rarely blooms indoors. Spadix with many staminate and pistillate flowers. Spathe sometimes colored. Some philodendrons and monsteras are much alike and are commonly confused. Technically, the two are separated by the net veins of *Monstera* and the lack of net

A

B

Figure 7-70 *Philodendron*, ivy. **A** *Philodendron* sp., common variety. **B** *Philodendron panduriforme*, fiddle leaf ivy, at top. (*B Merry Gardens, Camden, Me..*

veins in *Philodendron*, and by the differences in the floral structures of the two genera.

CULTURE: See *Monstera*.

PROPAGATION: Root stem cuttings in water or in soil.

Poa (Kentucky Bluegrass; *P. pratensis* L.)

USES: (1) For germination tests, seedling studies, and other physiological tests. (2) Use potted plants for food in animal cages and terraria. (3) To illustrate propagation by stolons. (4) Ornamental grasses of other species are used for dried bouquets.

DESCRIPTION: Perennial grass with stolons. Soft to firm stem, 1 to 2 ft high. Used as a soft, lawn grass in cooler, northern states. Life cycle from seed to seed is about 100 days. A fast-growing, annual bluegrass, *P. annua* L. is useful in physiological tests and chromosomal studies of developing pollen. Mixed seeds of ornamental grasses are available from seed companies.

CULTURE: Sow seeds in trays or pots. Use Standard culture method (p. 250), with moderate temperature and medium light.

PROPAGATION: Plant annual or perennial seeds, or transplant sections of perennial grass. Seed germination time varies among species from 2 to 3 days to 2 weeks.

Rhoeo (Moses-in-a-Boat; *R. discolor* Hance)

USES: (1) Hardy ornamental plant; can be used in hanging baskets. (2) Useful for many types of physiological and morphological studies. (3) To observe cy-

Figure 7-71 *Rhoeo discolor*, moses-in-a-boat. (*Merry Gardens, Camden, Me.*)

toplasmic flow in cells of hairs on staminate filaments. (4) To study meiotic divisions in the stained spore mother cells of pollen. See techniques for Acetocarmine Stains in Appendix A.

DESCRIPTION: Perennial, low herb with short stems, 8 to 10 in long (Fig. 7-71). Leaves dark green on top and deep purple beneath, sword-shaped, 6 to 10 in long and 1 to 3 in wide, in thick tufts on stem. Clusters of many small, white flowers, ½ in long, are crowded into boat-shaped bracts down among the dense leaves. Sepals and petals, three each. Sepals wither when flower opens. Six stamens, hairs on filaments; superior ovary with three chambers and one ovule in each chamber.

CULTURE: Standard method (p. 250), with low, medium, or bright light, and in moderate or warm temperature. Hardy and tolerates neglect. Will produce flowers most of year in medium or bright light.

PROPAGATION: Make stem cuttings; separate rootstock; or plant seed taken from the "boats."

Sansevieria (Mother-in-law-Tongue or Snake Plant)

USES: (1) To demonstrate asexual propagation by rooting portions of the leaves. (2) As an unusually hardy, ornamental, foliage plant that will produce fragrant white flowers when given optimum growing conditions.

DESCRIPTION: Long, slender, rigid leaves, 2 to 3 ft long and 1½ to 2½ in wide, ending in a sharp point (Fig. 7-72). Leaves grow in a cluster from underground, horizontal rhizome and are variously striped or marbled with yellow, white, or darker green. May bloom irregularly in bright light. Fragrant, small, greenish-white flowers borne on a slender stalk, 18 to 24 in high. Much confusion exists in nomenclature and description of cultivated types. *Sansevieria trifasciata* Prain var. *laurentii* is often named in gardening literature as a common type.

CULTURE: Standard culture (p. 250), with low, medium, or bright light, and moderate or warm temperature. Perhaps the hardiest of indoor plants. Produces

Figure 7-72 *Sansevieria* sp., mother-in-law-tongue.

good foliage even with neglectful treatment; consequently, the plant is often neglected. With regular watering and bright light, the foliage becomes much improved, and stalks of fragrant, white flowers appear irregularly throughout the year.

PROPAGATION: Slice through rhizome and separate leaves and rootstock. Studies of leaf cuttings can be fascinating for a class. The technique is described under *Asexual Propagation* in Chap. 6.

Scindapsus (Devil's Ivy or Pothos; *S. aureus* Engler)

USES: (1) Ornamental foliage plant. (2) For general physiological studies.

DESCRIPTION: Long, somewhat woody stems that climb with rootlets. Oblong, glossy, waxy, pointed leaves, sometimes with yellow or white markings (Fig. 7-73). Perfect flowers; but rarely, perhaps never, blooms indoors. Much like *Philodendron* and *Monstera*, and the three genera may be confused. *Scindapsus* is, in general, distinguished from the other two by the somewhat smaller leaves with unbroken, waxy, and variegated blades.

CULTURE: See *Philodendron*.

PROPAGATION: By stem cuttings in water or soil.

Sorghum (Common Sorghum; *S. vulgare* Pers.)

USES: (1) For seed germination tests. (2) Physiological tests with seedlings. (3) Genetic studies of plant color; obtain genetic seeds from biological supply compa-

Figure 7-73 *Scindapsus* sp., devil's ivy.

nies. (4) To study complete life cycle of a fast-growing plant; time from seed to seed is about 90 days.

DESCRIPTION: Erect, stout annual; 4 to 8 ft high, or more, outdoors. Grown as a crop plant for grain, syrup, and forage. Stem ½ to 1 in in diameter. Inflorescence a terminal panicle containing pairs of spikelets, with small, perfect florets. A grain fruit (caryopsis).

CULTURE: Ordinarily the mature plant is grown only as an outdoor crop. Seeds and seedlings are used for laboratory purposes. Germination time is 3 to 5 days.

Tradescantia (Wandering Jew; *T. fluminensis* Vell.)

USES: Same as for *Rhoeo*. *Tradescantia* is particularly useful for observing cytoplasmic movement in cells of staminate hairs, and for studying meiosis in the stained, developing pollen. Generally, N = 6, or a multiple of 6 (Handlos, 1970).

DESCRIPTION: Perennial herb; crawling stems to 3 ft long, readily rooting at nodes (Fig. 7-74). Oblong leaves to 3 in long. Terminal clusters of flowers; two long leaves (bracts) extending from under the cluster of flowers. Three green sepals that remain when flower opens; three petals, purple, white, or pink, and about ¼ in long; six stamens with hairs on filaments. Three-chambered ovary and two ovules in each chamber.

CULTURE: Standard method (p. 250), with low, medium, or bright light and moderate or warm temperature. Hardy plant that tolerates neglect.

PROPAGATION: Transplant portions of stems with rooted nodes.

Zea (Maize or Indian Corn; *Z. mays* L.)

USES: (1) Commonly used for studies of seed germination and seedlings. (2) For genetic studies. Obtain genetic seeds from biological supply companies.

Figure 7-74 *Tradescantia virginica*, wandering Jew. (*Courtesy of CCM: General Biological, Inc., Chicago.*)

DESCRIPTION: A stout, annual, cereal plant, grown for grain and fodder. Unbranched stem, 3 to 10 ft high. Long, narrow leaves, one at each node. Tassels of staminate florets at stem tip. Pistillate florets formed in 8 to 24 rows on a flower stalk (a rachis), which becomes the cob. Long styles are the silk of corn, one running from each pistil; fruit, a one-seeded grain. Pericarp of fruit and the seed coat are grown together, and are not easily separated except by special milling processes.

CULTURE: See Chap. 6 for general methods of seed germination and treatment of seedlings.

PROPAGATION: A forage crop plant, ordinarily not grown in the laboratory. Time of seed germination is 3 to 5 days.

Zebrina (Zebra Plant; *Z. pendula* Schnizl.)

USES: See *Tradescantia* and *Rhoeo*.

DESCRIPTION: Trailing stems, 2 to 3 ft long. No roots at nodes as in *Tradescantia fluminensis*. Oblong leaves, 1 to 3 in long, purple beneath and white and green striped above, accounting for "Zebrina" name. Small purple or pink flowers between two leaf bracts, one bract about one-half the size of the other. Three, each, of sepals and petals; six stamens; ovary with three chambers and two ovules in each chamber. Much confusion exists in distinguishing *Tradescantia* and *Zebrina*. Descriptions given here are based on Bailey (1949).

CULTURE: See *Tradescantia*.

PROPAGATION: By stem cuttings.

Figure 7-75 *Zebrina* sp., zebra plant.

REFERENCES

Abraham, G.: 1967, *The Green Thumb, Book of Indoor Gardening,* Prentice-Hall, Inc., Englewood Cliffs, N.J., 304 pp.

Bailey, L. H.: 1949, *Manual of Cultivated Plants,* The Macmillan Company, New York, 1116 pp.

Creighton, H. B.: 1965, "A Sustained Study of Plant Growth for an Introductory Course," *Am. Biol. Teacher,* **27** (2): 97–100.

DiPalma, J. R., R. McMichael, and M. DiPalma: 1966, "Touch Receptor of Venus Flytrap, *Dionaea muscipula,*" *Science,* **152:** 539–540.

Fondeville, J. C., H. A. Borthwick, and S. B. Hendricks: 1966, "Leaf Movements of *Mimosa pudica* L. Indicative of Phytochrome Action," *Planta,* **69:** 357–364.

Glass, B.: 1965, *Genetic Continuity,* BSCS Laboratory Block, D. C. Heath and Company, Boston. Student Manual, 154 pp.; Teacher's Supplement, 89 pp.

Graf, A. B.: 1970, *Exotica 3; Pictorial Cyclopedia of Exotic Plants: Guide to Care of Plants Indoors,* Roehrs Co., Rutherford, N.J., 1834 pp.

Handlos, W. L.: 1970, "Cytological Investigations of Some Commelinaceae from Mexico," *Baileya,* **17** (1): 6–33.

Hodgson, G. L.: 1970, "Effects of Temperature on the Growth and Development of *Lemna minor,* Under Conditions of Natural Daylight," *Ann. Botan.,* **34** (135): 365–381.

Hooft, J.: 1970, "Zamia From Seed," *Carolina Tips,* **33** (6): 21–22.

Jacobs, W. P., and C. E. Lamotte: 1964, *Regulation in Plants by Hormones, A Study in Experimental Design,* BSCS Laboratory Block, D. C. Heath and Company, Boston. Student Manual, 116 pp.; Teacher's Supplement, 27 pp.

Jaffe, M. J.: 1970, "Reversible Force Transduction in Tendrils of *Passiflora coerulea,*" *Plant and Cell Physiol.,* **11** (1): 47–53.

Jump, J. A.: 1968, "The Xerophytic Mimicry Plants," *Am. Biol. Teacher,* **30** (3): 201–205.

Keefe, A. M.: 1965, "Plants for a Window Garden," *Am. Biol. Teacher,* **27** (2): 118–123.

Lee, A. E.: 1963, *Plant Growth and Development,* BSCS Laboratory Block, D. C. Heath and Company, Boston. Student Manual, 88 pp.; Teacher's Supplement, 69 pp.

Middendorf, E. A.: 1968, "Plants with Blowguns," *Turtox News,* **46** (5): 162–164.

Mozingo, H. N., P. Klein, Y. Zeevi, and E. R. Lewis: 1970, "Venus's Flytrap Observations by Scanning Electron Microscopy," *Am. J. Botan.,* **57** (5): 593–598.

Plummer, G. L., and J. B. Kethley: 1964, "Foliar Absorption of Amino Acids, Peptides and Other Nutrients by the Pitcher Plant, *Sarracenia flava*," *Botan. Gaz.*, **125:** 245–260.
Plummer, G. L.: 1966, "Foliar Absorption in Carnivorous Plants—Part I," *Carolina Tips*, **29** (7): 25–26.
Plummer, G. L.: 1966, "Foliar Absorption in Carnivorous Plants—Part II," *Carolina Tips*, **29** (8): 29–30.
Postlethwait, S. N., and N. J. Enochs: 1967, "Tachyplants Suited to Instruction and Research," *Plant Sci. Bull.*, **13** (2): 1–5.
Reed, E. W., and E. A. Ihrig: 1968, "Common Plants Useful for Classrooms," *Plant Sci. Bull.*, **14** (2): 2–6.
Rhodes, L. W.: 1968, "The Duckweeds: Their Use in the Laboratory," *Am. Biol. Teacher*, **30** (7): 548–551.
Schwab, V. W., E. Simmons, and J. Scala: 1969, "Fine Structure Changes during Function of the Digestive Gland of Venus's-Flytrap," *Am. J. Botan.*, **56** (1): 88–100.
Sibaoka, T.: 1969, "Physiology of Rapid Movement in Higher Plants," *Ann. Rev. Plant Physiol.*, **20:** 165–184.
Stevenson, F. F.: 1965, "A Technique for Making Permanent Slides of Pollen," *Am. Biol. Teacher*, **27** (2): 124–126.
Tamhane, S. N.: 1970, "Plants as Teaching Aids, *Dieffenbachia picta* Schott," *Turtox News*, **48** (1): 35.
Webb, S. D.: 1968, *Evolution*, BSCS Laboratory Block, D. C. Heath and Company, Boston. Student Manual, 53 pp.; Teacher's Supplement, 55 pp.
West, W. R.: 1965, "Carnivorous Plants—Part I," *Carolina Tips*, **28** (8): 31–32.
West, W. R.: 1965, "Carnivorous Plants—Part II," *Carolina Tips*, **28** (9): 35–36.
West, W. R.: 1965, "Carnivorous Plants—Part III," *Carolina Tips*, **28** (10): 39–40.
West, W. R.: 1966, "Carnivorous Plants—Part IV," *Carolina Tips*, **29** (1): 1–2.
Willis, J. C.: 1966, *A Dictionary of the Flowering Plants and Ferns*, 7th ed., revised by H. K. Airy Shaw, Cambridge University Press, London, 1214 pp. and 53 pp. of keys.
Yurkiewicz, W. J., J. C. Parks, and G. L. Steucek: 1970, "Cockroach and Duckweed Show the Carbon Cycle," *Am. Biol. Teacher*, **32** (5): 296–297.

Other Literature

Bold, H. C.: 1967, *Morphology of Plants*, 2d ed., Harper & Row, Publishers, Incorporated, New York, 541 pp.
Esau, Katherine: 1960, *Anatomy of Seed Plants*, John Wiley & Sons, Inc., New York, 376 pp.
Mattoon, H. G. (ed.): 1958, *Plant Buyer's Guide*, 6th ed., Massachusetts Horticulture Soc., Boston. Supplements are issued regularly to keep information up-to-date.
Porter, C. L.: 1959, *Taxonomy of Flowering Plants.* W. H. Freeman and Company, San Francisco, 452 pp.
Savory, T.: 1962, *Naming the Living World, An Introduction to the Principles of Biological Nomenclature*, John Wiley & Sons, Inc., New York, 128 pp.
Scagel, Robert F., R. J. Bandoni, G. E. Rouse, W. B. Schofield, J. R. Stein, and T. M. C. Taylor: 1966, *An Evolutionary Survey of the Plant Kingdom*, Wadsworth Publishing Company, Inc., Belmont, Calif., 658 pp.

MOSSES, FERNS, AND RELATED PLANTS

Perhaps the living mosses and ferns, and their related plants are more neglected than any other plants in the biology laboratory. Nevertheless, many types can be easily maintained for indefinite periods within a greenhouse, in a glass terrarium, or in a growth chamber. This section discusses general uses for class studies and general methods for collecting, culturing, and preserving the plants. Included in the section are descriptions of several species that are easily cultured and that are appropriate for class studies or for individual student projects. The nomenclature is from Bold (1967a).

LIVERWORTS AND MOSSES

Special Uses

The following list contains only a few suggestions of many laboratory uses for liverworts and mosses. Other ideas can be gained from Anthony (1962), H.A. Miller (1962), Basile (1964), Matzke (1964), Bold (1965), and Bold (1967b).

1. The large cells of young moss leaves are excellent for introductory studies of green cells. The outline of cells in the single layer of tissue is considerably easier for beginning students to identify than is the multiple-layered tissue that we often

use, for example, those of *Anacharis* (Elodea). Although cytoplasmic streaming is more easily observed in cells of *Anacharis*, or in the cells of staminate hairs of *Tradescantia* and *Zebrina*, the large moss-leaf cells are excellent for demonstrating plasmolysis and for other osmotic tests. Leaves from liverworts are not as suitable as mosses for beginning studies, since most hepatics typically have a thickened cell wall. It must be noted, also, that some mosses develop several layers of cells as the leaves mature.

2. The two generations, the haploid gametophyte and diploid sporophyte, are more easily grown and observed in mosses than in any other plant, including the fern. Students (and the instructor) may be surprised to learn that living sporophytes of many mosses and hepatics are *green*. Thus, they are photosynthetic and are not completely dependent upon the gametophyte for nourishment.

3. The sexual reproductive organs of both groups are easily observed in living plants. The flagellated, swimming sperms and the nonmotile eggs may be found, with patient and careful study, under relatively low magnification (e.g., 10X). Mature antheridial and archegonial heads of *Mnium* and other mosses can be carefully dissected apart in several drops of water on a slide under a stereoscopic binocular microscope at magnifications of 10 to 30X. With the light reduced and with only a little search, the numerous biflagellated sperms will be seen swimming around the archegonia. Occasionally, with patient work and perhaps with several preparations, a student may be able to locate a mature egg still inside the archegonium and surrounded by sperms that have entered through the neck canal of the archegonium.

4. Sporogenesis, the formation of four spores by two meiotic divisions of a spore mother cell (sometimes called a sporocyte), is more easily traced in the capsule of the mosses than perhaps in any other plants. When sufficient numbers of plants are maintained or collected, the periodic observations of crushed and acetocarmine-stained capsules will allow the instructor and students to learn the particular outward appearance of capsules immediately before and during sporogenesis, and the season of the year to expect sporogenesis. For example, Steere (1964) states that meiosis in mosses is most likely to occur just when the annulus of a green capsule begins to turn red. The annulus is a ring of thick-walled cells between the mouth of a capsule and its lid, the operculum (Fig. 8-1). Capsules undergoing sporogenesis may be studied as fresh material, or they can be preserved in FAA solution (Appendix A) for later staining and study.

5. Information from fossils leads botanists to believe that bryophytes are not more ancient than vascular plants. Two quite different evolutionary series are apparent among nonseed plants: a dominantly haploid group, represented today by the liverworts, hornworts, and mosses, and a dominantly diploid series, represented by the vascular plants. Some evidence of what may have been parallel evolution can be discovered by students in the well-developed stomata on capsules of many mosses, and in the capsule of the hornwort, *Anthoceros*. This study, alone, may well be developed as an unstructured investigation in which students, themselves, find the stomates and are led (if necessary) to speculate upon the evolutionary significance of the presence of stomates in such "lowly" plants.

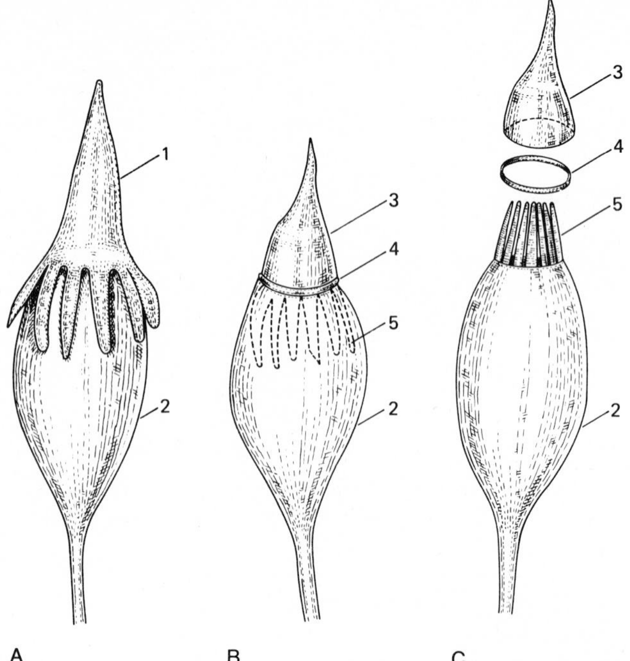

Figure 8-1 Generalized drawing of a moss sporangium; 1, Calyptra; 2, spore case; 3, operculum; 4, annulus; 5, peristome. **A** The intact sporangium with calyptra in place. **B** The calyptra has been removed. **C** The action when spores are released.

6. Because of their simple structure and small size, the plants are exceptionally well suited for physiological and environmental studies. For example, small clumps of plants may be placed in small vials that are then put in covered chambers on a layer of damp vermiculite or other material to retain humidity. In this manner, the plants can be treated for any of a number of conditions, such as variations in light, temperature, nutrient solutions, and plant hormones. Information and assistance with such tests can be gained from reports in botanical journals, particularly in the *Bryologist*, quarterly journal of the American Bryological Society. Several references to reports are given at the end of this chapter.

7. Chromosomal studies may be appropriate for individual student problems. For example, early research showed that an immature seta of a moss (the stalk of the sporophyte), when cut up and placed in nutrient solution, regenerates into a protonema which is then *diploid* and not haploid as is the protonema that develops from a haploid spore. Such a process is called *apospory*, or development of a gametophyte from a source other than the spore. Smith (1955) describes the studies of Marchal and Marchal (1911) and those of von Wettstein (1923) in developing polyploid mosses from the diploid gametophytes. Possibly, polyploidy can be induced, as well, with a colchicine solution (Appendix A).

8. The gemmae cups and their asexual gemmae offer a fascinating study of an unusual type of asexual reproduction. The cups are readily produced on *Marchantia polymorpha*, described later in this section.

9. Little study has been done of pigments in liverworts and mosses, where possibly the usual chromatographic procedures can be applied to pigments extracted from leaves. Beginning chromatographic techniques will be found in Stegner (1967); more complete information is given in Block et al. (1958), Hais and Macek (1964), and in Heftmann (1967). Rastorfer (1962) describes several tests with moss pigments.

Collecting

Mosses and liverworts are found over the world in moist land areas and in and near ponds and lakes. Collecting the plants involves little more than locating them and taking a small portion to the laboratory for a starting growth. Finding the plants, however, is not always an easy matter. Often they form a bright green or a dull, gray-green carpet along moist banks, on moist wood and stones, or they may be floating on or submerged in water. At other times, they occur in rock crevices, on tree bark, and under shrubs or in grass. Once a good collecting spot is located, several trips a year should yield both the gametophytes and sporophytes. The best procedure, however, for obtaining all phases of a life cycle is to maintain the growing plants in the laboratory or greenhouse. When collecting is not feasible, the living plants and spores can be obtained from most biological companies listed in Appendix B.

Necessary collecting equipment includes a small digging tool to remove plants with a portion of soil, a blade to pry plants loose from rocks and other substrata, and a dip net to obtain aquatic types. Be certain to remove the intact, rootlike rhizoids with the plant. Look for plants with fruiting bodies, and include them in the collection. Put the collections in plastic bags, damp newspapers, or other moisture-conserving containers. A good procedure is to put individual collections in a plastic bag that is then inflated by blowing into it and closed with a rubber band. The inflated bag prevents crushing the plants when the bags are stored in a transfer container. Label each bag with a number that refers to collecting data recorded in a field notebook. The plants may be retained in plastic bags at a cool temperature within the refrigerator for a week or longer, if they are not crowded and if they remain damp but not wet.

Identification of plants is most easily done after they are transferred to the laboratory. Conard (1956) provides an excellent key for beginners. Other useful references are those of Grout (1928–1940), and of Frye and Clark (1937–1947). Keys for the states of Washington, Oregon, West Virginia, Tennessee, Florida, Utah, and perhaps others, can be obtained by writing to the botany department of the state university or the state agricultural college. The list of mosses of the United States, compiled by Crum et al. (1965), may be another useful reference.

Culturing Liverworts and Mosses

A great many methods have been described for the indoor culture of mosses and liverworts. The variety of methods and occasional contradictory reports indicate that the culture, alone, may well provide interesting and worthwhile problems for individual students.

In general, put aquatic plants, such as *Riccia* and *Ricciocarpus*, in a well-illuminated freshwater aquarium. Grow land types on a substratum of soil, on vermiculite, or perlite with nutrient solution, or on an inorganic agar medium. Locate the land types in diffused light and in a cool temperature of 18 to 20°C. Keep the substratum moist but not excessively wet. A high atmospheric humidity must be maintained. If plants are kept in a laboratory, they must be maintained in covered chambers, as described below for the various methods.

1. *Terrarium culture.* This is probably the easiest method for maintaining a large quantity of plants continuously. Prepare the terrarium as described in Chap. 6, *Woodland Terrarium.* Arrange the plants over the substratum with the lower plant parts shallowly buried or pressed against the soil. Growth may be more successful if plants have been taken from the collecting area in a clump of soil that is now placed in a slight depression of the terrarium substratum. Every 3 or 4 weeks, sprinkle a dilute liquid plant food over the substratum. Place under medium light (500 to 1,000 ft-c) for 10 to 12 h daily at 18 to 20°C. The gametophytes can be *maintained* in light from a north window during the normal daylight hours, but they may not produce sporophytes. Cover the terrarium with glass or plastic film. Keep substratum moist but not wet.

2. *Soil culture.* The method for *Marchantia*, described by Anthony (1962), can be used for other liverworts and the mosses. He grows *Marchantia polymorpha* in flats with a soil mixture consisting of one-half neutral peat (pH 6.0 to 7.0) and one-half white sand. New flats are started (1) by dropping water on the gemmae cups of an established culture, thus causing the gemmae to splash into an adjacent unplanted flat; or (2) by transferring small portions of thalli directly onto the sand-peat mixture. A 20-20-20 water-soluble fertilizer is used in a concentration of 1 tablespoon per gallon of water. The flats are placed in the greenhouse, with a 150-W incandescent light bulb and a white porcelain reflector 2 ft above the surface of the flat. If a greenhouse is not available, a high humidity can be obtained by covering the box with glass or polyethylene. Anthony reports that with continuous illumination (at 55°F and high humidity) antheridial and archegonial discs begin to appear in 16 days and 21 days, respectively (Fig. 8-5). Mature sporophytes are obtained about 50 to 60 days after beginning the continuous light treatment.

3. *Vermiculite or perlite with a liquid mineral solution.* Schneider, Voth, and Troxler (1967) have described their method for growing *Marchantia polymorpha* on either vermiculite or perlite. Probably the system is equally effective for other liverworts and the mosses. Figure 8-2 illustrates their assembly, which consists of a 10-l earthen crock that contains nutrient solution. A porous, red-clay pot is snugly fitted into the lower crock. A fiber glass wick runs from the bottom of the lower vessel and through the drainage opening of the clay pot, where the unraveled ends of the fiber are spread out to radiate evenly through the top one-fourth layer of substratum. The substratum is then soaked with the nutrient solution and firmly pressed down. The plants are arranged over the top surface with their basal portions submerged. After planting, the substratum must again be pressed down into a firm condition. The planted assembly is put into a

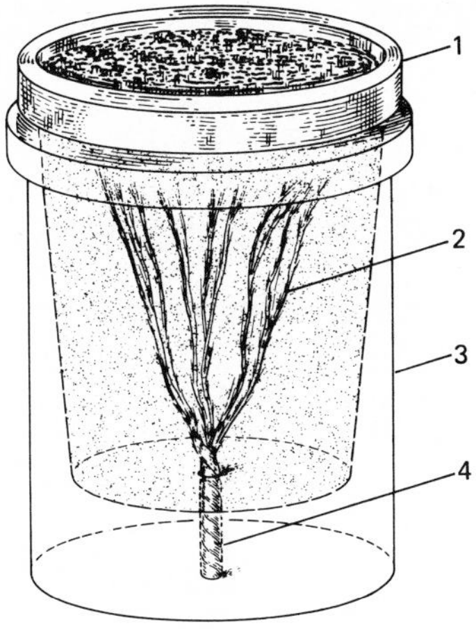

Figure 8-2 Apparatus for culture of *Marchantia* using nutrient solution; 1, clay pot; 2, unraveled ends of a fiber glass wick; 3, glazed, earthen crock; 4, unraveled fiber glass wick. (*After Schneider, Voth, and Troxler.*)

greenhouse or a growth chamber. To induce an early production of rhizoids, thoroughly wet the substratum several times each day for the first 3 days with the nutrient solution, diluted to one-half strength with distilled water. The authors used nutrient solution no. 5 of Voth (1943). Bold's basal medium (Appendix A) should serve equally well.

4. *Bold's basal medium.* Bold (1965) recommends using an inorganic medium solidified with 1.5% agar for the following plants: *Riccia crystallina, Riccia fluitans, Ricciocarpus natans, Sphaerocarpos* sp., *Anthoceros* sp., *Marchantia domingensis, Marchantia polymorpha, Fossombronia* sp. *Porella platyphylloidea, Notothylas orbicularis,* and *Funaria* (several species).

Prepare Bold's basal medium as described in Appendix A. Bold suggests that the nitrogen of this medium may be increased for such plants as *Marchantia* and *Anthoceros* (Appendix A). The agar medium may be used for germinating spores and gemmae, and for growing thallus transplants.

Although aseptic techniques are required for critical tests, as a rule the germinating spores of liverworts and mosses (and ferns) compete successfully with bacteria and fungi in an agar culture. Therefore, transfers of spores to a sterile agar plate can be accomplished simply by raising the plate cover and using tweezers to squeeze spores from a sporangium over the agar surface. Alternatively, a dry brush may be dipped into a mass of dry spores; the brush is then held over the agar surface, and a needle or forceps is used to spread the hairs in the brush, thus scattering the spores over the agar surface. Thallus fragments may be transferred by pressing the lower surface of fragments against the agar surface, where they often regenerate.

Schneider et al. (1967) describe a method for aseptic transfer of liverwort gemmae to agar medium. The following is a modification that generally reduces contamination sufficiently for most studies in introductory courses. Place a drop of 3% hydrogen perioxide in a gemmae cup. The effervescence causes the gemmae to float to the top of the liquid. Use an artist's brush to transfer the gemmae to a small dish of 1:9 household bleach:water solution for 30 sec. Run the solution through filter paper; rinse by running sterile water over the gemmae and through the filter paper. Transfer the gemmae from the paper to the agar medium. As a rule, no microbial growth appears if transfers are made quickly and if the medium is not exposed to the open air by completely removing the petri plate cover. For the teaching laboratory, probably satisfactory growth of gemmae can be obtained with less strictly aseptic techniques.

Put plates at 18 to 20°C under 500 to 1,000 ft-c light for a 12- to 18-h day. The wide range suggested here for each condition strongly indicates the need for more investigation to determine the optimum conditions for germination and growth of these plants. Such tests should be very appropriate for beginning biology students.

5. *Clay flowerpot germinator.* Klein and Klein (1970) describe a simple germinator that is appropriate for nonsterile culture of moss protonema and fern prothallia (Fig. 8-3). For this method, put water into a bowl and invert a clean, porous, clay pot in the water. After the water has diffused through the pot walls, sprinkle spores over the outer surface. Cover the assembly with a glass jar or closed, loose plastic bag. Klein and Klein recommend a photoperiod of 10 to 12 h with no more than 100 to 150 ft-c of white fluorescent light. When the spores have germinated, add Hyponex (diluted to one-tenth of the concentration recommended on the container) to the water in the lower bowl every week. Bold's basal medium (Appendix A) should be a good substitute for the water, with no Hyponex added.

Preserving Liverworts and Mosses

Dry preparations. The plants may be pressed dry and stored by the following procedure. Wash soil and debris from plants and blot them partially dry. Arrange the plants inside the sheets of several folds of newspaper, with branched portions spread to avoid overlapping of plant parts as much as possible. Conard (1956)

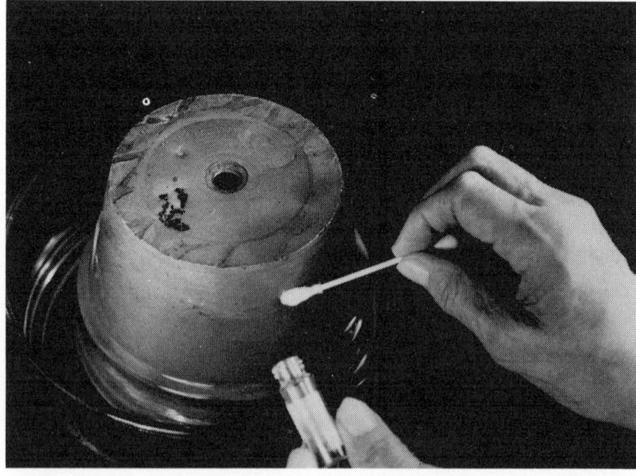

Figure 8-3 A clay flowerpot germinator for moss and fern spores. Spores from a vial are transferred with a cotton swab to surface of clay pot.

suggests that the newspapers be put into a stack no more than 12 in high with blotters between papers and with a weight of no more than 2 lb placed on top. When the plants are dry, place each specimen in a separate envelop or in a folded "pocket" with collecting data and identification on the outer flap. The envelope may be stored in an appropriately sized box, for example, a shoe box, or each envelope can be pasted onto a card and the cards then stored in a box or drawer.

When dried mosses are soaked in water for 10 to 15 min, or up to 10 to 15 h as may be required, or in boiling water for 5 to 10 sec, many plants reassume the normal appearance of a living plant, with even the chloroplasts in a normal position. This revival in appearance is very useful for furnishing mosses to classes when the fresh, living plants are unavailable.

Liquid preservation. Herbaceous plants may be preserved for indefinite periods in FAA solution (Appendix A). Various solutions with copper salts have been suggested for preserving the green color of plants; none has been formulated that retains the color indefinitely. The following methods will give successful plant preservation and will preserve the green color for different lengths of time, depending on the plant type.

1. FAA with $CuSO_4$ (Modified Miller and Blaydes, 1962)

Copper sulfate	0.2 g
Ethyl alcohol, 50%	90 ml
Formalin, 40% formaldehyde	5 ml
Glacial acetic acid	5 ml

Immerse specimens in solution for 3 or 4 days. Transfer to FAA solution (above solution without $CuSO_4$) for storage.

2. Keefe (1926) preservative

Ethyl alcohol, 50%	90 ml
Formalin, 40% formaldehyde	5 ml
Glycerin	2.5 ml
Glacial acetic acid	2.5 ml
Copper chloride	10.0 g
Uranium nitrate	1.5 g

Immerse specimens in solution for 3 to 10 days, depending on thickness. Transfer to FAA solution (Appendix A) for storage.

3. Copper lactophenol (Johansen, 1940)

Phenol, c.p.	20 g
Lactic acid, sp. gr. 1.21	20 g
Glycerin, sp. gr. 1.25	40 g

Distilled water	20 ml
Cupric chloride	0.2 g
Cupric acetate	0.2 g

Johansen recommends the solution for green algae. It is as effective for liverworts and mosses. Immerse specimens in solution for 3 to 10 days, depending on size of plant. Transfer to FAA solution (Appendix A) for storage.

Microscope-slide preparations. The following methods are appropriate for preparing slides of portions of liverworts and moss plants, or for an entire small plant.

1. Hoyer's fluid (Schuster, 1966)

Distilled water	50 ml
Gum arabic (U.S.P. flake)	30 g
Chloral hydrate	200 g
Glycerin	20 ml

Mix materials at room temperature, in the order given above. Use flaked gum arabic; the powdered material produces bubbles that do not disappear. Gum arabic dissolves slowly. Do not use heat. Allow mixture to stand for 24 h to disperse air bubbles formed during stirring. Final mixture should be clear, without sediment, and a light yellow color. Filter, if necessary.

Before mounting plants, soak them in water for several hours or longer, until thoroughly wetted. Spread portions of plant on slide. Cover with several drops of preservative. Add coverglass; avoid trapping air bubbles under coverglass. Plant parts may curl in medium, but they flatten within several minutes. Let slides lie flat for 2 to 3 days, or more, until medium has hardened. Schuster (1966) states that fragile plants deteriorate within 4 to 5 yr and that tougher tissue may shrink and deteriorate within 10 yr.

2. Semipermanent slide (Roehrick, 1964). A simple method for making a semipermanent slide is to place a plant portion in one or two drops of water on a slide. Add a coverglass, and gently heat the slide above a flame until the water begins to steam. Set slide aside to dry. Place a drop of glycerin at one edge of the coverglass so that the glycerin flows under the glass. If necessary, add a second drop of glycerin to fill the area beneath the coverglass. Allow the slide to dry for 48 to 72 h.

REPRESENTATIVE LIVERWORTS FOR THE LABORATORY

Representatives of two general types are described here: (1) ribbonlike, forking, thallose plants that grow flat on the soil or that float on water (*Marchantia* and *Ricciocarpus*); and (2) erect, prostrate, or hanging plants with parts that resemble leaves and stems (*Porella*).

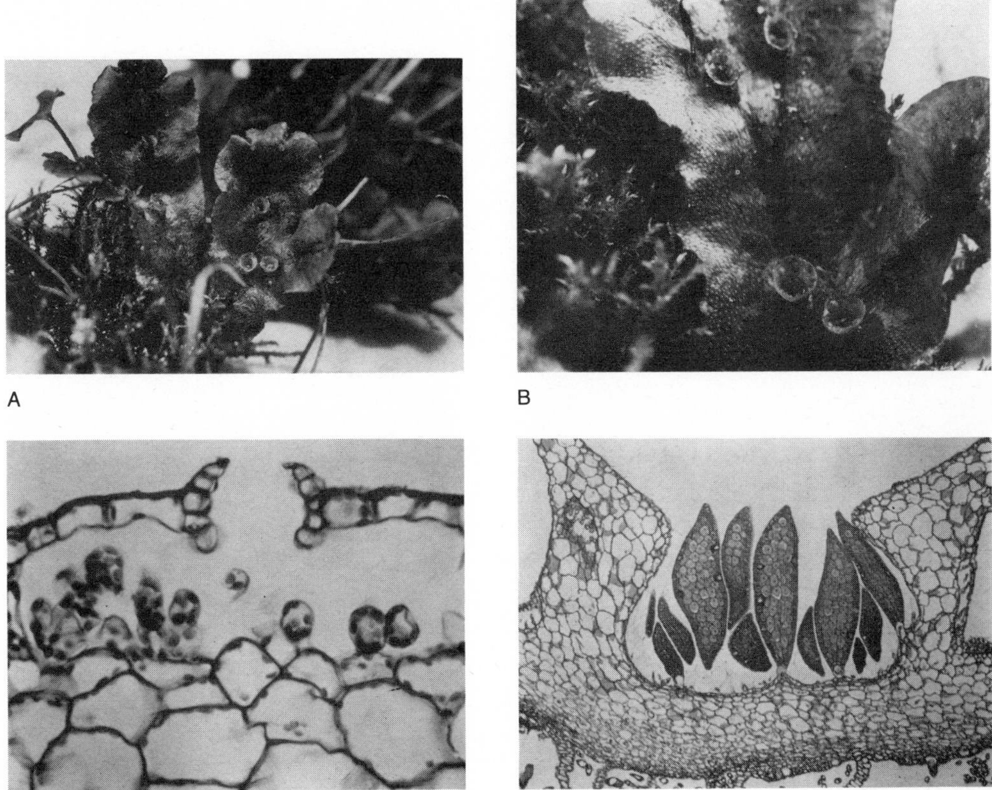

C *Courtesy of Carolina Biological Supply Company*

D *Courtesy of Carolina Biological Supply Company*

Figure 8-4 *Marchantia polymorpha.* **A** Thalli showing dichotomous branching and gemmae cups. **B** Close-up of thalli showing gemmae cups (several with gemmae). The surface outline is plainly visible of polygonal air chambers each with a central pore. The pores appear as bright dots in the photograph. **C** C.s. of thallus, showing pore and air chamber. **D** Median section of a gemmae cup showing gemmae and stalks.

Marchantia (Thallose Liverwort)

Probably *Marchantia* is the most commonly used liverwort in the laboratory. *Marchantia polymorpha* is found on moist soil particularly in northern United States, often appearing on burned areas. Other species like *M. domingensis* and *M. paleacea*, grow on moist rocks and clay banks in the southern United States. The dominant gametophyte generation consists of a green, flat, ribbonlike thallus (thalli, plural) that branches dichotomously and that may grow to a length of 3 or 4 in (Fig. 8-4). On the upper surface of a thallus can be seen the polygonal outlines of air chambers, each with a central pore and visible to the unaided eye. Internally, the thallus consists of a dorsal layer of air chambers, the walls of which are formed of cells containing many chloroplasts. Under the air chambers is a layer of

A *Courtesy of Carolina Biological Supply Company*

B *Courtesy of Carolina Biological Supply Company*

Figure 8-5 *Marchantia polymorpha.* **A** Archegoniophores with fingerlike lobes on disc. **B** Antheridiophores with scallop-edged discs.

parenchymatous, storage cells that contain only few plastids. On the undersurface of the thallus are found many rhizoids, and six to eight rows of small scales. Special asexual reproductive structures, called gemmae cups, form on the dorsal surface of thalli, and contain small two-notched, asexual reproductive bodies, called gemmae, that form continuously on short stalks at the bottom of a cup. When water falls into a cup, the gemmae are released and may float out of the cup. Those that become deposited on soil may grow into two thalli – one from each apical notch. Gemmae cups and the gemmae can be produced somewhat easily in the laboratory by following the procedures of Anthony, described earlier under culture method no. 2.

Anthony's procedures also produce numerous antheridia and archegonia that form on discs at the top of stalks (antheridiophores and archegoniophores), which rise from the dorsal surface of thalli (Fig. 8-5). The plants are unisexual, and antheridia and archegonia are formed on separate plants. Archegonia develop on the undersurface of an umbrella-shaped disc that is divided into nine, deeply cleft, fingerlike lobes. Antheridia form in the upper surface of a scallop-edged disc. Mature sperm swim to archegonia where eggs are fertilized. Small, yellow sporophytes develop under the archegonial disc. Mature sporophytes will be found after about 2 months of continuous light treatment by Anthony's method. Freehand, vertical sections of an archegonial disc, at a magnification of 40 to 100X, allow observations of the developing sporophyte (Fig. 8-6). Sporocytes (spore mother cells) form within the capsule and then undergo two meiotic divisions to form four spores. At the same time, rows of special long, pointed *elater* cells, with spiral thickenings in their walls, are formed between the rows of spores. Under damp conditions, elaters absorb water and swell; apparently they have a role in producing a gradual expulsion of spores. Under proper conditions, mature spores develop into thallose gametophytes.

If *Marchantia* cannot be collected locally, clumps of living plants can be

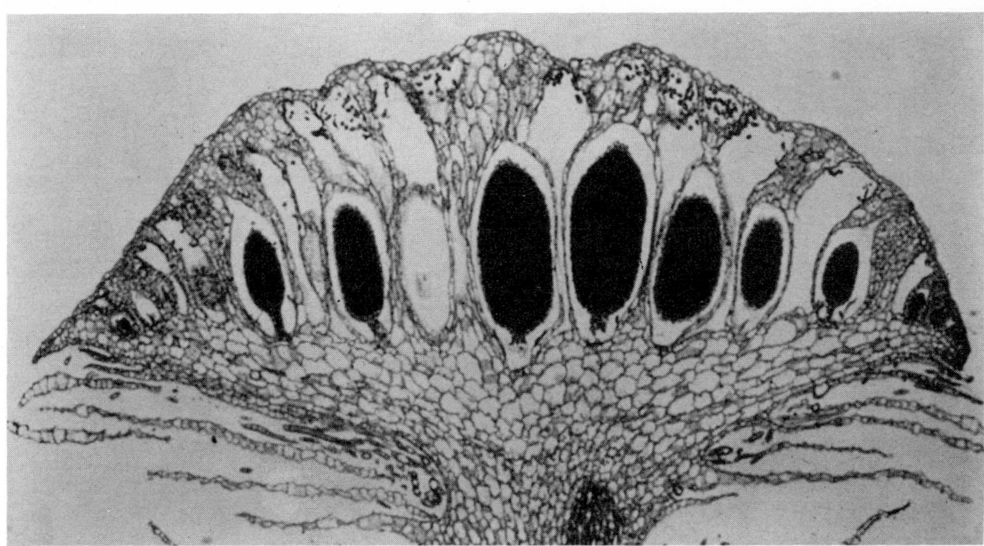

A *Courtesy of Carolina Biological Supply Company*

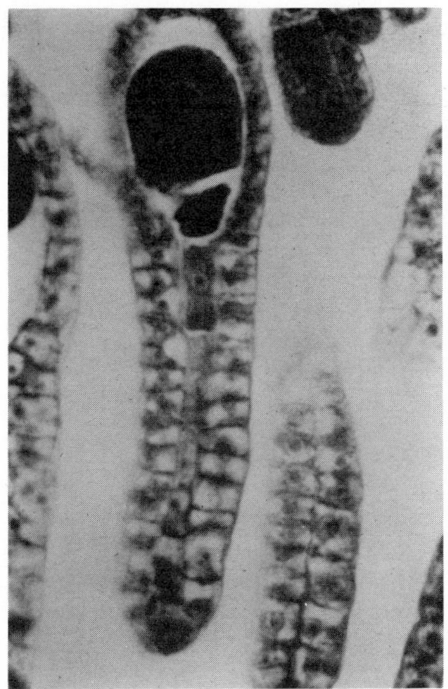

B *Courtesy of Carolina Biological Supply Company*

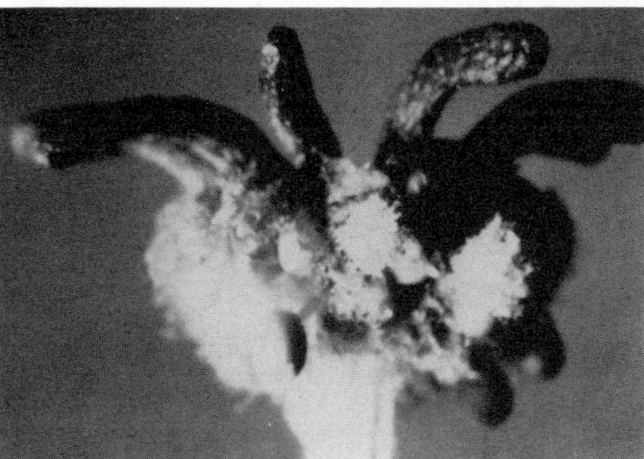

C *Courtesy of Carolina Biological Supply Company*

Figure 8-6 *Marchantia polymorpha.* **A** Median section of antheridial disc containing antheridia. **B** Median section of a portion of an archegonial disc with archegonia projecting downward under the disc. Only one archegonium is seen in full length. **C** Archegonial disc with mature sporophytes below the fingerlike lobes. Sporangia are in the process of discharging spores.

purchased from most biological supply companies listed in Appendix B. The plant is easily maintained from year to year by terrarium culture, described earlier in this section as culture method no. 1. When many fruiting bodies are wanted, transfer small portions of thalli (approximately 2 cm long) to one of the systems described under culture methods nos. 1, 2, 3, or 4.

Ricciocarpus (Thallose Liverwort; *R. natans* Corda)

Ricciocarpus natans forms a triangular, flat plant, 2 to 2.5 cm long, floating on quiet water or growing on very moist soil (Fig. 8-7). The thalli branch dichotomously and show conspicuous dorsal furrows. Rhizoids form on the undersurface of land plants; none are present on plants in water. The posterior portion of a thallus deteriorates continuously as the anterior portion grows. Upper layers of air chambers produce a spongy appearance that may be clearly observed under the stereomicroscope at 30 to 60X. Cells in the walls of air chambers contain many chloroplasts. The lower portion of a thallus consists of parenchymatous, storage cells, with few chloroplasts. Freehand, vertical sections of the thallus, made with a razor blade, will show these areas clearly when magnified at 30 to 60X.

Antheridia and archegonia develop, apparently during most of the spring and early summer months, on the dorsal surface of thalli, where they become sunken in the dorsal furrows. *Ricciocarpus* plants are bisexual, the male and female sex organs developing on the same thallus. Antheridia develop first at the apical notch and then become located posterior to the archegonia that develop later at the apical notch. The sex organs usually appear in floating plants but are found only occasionally in the land plants. Following fertilization, a small sporophyte develops within the archegonium. Sporocytes undergo two meiotic divisions, forming a tetrad of spores. Eventually the ripened spores are released into the water. Each spore, after a period of dormancy, may develop into a gametophyte.

Ricciocarpus may be collected locally or purchased from most biological supply companies listed in Appendix B. With introductory courses, probably the plant is most useful for the observation of the gametophytic thallus, and particularly for a

Figure 8-7 *Ricciocarpus natans.* The larger plants are about 5 cm across at widest point.

Courtesy of Carolina Biological Supply Company

Figure 8-8 *Riccia fluitans.* A large mass of plants is shown at top. At bottom a single dichotomously branching plant can be seen.

study of the air chambers in a cross section of the thallus. Because *Ricciocarpus natans* may grow "upside-down" at times in water, or the plant can be turned over by hand to produce this position of growth, Wagner (1961) has suggested that this characteristic might provide interesting studies. For example, what happens to the air chambers, which normally are in a dorsal position? Does the number of chloroplasts increase in the lower (now upper) layers, which normally contain few plastids? What happens when the container of plants is illuminated from below, with a black cover around the sides and top?

Riccia is closely related to *Ricciocarpus* and may be desirable for studying a submerged aquatic plant. *Riccia fluitans* may be grown in an aquarium, where it forms a mat of short, narrow, threadlike thalli in the water; or it can be grown on Bold's basal medium. See Fig. 8-8.

Porella (Leafy Liverwort; *P. platyphylloidea* Lindb.)

Although thallose liverworts represent only about 20 percent of all liverworts, the thallose types, particularly *Marchantia*, are studied in many biology courses to the total exclusion of the leafy types, such as *Porella*, *Frullania*, and *Scapania*. Perhaps the oversight of leafy liverworts occurs because of their superficial resemblance to the mosses, whereby a casual observation may result in a confusion of the two plant types. A closer examination, however, will show that in most mosses the leaves are all similar and are arranged around a rounded stem. On the other hand, those of leafy liverworts are in three ranks, usually consisting of two dorsal rows and a ventral row (sometimes lacking) on a dorsoventrally flattened stem. The position and structure of reproductive organs also differ. In mosses the reproductive structures generally develop in a terminal "bud." In *Porella* and other leafy liverworts the archegonia develop on short, lateral branches, that resemble vegetative branches until after fertilization, when the leaves surrounding an archegonium enlarge into a prominent sheathlike perianth; antheridia develop on specialized branches, in the axils of closely overlapping leaves.

Porella platyphylloidea is a relatively large liverwort growing to 7 or 8 cm long, and branching to form mats on moist rocks or trees, and occasionally on soil. Each leaf of the two dorsal rows bears a ventral lobe, giving the false appearance, then, of three rows of ventral leaves (Fig. 8-9). Individual plants are unisexual; the antheridial plant is somewhat more narrow than the archegonial plant. A single, relatively small, sporophyte develops from an archegonial head, although eggs of other archegonia may be fertilized. The sporophyte remains under the cover of the perianth leaves until spores and elaters are formed, when the seta lengthens carrying the capsule out of the cover. The capsule bursts along four vertical lines; all spores are released at once, and the fragile seta collapses within several hours.

Fertilization occurs in *Porella* during late summer and autumn, and sporophytes develop the following spring. The plants may be collected during the fall months and held in a terrarium through the winter. During this time the meiotic divisions of sporocytes can be studied from preparations of squashed acetocarmine-stained tissue (Appendix A), followed by a study of the later development of the sporophyte. Spores can be germinated on Bold's basal medium, described earlier in this section under culture method no. 4. A starting culture may be purchased from most biological supply companies listed in Appendix B.

REPRESENTATIVE MOSSES FOR THE LABORATORY

Sphagnum (Peat Mosses)

Sphagnum grows in bogs and swamps, and on the wet banks of lakes and ponds. The plants form many branches that make a dense mat which may accumulate and form peat bogs. In some rural areas, especially in Ireland, the dried, compressed peat is burned as a cheap fuel. Horticultural peat moss is the partially deteriorated moss and associated plant debris taken from bogs in the continental

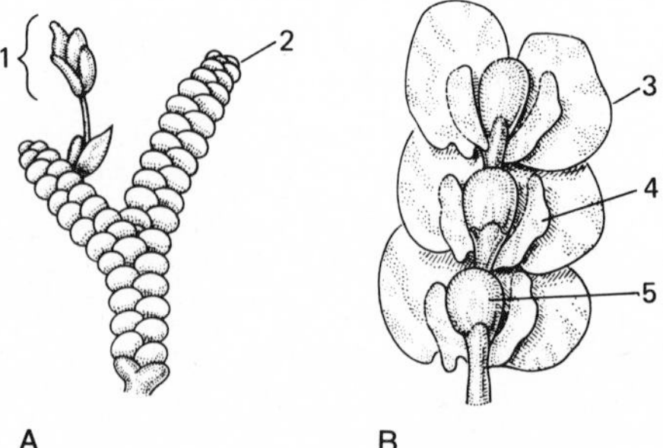

Figure 8-9 *Porella platyphylloidea*, a leafy liverwort. **A** Dorsal view showing lateral branch with an opened sporangium on sporophyte. **B** Close-up of ventral surface; 1, sporophyte; 2, dorsal leaves; 3, undersurface of dorsal leaf, enlarged; 4, ventral lobe of a dorsal leaf; 5, a ventral leaf.

A

B

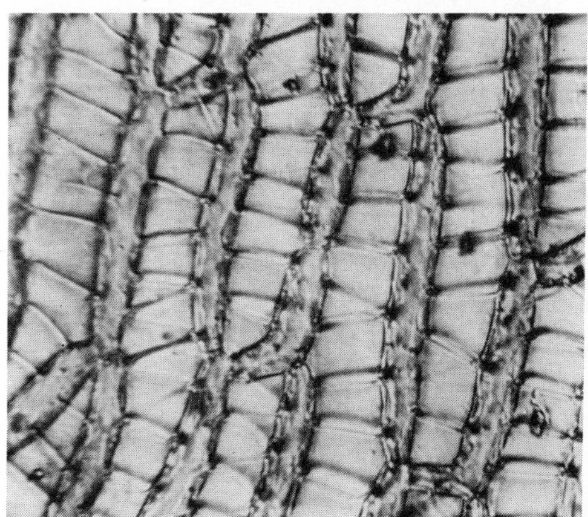

C

Figure 8-10 *Sphagnum* sp. **A** A mat of living plants on a rotted board. **B** A small group of separated plants that show types of branches. **C** Portion of a leaf, magnified to show long photosynthetic cells and dead, rectangular, storage cells.

United States and in other parts of the world. Horticultural *Sphagnum*, sold as "fibrous sphagnum" or as fine-textured "milled sphagnum" is the dried, fresh plant. Both peat and *Sphagnum* have a great capacity for absorbing and holding water, and the partially decayed plants produce an acid condition in a soil mixture. Both traits explain the horticultural usefulness of the plant.

When a single plant is separated from a mat of living plants, several kinds of branches can be distinguished: some that run horizontally, others that are curved and twisted, and many upright branches that bear antheridia and archegonia (Fig. 8-10). Some species have bisexual individual plants while in others the individual plants are unisexual. The relatively small sporophyte is borne on an elongated gametophytic stem, called a pseudopodium, that functions in place of the reduced seta to elevate the mature capsule. Spores are ejected from the capsule with an audible explosion (interesting to students). Under favorable conditions, a spore develops into a thalluslike protonema from which a single, leafy gametophyte grows.

Sphagnum is easily maintained in a greenhouse or in a covered glass terrarium, as described earlier in this section under culture method no. 1. Other than as a growth medium, perhaps the most important use for the plant is to study its modified leaf cells that are adapted for holding water. When a leaf is put in a water mount and magnified 100 to 400X, two distinct kinds of cells are seen: long, narrow, living photosynthetic cells; and large, rectangular, nonliving, water-storage cells with thickened walls (Fig. 8-10). The nonliving cells cause the light yellowish green color of adult plants.

Polytrichum (Haircap or Pigeonwheat Mosses)

This is a common moss distributed worldwide, forming dense green carpets on moist soil. The gametophyte of several species may grow to 5 or 6 in high (Fig. 8-11). A cross section of a mature leaf at 100 to 400X shows a somewhat complex histology, with an upper region of many parallel, photosynthetic lamellae, a lower region of several layers that contain parenchymatous and sclerotic cells, and a thickened midrib. The plant is anchored in the soil by a mass of false roots, rhizoids. Leaves are attached around the stem. Individual leafy gametophytes are unisexual, bearing antheridia and archegonia on separate plants at the apex of the main axis. Antheridial plants can be recognized by the cuplike arrangement of apical leaves around the antheridial head. Archegonial plants are not so easily recognized except by the general, dimorphic distinctions of the male and female

Figure 8-11 *Polytrichum* sp., gametophytes with sporophytes. Plants are 4 to 6 in high.

Courtesy of Carolina Biological Supply Company

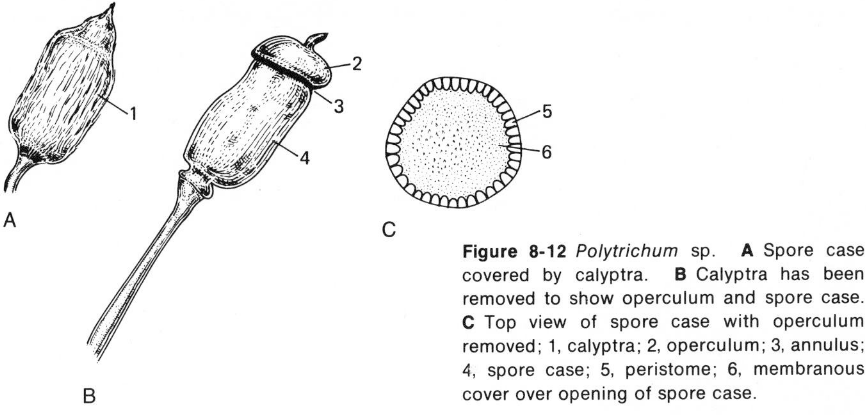

Figure 8-12 *Polytrichum* sp. **A** Spore case covered by calyptra. **B** Calyptra has been removed to show operculum and spore case. **C** Top view of spore case with operculum removed; 1, calyptra; 2, operculum; 3, annulus; 4, spore case; 5, peristome; 6, membranous cover over opening of spore case.

plants and by the fact that antheridial heads are absent. The motile sperms are splashed about by raindrops and swim to the archegonia, where fertilization occurs. The sporophyte develops into an elongated, needlelike structure (to 6 in long) that carries the top portion of the archegonium on the capsule as a hairy covering called the calyptra. Later, under the calyptra, the mature capsule consists of a lid called the operculum and a ring of thickened cells, the annulus, at the base of the lid. A ring of sharp teeth, the peristome, is fastened to a membranous layer that covers the mouth of the capsule under the operculum (Fig. 8-12). When the spores mature, the peristome gradually swells with moisture and lifts the operculum from the capsule. Spores are gradually released by movement of the seta after the calyptra and operculum have been shed. Under favorable conditions, a spore germinates into a branching filament, the protonema, on which several buds appear, each growing into a gametophyte.

The large *Polytrichum* is excellent for studying the alternation of generations in mosses, and for demonstrating the swimming sperms when an antheridial head is dissected in a drop of water and examined under 100 to 400X magnification. Maintain clumps of the moss in a covered terrarium, as described earlier in this section under culture method no. 1. Fertilization often occurs in *Polytrichum* during late autumn and early winter, and sporophytes mature in great numbers the following spring and early summer. If the plant cannot be collected locally, clumps of the moss may be purchased from a biological supply company (Appendix B).

Funaria (Cord Mosses)

This is a common genus of mosses, most often growing on moist limestone soil. *Funaria* is smaller than *Polytrichum*, but both are highly appropriate for laboratory culture and study. The individual gametophytes of *Funaria* (Fig. 8-13) are bisexual, but the archegonia and antheridia are in separate branches of a single plant. The antheridial head is easily identified by its cup shape and orange center. Sexual reproduction and sporophyte development are similar to those processes

Figure 8-13 *Funaria* sp., gametophytes with sporophytes; 1, calyptra; 2, spore case when calyptra has been removed.

in *Polytrichum;* the sporophyte rises 2 to 3 in high, has a peaked calyptra and an underlying operculum (Fig. 8-13). The peristome in most species is a double layer of toothlike segments. At maturity, spores are ejected, each developing, under favorable conditions, into a protonema. No elaters are formed. The leafy gametophytes develop as buds from the protonema.

Grow clumps of *Funaria* in a terrarium as described earlier in this section under culture method no. 1. Spores can be germinated on Bold's basal medium, described in culture method no. 4. They require light for germination.

Mnium

Mnium is widely distributed in the United States on very moist, often acidic soil, sometimes at the edge of water but not submerged. The plant is similar to *Funaria* and *Polytrichum*. The individual leafy gametophytes may be bisexual or unisexual, depending on the species. Sex organs are borne on special branches; male and female organs may occur together in the same head. Occasionally several sporophytes may grow from one head, each with a drooping capsule covered with a lid, the operculum (Fig. 8-14). When the operculum is removed, a double ring of long, slender teeth can be seen lying horizontally across the mouth. No elaters are formed in the capsule.

H. A. Miller (1962) describes a method for studying the hygroscopic movement of the peristome teeth. Other mosses may be used also. Cut the sporophyte from the gametophyte and run the seta through a small hole punched in a piece of

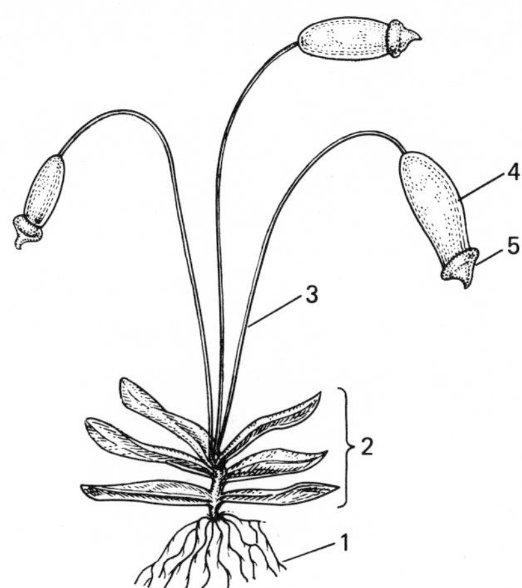

Figure 8-14 *Mnium* sp., gametophyte and sporophyte. Actual size is 4 to 5 in high. 1, Rhizoids; 2, gametophyte leaflike structures; 3, seta of sporophyte; 4, spore case; 5, operculum.

paper. Pull the sporangium down against the paper (Fig. 8-15). If the operculum has not already fallen loose, use a needle to remove it. Observe the sporangium on the paper under a stereomicroscope at 40 to 100X magnification, and locate the peristome ring. Then observe the swelling and movement of the teeth when a second person blows his breath, open-mouthed, over the peristome. The response to

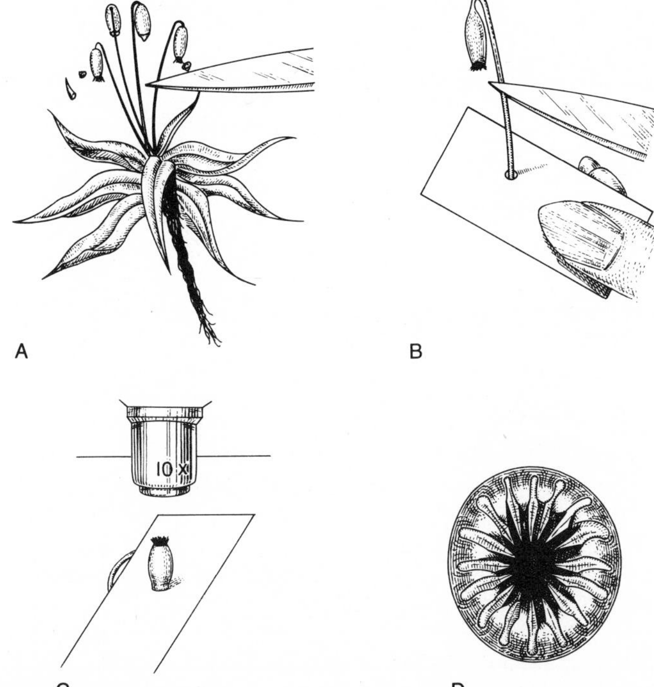

Figure 8-15 *Mnium* sp. Method of observing hygroscopic movement of peristome teeth. **A** Remove seta and sporangium. **B** Pull seta through hole in rigid paper. **C** Observe movement of teeth as described in text. **D** Top view of peristome consisting of long teeth that extend over opening of spore case. (*After H. A. Miller.*)

Courtesy of Carolina Biological Supply Company

the moisture is quite striking and will occur repeatedly. Dried specimens may react in the same way.

Mnium is easily maintained by the culture methods described earlier in this section. If plants cannot be collected locally, a starting clump of *Mnium* can be purchased from a biological supply company (Appendix B).

FERNS AND RELATED PLANTS

All locally collected plants generally grow quite satisfactorily in the laboratory or greenhouse. So far as is practical, however, the plants must be given indoor conditions that approximate those of their outdoor environment, particularly in regard to soil, moisture, and humidity. If greenhouse facilities are not available, a glass terrarium or other similar equipment will be necessary to provide a high humidity for woodland ferns. A terrarium is easily established, and the ferns can be grown during the entire school year for class studies and as ornamental plants. Also an aquatic, floating fern, *Marsilea*, is included in this section. This plant usually grows and multiplies in magnificent fashion when put into a freshwater aquarium under bright light. As well, the sporophytes of many horticultural ferns grow well within the laboratory in pots of soil. Their maintenance is described below under *Sporophyte Culture*. Conservation practices need to be reemphasized here regarding field collecting, in that a person should collect only enough plants to start an indoor collection. In no case should more than one-third of a plant stand be removed from the field, thus allowing the remaining plants to fill in the empty area. All plants described in this section may also be obtained from plant nurseries, locally or by catalog, or from biological supply houses. The shipment of large plants is costly and is likely to result in plant damage. Therefore, a person may wish to purchase the rooted rhizomes in place of mature sporophytes. A source list of suppliers is given in Appendix B.

SPOROPHYTE CULTURE

Soil and potting. A standard soil mixture for ferns consists of 1 part sharp sand, 1 part peat moss, and 2 parts "rich" loam (with humus matter) or 2 parts leaf mold. Place ½ to 1 in of shard (broken pieces of clay pottery or small rocks) in bottom of pot, which must have drainage openings. Some persons mix charcoal with the shard. Place the roots and most of the rhizome under the soil, with the new growing tips from the rhizome and the base of fronds above the soil level. The plants grow better when slightly potbound and should be put in a pot as large as the rhizomes but with little extra space. Repot every 3 yr or sooner if plant roots grow to fill a pot.

Light. Place pots in an east or south window, or 12 to 14 in below low to medium light (500 to 800 ft-c) during 12 to 14 h of each 24-h period.

Water. For most plants keep soil slightly moist, with good drainage. Xerophytic ferns tolerate less moisture. If the atmosphere becomes very dry, stand the pot on gravel with water in a tray. Be certain that the bottom of the pot rests above the water level, although certain ferns, indicated later, will produce better growth when the pot stands in water. Alternatively, thoroughly water the soil and place a plastic bag over the plant and pot. Under these conditions, take care that adequate light reaches the plant. Unless a plant is too large, every 2 or 3 weeks immerse the pot and total plant in a container of water for 10 to 15 min. This practice helps to clean the foliage and wash off pests.

Temperature. Moderate to warm temperature of 18 to 23°C.

Fertilizer. Liquid plant food every 3 to 4 weeks. Many ferns rest during winter months, as indicated by a slow growth or no growth. Give no food during this time.

GAMETOPHYTE CULTURE

The spores of most ferns of the family Polypodiaceae, as well as those of other particular species such as the various species of *Osmunda*, readily germinate and develop into gametophytes by one or more of the procedures described below. Note that spores with chlorophyll and thin walls have a short viability period; thick-walled, brown spores are more resistant.

Collect spores by placing fronds with newly matured sori on white glazed paper, spore-side-down. Put in a protected location where the frond and spores will not be disturbed by air movements. Within a few hours the spores will be released on the paper in the shape of the frond (Fig. 8-16A). A good way to determine if sori are mature is to place a single pinna, sorus side up, on a dry glass slide with no coverglass. Examine at 60 to 100X with the stereomicroscope or a compound microscope. The ejection of spores from sporangia can be clearly seen, making a vivid demonstration for students.

The time required for the development of a gametophyte varies among species. As a rule, antheridia and archegonia of polypodiaceous ferns develop within 6 weeks. Their production may continue for several months if gametophytes are maintained on agar medium that is kept slightly moist with an inorganic salt solution (e.g., Bold's basal medium, Appendix A). Eventually, young sporophytes develop (Fig. 8-17); these can be transferred, still on the gametophyte, to small bowls of soil moistened with inorganic salt solution. When roots have developed, transfer the sporophytes to pots of soil. Bold (1967b) reported that embryonic sporophytes of *Dryopteris*, ultimately transferred to pots of soil in the greenhouse, matured and produced spores in 3 yr.

A number of methods are found in the literature for germinating spores and growing prothallia. Several of these are described below.[1]

1. *Bold's basal medium.* For *Equisetum*, *Selaginella*, and ferns of the family Polypodiaceae. Prepare the inorganic liquid medium according to instructions given in Appendix A. The sterilized medium is dispensed into sterile petri dishes or small flasks and inoculated with a thin layer of widely dispersed spores, or the liquid medium may be used in one of the procedures that follow.

[1]Also see methods described by Davis and Postlethwait (1966).

Figure 8-16 Spore prints. **A** At left, print of spores at the end of 24 h. At right, a frond that has not yet been lifted from spore print. **B** Close-up of spore print after frond in **A** is lifted.

2. *Bold's basal medium with 1.5% agar.* This is a preferred medium for aseptic culture of most types of spores. The preparation of the agar medium is described in Appendix A. Dispense the sterile agar medium into petri plates, or into large tubes for agar slants.

Use a small, soft brush to spread spores thinly over the agar medium. If the later prothallia become crowded into clumps, use tweezers or a transfer loop to carefully spread them out over the medium. The agar medium is preferred because the plates or tubes of agar are easier to handle than are those with liquid medium, and because the gametophytes apparently develop more satisfactorily on the solid medium.

3. *Agar medium method.* The technique is similar to method no. 2. In place of inorganic salts, listed for no. 2, add 1 g of Vigoro (or a houseplant food) and 15 g of agar to 1 l of water. Heat with constant stirring until agar is dissolved. Remove the heat and add 2 to 3 drops of 1% ferric chloride solution (1 g of ferric chloride in 99 ml of water). Autoclave or put in a pressure cooker for 15 min at 15 lb pressure. Dispense into containers and inoculate with spores as described in method no. 2. It may be necessary to experiment with different brands of commercial plant foods to find an appropriate type for a particular fern. The method should be much simpler than others and could well provide an appropriate series of tests for students.

4. *Plaster of paris blocks for Selaginella.* Select cans, bowls, or small cardboard

Figure 8-17 Germination of fern spores and growth at end of 8 months of gametophytes and young sporophytes on Bold's basal medium with 1.5 percent agar. **A** View of entire agar culture showing a mass of prothallia (gametophytes) and several young sporophytes. Outer ring is the wall of the petri dish. Inside the outer ring is the ring of agar medium. **B** Close-up view of many prothallia that overlap at edges obscuring the outline of individual plants. Note distinct heart shape and apical notch of gametophyte in lower left corner. Several young sporophytes are seen as a light-colored cluster at top, left of center. **C** Close-up of young sporophyte and prothallia in 8-month culture. **D** Close-up of young sporophytes in 9-month culture. At this stage, the gametophytes and attached sporophytes can be transferred to a bowl of soil.

boxes to serve as molds for plaster of paris blocks that will fit into a covered culture bowl. A mold at least 1 in deep and 3 to 4 in in diameter is desirable. Coat the inner surface of the mold with petroleum jelly. Mix enough plaster of paris (from a drugstore or hardware store) with water to make a thick, creamy mixture that is still thin enough to pour. Stir the mixture slowly to avoid introducing air bubbles. Pour the mixture into the mold, taking care that air bubbles do not form. When the plaster has hardened, remove the block from the mold.

Before using the plaster blocks, soak them for 24 h in several changes of distilled water. Then place a block in a culture bowl containing enough of Bold's basal medium (Appendix A) to submerge the lower half of the block. Collect *Selaginella* spores by putting mature strobili in a dry covered petri dish for about 24 h. The spores will be ejected from sporangia and are then brushed over the top surface of the plaster block. Cover the bowl with a glass plate or plastic film. If necessary, add medium to replace for evaporative loss. Store bowls at a cool to moderate temperature, 18 to 21°C, in low to medium light (200 to 600 ft-c). As a rule, *Selaginella* spores produce female gametophytes in about 90 days and male gametophytes in less time.

5. *Clay pot culture with Bold's basal medium.* Select a clean, porous clay pot and fill with *Sphagnum* or peat moss. Invert the pot into a culture bowl so that the lower part of the pot is covered with about an inch of Bold's basal medium (Appendix A). After the solution has soaked through the clay walls, scatter spores over the wet surface of the pot. Cover the bowl with a battery jar to prevent evaporation and to maintain a high humidity. If mildew or "damping off" occurs, water the plants with a weak solution of potassium permanganate, made by placing one small crystal of the compound in a quart of water. Add medium to replace for evaporative loss.

6. *Soil culture.* Some persons prefer to germinate spores and grow the gametophytes on soil in a petri dish or in a pot. A mixture of 1 part, each, of loam and sand, and 2 parts of peat moss is a suitable medium. Sterilize the soil and container in the oven at 180°F for 30 min. After the soil has cooled, moisten with Bold's basal medium (Appendix A) and sprinkle spores thinly over the soil surface. Cover with a glass plate or plastic film.

REPRESENTATIVE FERNS AND RELATED PLANTS FOR THE LABORATORY

Plants of this section are arranged alphabetically by generic names so that their descriptions may be located more easily. However, to understand the relationships among the plants, a person should consult the outline given below. Division and family names are from Bold (1967a); other nomenclature is based primarily on Bailey (1949).

1. Division Microphyllophyta[1] (club "mosses"): *Selaginella*
2. Division Arthrophyta: *Equisetum* (horsetails)
3. Division Pterophyta (ferns)
   a. Family Osmundaceae: *Osmunda*
   b. Family Polypodiaceae (common fern family): *Polypodium, Platycerium, Adiantum, Asplenium, Nephrolepsis, Dryopteris*
   c. Family Marsileaceae: *Marsilea*

[1]Division Microphyllophyta is also named Lycopodophyta (Scagel et al., 1965), and Lepidophyta (Smith, 1955).

Adiantum (Maidenhair Fern; *A. capillus-veneris* L.)

USES: (1) For studying all plant parts. (2) To germinate spores and grow gametophytes.

DESCRIPTION: Slender, hairy, horizontal rhizomes that form a matted underground growth. Slender, black, wiry stem, 12 to 20 in high. Curved fronds, divided once or twice into small wedge-shaped and notched pinnae, ½ in across. Sori on undersurface, near margins of pinnae with edges of pinna folded back and over sori to make indusia (Fig. 8-18). *Adiantum pedatum* L. is also suitable for indoor culture, although more difficult to grow. The plants are found on moist, limestone banks of streams and are almost hydrophytic.

CULTURE: See culture of sporophytes and gametophytes at beginning of this section. Plant needs high humidity and slightly moist soil. Let pot stand in ½ in of water and in sunlight. If soil lacks calcium, add 1 teaspoon of powdered limestone for each quart of soil mixture. Repot every 3 yr, or more often if soil becomes filled with roots. Plants need a rest period during winter months.

PROPAGATION: Separate rhizomes with a sharp blade, or propagate by spores.

Asplenium (Bird Nest Fern or Spleenwort; *A. nidus* L.)

USES: (1) For study of plant parts. (2) To study asexual reproduction by offset plantlets. (3) As an ornamental foliage plant.

DESCRIPTION: Rosette of tongue-shaped, stiff fronds, 2 to 3 ft long, that unfold from a center nest of coarse, brown hairs. Fronds are not divided in the typical fern fashion. Humus collects in nest, and roots grow throughout this for food and water. Sori are arranged along the pinnate veins.

Another suitable species, *A. bulbiferum* or Mother Spleenwort, bears plantlets on fronds that can be removed with a portion of the frond and put into a terrarium until the roots are developed, at which time they may be potted in soil. A. E. Allgrove (see Appendix B) sells a "fernarium," an educational kit which contains *A. platyneuron* (Ebony Spleenwort), a small rock fern for terraria.

CULTURE: See *Adiantum*. Do not let water collect in center crown or lower plant will rot. An interesting and attractive plant, which may, however, become too large for indoor use. Obtain from local nurseries or by catalog.

Figure 8-18 *Adiantum* sp., maidenhair fern.

PROPAGATION: Propagate by cutting through rootstock, a difficult task because of size and toughness of plant.

Dryopteris (Shield Fern; *D. dentata* C. Chr.)

USES: (1) For studying all plant parts. (2) To germinate spores and grow gametophytes.

DESCRIPTION: A common, hardy greenhouse and laboratory fern. Other species are as hardy and as commonly used, for example, *D. goldiana* and *D. spinulosa*.

Underground rhizome, covered with adventitious roots that rise above ground among the leaf bases. Pinnate fronds, 1 to 2 ft long, with blades 6 to 10 in across. Hairs on petioles and blades. Long, narrow pinnae, deeply cut at margins to make narrow, closely packed lobes. Small sori on undersurface of pinnae in a line along each side of midvein. Sporangia, each with an annulus, arise from lower leaf surface and are covered with a shieldlike indusium that rises as a flap of tissue from lower leaf surface (Fig. 8-19).

CULTURE: See *Adiantum*.

PROPAGATION: Separate plant parts by cutting through rhizome with a sharp blade. Obtain plants locally or by catalog (see Appendix B).

Equisetum (Horsetails, Scouring Rushes)

USES: (1) For study of free-hand cross sections and long sections of stem, stained with acetocarmine. (2) To study meiosis in the spore mother cells by collecting a series of sporangia from tip of strobilus to base, which are then crushed on slides and stained with acetocarmine as described in Appendix A. (3)

Figure 8-19 *Dryopteris* sp., shield fern. **A** Entire sporophyte. **B** Close-up of a second species, showing circular, shieldlike indusia. (*A and B Courtesy of CCM: General Biological, Inc., Chicago.*)

A

B

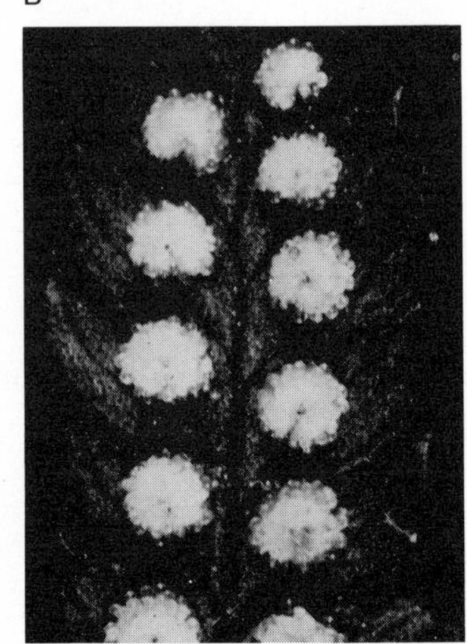

To study stem surface and stomata with a stereomicroscope. (4) To germinate spores and grow gametophytes.

DESCRIPTION: Rushlike plants with underground, perennial, jointed rhizome. Erect, jointed aerial stems sometimes branching and often hollow. Stems are 2 to 4 ft high and with long grooves and ridges. Small photosynthetic leaves develop in a ring at stem nodes and soon become dry, toothlike scales (Fig. 8-20). Internodes of most species (*E. scirpoides* is an exception) contain a large central canal, a medium-sized canal (vallecular canal) under each groove to which the stomata connect, and a small canal under each ridge (carinal canal). To study canals prepare a microscopic water mount of a thin cross section from an internode.

Sporangia are formed in a strobilus at the tips of vegetive stems (*E. hyemale*), at tip of special nonchlorophyllous stems (*E. arvense*) that die after spores are shed, or on nongreen fertile stems that become chlorophyllous after spores are shed (*E. sylvaticum*).

Equisetum arvense is common to much of United States, except the southern states, in both marshy and dry areas. *Equisetum telmateia* and the evergreen *E. hyemale* are found in moist woodlands of eastern and southern United States and along streams of the western coastline. *Equisetum sylvaticum* is native to woodland areas of lower Canada from Alaska to Quebec, and northcentral and northeastern United States. Other species are distributed over much of United States.

CULTURE: Plants collected locally can be transferred, generally with little difficulty, to outdoor beds or to the greenhouse and laboratory. Follow instructions given earlier in chapter for sporophyte culture. Bierhorst (1964) recommends planting *E. arvense* in an outdoor bed, where the plant forms nearly mature strobili below ground by the fall months. A clump may be dug up during the fall and transferred to an indoor pot where fertile shoots will appear within several weeks. The developing spores of the strobili may be studied, or mature spores can be germinated on Bold's basal medium with 1.5% agar (Appendix A). To inoculate agar plates with spores, sterilize the strobili by rapidly passing them through a flame, or by soaking them for 30 sec in 1 part household bleach mixed with 9 parts water. Crush the strobili over the Bold's basal medium to expel spores onto the medium surface. Use a small, soft brush to separate and spread spores evenly. Bold (1967b) states that gametophytes of *E. arvense* and *E. hyemale* grow luxuriantly on the Bold's basal medium with 1.5% agar. When placed at 22°C, with 12-h periods of 150 ft-c of light each day, antheridia may be produced in 30 days. Archegonia are produced less regularly under these conditions.

Spores of *E. telmateia* may be obtained during March or April from Carolina Biological Supply Co. at their Powell Laboratory Division, Gladstone, Oregon 97027. *Equisetum* spores are viable for only several weeks and must be obtained fresh and germinated at once.

PROPAGATION: Separate rootstock with a sharp blade. Obtain plants from a biological company or from A. E. Allgrove (see Appendix B).

Marsilea (Clover Leaf Fern; *M. vestita* Hook. and Grev.)

USES: (1) For study of sporocarps. (2) For study of stomates on upper surface of floating leaves. (3) As a freshwater aquarium plant.

A *Courtesy of Carolina Biological Supply Company*

Figure 8-20 *Equisetum* sp., horsetails. **A** *E. arvense;* strobili are shedding spores. **B** *E. telmateia.* **C** *E. hyemale.*

C *Courtesy of Carolina Biological Supply Company*

B *Courtesy of Carolina Biological Supply Company*

DESCRIPTION: A heterosporous fern that resembles the four-leaved clovers (Fig. 8-21). Worldwide distribution in warm fresh water. Leaf has four leaflets, each ½ to ¾ in across, that rise on a long petiole above shallow water or that float in deeper water.

Elongated and branched stolon grows on or in mud. Alternate leaves arise in two rows, one from each side of stolon. Roots from stolon may form a mat in the mud. Occasionally sporocarps form on short branches at base of petioles. A sporocarp contains both the megasporangia, each with one mature megaspore, and the microsporangia, each with many microspores. The mature sporocarp, about ⅛ to ¼ in in diameter, resembles a small brown nut with an outer stony covering. Mature sporocarps remain in mud after vegetative portions die. Eventually the hard covering deteriorates, and absorbed water causes an inner gelatinous material to swell and protrude from sporocarp, forming an outside ring of jelly. The sori with enclosed clusters of sporangia are carried along on the ring of jelly. Eventually microspores germinate into microgametophytes, and megaspores develop into macrogametophytes. Megaspores are visible to the naked eye; the female gametophyte may be visible at a magnification of 60X as a protruberance from the megaspore. Microspores and their male gametophytes must be observed under magnifications of about 200X. The mature sperm swim to archegonia, where fertilization occurs. Development of gametophytes and fertilization may occur within a 24-h period. The sporophyte plant grows during the summer months and forms sporocarps during the late fall months.

Marsilea sporocarps can be obtained from biological companies and germinated in an inorganic salt medium by first cutting away a portion of the stony wall. The study can be attractive to students who are particularly interested in ferns and in the evolutionary development of heterospory.

CULTURE: Put plants in pots of soil and submerge them in a freshwater aquarium. Obtain sporophyte plants from biological suppliers.

PROPAGATION: Separate stolons and root mats.

Nephrolepis (Sword Fern; *N. exaltata* Schott)

USES: (1) For study of all plant parts. (2) To germinate spores and to grow gametophytes.

DESCRIPTION: Many horticultural and mutant types have been developed from *N. exaltata*, and the basic species is now rarely found in greenhouses. Catalogs from suppliers describe many cultivars and varieties; for example, *N. exaltata* var. *bostoniensis* is the Boston fern, a common houseplant.

Nephrolepis exaltata, the parent plant, has stiff, erect, compound-pinnate fronds rising from an underground rhizome (Fig. 8-22). Fronds are 2 to 4 ft long; pinnae 2 to 3 in long. Rachis is generally hairy. One line of sori on each side of ventral midvein midway between rib and margin with kidney-shaped indusium attached at one edge, accounting for name (Greek: *nephro* means kidney).

CULTURE: See *Sporophyte Culture* and *Gametophyte Culture*, earlier in chapter. Obtain plants locally or by catalog.

PROPAGATION: Separate rhizome with sharp blade.

A

B

C

Figure 8-21 *Marsilea* sp., water fern. **A** Group of plants with floating leaves. **B** Close-up of plants in aquarium. **C** Close-up of leaves (or fronds). (*A Courtesy of CCM: General Biological, Inc., Chicago.*)

Figure 8-22 *Nephrolepis* sp., sword fern.

Osmunda (Interrupted Fern; *O. claytoniana* L.)

USES: (1) To study meiosis and the large chromosomes in stained spore mother cells ($n = 22$). (See Appendix A for spore stains.) (2) To cultivate and study the gametophytes. (3) As an attractive ornamental plant.

DESCRIPTION: Vertical, underground rhizomes with thick covering of wiry roots and leaf bases that remain long after the annual, green fronds die. Pinnately divided, coarse fronds, 2 to 4 ft long. Some fronds bear three to five pairs of brown fertile pinnae near middle of leaf and with sterile pinnae above and below; hence the name "interrupted" fern. Sporangia with no indusia are formed on undersurface of fertile pinnae. The sporangia split open, releasing mature spores that fall to the ground and develop into a chlorophyllous gametophyte with the sex organs on underside. Antheridia form near margins and archegonia near the midline. Growth of embryo and young sporophyte is slow compared to related ferns. This species, as well as *O. regalis* (bears sporangia on distal pinnae), and *O. cinnamomea* (bears sporangia on separate fertile fronds) are widely distributed over North and South America.

Garden stores supply "osmunda fiber," the fibrous roots, in bags or bales for a potting medium, particularly for orchids and other epiphytes.

CULTURE: As a rule, sporophyte plants collected locally will grow easily indoors. See methods described earlier in chapter for culture of sporophytes and gametophytes. Spores are viable for only a short time and should be put on media shortly after they are collected. Antheridia and archegonia may develop within 2 to 3 months. Gametophytes can be retained in cultivation if transferred periodically to fresh medium.

PROPAGATION: Separate rootstock with portions of the upper fronds. Obtain plants from nurseries or by collecting in the field. See source list of suppliers in Appendix B.

Figure 8-23 *Platycerium bifurcatum*, staghorn fern. (*Merry Gardens, Camden, Me.*)

Platycerium (Staghorn Fern; *P. bifurcatum* C. Chr.)

USES: (1) As a novelty fern because of bizarre foliage that resembles antlers. (2) To observe epiphytic growth.

DESCRIPTION: An attractive, epiphytic fern, with two types of fronds: flat, low-growing, rounded, sterile fronds, covered with hair and creeping over soil somewhat like a liverwort thallus; and erect, forked, fertile fronds, 2 to 3 ft long, that may become pendant (Fig. 8-23). Naked sori on fertile fronds in large masses on undersurface of forked ends. Underground rhizome from which fronds and roots arise.

CULTURE: See *Sporophyte Culture*, earlier in chapter. Double the proportion of humus in the soil mix. To best demonstrate epiphytic habit, grow the fern on a "plant plank" described in Chap. 6, or on a large, bark slab with *Osmunda* or *Sphagnum* as the growth medium. For gametophyte growth, try methods described under *Gametophyte Culture*, given earlier in chapter.

PROPAGATION: Separate rootstock. Obtain plants at local nurseries, if possible, or by catalog. See Appendix B.

Polypodium (Common Polypody Fern; *P. vulgare* L.)

USES: (1) As a terrarium plant. (2) To study all plant parts. (3) To germinate spores and grow gametophytes.

DESCRIPTION: Found in western North America. Many horticultural types have been developed. *Polypodium virginianum* is the common polypody of eastern North America.

Branched, creeping, rhizome from which fronds and roots arise. Frond is pinnately divided and 6 to 12 in high (Fig. 8-24). Many large naked sori on undersurface of pinnae in a single row on each side of midrib. *Polypodium polypodioides*

A

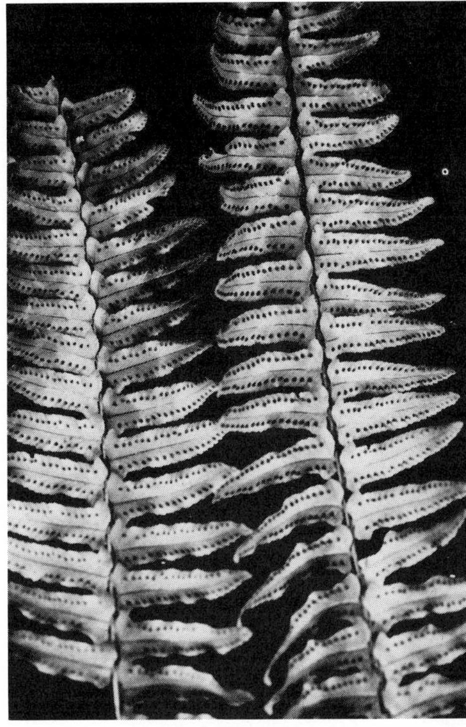

B

C

Figure 8-24 *Polypodium vulgare*, common polypody fern. **A** Entire plant. **B** Undersurface of fronds showing sori. **C** Close-up of pinnae showing naked sori.

is a xerophyte that is found in cool desert areas of South America and Caribbean islands.

CULTURE: See methods for culturing sporophytes and gametophytes, given earlier in chapter. A hardy plant that is often grown indoors.

PROPAGATION: Separate rhizomes with sharp blade. Plants may be purchased from biological suppliers and plant nurseries. See Appendix B.

Polystichum (Christmas Fern; *P. acrostichoides* Schott)

USES: (1) For studying all plant parts. (2) To germinate spores and grow gametophytes.

DESCRIPTION: Species of *Polystichum* are widely distributed in woodland areas over the United States. Many of these ferns were formerly placed in the genus *Aspidium*, and gardeners may still use that name. Perennial and evergreen plant. Stalked rhizome covered with roots and leaf bases. Fronds 10 to 30 in long and 2 to 3 in wide; divided pinnately (Fig. 8-25). Pinnae with slightly toothed margins. Large sori, covered with a central indusium, are scattered over most of undersurface of distal pinnae.

CULTURE: See methods for growing sporophytes and gametophytes, described earlier in chapter. Obtain plants at local nurseries or by catalog. See Appendix B.

PROPAGATION: Separate rootstock.

Selaginella (Spike Moss; *S. kraussiana* Braun.)

USES: (1) To demonstrate the more advanced characteristics of these plants, including the strobili, heterospory, and the development of unisexual gametophytes within the spores and sporangia. (2) Use *S. lepidophylla* (Resurrection Plant) as a novelty to demonstrate plant adaptations to arid conditions. (3) As a terrarium plant.

Figure 8-25 *Polystichum* sp., Christmas or holly fern. (*Courtesy of CCM: General Biological Inc., Chicago.*)

DESCRIPTION: The genus *Selaginella* includes a large group of plants of approximately 700 species, native mostly to very moist tropical regions. *Selaginella kraussiana* is a mosslike, creeping plant, commonly grown in greenhouses as a soil cover around other potted plants and as a planting under the benches. The plant roots freely from the jointed, branching stems. Bright green, minute, scalelike leaves in four rows, with larger leaves (⅛ in long) on either side of stem and a pair of smaller leaves on dorsal surface of stem. Heterosporous. Sporophylls in terminal strobili (strobilus, singular), containing microsporangia and megasporangia in axils of leaves. Microspores develop into male gametophytes and megaspores into female gametophytes. Development of both types of gametophytes occurs in spores while they remain within the sporangia. When gametophytes are nearly developed, spores are shed with the greatly reduced gametophyte. Gametophytes are without chlorophyll and are nourished by materials within spores.

Most species are tropical and require moist, humid surroundings. The "resurrection plant" (*S. lepidophylla*), however, can live in very dry soil or removed from soil, at which times it curls into a ball. When wet, the plant expands, exposing the green upper surfaces of foliage. See Fig. 8-26.

CULTURE: Put plants in a terrarium, prepared as described in Chap. 6 under *Woodland Terrarium*. Add one teaspoon of dry bone meal to each quart of soil mixture. Continue to add small amounts of bone meal in water solution every 2 or 3 months. Place terrarium at 21 to 24°C, in a location for sunlight during several hours of the day. Do not allow soil to become dry. A glass or plastic cover on terrarium should hold moisture so that watering is required only every 4 to 6 weeks. If much moisture condenses on undersurface of cover, open terrarium for several

Figure 8-26 *Selaginella* sp., spike "moss." **A** *S. lepidophylla*, resurrection plant when dry and brown. **B** Resurrection plant after soaking overnight. The plant is now a bright green.

A

B

hours to prevent growth of molds. Ordinarily the plants thrive and produce strobili with spores.

Gametophytes may be cultured on plaster of paris blocks as described under *Gametophyte Culture* earlier in chapter.

PROPAGATION: Separate sporophyte stems with roots, and transfer portions to a new container. Obtain sporophyte plants from biological supply companies or from plant nurseries. See sources in Appendix B.

REFERENCES

Anthony, R.E.: 1962, "Greenhouse Culture of *Marchantia polymorpha* and Induction of Sexual Reproductive Structures," *Turtox News*, **40** (1): 2–5.

Bailey, L.H.: 1949, *Manual of Cultivated Plants*, The Macmillan Company, New York, 1116 pp.

Basile, D.V.: 1964, "New Procedures of Bryophyte Culture which Permit Alternation of the Culture Medium During the Life Cycle," *Bryologist*, **67**: 141–146.

Bierhorst, D.W.: 1964, "Suggestions and Comments on Teaching Materials of the Non-Seed Bearing Vascular Plants," *Am. Biol. Teacher*, **26** (2): 105–107.

Block, R.J., E.L. Durrum, and G. Zweig: 1958, *A Manual of Paper Chromatography and Paper Electrophoresis*, 2d ed., Academic Press, Inc., New York, 710 pp.

Bold, H.C.: 1965, "The Neglected Cryptogams," *Am. Biol. Teacher*, **27** (2): 101–103.

Bold, H.C.: 1967a, *Morphology of Plants*, 2d ed., Harper & Row, Publishers, Incorporated, New York, 541 pp.

Bold, H.C.: 1967b, *A Laboratory Manual for Plant Morphology*, Harper and Row, Publishers, Incorporated, New York, 123 pp.

Conard, H.S.: 1956, *How to Know the Mosses and Liverworts*, Wm. C. Brown Company, Publishers, Dubuque, Iowa, 226 pp.

Crum, H., W.C. Steere, and L.E. Anderson: 1965, "A List of the Mosses of North America," *Bryologist*, **68** (4): 377–434.

Davis, B.D., and S.N. Postlethwait: 1966, "Classroom Experimentation Using Fern Gametophytes," *Am. Biol. Teacher*, **28** (2): 97–102.

Frye, T.C., and L. Clark: 1937–1947, *Hepaticae of North America*, parts I-V, University of Washington Press, Seattle.

Grout, A.J.: 1928–1940, *Moss Flora of North America, North of Mexico*, vols. I, II, and III, published by the author, Newfane, Vermont. May be obtained from Chicago Natural History Museum.

Hais, I.H., and K. Macek (eds.): 1964, *Paper Chromatography*, Chemical Rubber Co., 2310 Superior Ave., Cleveland, Ohio.

Heftmann, E.: 1967, *Chromatography*, 2d ed., Reinhold Publishing Corporation, New York, 851 pp.

Johansen, D.A.: 1940, *Plant Microtechnique*, McGraw-Hill Book Company, New York, 523 pp.

Keefe, A.M.: 1926, "A Preserving Fluid for Green Plants," *Science*, **64**: 331.

Klein, R.M., and D.T. Klein: 1970, *Research Methods in Plant Science*, Natural History Press, Garden City, N.Y., 756 pp.

Matzke, E.B.: 1964, "The Aseptic Culture of Liverworts in Microphytotrons," *Bryologist*, **67**: 136–141.

Miller, D.F., and G.W. Blaydes: 1962, *Methods and Materials for Teaching the Biological Sciences*, 2d ed., McGraw-Hill Book Company, New York, 453 pp.

Miller, H.A.: 1962, "Bryophytes in the Biology Class," *Carolina Tips*, **25** (8): 29–31.

Rastorfer, J.R.: 1962, "Photosynthesis and Respiration in Moss Sporophytes and Gametophytes," *Phyton*, **19:** 169–177.

Roehrick, C.G.: 1964, "Bryology for the High School," *Am. Biol. Teacher*, **26** (2): 124–130.

Scagel, R.F., R.J. Bandoni, G.E. Rouse, W.B. Schofield, J.R. Stein, and T.M.C. Taylor: 1965, *An Evolutionary Survey of the Plant Kingdom*, Wadsworth Publishing Company, Inc., Belmont, Calif., 658 pp.

Schneider, M.J., P. D. Voth, and R.F. Troxler: 1967, "Methods of Propagating Bryophyte Plants, Tissues, and Propagules," *Botan. Gaz.*, **128** (3–4): 169–174.

Schuster, R.M.: 1966, *The Hepaticae and Anthocerotae of North America*, vol. I, Columbia University Press, New York, 802 pp.

Smith, G.M.: 1955, *Cryptogamic Botany, Vol. II: Bryophytes and Pteridophytes*, McGraw-Hill Book Company, New York, 399 pp.

Steere, W.C.: 1964, "The Use of Living Bryophytes in the Teaching of Botany," *Am. Biol. Teacher*, **26** (2): 100–104.

Stegner, R.W.: 1967, *Plant Pigments*, Rand McNally & Company, Chicago, 71 pp.

Voth, P.D.: 1943, "Effects of Nutrient-Solution Concentration on the Growth of *Marchantia polymorpha*," *Botan. Gaz.*, **104:** 591–601.

Wagner, K.A.: 1961, "A Key to Common Liverworts," *Carolina Tips*, **24** (6): 22–23.

Other Literature

Andrews, S., and P.L. Redfearn: 1965, "Observations on the Germination of the Gemmae of *Hyophila tortula* (Schwaeger) Hampe," *Bryologist*, **68** (3): 345–347.

Cobb, B.: 1956, *A Field Guide to the Ferns*, Houghton Mifflin Company, Boston, 281 pp.

Freeberg, J.A., and R.H. Wetmore: 1957, "Gametophyte of *Lycopodium* as Grown *in vitro*," *Phytomorphology*, **7** (2): 204–217.

Hoshizaki, B.J.: 1970, "The Genus *Adiantum* in Cultivation [Polypodiaceae]," *Baileya*, **17** (3): 97–144.

Milton, M.W., and J. Colaiace: 1969, "The Induction of Sexual Reproductive Structures of *Marchantia polymorpha* Grown Under Aseptic Culture Conditions," *Bryologist*, **72** (1): 45–48.

Pohl, R.W.: 1971, Dept. of Botany and Plant Pathology, Iowa State University, Ames. Unpublished list of living plant materials for taxonomic instruction.

Register, T.E., and W.R. West: 1971, "Marchantia Life Cycle," *Carolina Tips*, **34** (5): 17–20.

PART

three

MICROORGANISMS IN THE LABORATORY

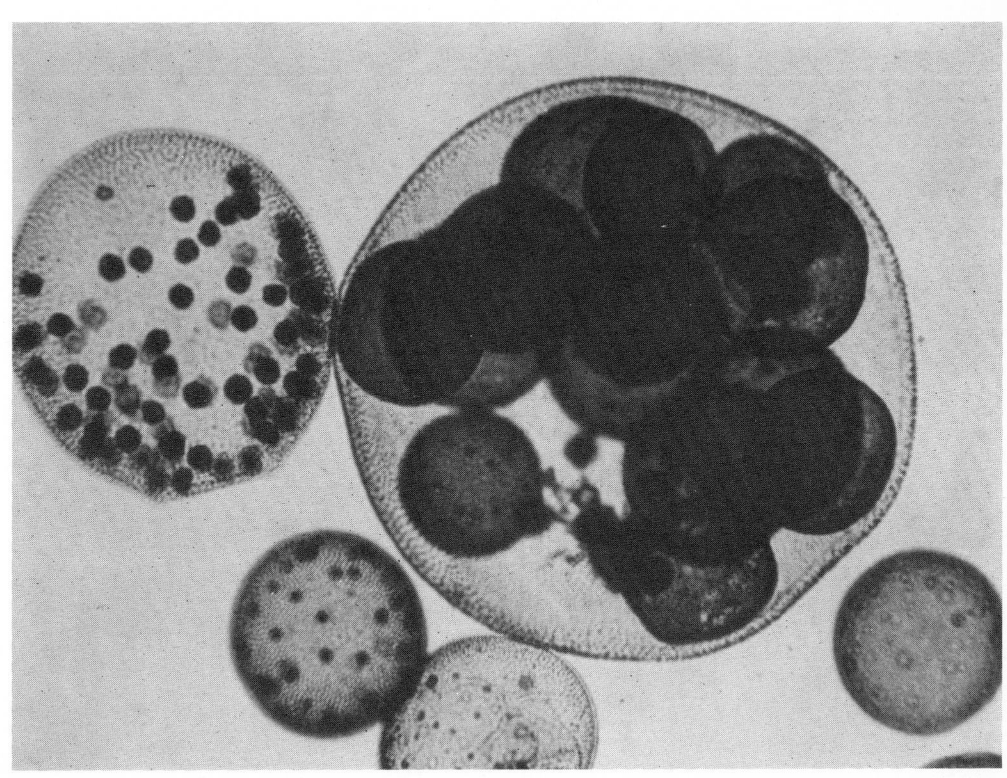

The major groups of microorganisms are generally designated as the viruses, rickettsiae, bacteria, fungi, algae, and protozoa. In reality, various taxonomic categories have been suggested for these organisms, including the kingdoms of Monera and Protista, and various systems of phyla and divisions for animallike and plantlike microorganisms. For our purposes, we shall follow the usual practice of assigning the name "microorganism" to all these groups. Despite the fact that many algae and fungi are not microscopic, all organisms of these groups share a relationship in their morphological simplicity that sets them apart from the more complex, "higher" plants and animals.

The discussion of this section is divided into four chapters that are concerned with: (1) the viruses and bacteria; (2) the fungi and lichens; (3) the algae; and (4) the protozoa.

As a rule, microorganisms are especially interesting to students, possibly because the studies reveal an entirely new world to them. With few exceptions, microorganisms are easily collected and maintained. They occur abundantly in fresh and salt water, in the soil and air, and as parasites of both plants and animals. Except for the larger fungi and algae, conservation of these organisms is of little, or no consequence, and no legislation is required to control collection.

VIRUSES AND BACTERIA

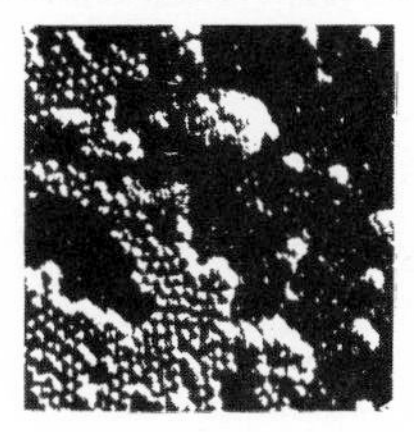

The first portion of this chapter contains a description of the equipment and techniques that may be employed in microbiological work, including that of the fungi discussed in the next chapter. A second portion describes techniques for studying viruses. A last portion consists of descriptions of selected genera of bacteria that are appropriate for the biology laboratory, and of special techniques for bacteriological studies.

The precautions given below should be observed when working with non-pathogenic microbes. More stringent rules, of course, must be observed with pathogens; it seems unnecessary to state that live pathogens are not appropriate for an introductory biology course.

1. Do not eat or drink when working with these microbes.

2. Keep hands away from the mouth and eyes.

3. Do not put pencils or other objects in the mouth.

4. When removing covers from a culture and when making transfers, place the opening of a tube or plate at an angle away from the face. Do not inhale or exhale directly over the open vessel.

5. When making transfers, disturb the culture as little as possible to avoid unnecessary dispersal of organisms.

6. To avoid the inhalation of microbes and the contamination of a culture by exhalation, do not talk when making a transfer.

7. When flaming a loop or needle after a transfer, avoid spattering of microbes by first holding the needle or loop above the flame to slowly burn off any remaining clumps of microbes. Then hold the needle or loop at an almost vertical slant in the hottest portion of the flame until the wire becomes red from the heat.

8. If a culture is spilled, immediately cover the surface with a disinfectant solution, for example, 1% Lysol solution. After several minutes, rinse the surface with water.

9. Wash hands thoroughly with soap when the work is finished.

10. Autoclave cultures before disposing of them, or, alternatively, cover a culture with 40% formaldehyde and allow the culture to set for 24 h before disposal. Cultures in plastic plates can be incinerated, in which case neither of the preceding treatments is necessary.

EQUIPMENT AND TECHNIQUES

Standard equipment for the cultivation of microbes usually includes the following items (Fig. 9-1): inoculating needle and transfer loop; Bunsen burner, alcohol lamp, or propane cylinder; autoclave or pressure-cooker; dry-oven sterilizer; incubator; glassware, including test tubes, petri dishes, flasks, pipettes, and glass spreaders.

Tubes and flasks are plugged with nonabsorbent cotton, with screw-on or plastic caps, or, if not lipped, with metal-spring caps (Fig. 9-2). Often a cheesecloth covering is placed over the bottom and sides of a cotton plug to prevent plugs from sticking to a film of medium in the mouth of the flask. Figure 9-3 shows one method for preparing cotton plugs. The plug should extend about 2.5 cm into a flask with about 2 cm above the mouth opening. The usual "rules-of-thumb" for determining if a plug is of proper size are: (1) the empty tube or flask does not fall loose when lifted by grasping only the plug; and (2) the plug does not squeak when rotated in the mouth of the tube or flask, indicating that the plug is not too large and too tight.

Cleaning Glassware

Glassware should be washed thoroughly in hot, soapy water, rinsed in running tap water, then dipped into 1% HCl to neutralize any remaining alkali of the soap, and finally rinsed in distilled water. Allow glassware to dry completely in an oven before reuse or storage. Slides and coverglasses require a more thorough cleaning with a soft polish, for example with Bon Ami, or by soaking in concentrated nitric acid for 10 min. Either treatment is followed by thorough rinsings in running water for 10 min.

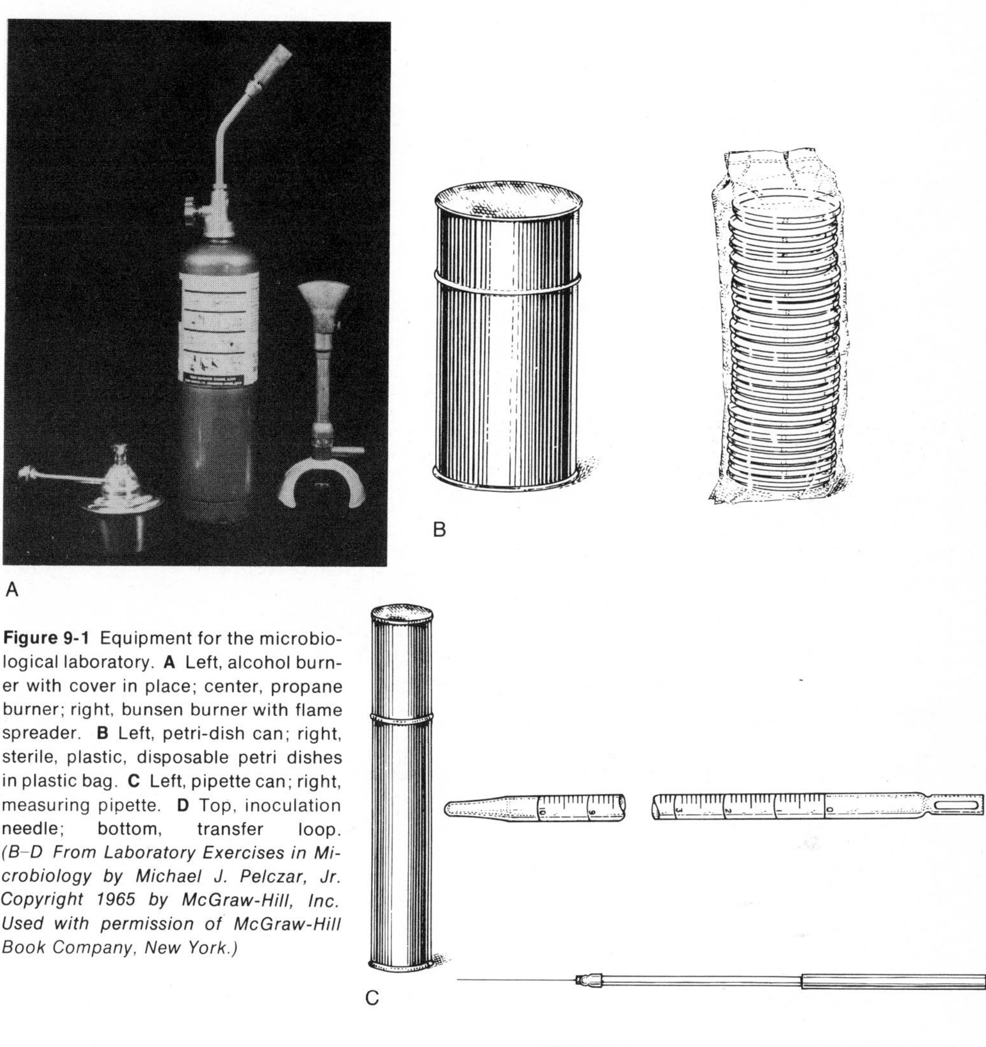

Figure 9-1 Equipment for the microbiological laboratory. **A** Left, alcohol burner with cover in place; center, propane burner; right, bunsen burner with flame spreader. **B** Left, petri-dish can; right, sterile, plastic, disposable petri dishes in plastic bag. **C** Left, pipette can; right, measuring pipette. **D** Top, inoculation needle; bottom, transfer loop. *(B–D From Laboratory Exercises in Microbiology by Michael J. Pelczar, Jr. Copyright 1965 by McGraw-Hill, Inc. Used with permission of McGraw-Hill Book Company, New York.)*

Many kinds of new slides are coated with a thin layer of oil and need to be washed before use. To test for oil on a new slide, transfer a loop of water to the slide. If the water remains in a rounded drop, the slide is coated with oil and needs washing. If the drop spreads out into a thin layer, the slide is clean.

Sterilizing Glassware and Media

Glassware is most easily sterilized in an oven, which may be the oven of an ordinary cooking stove or a special laboratory oven. Petri plates may be sterilized in

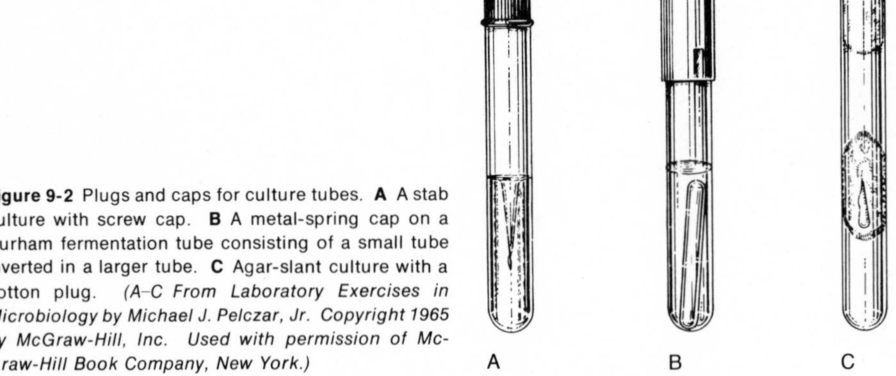

Figure 9-2 Plugs and caps for culture tubes. **A** A stab culture with screw cap. **B** A metal-spring cap on a Durham fermentation tube consisting of a small tube inverted in a larger tube. **C** Agar-slant culture with a cotton plug. *(A–C From Laboratory Exercises in Microbiology by Michael J. Pelczar, Jr. Copyright 1965 by McGraw-Hill, Inc. Used with permission of McGraw-Hill Book Company, New York.)*

metal cans, paper bags, or in lots of four or five plates wrapped in paper. Pipettes are best sterilized in special metal pipette boxes, or in devised cylinders or cans of appropriate length. Otherwise, each pipette may be rolled separately or in small groups in paper, with the outside paper marked to indicate the mouth end of a pipette, thus ensuring the safe removal of the pipette at the time of use.

Place the glassware in a cold oven, taking care not to allow paper wrapping or plugs to touch against the inner walls. Set the oven at 180°C; when that temperature has been reached, begin the timing and continue heating for 2 h. A higher temperature or a longer time may cause the wrapping to ignite. To prevent glass

Figure 9-3 A technique for making cotton plugs. **A** Roll of cotton. **B** A strip is cut from the roll. **C** The strip is cut into halves. **D** Each half is rolled tightly and folded. **E** A tube with cotton plug. If plug is too thick, remove a top layer from the cotton square at **C**. *(From O. B. Williams and Orville Wyss.)*

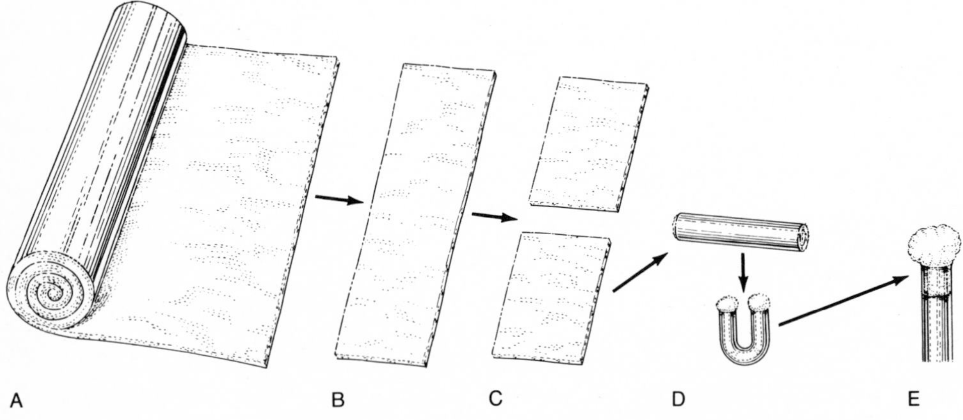

breakage from sudden cold air, allow the oven to cool to below 90°C before opening the door.

Dispense culture medium in flasks or tubes for sterilization, with each container not more than half full to allow space for air expansion during sterilization and to prevent wetting of cotton plugs. Plug or cap the containers and sterilize in an autoclave or household pressure-cooker, generally at 15 lb pressure for 15 min. An electric hot plate of at least 1,200 W, or the burner of an ordinary gas or electric range, is required to gain the required pressure in a pressure-cooker. Before using an autoclave or pressure-cooker for the first time, a person *must* read carefully the accompanying manual or *must* receive instruction from an experienced person. When sterilization time is completed, disconnect the sterilizer and allow the pressure to decrease to zero before opening the sterilizer. A quick lowering of pressure by fully opening a steam outlet may cause plugs to blow out of the flasks and tubes, and may give a person a serious steam burn.

Allow the sterile medium in flasks to cool to 45 to 50°C (warm but not hot to touch), and then pour into sterile petri dishes or test tubes with proper attention to aseptic techniques. Agar medium solidifies at about 40 to 42°C, and the pouring must be done promptly. For this reason, it is better to sterilize agar media in flasks of no greater volume than 500 ml (each flask half-filled), as a greater quantity of medium may solidify before transfers can be completed. If the medium is at a higher temperature when poured, the heat produces an undesirable condensation that accumulates on the agar surface or on the inner surface of the plate cover.

Slant-agar tubes are made by placing the upper end of the freshly poured tube on a book or other object to raise that end about one-half inch. When the agar medium in a plate has solidified, invert the plate and store in that position to prevent accumulation of condensed water on the agar surface.

Sterilization by Filtration

Certain types of organic substances, for example, antibiotics, hormones, and some sugar solutions, are destroyed by heat sterilization. Therefore, when such substances are contained in a medium, sterilization is accomplished by filtration through microporous discs, generally of asbestos or cellulose compounds. Microfilter apparatus is available from many biological suppliers. A simple inexpensive device, useful for only several milliliters of material, is the Swinney adapter, which with its filter disc is attached to any Luer-Lok type of syringe and needle. The entire unit of adapter, needle, and syringe is sterilized in an autoclave before use.

Transfer Techniques

Figures 9-4 through 9-7 illustrate the techniques for various types of transfers.

Dilutions and Plate Counts

Often a count of the number of microorganisms in a sample is necessary, for example, in testing water or milk, in genetic studies of mutation frequency, and in various biochemical analyses. A common technique for determining the number

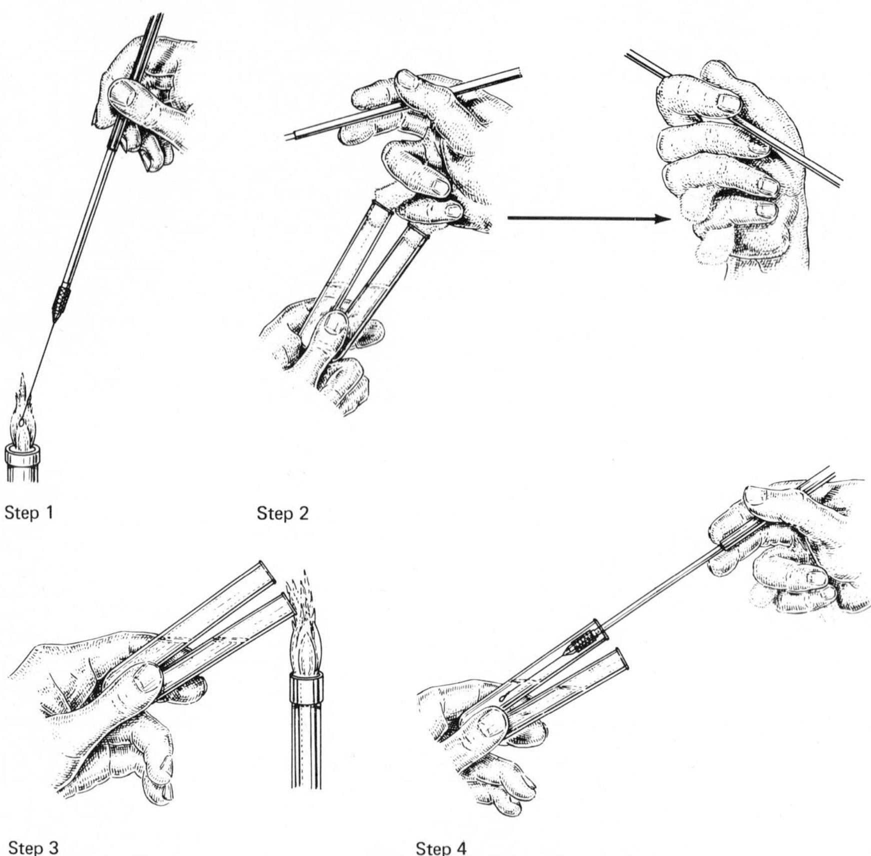

Figure 9-4 Techniques for a tube to tube transfer. *Step 1.* Sterilize the transfer loop in a flame. *Step 2.* Remove the plugs from tubes as shown. See inset for method of holding plugs. *Step 3.* Pass mouth of tubes through the flame. *Step 4.* Insert the transfer loop into the tube containing the culture, and then place the loop in the tube to be inoculated. Following this, flame the tube mouths again, replace cotton plugs, and flame the transfer loop again. *(Based on Laboratory Exercises in Microbiology, by Michael J. Pelczar, Jr. Copyright 1965 by McGraw-Hill, Inc. Used with permission of McGraw-Hill Book Company, New York.)*

of organisms is to inoculate an agar plate with a known amount of sample and to count the number of discrete colonies that appear in 24 to 48 h of incubation. Each colony is assumed to have started from one organism, and thus a count of the colonies gives a count of organisms in a known quantity of the original sample at the time when plates were inoculated. Generally, however, a sample must first be diluted so that colonies will be separated on a plate and, thus, can be counted individually. A desirable dilution is one that produces between 30 and 300 colonies on one plate.

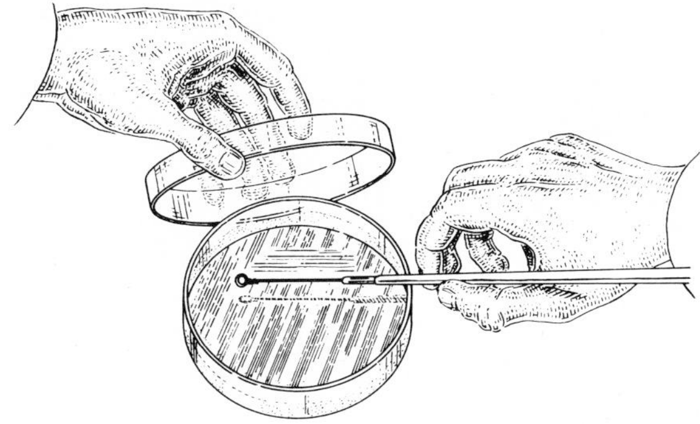

Figure 9-5 Techniques for streaking an agar-medium plate. Repeat Steps 1, 2, and 3 of Fig. 9-4, using only the one tube that contains the culture to be transferred. Flame the mouth of the tube and replace the plug. Raise the lid of the petri plate slightly and move the loop back and forth over the medium, as shown, to gain a good distribution of organisms over the surface.

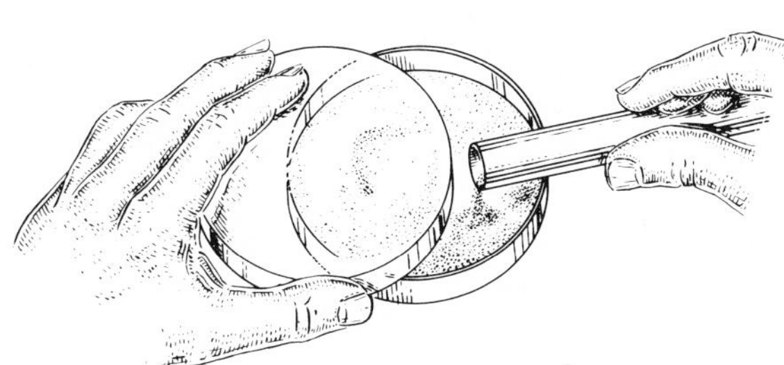

Figure 9-6 Technique for pouring a plate. Remove plug from tube, and flame tube mouth, as shown in Fig. 9-4. Raise one edge of the plate cover, and pour medium into lower plate. *(Based on Laboratory Exercises in Microbiology, by Michael J. Pelczar, Jr. Copyright 1965 by McGraw-Hill, Inc. Used with permission of McGraw-Hill Book Company, New York.)*

Figure 9-7 Technique for using a pipette. **A** Use mouth or bulb suction to draw liquid into a pipette. Control flow from pipette with forefinger, as shown. **B** Technique for pipetting liquid into a petri dish.

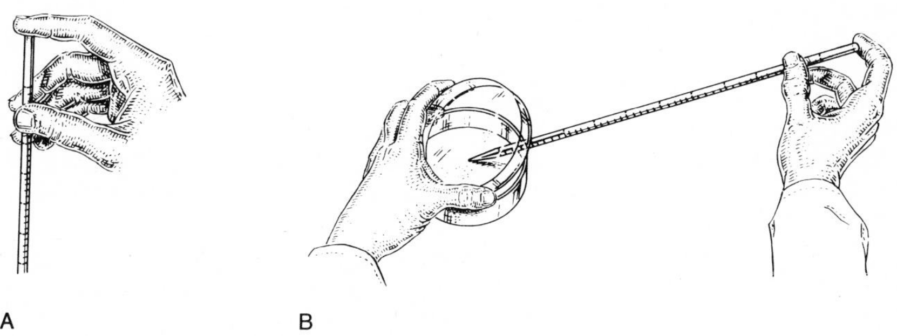

A B

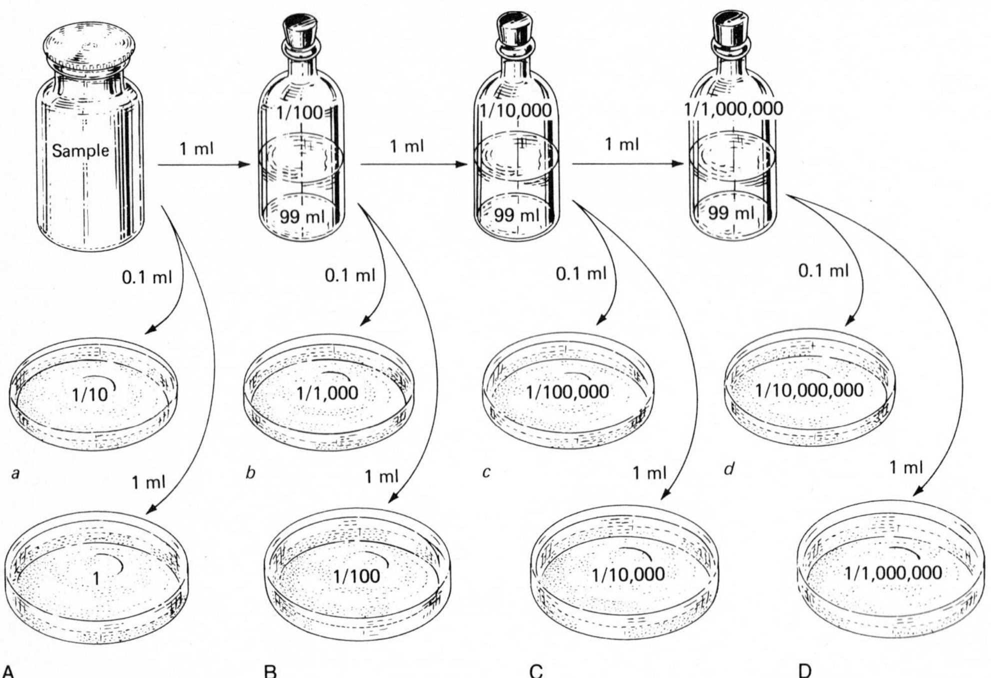

Figure 9-8 A quantitative dilution method. *(From Microbes in Action, 2d ed., by H. W. Seeley, Jr., and P. J. Van Demark. Copyright ©1972 by W. H. Freeman and Company, San Francisco.)*

An example of a series of dilutions is illustrated in Fig. 9-8. Stoppered tubes, flasks, or bottles containing water, nutrient broth, or 0.85% NaCl solution are sterilized and used for the dilutions. Eight-ounce capped medicine bottles, from a drug store, are convenient and inexpensive substitutes for the above containers. Prior to sterilization, however, the caps of the bottles should be loosened slightly. As shown in Fig. 9-8, 99 ml of diluting medium is placed in each of a series of flasks or bottles. Because the number of required dilutions cannot be known until after colonies have appeared on plates, it is best to prepare plates from at least two or three dilutions. One milliliter of sample is transferred with a sterile pipette to the first dilution bottle. The used pipette is then placed in a tray of disinfectant solution. The first dilution is thoroughly mixed by shaking the tightly capped bottle at least 25 times. The dilution procedure is repeated by transferring 1 ml of the first dilution to the second container, followed by thoroughly shaking the second container. A total series of three to six dilutions is made, each time with a fresh pipette.

Prepare poured-plates from the last two or three dilutions by pipetting 1 ml of dilution to each of a desired number of sterile petri plates. A tube of melted and cooled agar is then poured into each plate over the suspension, and the covered plates are rotated gently to disperse the cells throughout the medium. Invert plates and incubate at room temperature for 48 h, or in an incubator for 24 to 48 h at a temperature indicated later for particular organisms.

Make counts of colonies in a plate by placing the plate over a sheet of graph paper. If many colonies are present, count the number in one-fourth of the plate and multiply the count by four for a total plate count. Then calculate the total number of organisms in a milliliter of sample by multiplying the plate count by the dilution number. For example, in Fig. 9-8, the count of plate D is multiplied by one million to obtain the number of organisms in 1 ml of original sample at the time of transfer.

THE VIRUSES

Although bacteria are commonly studied in the biology laboratory, the viruses receive only scant attention. Obviously, the viruses lend themselves to little study in a beginning course, since all viruses are obligate parasites and many are pathogenic to their hosts. Nevertheless, viruses that parasitize nonpathogenic bacteria can be appropriately incorporated into laboratory studies, thus enabling students to gain at least a beginning understanding of viral action.

Viruses are not cellular in structure and have no true nucleus, cytoplasm, cell membrane, or cell wall. The fact that they multiply only when inside a host cell, and are inert particles when outside the host cell, remains intrinsically interesting to students and researchers alike. Thus, it seems that viruses approach the borderline that separates life from nonlife. Electron microscopy has shown that viruses may have the shape of a rod, a sphere, a cube, or a polyhedron (Fig. 9-9). Electron microscopes have also shown that viruses range in size from about 10 nm (0.01 micron) to nearly 300 nm (0.3 micron), a particulate size that cannot be viewed with the ordinary light microscope. In contrast, the smallest bacteria measure about 0.75 to 1.0 micron in diameter, and are visible under the light microscope.

The simplest type of virus consists of molecules of nucleic acid inside a protein coat. The most complex viruses contain other compounds, including nucleoproteins, lipids, carbohydrates, and occasionally other substances. Strangely, it seems, RNA is the only nucleic acid present in viruses that parasitize plant cells, and either RNA or DNA (but not both) are present in viruses that parasitize animal cells. As well, either DNA or RNA may be present in bacterial viruses, called bacteriophages, although most bacterial viruses contain DNA. Figure 9-10 illustrates the relative sizes of some viruses and bacteria in comparison to a yeast cell.

Although several theories have been offered regarding the mechanism that triggers the reproduction of viruses inside a host cell, no theory appears completely correct, and much probing and questioning remain for future investigators.

Several systems of classification are used for viruses. In a broad sense, they are grouped as plant, animal, or bacterial viruses. Mammalian viruses are often grouped according to the type of tissue attacked, for example, neurotropic (nerve cells), dermatotropic (skin cells); and viscerotropic (cells of the viscera and other internal organs). Most commonly, however, viruses are identified by their specific

A

Figure 9-9 Electron micrographs of viruses. **A** Tobacco rattle virus, magnified about 23,600 times. Both the short and long rods are required for infection. **B** Human polio virus magnified about 34,000 times. **C** Influenza virus magnified about 55,000 times. *(A Courtesy of M. K. Corbett. B and C Courtesy of Parke, Davis and Company.)*

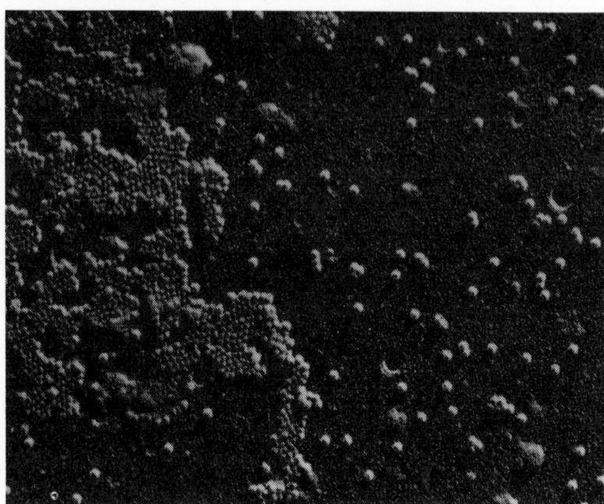

B

C

hosts and also by a trivial name, that may be the name of the researcher who isolated the strain, or that may be simply a label consisting of letters and numbers assigned to a strain by the isolator. Thus, the classification of viruses and the recognition of viruses by their trivial names becomes a situation that is comprehensible only to an experienced virologist. Possibly many virologists gain a working knowledge of only several restricted and specialized groups of viruses. Therefore, to beginning biology students, the names of viruses may be strange and confusing, for example, when students see illustrations of the TMV particles, or when they work with the T_2 virus, one of the bacteriophages that parasitizes *Escherichia coli* and that is used most often in beginning courses (Fig. 9-11).

Demonstration of Bacteriophage Plaques

A bacteriophage is a virus that infects a bacterium. Infection occurs when the tip of the virus tail becomes attached to the bacterial wall, and the viral nucleic acid is injected into the bacterial cell. In some manner, not completely understood, the viral nucleic acid, mostly DNA, "takes over" the bacterial metabolic activities in such a way that many new complete viruses are formed. One infected bacterial

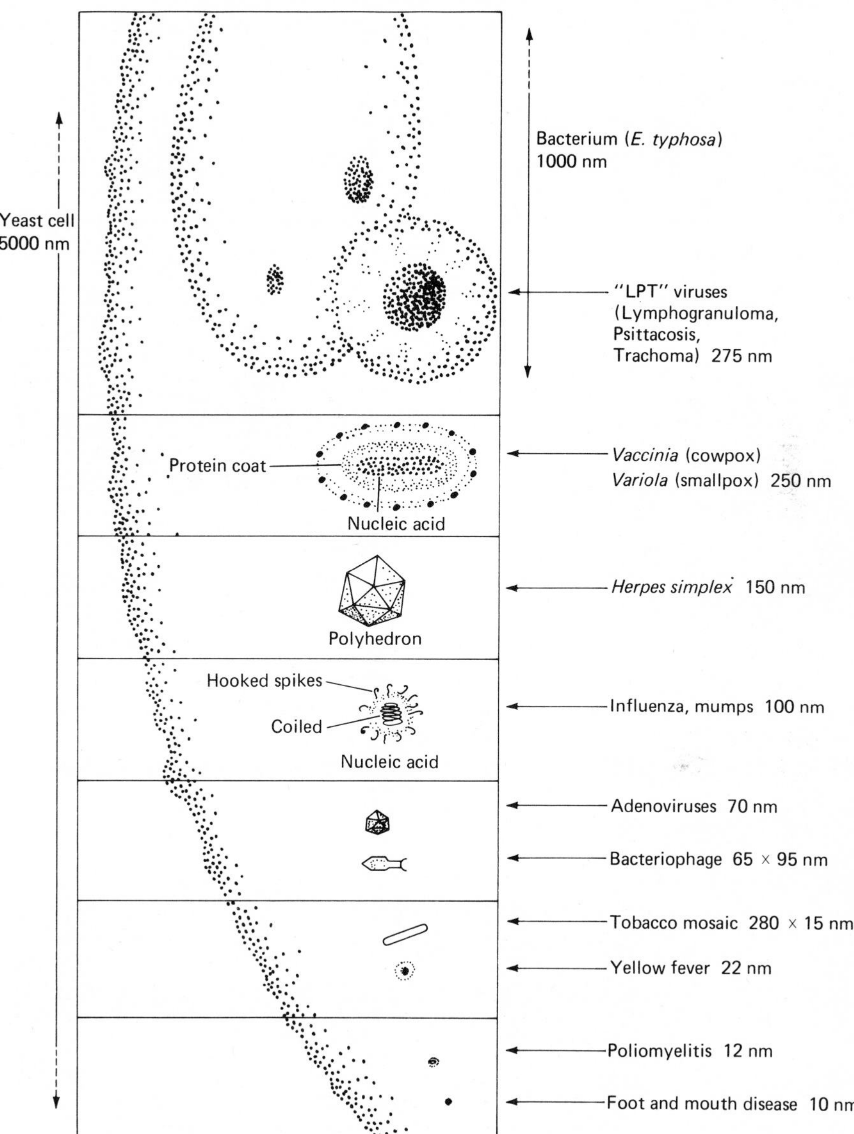

Figure 9-10 Relative size of yeast, bacteria, and viruses. *(From Microbiology, 2d ed., by Michael J. Pelczar, Jr., and Roger D. Reid. Copyright 1965 by McGraw-Hill, Inc. Used with permission of McGraw-Hill Book Company, New York.)*

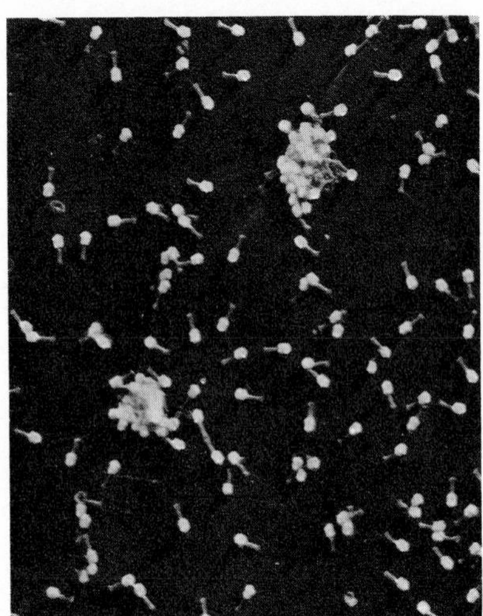

Figure 9-11 T_2 coliphage (X 13,900). *(From R. M. Herriott and J. L. Barlow, 1952, in J. Gen. Physiol.,* **36:** *17–28.)*

cell may produce as many as 300 new virus particles in about 30 min. The bacterial wall then ruptures, a process called lysis, and the new phages are released to enter other bacterial cells. Lysis, the disintegration of bacteria, becomes apparent in a tube of broth when the broth begins to lose its cloudiness and becomes clear. Agar plates, spread with a film of *E. coli*, show the effects of lysis when clear spots, the plaques, appear on the agar surface (Fig. 9-12).

The T strains of viruses, including T_1, T_2, T_3, T_4, T_5, T_6, and T_7, infect specific strains of the bacterium *E. coli*. The T_2 strain is used most often with beginning courses, and the *E. coli* strain B is the host selected most often for the T_2 virus (and

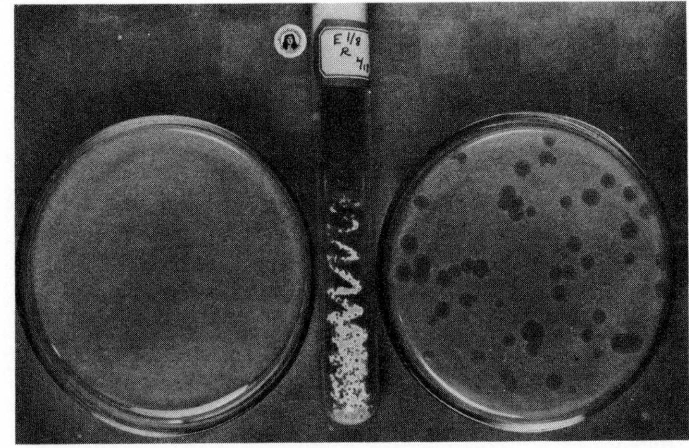

Figure 9-12 Virus plaques. Plate at left contains a culture of bacteria uninfected by bacteriophage. Center slant tube and plate at right contain cultures of bacteria infected by phage. Clear spots in tube and right plate are plaques produced by phage lysis of bacteria. *(From E. C. Saudek and D. R. Collingsworth, 1947.)*

for other T strains). T_1 and T_2 are supplied for educational purposes by the American Type Culture Collection (ATCC; see Appendix B). T_1 is more resistant to sterilization and is therefore more difficult to eliminate from glassware after the exercise is completed. The more desirable T_2 strain can be obtained in a freeze-dried preparation from ATCC, or in suspension with *E. coli* from Carolina Biological Supply Company. Occasionally, a local research laboratory will be able to furnish a starting culture.

Seeley and Vandemark (1972) have outlined the procedures for isolating bacteriophage from raw sewage. Such work could well lead to individual research by students who wish to investigate bacteriophages further than the demonstration described below. The following procedure is a common one that uses the T_2 phage and *E. coli* B, both obtained from commercial sources or from a university microbiology department.

Preparation of stock culture of T_2 bacteriophage. Aseptic techniques and sterile equipment are required for all procedures described here. Instructions for rehydrating the freeze-dried preparation are sent with the order from a commercial supplier. Briefly, rehydration is accomplished by adding 0.5 ml of trypticase broth (Appendix A) to the ampule of dried material. Then transfer the rehydrated phage suspension to a tube or flask of trypticase broth containing *E. coli* strain B that is at 6- to 12-h growth. Incubate the tube containing T_2 and *E. coli* at 37°C until lysis of the bacteria occurs (12 to 16 h), as indicated by a clearing of the broth. This procedure, then, produces a dense stock culture of the bacteriophage that must be separated from the bacterial cells in order to obtain a bacteria-free phage filtrate.

To separate the filtrate, centrifuge the bacterial-viral suspension for 10 min at 2,000 rpm, or allow the suspension to set until bacterial clumps have settled to the bottom of the tube. Remove the clear fluid at the top with a sterile pipette and transfer the fluid to a sterile vial or test tube. Store this bacteriophage stock at 40°C for later use. In refined work the bacteriophage extract must be passed through a bacteriological filter, for example, through a micropore filter described earlier. Filtration is not required when the purpose is only to demonstrate plaques, as in this procedure.

Demonstration of plaques. The following materials are required: (1) a flask containing 150 ml of sterile trypticase agar; (2) 10 plugged tubes, each containing 9 ml of sterile tap water and labeled 1 through 10; (3) 10 plugged tubes each containing 3 ml of soft trypticase agar with tubes labeled 1 through 10; (4) 10 empty sterile petri dishes labeled 1 through 10 on the outer, lower plate; (5) ten 1 ml pipettes; (6) a 6- to 12-h old culture of *E. coli* strain B in trypticase broth; and (7) the stock culture of *E. coli* bacteriophage prepared as above. See Appendix A for all media. Aseptic techniques and sterile equipment are required for all procedures described below.

1. Pour 15 ml of trypticase agar into each of the 10 labeled petri dishes. When the agar has solidified, place plates in the incubator at 35°C to dry the agar surface. They are to be removed when steps 2 and 3 are completed.

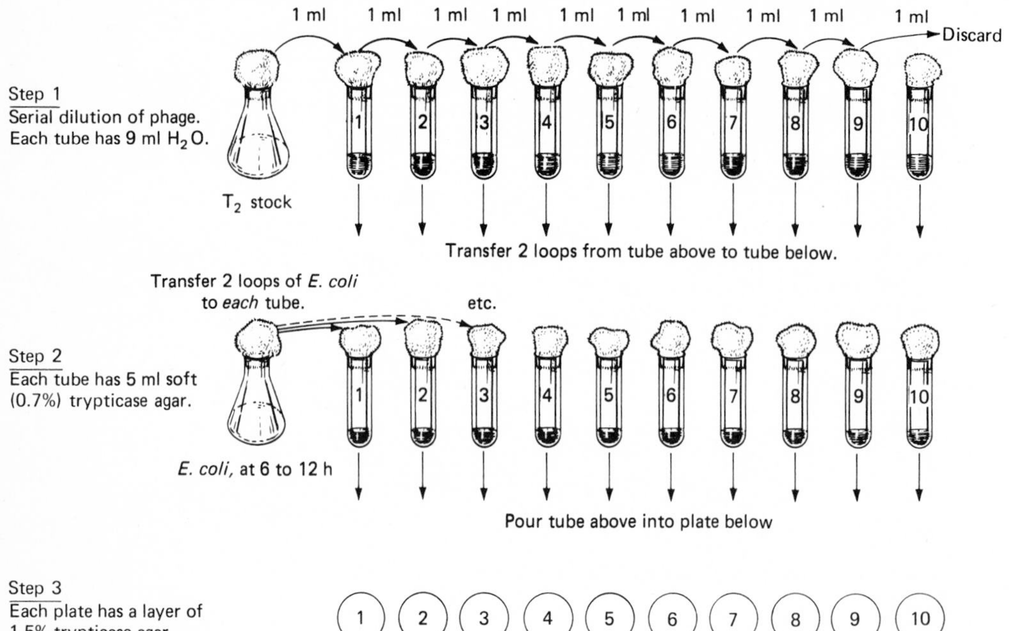

Figure 9-13 Procedures for demonstrating virus growth. (See discussion in text.)

2. Dilute the T_2 stock culture through a series of 10 water tubes, as follows (Fig. 9-13, Step 1): Using a pipette, transfer 1 ml of the phage extract to Tube 1. Discard the used pipette into a tray of disinfectant. Roll the plugged tube between the hands to distribute the phage evenly through the water. In the same manner, use a fresh pipette to transfer 1 ml of suspension from Tube 1 to Tube 2. Repeat the transfers through Tube 9, each time with a fresh pipette and with a thorough mixing of the suspension by rolling the tube. Remove and discard 1 ml of suspension from Tube 9, in place of transferring the suspension to Tube 10. Tube 10 will serve as a control.

3. Place the 10 tubes of soft trypticase agar in boiling water until the medium is liquefied. Cool the medium to 45°C and store the tubes in a water bath at that same temperature. To each tube of soft agar transfer two loops of the *E. coli* culture, returning the tube to the warm water bath after the transfer.

4. Remove the 10 plates of agar from the incubator.

5. Transfer two loops of the phage from Dilution Tube 1 to Tube 1 of *E. coli* in soft agar (Fig. 9-13, Step 2). Roll the tube between the hands to thoroughly mix the

suspension. Immediately pour the mixture into Petri Plate 1 (Fig. 9-13, Step 3). Rotate the plate carefully to spread the mixture over the soft agar.

Repeat the procedure by transferring two loops of phage from Dilution 2 to Tube 2 of soft agar containing *E. coli*. Mix the tube contents and pour into agar Plate 2. Repeat by transferring two drops from each of Dilution Tubes 3 through 10 to the corresponding numbered tube of *E. coli* in soft agar, and pouring each mixture into the plate of the same number. In this manner, Plate 10 contains no phage and becomes the control plate.

6. When the second layer of agar has solidified, invert the plates and incubate them at 35°C.

7. Observe the plates after 6 to 8 h of incubation and again after 12 to 18 h. The *E. coli* will appear as a white, creamy growth. Clear, circular areas are the plaques that have formed in spots where the bacteriophage has destroyed (lysed) the bacteria.

THE BACTERIA

As with the classification of other organisms, the placement of bacteria into a division or a class becomes a troublesome matter for a general biologist. The situation may be more confounding when a person discovers that within almost all areas of biology, the specialists, themselves, accept no common system of classification. Nevertheless, the Seventh Edition of Bergey's *Manual of Determinative Bacteriology* (1957) is used extensively among bacteriologists and therefore is the basis for the taxonomy of these discussions. According to Bergey's Manual, all bacteria are placed in ten orders of the Class Schizomycetes.

This section describes the general techniques for culturing and studying bacteria, and also describes special studies that are particularly appropriate and interesting for general biology courses. The teacher or student who wishes to obtain information about certain genera is referred to Table 9-1 and to the multitude of publications on bacteriology, including those given at the end of this chapter.

General Characteristics of Bacteria

Most bacteria range in size from about 0.75 or 1.0 micron to about 4.0 or 5.0 microns. However, some of the filamentous bacteria may reach a length of 100 microns or more. Strangely, it seems, almost all bacteria assume one of three general forms, cylinders or rods—the *bacillus* (bacilli, plural); a spherical form, the *coccus* (cocci, plural); a spiral form with one or more spirals, the *spirillum* (spirilla, plural). Due to particular types of cell division, coccus bacteria often display characteristic arrangements: pairs, or *diplococcus;* chains, or *streptococcus;* four cells arranged into a square, a *tetrad;* clusters somewhat like a bunch of grapes, or *staphylococcus;* and a cuboidal pattern of eight or more cells, a

TABLE 9-1. SUMMARY OF CHARACTERISTICS, NATURAL SOURCES, AND CULTURE OF COMMON BACTERIA

GENUS AND SPECIES (ORDER)	CHARACTERISTICS OF GENUS	NATURAL SOURCES	CULTURE
Acetobacter aceti (Pseudomonadales)	Ellipsoid to rod-shaped. Single, pairs, or chains. Motile with flagella at one or both ends, or nonmotile. Obligate aerobes. Young cells gram negative; old cells gram variable. Oxidizes various organic chemicals to organic acids, including ethyl alcohol to acetic acid	Fermenting plant tissue. Film on vinegar	Mannitol agar; also see section on "Bacteria in Vinegar," later in chapter, 26°C
Achromobacter liquefaciens (Eubacteriales)	Rod-shaped. Motile with flagella over entire cell, or nonmotile. Gram negative. Aerobic	Salt water. Fresh water. Soil	Nutrient agar, 30°C
Aerobacter aerogenes (Eubacteriales)	Short rods. Motile with flagella over entire cell, or nonmotile. Gram negative. Ferments glucose and lactose. Aerobic and facultatively anaerobic	Air, soil, water	Nutrient agar, 30°C
Arthrobacter globiformis (Eubacteriales)	*Young cells:* rods that are straight, bent, curved, swollen, or club-shaped; gram negative or positive. *Older cells:* Coccoid; gram negative or positive. Nonmotile. Aerobic	Soil	Nutrient agar, 26°C
Azotobacter chroococcum (Eubacteriales)	Large rods, sometimes coccoidal and yeastlike. Motile. Gram negative. Obligate aerobes. Nonsymbiotic, nitrogen-fixing bacteria	Soil. Water	N-free Mannitol solution. Mannitol agar. See "Soil Bacteria," later in chapter, 26°C
Bacillus subtilis (Eubacteriales)	Rods, sometimes in chains. Endospores. Motile by flagella over entire cell, or nonmotile. Gram positive or variable. Aerobic or facultatively anaerobic	Soil. Dust. Hay infusions	Nutrient agar, 30°C

TABLE 9-1. (Continued)

GENUS AND SPECIES (ORDER)	CHARACTERISTICS OF GENUS	NATURAL SOURCES	CULTURE
Brevibacterium linens (Eubacteriales)	Short rods. Motile or nonmotile. May contain red, orange, yellow, or brown pigments. Aerobic or facultatively anaerobic. Used for making limburger cheese; forms yellow-orange pigment on ripening cheese, and softens and flavors the cheese	Dairy products	Nutrient agar, 30°C
Escherichia coli (Eubacteriales)	Short rods. Gram negative. Motile or nonmotile. Ferments glucose and lactose. Aerobic or facultatively anaerobic. Colonies produce a fecal odor	Widely distributed. In feces of man and may be pathogenic, producing enteritis and peritonitis	Nutrient agar with 0.5% NaCl, 37°C
Lactobacillus casei (Eubacteriales)	Rods, generally long and slender. Gram positive. Nonmotile. Ferments carbohydrates to lactic acid. Generally anaerobic. Important in cheese ripening	Widely distributed in dairy products, silage, and manure	Lactobacillus medium, 37°C
Micrococcus luteus (Eubacteriales)	Spherical cells in irregular masses. Gram positive. Motile or nonmotile. Some species with yellow, orange, or red pigment. Saprophytic or facultative parasitic. Aerobic	Soil. Decaying materials. Milk	Nutrient agar, 26°C
Photobacterium fisheri (Pseudomonadales)	Slightly curved rods. Gram negative. Aerobic. Motile. Bioluminescent	Seawater. Fresh and decaying saltwater fish	Photobacterium agar. See "Bioluminescent Bacteria," given later in chapter, 26°C
Proteus vulgaris (Eubacteriales)	Rods. Gram negative. Motile by flagella over entire body. May produce swarming colonies of amoeboid cells on solid media. Aerobic	Human and other animal feces. Decaying materials	Nutrient agar, 37°C

TABLE 9-1. (Continued)

GENUS AND SPECIES (ORDER)	CHARACTERISTICS OF GENUS	NATURAL SOURCES	CULTURE
Pseudomonas fluorescens (Pseudomonadales)	Short rods. Generally motile with flagella at one or both ends. Gram negative. May develop fluorescent pigments of bluish-green or other colors. May reduce nitrates in soil to nitrites, ammonia, or free nitrogen. Aerobic	Widely distributed. Soil, fresh water and saltwater	Nutrient agar, 37°C
Rhizobium leguminosarum (Eubacteriales)	Small rods. Motile when young. Gram negative. Aerobic. Causes formation of nodules on roots of legumes where it lives symbiotically. Fixes atmospheric nitrogen	In nodules of clover and other legumes. Other species in alfalfa, peas, and beans	*Rhizobium* medium. See culture of *Rhizobium* later in chapter under "Soil Bacteria," 26°C
Sarcina lutea (Eubacteriales)	Spherical. Cells in cubical packets of 8 cells each due to cell division along 3 perpendicular planes. Generally nonmotile. Gram positive. Aerobic. Usually yellow, orange, or red pigment	Air. Water	Nutrient agar, 30°C
Serratia marcescens (Eubacteriales)	Small rods. Motile by flagella over body. Gram negative. Aerobic. Generally, a bright red pigment. White, pink, and rose-red strains are common. Saprophytic on decaying animals	Decaying vegetation	Nutrient agar, 26°C
Staphylococcus aureus (Eubacteriales)	May be pathogenic. May cause food poisoning, and is a common cause of boils and carbuncles when the bacterium enters skin through lesions. Because of potential pathogenicity, the organism is not recommended for biology laboratory studies. It is included here only because the bacterium is often used without the knowledge of the possible pathogenicity		Culture is not recommended

TABLE 9-1. (Continued)

GENUS AND SPECIES (ORDER)	CHARACTERISTICS OF GENUS	NATURAL SOURCES	CULTURE
Streptococcus lactis (Eubacteriales)	Small and ovoid or ellipsoid cells. In pairs or short chains. Capsules on cells of some species. Nonmotile. Gram positive. Ferments lactose to lactic acid	Milk, silage, dairy equipment	*Lactobacillus* medium, 37°C
Streptomyces albus (Actinomycetales)	Funguslike bacterium. Cells form a much-branched mycelium. Chains of conidia at ends of erect hyphae. Nonmotile. Aerobic. Gram positive. Many species of *Streptomyces* produce antibiotics	Soil. Decaying vegetation	Sporulation agar, or nitrate sucrose agar. Also see "*Streptomyces:* Antibiotic Production," later in chapter

Taxonomy and characteristics are based on Bergey's Manual, 7th ed. Optimum temperature is given. All may also be grown at room temperature with longer time. See Appendix A for all media.

sarcina arrangement (Fig. 9-14). Occasionally, bacillus bacteria may form in pairs or in a long chain. Spirilla rarely form in groups, although they exhibit great differences in individual shape that may be a long, many-spiraled cell, a short, tightly coiled cell, or a cell with only one twist that produces a comma shape. The comma-shaped bacteria are often called vibrio bacteria. See Figs. 9-15, 9-16.

The bacterial cell consists of a cell wall, a cell membrane, a nuclear area of DNA, and a cytoplasm that contains nucleoprotein particles, food granules, and RNA. The electron microscope reveals that the nuclear area contains fibers similar to strands of DNA. A discrete nucleus with a nuclear membrane is not present, however. Flagella of definite numbers, from one to many, and of definite arrangement often occur in bacilli and occasionally in spirilla. Cocci rarely possess flagella. Some bacteria form an outer coat of slime or a tough coat called a capsule. All species of the genera *Bacillus* and *Clostridium* are able to form highly resistant and characteristic endospores. Spore formation occurs in only a few other species of the true bacteria. See Fig. 9-17.

Table 9-1 describes selected bacterial species commonly studied in the biology laboratory.

Microscopic Study of Bacteria

Living bacteria may be studied in a wet mount prepared by placing a drop of bacteria in suspension on a clean glass slide with a coverglass. Generally the wet mount begins to dry under the coverglass after about 10 min. If the margin of the

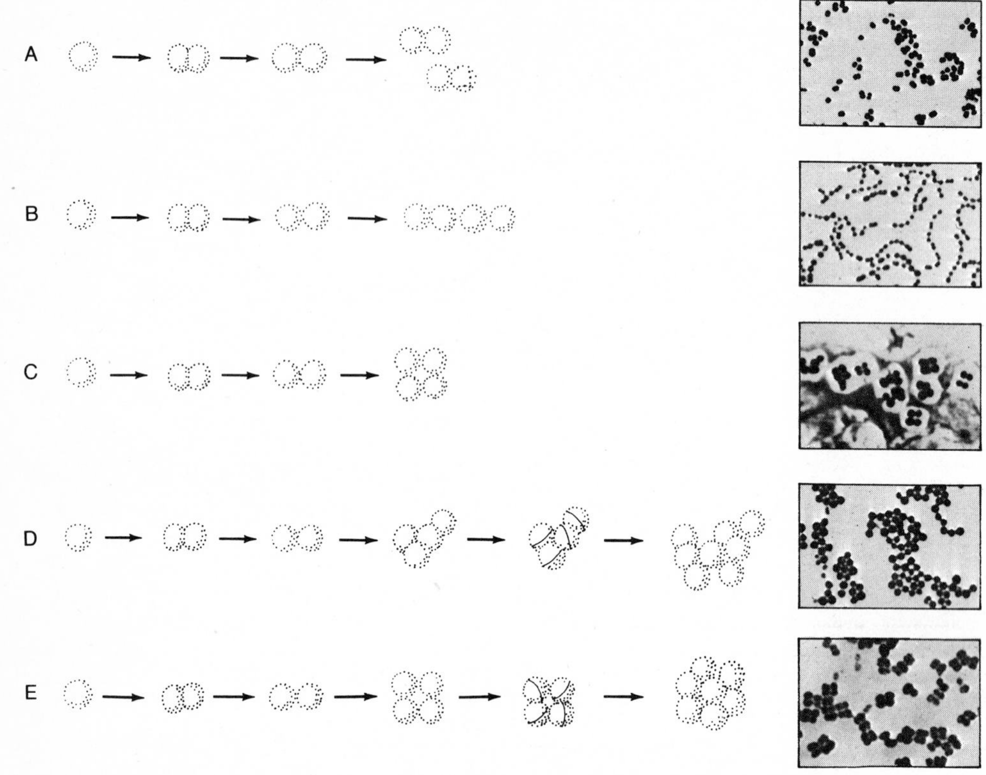

Figure 9-14 Characteristic arrangements of coccus bacteria. **A** Diplococcus. **B** Streptococcus. **C** Tetracoccus. **D** Staphylococcus. **E** Sarcina. The diagram at left of each figure illustrates the pattern of bacterial division. *(From Microbiology 3d ed. by Michael J. Pelczar, Jr., and Roger D. Reid. Copyright 1972 by McGraw-Hill, Inc. Used with permission of McGraw-Hill Book Company, New York.)*

coverglass is sealed to the slide with Vaseline or clear nail polish, the temporary mount will remain in satisfactory condition for several hours. Probably a more satisfactory procedure is to prepare a hanging drop over the depression of a concave or deep-well slide. Prepare the hanging drop by first ringing the slide concavity or well with Vaseline. Place a drop containing bacteria in the center of a coverglass and lower the inverted slide onto the coverglass, pressing the Vaseline ring against the coverglass below. Then carefully turn the slide over, carrying the coverglass to an uppermost position. As a rule, a hanging drop will remain in good condition for at least 3 or 4 h or longer. A temporary mount or a hanging drop may be examined with both the low-power and high-power objectives.

Observation with the oil immersion lens. Ordinarily the oil immersion lens, with a magnification of approximately 1,000X, is used to examine stained bacteria, either on a fixed-smear slide without a coverglass or with a permanent slide containing a coverglass. To use the lens, place a drop of immersion oil in the center of

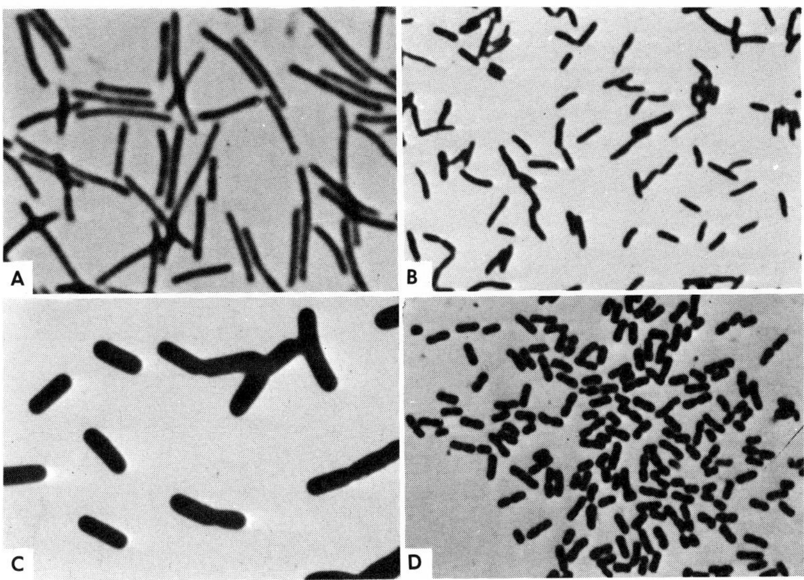

Figure 9-15 Typical rod-shaped bacteria, showing differences in length and width. **A** *Clostridium sporogenes.* **B** *Pseudomonas* sp. **C** *Bacillus megaterium.* **D** *Salmonella typhosa. (From J. Nowak, "Documenta Microbiologica," Erster Teil, "Bakterien," Gustav Fischer Verlag, Tena, 1927.)*

a fixed smear or on the coverglass of a permanent slide. While closely observing from the side of the microscope, lower the oil immersion objective barely into the drop of oil taking care not to press the objective against the coverglass or slide. Then use the fine adjustment to bring the bacteria into view. Because the opening of the oil immersion objective is quite small, the oil is required for the passage of a maximum amount of light into and through the lens. When the observations are completed, wipe the lens and the slide with lens paper to remove the oil film. Oil that remains on the lens will thicken and may require cleaning with a *small* amount of xylol on lens paper. The xylol must then be removed quickly with dry lens paper. Excessive xylol will loosen and damage the lens.

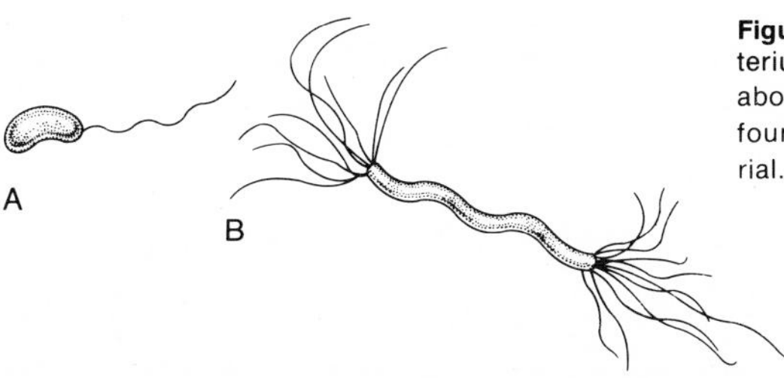

Figure 9-16 Spiral bacteria. **A** *Vibrio cholerae,* a bacterium associated with Asiatic cholera. Actual length is about 2 microns. **B** *Spirillum volutans,* a saporophyte found in stagnant or polluted water and in decaying material. Actual length is 20 to 40 microns.

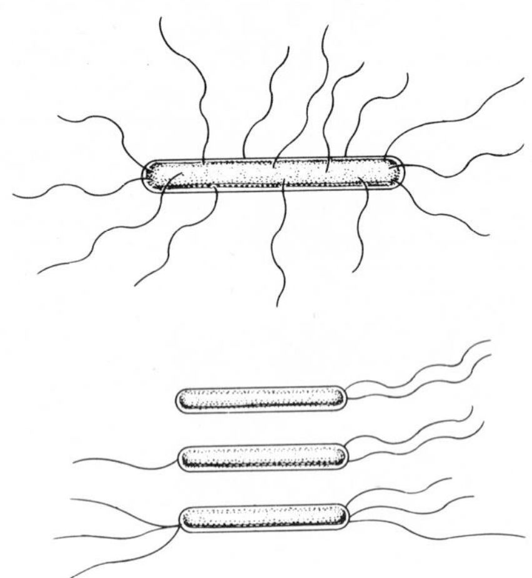

Figure 9-17 Flagellation in bacteria. Top, peritrichous; flagella cover the cell. Bottom, polar; flagella are at one or both ends of the cell.

Preparing a Stained Smear of Bacteria

Ordinarily bacteria are prepared for staining by simply drying a drop of suspension on a clean slide and fixing the bacteria with heat. Although the process is simple, this is a routine procedure for most bacterial stains. Because of their minute size and their stiff walls, bacteria do not become greatly distorted when dried and heated. Therefore, a coagulation of tissues with a fixing compound is not necessary as with tissues of higher organisms, although fixatives other than heat are required for particular cytological studies of bacteria and for accurate measurements of their cell size and shape.

The best smears are made when a small quantity of surface growth is removed from a solid medium and mixed with tap water, although a drop of bacteria from a liquid medium is often used and generally is satisfactory. The steps for preparing and staining a smear of bacteria are listed below and are shown in Fig. 9-18.

1. Place a drop of bacteria on a *clean* slide. If the bacteria are on a solid medium, place a *small* amount of surface growth in a drop of tap water on a slide.

2. Use an inoculating loop or needle to disperse the solid material throughout the water. The water should be only faintly turbid and not milky. A thick film produces an unsatisfactory slide that is of little or no use.

3. Allow the film to dry in the air, or dry by holding the slide several inches above a flame.

4. Fix the dry smear to the slide by quickly passing the slide, with smear side upward, through a flame three times. Excessive heat must be avoided.

5. Cover the smear with stain for 5 to 30 sec. The staining time for common stains is given in Table 9-2. A screen over a pan or other container makes a satisfactory staining equipment.

6. Rinse off the stain with clear water.

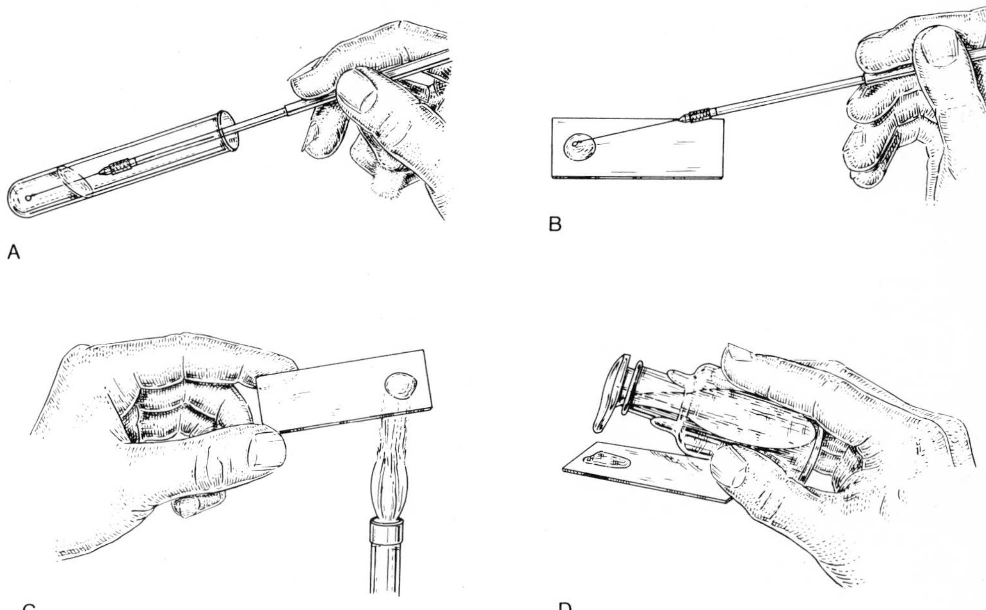

Figure 9-18 Staining a bacterial smear. **A** Remove organisms with a transfer loop. **B** Spread materials on a clean slide, over an area about the size of a dime. Allow the smear to dry. **C** Fix the smear by quickly passing the slide (smear side up) through a flame three times. **D** Stain the smear for the appropriate time indicated for the stain. Wash and dry the slide. *(Based on Laboratory Exercises in Microbiology, by Michael J. Pelczar, Jr. Copyright 1965 by McGraw-Hill, Inc. Used with permission of McGraw-Hill Book Company, New York.)*

7. Allow the smear to dry and examine with the oil immersion lens.

8. The stained and uncovered smear will remain in good condition for an indefinite period, although in time the dye will fade. If a person wishes to add a coverglass, however, a drop of balsam or other mountant is placed on the dry smear, and a coverglass is pressed evenly against the mountant.

Negative Staining with Nigrosin[1]

Mix a loop of bacteria and a loop of nigrosin stain (Appendix A) on a slide; spread to a thin film. Allow film to dry for about 10 min. Do not fix bacteria in heat as is done with other stains described above. Examine the dry, stained smear with oil immersion lens. The cells appear white or clear against an evenly dark-gray background of nigrosin. Because the cells are not heated, they retain their original size and shape, and the mounts can be used for determining the size and shape of the bacteria. The bacteria in scrapings from around the teeth are demonstrated es-

[1]Black or blue India ink is often used in place of nigrosin. The ink may contain bacteria, however, and cannot be used when staining a pure culture unless it is first examined for contamination.

TABLE 9-2. GENERAL STAINS FOR BACTERIA*

STAIN	STAINING TIME	COMMENTS
Crystal violet	10 sec	A general purpose stain for bacterial smears
Ammonium oxalate crystal violet	10 sec	General-purpose stain
Methylene blue	1–2 min	Stains chromatin selectively and other cell parts lightly
Loeffler's alkaline methylene blue	1–2 min	Intense stain for chromatin granules. Other cell parts stain lightly
Carbol fuchsin (Ziehl's)	5 sec	An intense stain and not suitable when much debris is present

*See Appendix A for preparation of stains.

pecially well by this method. The scrapings may contain spirilla bacteria which are difficult to find in other ordinary laboratory preparations.

Gram's Stain (Hucker Modification)

Most bacteria are classified as gram-positive or gram-negative according to their reaction to Gram's stain. A bacterial species is gram-positive if it retains a blue color from the first stain of the procedure. A gram-negative organism loses the first stain and becomes a red color from the second stain. Note: Bacteria are variable in reaction to Gram's stain particularly because of the age of a culture. For reliable results, the stain must be done on 18- to 24-h cultures and on *thin* smears.

Solutions. (1) Ammonium oxalate crystal violet; (2) iodine stain, Gram's; (3) safranin stain (see Appendix A).

Procedure.

1. Prepare a smear of bacteria and fix with heat as described above. Cover the smear with ammonium oxalate crystal violet and allow to set for 1 min.

2. Dip in tap water for 1 to 2 sec.

3. Cover with iodine stain (Gram's) for 1 min.

4. Gently wash slide in tap water and dry by blotting.

5. Cover smear with safranin for 10 sec.

6. Wash in tap water and allow to dry. Examine with oil immersion lens. A violet or blue color = gram-positive. Red or pink color = gram-negative.

Preserving and Storing Bacteria

Stock cultures of most bacteria described in this chapter may be preserved on nutrient agar slants. Duplicate sets of cultures should be maintained, and tubes with screw-on caps are preferable. The freshly transferred slants are incubated at 30 to 35°C for 12 to 18 h, or at room temperature for 24 h. After the incubation period, tighten the caps and store at 6 to 10°C.

Transfers to fresh medium must be done at regular intervals, generally about every 2 months. The storage time between subcultures can be increased to about 1 year by covering a slant with sterile mineral oil to a depth of 1 cm above the agar slant. Transfers to fresh medium may then be done by removing a loopful of bacterial growth from under the oil and resting the loop against the inner wall of the culture tube to drain off oil before completing the transfer.

Prepare the mineral oil (medicinal grade) by sterilizing in an autoclave or pressure cooker at 15 lb pressure for 60 min. After sterilization, place the oil in an oven at 110°C for 1 h to remove moisture that has accumulated on the inner wall of the container above the oil.

SPECIAL TECHNIQUES FOR STUDYING BACTERIA

Pure cultures of bacteria may be obtained from most biological supply companies. The American Type Culture Collection (Appendix B) supplies certain cultures at a reduced cost for educational use. In place of pure cultures, a study of bacteria from their natural sources is perhaps more desirable for general biology courses. Bacteria are abundant, of course, in soil, air, water, and food. The following discussions describe techniques for obtaining and studying bacteria in their normal environment. See Table 9-1 for general characteristics of the various groups.

AIR BACTERIA

Bacteria and other microbes do not live in the air but are carried about on dust particles and on droplets of moisture, perhaps those expelled from the human respiratory tract by sneezing or coughing. Airborne microorganisms may be transported a few feet or many miles, depending on air currents, temperature, and other factors, such as the formation of resistant spores by certain species. Bacteria expelled into the air may live for only a few seconds or, in the case of spore formers, they may live for years and may be carried great distances by air currents. Several methods are described below for obtaining air microorganisms.

Open Hay or Beef Infusions

The simplest method for obtaining bacteria is to allow the organisms to settle from the air into open containers of medium. The medium may be a small piece of cooked or raw beef in water, or a small amount of dry hay or grass in water. When the medium is left open and exposed to the air, a film of bacteria will form on the organic material and on the water surface within 24 to 48 h. Since sterile techniques are not employed, the bacteria are not only those on air particles but also the microbes in the container, on the food particles, and from the laboratory personnel as well. A small portion of the bacterial film on the water is then used to make a stained smear according to the techniques described earlier in the chapter. As a rule, many bacillus and coccus forms are obtained in this manner. The procedure can serve as a means for obtaining bacteria easily and at practically no cost.

Exposure of Nutrient Agar Plates to Air

Prepare several petri plates with nutrient agar. Raise the cover and expose the surface for a determined time period, perhaps 10 to 15 min. During this time, place the inverted cover on the table.

Plates may be exposed to different situations, for example, in a classroom during early morning before classes begin; in the same room immediately after classes; the room during early fall or late spring with windows raised; and the same room during midwinter when windows are closed. Many other types of tests can be devised for comparing air samples, for example, outdoors on a windy day and again on a day when the air is still. As a rule, a surprising number of bacteria (and molds) are collected in the dressing room of the football team.

After exposure, incubate the plates at 37°C for 24 h, or at room temperature for 48 h. Each colony appearing on the plate represents one air particle with microbes that has fallen on the agar. This technique is somewhat crude, because only the larger and heavier particles fall out of the air during a given time. Also the volume of an air sample is unknown. Nevertheless, the technique provides a rough estimate of the relative degree of air contamination and gives information about the kinds of airborne microbes.

Quantitative Analysis of Air Bacteria

Special equipment is required for precise bacterial analysis of air. Two types of equipment, shown in Fig. 9-19, are the sieve sampler (a solid impingement device), and a suction flask of broth (a liquid impingement device). With either equipment, a known quantity of air is drawn through the medium for a known length of time. In this way a quantitative analysis of air bacteria is gained.

Figure 9-20 shows a simpler and less precise device by which air is drawn through a sterile tube containing a sterile filter of wet sand or cotton. After a set time, for example, 1 min or less, transfer the filter to a tube of nutrient broth by aseptic techniques. Shake the plugged tube vigorously to disperse microorganisms throughout the broth. Use a sterile pipette to transfer 0.1 ml of the broth to each of several petri plates containing nutrient agar. Spread the drop evenly over the agar surface of each plate. Incubate at 37°C for 24 h or at room temperature for 48 h. Each colony that appears on the agar represents one air particle with microbes. If more than 300 colonies have grown on the plate, another air sample must be filtered, followed by serial dilutions of the broth before transfers are made to plates. See dilution techniques described earlier in this chapter.

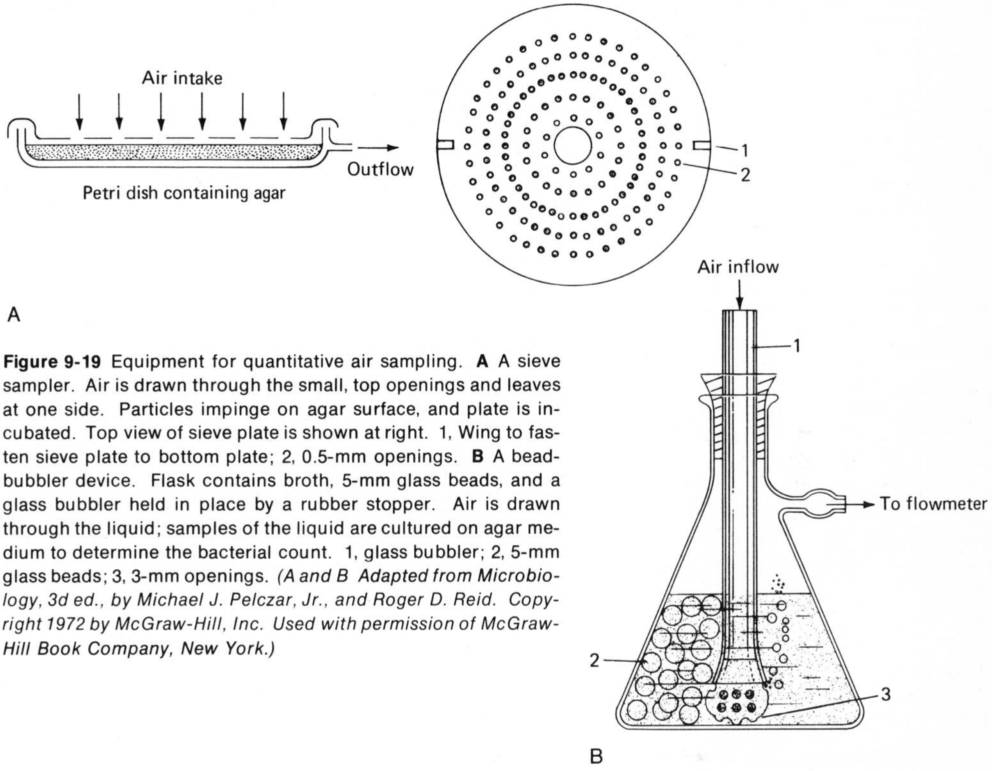

Air intake

Outflow

Petri dish containing agar

1

2

A

Air inflow

1

To flowmeter

2

3

B

Figure 9-19 Equipment for quantitative air sampling. **A** A sieve sampler. Air is drawn through the small, top openings and leaves at one side. Particles impinge on agar surface, and plate is incubated. Top view of sieve plate is shown at right. 1, Wing to fasten sieve plate to bottom plate; 2, 0.5-mm openings. **B** A bead-bubbler device. Flask contains broth, 5-mm glass beads, and a glass bubbler held in place by a rubber stopper. Air is drawn through the liquid; samples of the liquid are cultured on agar medium to determine the bacterial count. 1, glass bubbler; 2, 5-mm glass beads; 3, 3-mm openings. *(A and B Adapted from Microbiology, 3d ed., by Michael J. Pelczar, Jr., and Roger D. Reid. Copyright 1972 by McGraw-Hill, Inc. Used with permission of McGraw-Hill Book Company, New York.)*

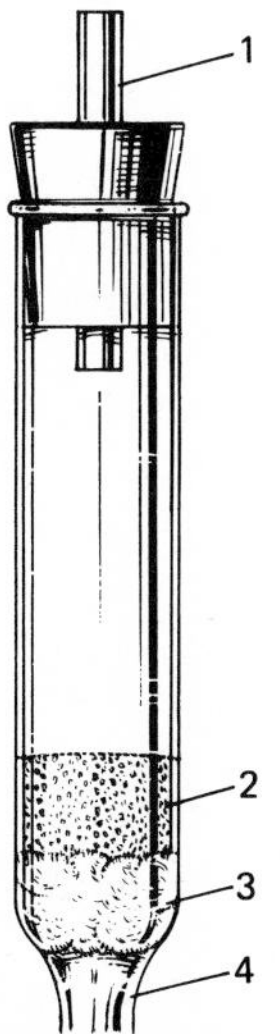

Figure 9-20 Tube filter for air sampling. A simple device, consisting of air inlet (1), sand (2), cotton (3), and air outlet (4). See text for procedures. *(Based on Fundamentals of Microbiology, by Martin Frobisher. Copyright 1968 by W. B. Saunders Company, Philadelphia.)*

Count the number of colonies on each plate and calculate the average number of air particles for each milliliter of broth for the unit of time that the suction filter was operated. Most likely all air samples will contain mold and yeast as well as bacteria.

BIOLUMINESCENT BACTERIA

The emission of light by a living organism is called bioluminescence. The majority of luminous microorganisms occur in ocean water. Because little heat is produced by luminous organisms, bioluminescence is often called "cold light."

The phenomenon is easily demonstrated in *Photobacterium fischeri*, a marine curved-rod bacterium. The demonstration of luminescence by these organisms can be a fascinating study for students and may lead to tests beyond the demonstration described here. A culture may be obtained from ATCC, Carolina Biological Supply Company, General Biological, Inc., and other sources.

A maximum luminescence is displayed in a fresh culture. Therefore, upon receiving the organism, make transfers to slant tubes or petri plates containing *Photobacterium* agar (Appendix A). Inoculate the fresh plates and tubes heavily with two or three loops of the stock culture. Incubate at room temperature for 18 h. Examine the plates in a dark room or closet, after allowing time for the eyes to adjust to the darkness. Record data each day for about 7 days regarding the persistence of luminescence.

BACTERIA IN VINEGAR: *ACETOBACTER ACETI*

Bacteria and fungi form a sediment called "mother of vinegar" in *unpasteurized* cider. The organisms may change the cider into vinegar. *Acetobacter aceti* can be isolated from untreated cider vinegar, or it can be first grown in apple cider that becomes vinegar as the bacteria accumulate. For the second procedure, obtain unpasteurized apple cider at a grocery store. Place enough cider in sterile flasks or bottles (preferably flat bottles) to make a ½-in layer when the container is laid on one side. Generally, a film of *A. aceti* will form on the surface of the cider within 1 or 2 weeks. Using aseptic techniques, transfer loops of the bacterial film to petri plates containing agar medium prepared as described below.

Add 0.1 g yeast extract and 1.5 g agar to 98 ml of distilled water. Autoclave at 15 lb pressure for 15 min. By aseptic transfer, add 1 g of $CaCO_3$ that has been sterilized in an oven at 160°C for 2 h, and 2.0 ml 95% ethyl alcohol. Pour plates in the usual manner, described earlier in this chapter.

When the agar has solidified, streak each plate with a loop of surface film from the cider culture. Incubate at 26°C or at room temperature for 24 to 48 h, or longer. Examine plates for a clearing of the $CaCO_3$ which indicates that the organisms are producing acetic acid from the alcohol, and that the acid is reacting with the $CaCO_3$. Alternatively, the organism can be maintained and stored on Mannitol agar (Appendix A).

SOIL BACTERIA

Although many persons may consider soil to be a nonliving, practically inert substance, most soil is very much alive. Bacteria and also fungi, algae, and protozoans transform soil into a living, metabolizing system which has been described as a "sort of Lilliputian Zoo." Counts made by A. Burgess (1958) of soil organisms in *one gram* of fertile agricultural soil produced the results shown below:

Bacteria:	
Direct count	2,500,000,000
Dilution plate	15,000,000
Actinomycetes	700,000
Fungi	400,000
Algae	50,000
Protozoa	30,000

Soil microorganisms, other than bacteria, are described in the chapters on protozoa, algae, and fungi.

The following procedures permit students to isolate and examine some typical soil bacteria, in particular: *Streptomyces*, a branching, funguslike bacterium; *Azotobacter*, a nonsymbiotic, nitrogen-fixing bacterium; and *Rhizobium*, a symbiotic, nitrogen-fixing bacterium.

Streptomyces and Other Actinomycetes: Antibiotic Production

Streptomyces and a number of other genera are grouped together in the order Actinomycetales. Figure 9-21 contains illustrations of *Streptomyces* and other genera of the Actinomycetales, all of which are commonly referred to as actinomycetes.

Add 1 g of soil to 9 ml of sterile water (a 1/10 dilution). Mix thoroughly by shaking at least 25 times. Aseptically transfer 1 ml to 9 ml of sterile water, thus producing a 1/100 dilution. Mix thoroughly as before, and continue the transfers to make dilutions of 1/1,000, 1/10,000, 1/100,000, and 1/1,000,000 (see Fig. 9-8).

Transfer 1 ml of the 1/100,000 and 1/1,000,000 dilutions to separate petri plates containing nitrate-sucrose agar (Appendix A). Use a sterile, glass spreader to spread the suspension over the agar surface of each plate. Sterilize glass spreader by dipping in alcohol and passing through a flame. Allow the blue flame of the alcohol to disappear. Cool the spreader before use by holding in air for 30 sec and then touching it at edge of agar surface. Label the dilution on each plate. Incubate at room temperature for 3 to 7 days.

If present, *Streptomyces* and other actinomycetes will appear as small, powdery, fuzzy edged colonies. Examine plates under a stereomicroscope and under the low-power objective of a compound microscope. Carefully transfer a portion of a well-isolated colony to a dry slide and examine under the high-power objective. Take care not to lower the objective into the uncovered preparation. Exam-

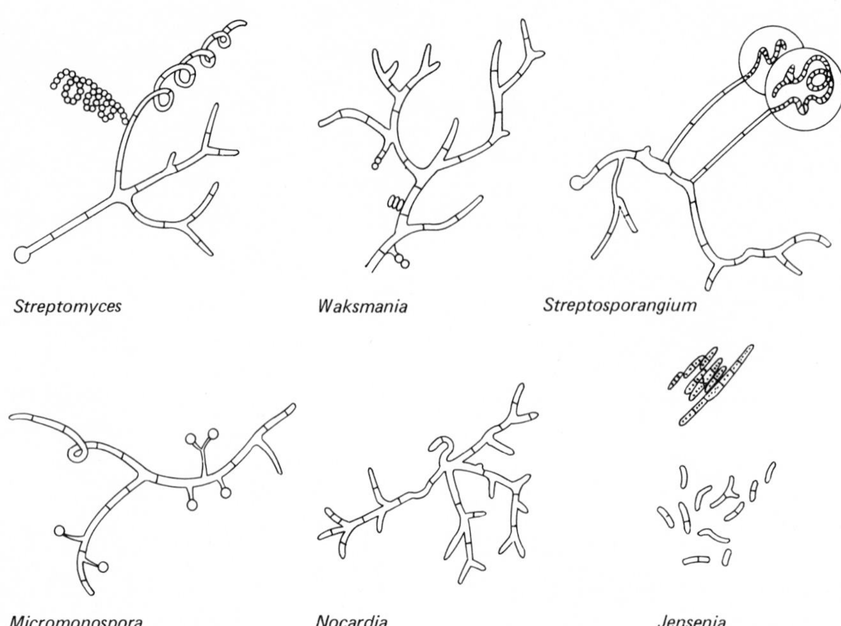

Figure 9-21 *Streptomyces* and other actinomycetes. *(After V. B. D. Skerman, from "The McGraw-Hill Encyclopedia of Science and Technology," McGraw-Hill Book Company, New York, 1961, vol. 1, p. 56.)*

ine for branching mycelium, aerial hyphae, and spores. Remove a small portion of a colony to a drop of water on a slide. Prepare a smear according to techniques described in the introduction to this chapter. Stain with crystal violet. Examine with an oil immersion lens. Look for branching mycelium and spiral chains of spores in *Streptomyces*.

Isolation of antibiotic producers. Although antibiotic producers appear only rarely, the plates should be examined carefully to detect clear zones or halos around a moldlike colony, indicating that bacteria in the area of the actinomycete have not grown or have been destroyed by an antibiotic. Isolate organisms from colonies with clear zones by aseptically transferring a portion of each colony to a separate slant tube of nutrient agar. Incubate at room temperature until a good growth forms. See Fig. 9-22, Step 1.

Test the organisms in the slant tubes for antibiotic production with several types of bacteria, for example, *Sarcina lutea, Serratia marcescens*, or any other bacteria in stock. Prepare a water suspension of each isolated organism by aseptically pipetting about 5 ml of sterile water into a slant tube and scraping the colony loose into the water with a transfer loop. Streak nutrient agar plates with each actinomycetes, making only one broad streak across the middle of a plate (Fig. 9-22, Step 2). Incubate at room temperature for 3 to 7 days. When a good growth has formed, streak water suspensions of each bacteria (prepared as above for the actinomycetes) at right angles and into the actinomycetes line of growth (Fig. 9-22,

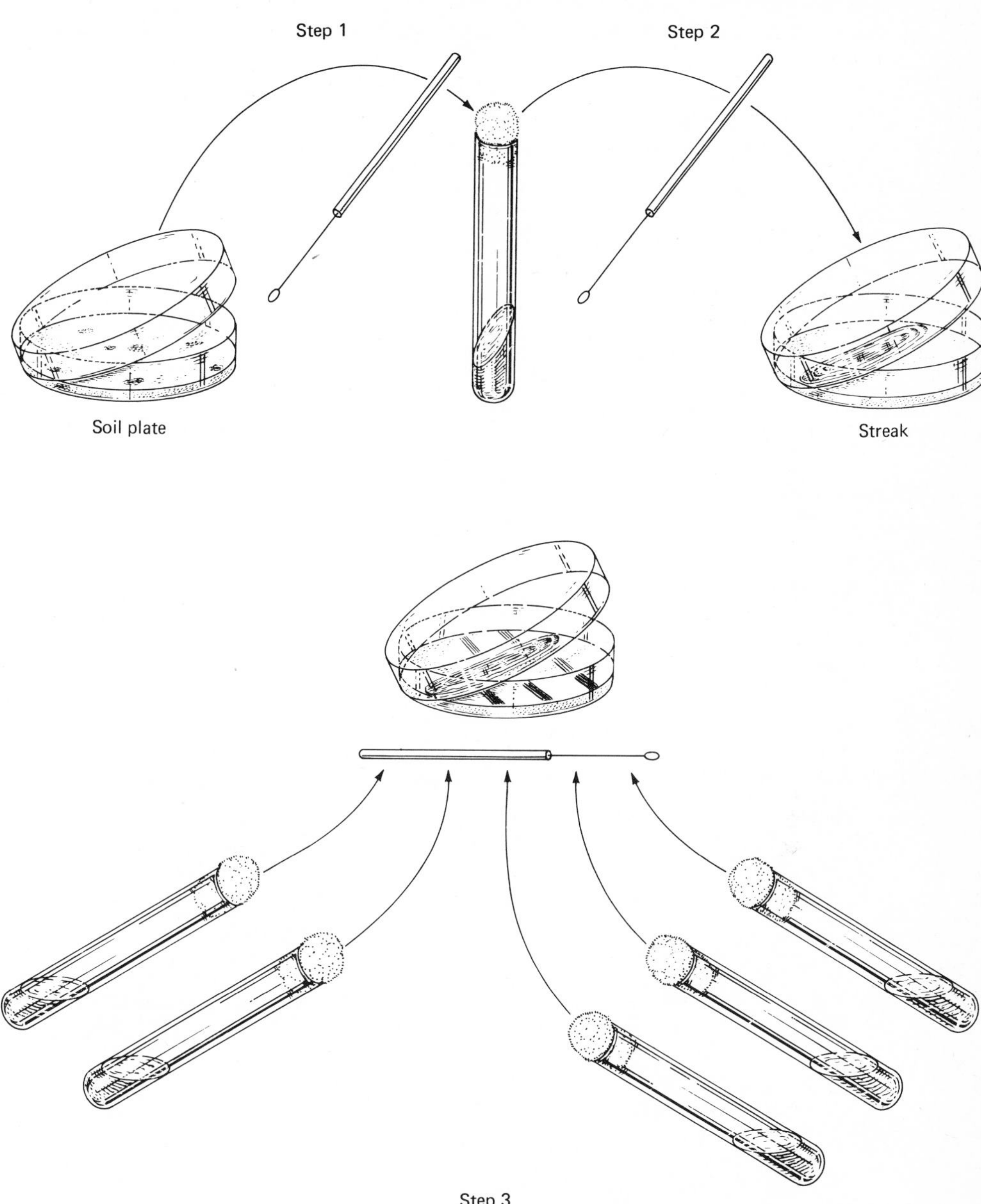

Figure 9-22 Procedures for testing antibiotic production of actinomycetes. See text for description. *(From Life in the Soil, by David Pramer. Copyright 1964 by Biological Sciences Curriculum Study.)*

Step 3). Incubate the plates at room temperature for 48 h. Examine for antibiotic inhibition indicated by clear areas in the bacterial streaks at points adjacent to the actinomycetes (shown in center plate of Fig. 9-22, Step 3).

Cultivation and Examination of *Azotobacter:* Nonsymbiotic N-Fixing Bacteria

To obtain *Azotobacter*, put 0.5 g of fertile garden soil in 20 ml of Mannitol solution, N-free (Appendix A). Mannitol is used readily by *Azotobacter* but is not metabolized so readily by competing bacteria in the soil. Incubate at room temperature for 7 to 10 days. Do not move or shake the culture as bacteria form in a surface film that turns brown with age. When a surface film is well formed, transfer loops of the film to streak plates containing Mannitol agar (Appendix A). Incubate for 7 to 10 days at room temperature. Observe for large mucoid colonies. Examine heat-fixed smears, stained with crystal violet. See Table 9-1 for descriptions.

Rhizobium: Symbiotic N-Fixing Bacteria

Species of the genus *Rhizobium* have the ability to penetrate the root-hair tips of legumes and to develop threads of cells that grow into the roots. The "infection threads" swell and burst, releasing bacteria within the plant cells (Fig. 9-23). The free bacteria within the root cells assume a peculiar shape called a bacteroid that according to species may be club-, pear-, T-, or Y-shaped (Fig. 9-23). The presence of the bacteria stimulates the multiplication of plant cells, resulting in the development of a nodule. Nodules vary in size and shape according to the species of legumes. The nodules of the velvet bean may grow to 1 in or more in diameter, whereas those of clover may be only 5 to 10 mm in diameter (Fig. 9-24).

The bacteria are nourished by plant juices, and the plant gains nitrogen compounds, made available by the bacteria which take nitrogen directly from the air. The end result is that atmospheric nitrogen, which cannot be utilized directly by a plant, is converted by rhizobial bacteria to nitrogen compounds that remain in the soil for plant use or that are returned to the soil when the legumes die and decay. Nonleguminous plants will die in a soil that contains no nitrogen compounds, whereas legumes, in association with *Rhizobium*, will flourish in a nitrogen-deficient soil.

Not all genera and species of legumes develop nodules, although a majority of those that have been investigated do nodulate, including principally the crop legumes. Furthermore, a certain amount of specificity exists between legumes and species of *Rhizobium*, and apparently each rhizobial bacterium has a limited number of leguminous hosts. Crop legumes are divided into several cross-inoculation groups, wherein legumes of a group develop nodules when exposed to the rhizobial bacteria taken from nodules of any other legume within the group.

Observations of nodules. Collect nodulated legumes, such as clover, alfalfa, soybeans, peas, and beans. Wash roots and examine for nodules. Slice a thin, transverse section of a large nodule, and examine under the low-power objective. Note the shape of the nodule, the bacteroidal tissue in the plant cortex, and the vascular tissue of the root.

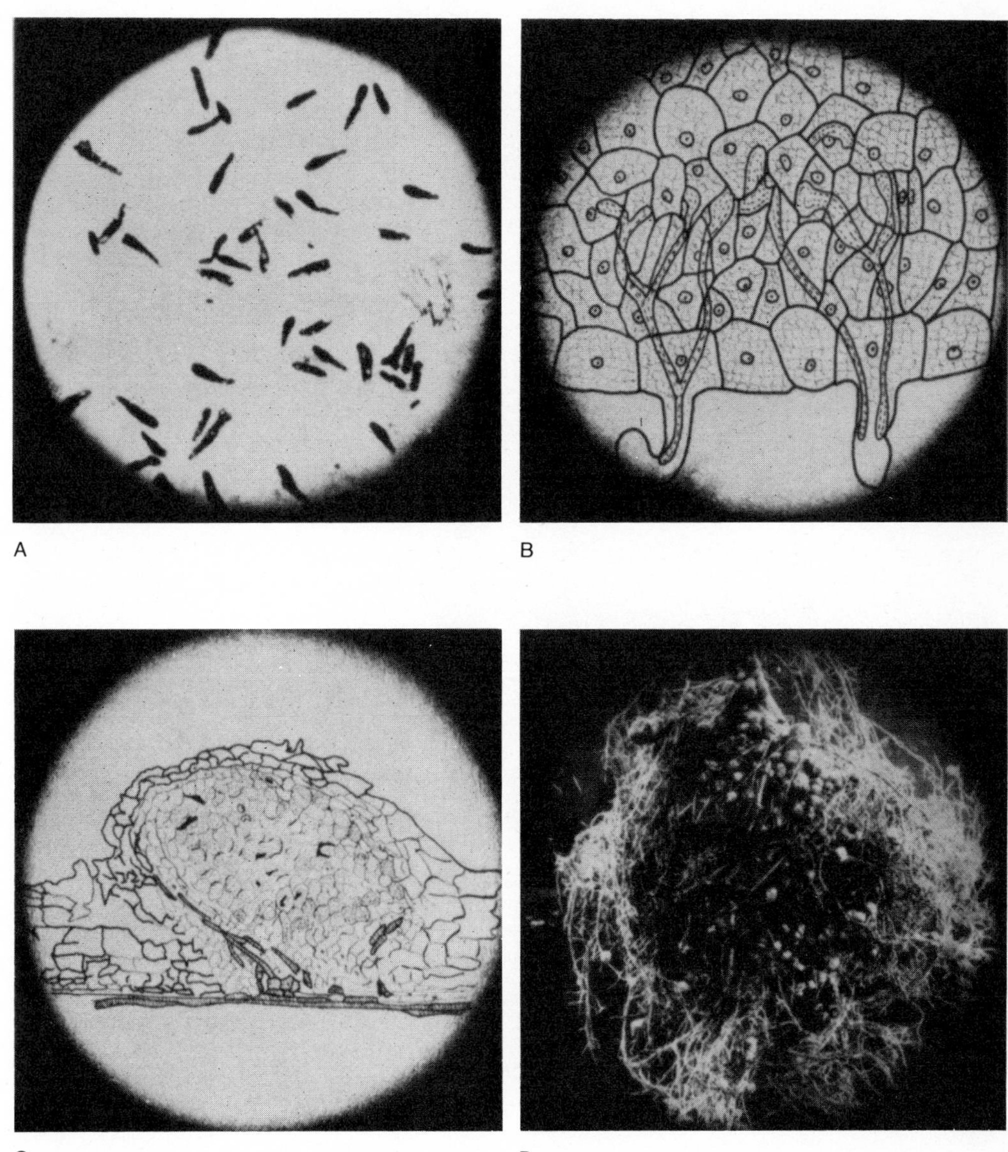

Figure 9-23 *Rhizobium,* symbiotic N-fixing bacteria. **A** Bacteria (bacteroids) from a root nodule of a legume. **B** Infection threads are produced by bacteria through root hairs and into root tissue. **C** The threads break open and release bacteria. As bacteria multiply the cells swell forming a nodule on the root. **D** Root of a legume with many nodules. *(A–D From Agricultural Research, USDA, 1953. Courtesy of L. W. Erdman.)*

Figure 9-24 Roots with nodules from several legumes. **A** Lespedeza. **B** Velvet bean. **C** Cowpea. **D** Crimson lover. *(From USDA, Farmer's Bulletin 2003, 1953. Courtesy of L. W. Erdman.)*

Crush a small nodule in a drop of water on a glass slide. Prepare a heat-fixed smear as described in the introduction to this chapter. Stain with nigrosin negative stain (Appendix A), and examine for bacteroids with oil immersion.

Culture techniques. Rhizobium species are easily cultured on artificial media, although the organism does not use atmospheric nitrogen when removed from the legume, and therefore a nitrogen compound must be added to the medium. For the procedure, soak a large nodule in a 1:9 solution of household bleach and water for 3 min to destroy surface microorganisms. Rinse thoroughly in sterile water. Transfer the nodule to a sterile petri plate, and cut the nodule into small pieces with a sterile razor blade. Add about 0.5 ml of sterile water, crush nodule pieces, and streak the suspension on plates of Mannitol agar (Appendix A). Incubate at room temperature for 4 to 7 days. Colonies will be moist and glistening. Prepare a negative stain of the bacteria with nigrosin stain (Appendix A). Examine with the oil immersion lens. When grown on artificial medium, *Rhizobium* assumes a rod shape and is not bacteroidal.

Larson, Clary, and Lwanga (1969) describe a laboratory experiment that should be very useful for illustrating symbiotic nitrogen fixation and the availability of the fixed nitrogen to other plants that are grown with a nodulating legume.

MILK BACTERIA

Species of *Streptococcus* and *Lactobacillus* ferment the carbohydrates of milk, producing mainly lactic acid. When milk is produced and maintained under sanitary conditions, the bacterial flora will consist primarily of species of these two genera. Such milk, when placed under conditions that permit the bacteria to grow, will contain a desirable sour flavor and odor. Milk of unsanitary quality, however, contains a broad spectrum of other bacteria that may produce undesirable changes in the consistency, flavor, and odor of the milk. Milk and milk products, sold for human consumption, are regularly analyzed for bacterial counts under the supervision of government officials. A high bacterial count indicates unsanitary handling or improper processing and storage.

Two standard tests for milk are described here: the standard plate count (SPC), and the reductase test.

Standard Plate Count of Milk Bacteria

1. Obtain various samples of milk and refrigerate milk until the time of testing but not longer than 12 h.

2. Serially dilute each sample, first shaking the sample thoroughly, and then aseptically transferring 1 ml of milk to 9 ml of sterile water, thus producing a 1/10 dilution. Repeat dilutions, preparing 1/100, 1/1,000 and 1/10,000 dilutions (see dilution techniques in Fig. 9-8).

3. Transfer 1.0 ml of each dilution from a sample to separate empty, sterile petri plates, and to each add 15 ml of sterile tryptone glucose extract agar (Appendix A) cooled to 45°C. Distribute the organisms evenly by gently moving the plate in a circular pattern.

4. Incubate for 48 h at 37°C. Count the colonies and calculate the number of microbes in each milliliter of the original sample. See *Dilutions and Plate Counts*, given earlier in the chapter. Compare the counts of each type of milk.

Reductase Test for Milk

Methylene blue is used as an indicator compound for the reductase test.

1. Obtain various samples of milk, for example, raw, pasteurized, and certified milk. Refrigerate milk until the time of tests but not longer than 12 h.

2. Label a test tube for each sample of milk, and transfer 10 ml of milk into each corresponding tube. To each tube add 1 ml of methylene blue solution (1 part of dye in 20,000 parts of distilled water; see preparation of methylene blue in Appendix A).

3. Place tubes in a 37°C water bath. Observe tubes for the time required for a complete loss of the blue color.

Actively growing bacteria consume oxygen in milk. As the oxygen content diminishes, methylene blue changes from an oxidized form to a colorless, reduced form. The time required for the color loss is an indication of the relative bacterial count.

Use the standards given in Table 9-3 to establish the quality of each sample.

BACTERIA IN WATER

Bacteria are present in practically all types of water, including, of course, the tap water that we drink. Atmospheric water of rain and snow may be relatively free of bacteria after the dust particles have been removed from the air during the early part of the rain or snow period. Snow and rain become contaminated, however,

TABLE 9-3. STANDARDS FOR METHYLENE BLUE—REDUCTASE TEST OF MILK

TIME OF DECOLORIZATION OF METHYLENE BLUE	QUALITY OF MILK	BACTERIAL CONTENT
More than 6 h	Good	Less than 500,000/ml
Less than 6 h but not less than 2½ h	Fair	500,000 to 4,000,000/ml
Less than 2½ h but not less than 30 min	Bad	4,000,000 to 20,000,000/ml
Less than 30 min	Very bad	More than 20,000,000/ml

as soon as they touch the earth surface. Normally the bacterial count decreases in stored water, primarily because of sedimentation and other factors including temperature and limited food supply. Ground water from wells and springs may be relatively free of bacteria because of filtration as the water passes through the soil. However, ground water becomes contaminated, the same as atmospheric water, as soon as it reaches the earth surface.

Although most water bacteria are harmless, the contamination of water with pathogenic bacteria is always a possibility. For this reason, water destined for household and recreational purposes is tested regularly. The pathogens most commonly carried in water are those of intestinal diseases, including typhoid fever, paratyphoid fever, dysentery, and cholera, which are introduced into the water as a result of contamination with human sewage. Since the sewage from one or even more sources of infectious diseases becomes greatly diluted in a large body of water, attempting to detect the pathogens in small samples is a futile task. Nevertheless, only a few pathogens may well infect a great many humans who drink or swim in the contaminated water.

Because the direct detection of intestinal pathogens is an unrealistic procedure, tests are made to detect other intestinal bacteria, in particular *Escherichia coli*, and other coliforms, which are excreted in great number in all human feces. The coliform group includes all bacteria that are aerobic or facultative anaerobic, gram-negative, nonspore-forming bacilli that ferment lactose with gas formation. The presence of coliform bacteria in water means that the water may be contaminated with human feces, which then means that intestinal pathogens may also be present in the water.

Standard methods for bacteriological determination of water are described jointly by the American Public Health Association, the American Water Works Association, and the Federation of Sewage and Industrial Wastes Associations. These procedures, published in *Standard Methods for the Examination of Water and Wastewater Including Bottom Sediments and Sludges* (1971), are used in public health laboratories. Three types of general tests are described below: (1) the total plate count, used to determine the relative number of bacteria in a water sample; (2) a series of three tests used routinely to detect coliform bacteria; and (3) the membrane-filter technique. The tests do not distinguish between coliforms of fecal origin and those of non-fecal origin (for example, the *Aerogenes*). The American Public Health Association considers either type of contaminated water unsanitary and unsafe for human use.

Total Plate Count of Water Bacteria

Materials. (1) A clean and sterile, 100-ml, glass-stoppered or screw-capped bottle. Before sterilization, wrap the cap and neck of the bottle with paper and tie with string. Sterilize the bottle for 30 min at 15 lb pressure. (2) Flasks or tubes of nutrient agar. (3) Sterile, petri plates. (4) Sterile, 1-ml pipettes.

Procedures.

1. *To collect tap water,* allow the water to run for 5 min to clean the faucet opening and pipes of accumulated bacteria. Remove the stopper from the bottle,

holding the stopper by the paper. Fill the bottle to a level that will be about $^1/_2$-in below the stopper. Replace the stopper and the paper cover; tie the cover with string. Do not allow the fingers to touch the bottle opening and neck, or the stopper.

2. *To collect a water sample from a body of still water*, remove the stopper with the paper covering as described above for tap water. Put the bottle in the body of water, mouth down, to a depth of about 1 ft. Turn bottle upward and, when filled, remove and replace the stopper as described for tap water.

3. Because the bacteria multiply rapidly, make tests immediately or store the bottles at 6 to 10°C.

4. At the time of tests, shake the sample vigorously at least 25 times. Using aseptic techniques, transfer 1 ml of water sample to a sterile petri plate. Add nutrient agar, melted and cooled to about 45°C. Mix the agar medium and water sample by gently rotating the plate. When the agar has solidified, invert the plates and incubate at 35°C for 24 h, or at room temperature for 48 h. If more than 300 colonies develop on a plate, the procedure must be repeated with the addition of serial dilutions of the water sample before plating. Serial dilutions are described in the introduction to the chapter.

5. Count the colonies and determine the average number of organisms for each milliliter of water.

Since many types of bacteria are unable to grow on nutrient agar or under aerobic conditions, the total-plate method does not give a count of all microorganisms in a water sample. The plate method is useful, however, in determining a relative count of bacteria in water from different sources, or in water of the same source sampled at different times.

Presumptive Test for Coliforms

Materials. (1) Water samples. (2) Sterile water in tubes or bottles for serial dilutions. (3) Five fermentation tubes, each with 30 ml of lactose-peptone broth (Appendix A). (4) Ten fermentation tubes, each with 10 ml of lactose-peptone broth (Fig. 9-25). (5) Sterile, 1-ml and 10-ml pipettes.

Procedures.

1. Place 10 ml of the water sample into each of the five fermentation tubes containing 30 ml of medium.

2. Place 1 ml of water sample into each of five fermentation tubes containing 10 ml of medium.

3. Place 0.1 ml of water sample into each of the other five fermentation tubes containing 10 ml of medium.

4. Incubate tubes at 35°C for 48 h. The formation of any gas in any of the tubes is considered a positive presumptive result for coliforms. The absence of gas in any tube at the end of 48 h is considered a negative test for coliforms.

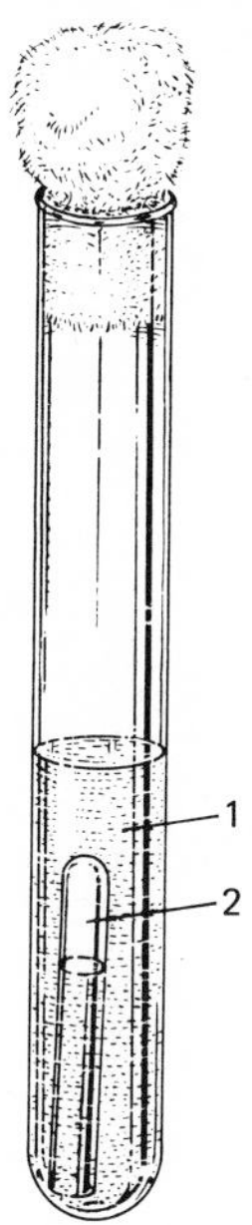

Figure 9-25 Fermentation tube for coliform presumptive test. 1, Lactose peptone broth to which has been added a measured amount of water sample. 2, Gas at tip of small inverted tube indicates presence of coliform bacteria. The inverted tube was filled with medium at beginning of test.

Confirmation Test for Coliforms

Materials. (1) Fermentation tubes from the presumptive tests, and which contain the least amount of water sample (10 ml, 1 ml, or 0.1 ml) that produced gas. (2) Eosin Methylene Blue agar (EMB) plates, one for each positive fermentation tube that is to be tested (see Appendix A for EMB agar).

Procedures.

1. Transfer a loopful of culture from each positive-fermentation tube to a separate EMB plate.

2. Incubate plates at 35°C for 24 h.

3. *Escherichia coli* produces small, flat colonies with a dark, metallic sheen. Other coliforms, for example, *Aerogenes*, produce raised, pink colonies with little or no sheen. Consider either type of colony a positive result.

Completed Test for Coliforms

Materials: (1) Plates from the confirmation test that contain typical coliform colonies. (2) Slant tubes of nutrient agar. (3) Fermentation tubes of lactose-peptone broth (Appendix A).

Procedures.

1. Select several typical coliform colonies on the EMB plates from the confirmation tests, and make transfers of each to a fermentation tube of lactose-peptone broth and to a slant tube.

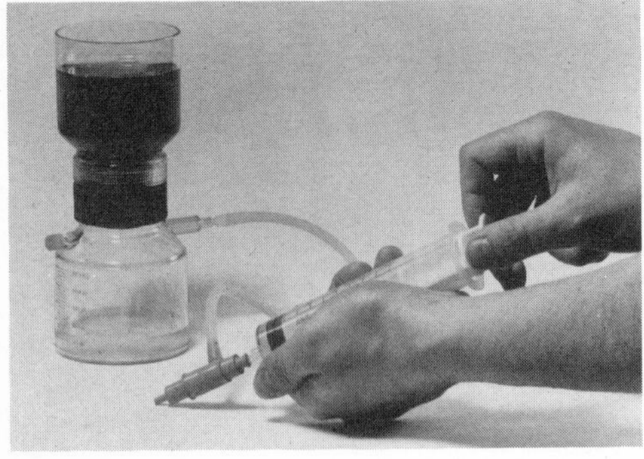

A

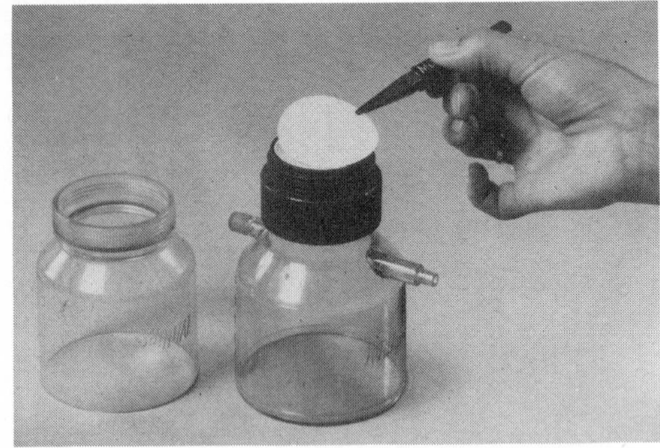

B

Figure 9-26 Membrane-filter technique for determination of coliforms in water. **A** The Millipore Sterifil System is made of unbreakable polycarbonate plastic and is used in pharmaceutical and hospital laboratories throughout the world for microbiological and particulate analysis. For educational purposes, the system can be simplified by substituting a plastic syringe and two-way valve for the electrical vacuum pump employed in most analytical laboratories. The funnel and base assembly holds a 47-mm Millipore bacterial retentive membrane filter, which retains on its surface all particles or microorganisms larger than the specified filter pore size (typically 0.45 microns). **B** For microbiological analysis, the Sterifil System is first sterilized, either by boiling or autoclaving. The bacterial retentive Millipore membrane filter is then placed aseptically on the support base, using sterile smooth-tipped forceps. The funnel is then screwed down over the base. **C** Premeasured ampules of nutrient medium are available for culturing coliform bacteria, total bacteria, yeast, mold, and many others. **D** After filtering a sample to isolate any microorganisms on the Millipore filter surface, the filter is transferred to an absorbent pad that has been saturated with the proper nutrient medium. The growing organisms are nourished by capillary action through the filter pores. **E** Shown is a mixed culture of coliform and other bacteria growing on a Millipore membrane filter after 24 h of incubation at 32°C (MF-Endo Medium). When the filter is dried, coliform colonies will exhibit a distinctive green metallic sheen that makes identification easy. Coliform bacteria are the indicator organisms used by most public health laboratories to measure the sanitary quality of drinking water. *(A–E Courtesy of the Millipore Corporation, Bedford, Mass.)*

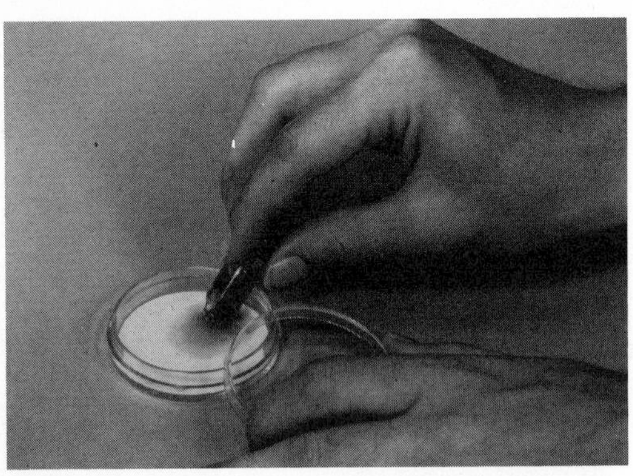

C

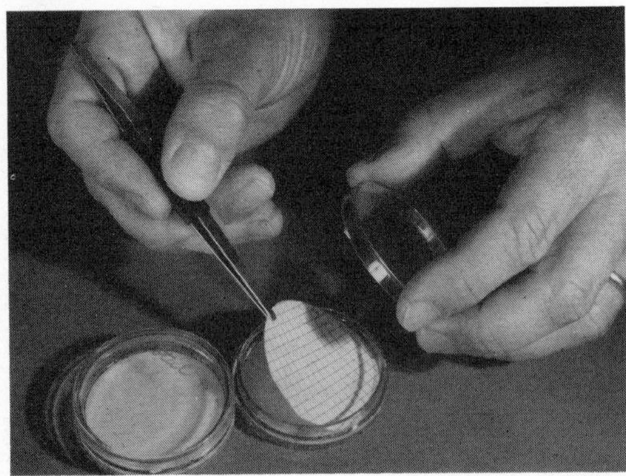

D

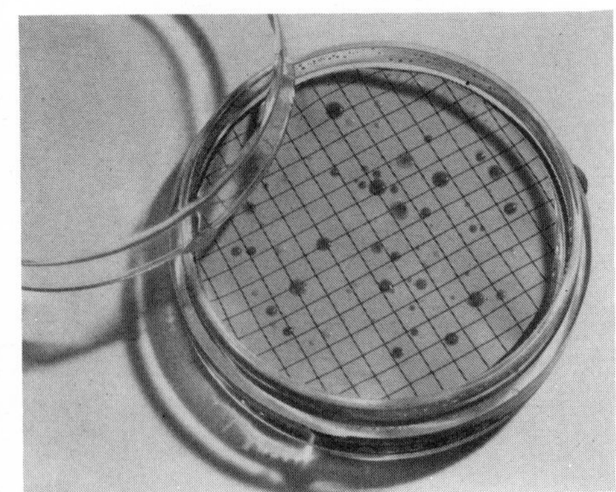

E

A

B

Figure 9-27 Other tests with the membrane-filter technique. **A** This student at Weston High School, Weston, Mass., is using the top part of a Millipore Sterifil Filter System and a small vacuum cleaner to filter exhaust smoke through a Millipore membrane filter. Particulate matter is trapped on the filter surface, where it can be examined under a microscope, analyzed using chemical spot tests, or weighed to determine the level of solid contaminants in a measured volume of air. **B** The Millipore Sterifil System is an ideal tool for studying solid contaminants in cigarette smoke. Filters can be analyzed gravimetrically to compare the weight of solids obtained from different brands of cigarettes. The test also demonstrates the filter's ability to retain bacteria, since particles in a carbon aerosol are approximately the size of a small bacterium (about half a micron). (*A and B Courtesy of the Millipore Corporation, Bedford, Mass.*)

2. Incubate tubes at 35°C for 48 h.

3. Observe fermentation tubes for the production of gas.

4. Make a Gram stain of organisms growing on the slant tube.

5. Consider the results as a completed positive test for coliforms if gas is produced from the lactose broth and if the organisms of the slant tube are gram-negative, nonspore-forming bacilli.

Membrane-filter Technique

In 1965 the American Public Health Association adopted standard methods for confirming the presence of coliform bacteria with filter membranes. The technique allows the completion of a test within 24 h. The procedure involves drawing a volume of water through a sterile filter disc placed in a filtration unit. The disc is then removed and placed in a petri plate with appropriate medium, and incubated at 35°C for 20 h. See Fig. 9-26. For more information, order a brochure from the Millipore Corporation, Bedford, Mass. 01730, which describes their filtering equipment, and their manual *Millipore Experiments in Environmental Microbiology* (price 50¢). The equipment is relatively inexpensive and has versatile uses for other tests in addition to water analysis. (See Fig. 9-27.)

REFERENCES

American Public Health Association, Inc.: 1971, *Standard Methods for the Examination of Water and Wastewater, Including Bottom Sediment and Sludges*, 13th ed., New York.

Breed, R.S., E.G.D. Murray, and N.R. Smith (eds.): 1957, *Bergey's Manual of Determinative Bacteriology*, 7th ed., The Williams & Wilkins Company, Baltimore, 1132 pp.

Burgess, A.: 1958, *Microorganisms in the Soil*, Hutchinson & Co. (Publishers), Ltd., London.

Committee on Bacteriological Technic, Society of American Bacteriologists: 1957, *Manual of Microbiological Methods*, McGraw-Hill Book Company, New York, 315 pp.

Larson, L.A., S. Clary, and K. Lwanga: 1969, "Nitrogen Fixation and Availability," *Am. Biol. Teacher*, **31** (9): 587–589.

Pelczar, Michael J., Jr., and Roger D. Reid: 1972, *Microbiology*, 3d ed., McGraw-Hill Book Company, New York, 948 pp.

Rhodes, A.J., and C.E. van Rooyen: 1968, *Textbook of Virology*, 5th ed., The Williams & Wilkins Company, Baltimore, 996 pp.

Seeley, H.W., Jr., and P.J. Vandemark: 1972, *Microbes in Action, A Laboratory Manual of Microbiology*, 2d ed., W.H. Freeman and Company, San Francisco, 225 pp.

Stanier, R.Y., M. Doudoroff, and E.A. Adelberg: 1970, *The Microbial World*, 3d ed., Prentice-Hall, Inc., Englewood Cliffs, N.J.

Other Literature

Adams, M.H.: 1959, *Bacteriophages*, John Wiley & Sons, Inc., New York, 592 pp.

Baker, F.J.: 1967, *Handbook of Bacteriological Technique*, 2d ed., Appleton-Century-Crofts, Inc., New York, 482 pp.

Barthelemy, R.E., J.R. Dawson, Jr., and A.E. Lee: 1964, *Innovations in Equipment and Techniques for the Biology Teaching Laboratory*, D.C. Heath and Company, Boston, 116 pp.

Berman, D.: 1968, "The Enumeration of Bacterial Viruses by the Plaque Technique," *Am. Biol. Teacher*, **30** (6): 486–487.

Brock, T.: 1970, *Biology of Micro-organisms*, Prentice-Hall, Inc., Englewood Cliffs, N.J., 740 pp.

Cairns, John: 1966, "The Bacterial Chromosome," *Sci. Am.*, **214** (1): 37–44.

Carpenter, P.L.: 1967, *Microbiology*, 2d ed., W.B. Saunders Company, Philadelphia, 476 pp.

Collins, C.H., and P.M. Lyne: 1970, *Microbiological Methods*, 3d ed., Butterworth & Co. (Publishers), Ltd., London, and University Park Press, Baltimore, 454 pp.

Edgar, R.S., and R.H. Epstein: 1965, "The Genetics of a Bacterial Virus," *Sci. Am.*, **212** (2): 70–78.

Fagle, D.L., and L.S. McClung: 1960, "Bacteriology Inoculating Devices Useful in the Biology Laboratory," *Am. Biol. Teacher*, **22** (6): 337–338.

Grassmick, R.A.: 1969, "An Inexpensive Pipetting Assembly," *Am. Biol. Teacher*, **31** (9): 590–592.

Horne, R.W.: 1963, "The Structure of Viruses," *Sci. Am.*, **208** (1): 48–56.

Kellenberger, E.: 1966, "The Genetic Control of the Shape of a Virus," *Sci. Am.*, **215** (6): 32–39.

Kirchen, R.V.: 1967, "Bacteriophage Plaque Morphology," *Carolina Tips*, **30** (7): 26.

Lechevalier, H.A., and D. Pramer: 1971, *The Microbes*, J. B. Lippincott Company, Philadelphia, 507 pp.

Luria, S.E., and J.E. Darnell: 1967, *General Virology*, 2d ed., John Wiley & Sons, Inc., New York, 512 pp.

McClung, L.S.: 1960, "Microbiology Teaching Aids," *Am. Biol. Teacher*, **22** (6): 352–385.

McClung, L.S., and H.R. Arthur: 1960, "Simple Procedure for Making Pipettes for Bacteriology," *Am. Biol. Teacher,* **22** (6): 338–340.

Skerman, V. B. D.: 1967, *A Guide to the Identification of the Genera of Bacteria,* 2d ed., The Williams & Wilkins Company, Baltimore, 303 pp.

Walter, W. G.: 1968, "The Use of Wooden Applicator Sticks," *Am. Biol. Teacher,* **30** (6): 473–475.

THE SLIME MOLDS, FUNGI, AND LICHENS

The fungi consist of a large number of achlorophyllous organisms that usually possess definite cell walls. Most are composed of filaments called hyphae, which when in mass form a mycelium. The cells of fungi contain definite nuclei that undergo the usual phases of mitosis. Many fungi exhibit asexual reproduction by spores, and sexual reproduction by the fusion of nuclei of opposite mating types. For another large group of fungi, the sexual phase has not been observed; these organisms are classed as imperfect fungi or the Fungi Imperfecti.

The present chapter describes special techniques for culturing and studying certain types of (1) slime molds, (2) the true molds, (3) the yeasts, and (4) the mushrooms. A last portion of the chapter deals with several techniques for studying the lichens.

TAXONOMIC SYSTEM FOR FUNGI OF THIS CHAPTER

The organisms discussed in this chapter are arranged into five large groups: Myxomycetes, Phycomycetes, Ascomycetes, Basidiomycetes, and Deuteromycetes or the Fungi Imperfecti. For easier reference, the organisms within each group are arranged alphabetically by generic name. An alphabetical system can be misleading, however, since the evolutionary sequence and the relationships among genera become obscured. Therefore, the taxonomic outline presented below should be used when studying any fungi of this chapter.[1]

Division Myxomycota

Class Myxomycetes
Physarum polycephalum

Division Phycomycota

Class Phycomycetes
Allomyces macrogynus
Saprolegnia
Achlya
Rhizopus stolonifer
Pilobolus crystallinus

Division Ascomycota

Class Ascomycetes
Saccharomyces cerevisiae
Aspergillus
Penicillium
Neurospora crassa
Sordaria fimicola
Peziza vesiculosa
Sarcoscypha coccinea
Scutellinia scutellata
Pyronema omphalodes

Division Basidiomycota

Class Heterobasidiomycetes
Puccinia graminis

Class Homobasidiomycetes
Agaricus bisporus

Division Deuteromycota
(Fungi Imperfecti)

Class Deuteromycetes
Alternaria tenuis
Dactylella dreschsleri
Arthrobotrys dactyloides
Arthrobotrys conoides
Rhizoctonia solani

COLLECTING AND PURCHASING FUNGI

Pure cultures may be obtained from most biological supply companies. Most species described here are also available from the American Type Culture Collection (ATCC). Many types of mixed cultures are easily obtained from water, soil, and the air.

Small fungi on bark, soil, or wood are collected on a portion of the substratum and transferred to the laboratory in covered, rigid containers, such as small cardboard or plastic boxes that may be carried in a basket. Plastic or paper bags are not satisfactory, as specimens may be crushed during transfer. Fungi growing under moist conditions must not become dry during transfer to the laboratory. Mushrooms and other higher fungi are taken with a trowel that is inserted into the soil below the stalk so as to remove the mycelium in the soil with the fruiting body above the soil. When handled carefully, the small woodland mushrooms and the cup fungi often can be maintained successfully, at least temporarily, in a covered terrarium. (See techniques for terrarium and bottle cultures in Chap. 6.)

Leaves and stems from plants with rusts and smuts are best collected by immediately transferring individual specimens to a folded newspaper that is placed in a plant press, thus preventing the dispersal and mixing of parasitic spores. Such collections are pressed until dry and retained and isolated within the folded and

[1]Division names are based on the system of Bold (1967a).

sealed sheets until their later use for class studies. The sheets may be placed in sealed and labeled envelopes of appropriate size for storage.

MAINTAINING STOCK CULTURES

Methods for maintaining molds and yeast in stock cultures are much the same as those described in Chap. 9 for maintaining bacterial cultures. As with bacteria, molds and yeasts are transferred to slant tubes of appropriate agar medium. After 24 to 48 h of growth at room temperature, the tubes may be stored in a refrigerator at 6 to 10°C for 6 months or longer, at which time the culture should be transferred to fresh medium. Cultures can be stored for several years when the agar slant is covered with sterile mineral oil (see Chap. 9), or when cultures of sporulating molds are stored in a freezer at −20°C (Carmichael, 1956). Also see Raper and Alexander (1945) for a freeze-drying method of preserving sporulating molds.

PRESERVING FUNGI

Liquid Preservation

Fleshy fungi such as mushrooms, bracket fungi, and cup fungi, and plant parts with fungal parasites can be preserved indefinitely in either solutions 1 or 2, given below. Solutions 3 and 4 are used for colored fungi.

1. 5% *formalin.* Add 5 ml of 40% formaldehyde solution to 95 ml of distilled water.

2. *Formalin, acetic acid, alcohol solution.*

Formaldehyde (40% sol.)	50 ml
Glacial acetic acid	50 ml
Alcohol, 70% ethyl or isopropyl	900 ml

3. *Fungi with water-soluble pigment.*

Glacial acetic acid	10 ml
Mercuric acetate	1.0 g
Neutral lead acetate	10.0 g
Ethyl alcohol, 90%	1,000 ml

4. *Fungi with water-insoluble pigment.*

Glacial acetic acid	5 ml
Mercuric acetate	10 g
Distilled water	1,000 ml

Dry Preservation

Plant parts with a fungal parasite, such as wheat infected with the rust *Puccinia*, may be dried in a plant press and stored for later use as described earlier in the chapter. Fleshy fungi, including the mushrooms, puffballs, cup fungi, and others, may be dried in sunlight or under a 100-W lamp. The dried specimens are then wrapped with tissue and stored, uncrowded, in covered, rigid containers. Paradichlorobenzene crystals are added to the container to prevent insect contamination. (See Chap. 5 for use of paradichlorobenzene in insect collections.)

MEDIA FOR FUNGI

Whereas bacteria generally require a proteinaceous medium of approximately pH 7, most fungi require a carbohydrate medium and a slightly acid pH of 5 to 6. Satisfactory media for many fungi include Sabouraud's agar, potato dextrose agar, oatflake agar, and cornmeal agar. (All formulas are given in Appendix A.) Special media for certain species are indicated with the descriptions of genera later in the chapter.

Many types of ordinary organic materials, including common foodstuffs, provide excellent and inexpensive media for growing molds and yeasts. Practically all grapes have wild yeasts on their surfaces. When several grapes are crushed in a culture bowl that is then covered and left at room temperature, a variety of relatively large, ovoid or rectangular wild yeasts multiply rapidly to form a scum within several days. The budding of wild yeast cells into branched chains gives a more spectacular display than the ordinary baking yeast that is used most often in the laboratory.

Penicillium, *Aspergillus*, and other molds develop within 3 to 7 days at room temperature on crushed citrus fruit or tomatoes that are first exposed to the air for several hours before covering the culture bowls. A better and quicker method for obtaining these molds is to enlist the assistance of the manager of a vegetable counter in a grocery store. Most often he will cooperate generously in saving rotted and molded fruits and vegetables that often present a veritable "garden" of microorganisms. Molded cheese, and cheeses flavored with particular molds (Camembert, Roquefort, and others that show dark streaks of mold spores), old cottage cheese, molded bacon and ham, as well as bread — all represent inexpensive and interesting sources for molds, yeast, and bacteria. In every case, a small portion of the mold growth may be transferred to a drop of water, or preferably to a drop of alcohol or lactophenol (Appendix A), on a slide for microscopic study.

SPECIAL TECHNIQUES FOR STUDYING FUNGI

The equipment and general techniques for handling fungi are the same as those described for bacteria in Chap. 9. Several special techniques are described here; others are given later in the chapter with individual genera and species.

Microscopic Examination of Fungi

Preparing slide mounts. Temporary mounts of small portions of fungi are prepared in the same manner as described in Chap. 9 for bacteria. In place of water mounts, however, better mounts are obtained with one of the following solutions: (1) lactophenol mountant with aniline blue stain or other stain (Appendix A); (2) a weak detergent-water solution to reduce air bubbles; (3) a 20% glycerol solution in water; or (4) a weak, aqueous solution of KOH (2 to 10%) for small portions of dried fungi.

Observing fungal growth. About 8 to 12 h before the laboratory session, spread a few spores on a thin layer of agar medium in a petri plate. Place the plate in a dry, cold oven and allow the temperature to gradually rise to near 70°C during a 30-min period. Turn off the heat at once and leave the plate in the oven until the temperature drops to 30°C or lower. Then transfer the plate to room temperature until class time. At the time of use, remove the cover of the petri dish, and place a thin coverglass over the germinating spores. During a 2-h period, examine the growth every 15 min under 100X magnification to observe the development of mycelium. This method may be used also for observing the development of the large ascospores of *Sordaria* and related molds.

Culturing Fungi on Slides (Nonsterile Technique)

This technique is recommended for observing a fungus in a natural and undisturbed growing condition. The procedure is relatively easy and gives excellent "mini-cultures" for student observations. See Fig. 10-1.

1. Prepare 1,000 ml of nutrient agar medium (Appendix A) to which are added, before sterilization, 2.5 g tartaric acid and 25 g dextrose.

Figure 10-1 Slide-coverglass culture chamber. **A** Coverglass with wax on three sides. **B** Agar medium, with mold spores, has been added to the coverglass. **C** The coverglass has been inverted onto a glass slide. More wax has been added around the three sides. **D** Side view of culture chamber; 1, coverglass; 2, wax; 3, agar medium; 4, glass slide.

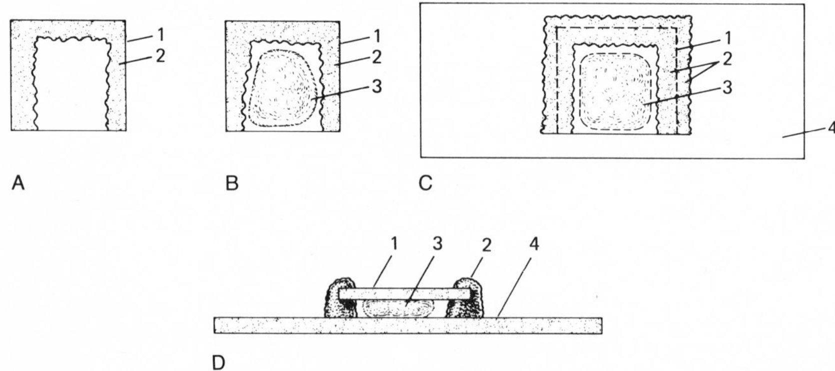

2. Convert a coverglass into a small growth chamber by building a wall of wax about 2 mm high around three edges of the coverglass. The wall is most easily constructed by moving a lighted candle around the edge of the coverglass in such a way that the melted wax drips onto the edges of the coverglass.

3. Transfer several loops of fungus spores to 5 ml of the melted medium (at about 45°C). Mix well, and transfer several loops of the mixture to the wax chamber on the coverglass.

4. Allow the medium to solidify and invert the coverglass onto a clean slide. Seal the coverglass to the slide by adding more melted wax to the three walls of the chamber.

5. Prepare a moist chamber by fitting several circles of paper toweling or filter paper into the bottom of a petri dish or culture bowl. Add 20% glycerol (20 ml glycerol in 80 ml water) to thoroughly dampen the paper and to form a shallow layer of solution above the paper. Place short strips of ¼-in glass tubing on the paper.

6. Lay the slide-coverglass culture on the glass tubing with the slide suspended above the layer of glycerol. Add the top cover of the petri dish, or cover a culture bowl with a glass plate or plastic film pressed tightly around the edge of the bowl. Several slide preparations can be placed in one large culture bowl.

7. Incubate at room temperature for 2 or 3 days, or longer, observing at regular intervals for mycelial growth and sporulation.

8. When sporulation is evident, remove the slide for microscopic examination. If necessary, wipe moisture from the lower slide surface and the upper coverglass surface (carefully). Examine the fungus through the coverglass with 30 to 100X or greater magnification.

The tartaric acid medium is not conducive to bacterial growth; however, airborne mold and yeast spores may contaminate the culture. Such contaminants can be held to a minimum if all glassware is clean and if a person works quickly. On the other hand, the contaminants are not necessarily undesirable if the purpose is simply to observe fungal growth.

Soil Fungi and Antibiotic Production

To obtain soil fungi that may produce antibiotics, follow the method described for soil bacteria in Chap. 9, including the discussion on isolation of antibiotic producers.

Fungi from Airborne Spores

Several techniques for obtaining airborne microorganisms are described in Chap. 9 under "Air Bacteria." The methods are equally satisfactory for obtaining molds or yeasts when a carbohydrate medium is substituted for the bacterial medium.

THE MYXOMYCETES (SLIME MOLDS)

Because the Myxomycetes exhibit both plant and animal characteristics, they are often included with the protozoa. The animallike stage, the plasmodium, definitely resembles amoeboid protozoa. The reproductive stage is plantlike, and the production of spores in fruiting bodies is comparable to that of the molds.

Physarum (Slime Mold; *P. polycephalum* Schw.)

USES: (1) To culture and observe the stages of the life cycle. (2) To demonstrate protoplasmic flow in the plasmodium stage. (3) To demonstrate tropic responses of the plasmodium. For other suggestions, see Alexopoulos (1963), Alexopoulos and Koevenig (1964), Kerr (1965), Collins (1966, 1969), and Vogel (1969).

DESCRIPTION: *Physarum* and other slime molds grow on moist, decaying vegetation, often on logs of dead tree stumps where, under good conditions, they may form a large protoplasmic mass that on rare occasions covers an area of several square feet. This protoplasmic mass is a plasmodium, the amoeboid stage of the life cycle. More often the plasmodia are buried in the substratum and thus are easily overlooked. If small portions of damp, decaying vegetation are brought into the laboratory and maintained under low light and high humidity, the plasmodia, if present, may emerge and move onto the surface of agar or paper toweling. The plasmodia can then be maintained by the procedures described below under Culture.

The plasmodium of *Physarum polycephalum* is a yellow mass that can be observed without magnification, moving slowly across the substratum at the rate of perhaps 2 to 3 cm/h (Fig. 10-2). As it moves, the plasmodium ingests bacteria, yeast, and other microorganisms, as well as particles of decaying organic material. Under a magnification of 100X the bidirectional flow of fluid is easily observed within a network that extends throughout the gelatinous protoplasm (Fig. 10-2).

Figure 10-2 *Physarum polycephalum*, a slime mold. **A** Plasmodium on agar medium in petri dish. **B** Close-up view of a portion of the plasmodium.

A

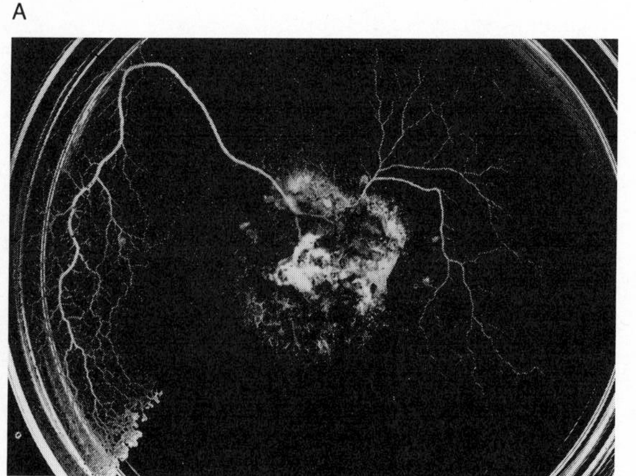

B

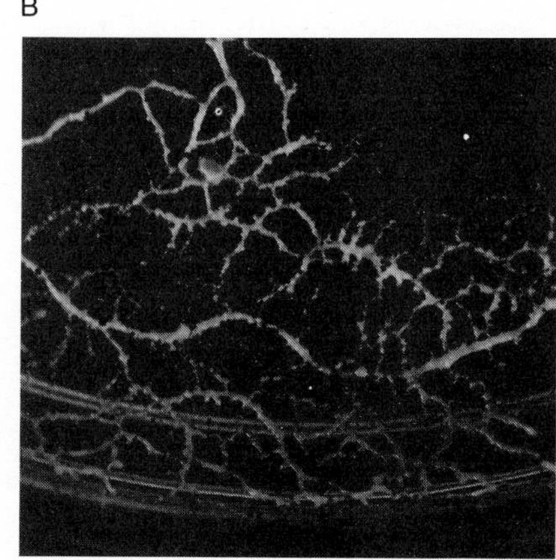

The liquid can be seen to move at a relatively rapid rate in one direction, to stop for a moment, and then to flow in the reverse direction.

For best observations of the flow, aseptically transfer a small portion of plasmodium to a petri dish containing a very thin layer of weak maltose agar. Prepare the agar medium by adding 3 g of maltose, 1 g of peptone, and 20 g of plain agar to a liter of water. Sterilize the medium and pour a very thin, almost transparent, layer in a sterile petri dish. The plasmodium will spread over the agar within 12 to 24 h. Observe the movement of fluid protoplasm in the network of plasmodial veins by inverting the plate under the low-power objective of the microscope.

If sufficient moisture and food are present, the organism may remain in the plasmodial stage for several weeks. If the substratum becomes dry, the plasmodium generally forms a sclerotium consisting of a hardened, multinucleated, irregular and dormant mass. If drying does not occur, the protoplasm may eventually become concentrated at various spots, forming stalks with sporangia. Each sporangium contains nuclei whose descendants occur in the many spores that develop. The spores are discharged and may remain dormant for several years. Under favorable conditions, a spore develops into one or two small amoeboid gametes (myxamoebae) that soon become biflagellated. The biflagellated forms fuse in pairs to make amoeboid zygotes which then develop into multinucleated plasmodia, thus completing the life cycle.

Occasionally, other slime molds may be collected from moist decaying vegetation, for example, species of *Dictyostelium*, *Fuligo*, and *Stemonitis* (Fig. 10-3). Cultures of several of these genera can be obtained from most biological suppliers.

CULTURE: Method 1. Put portions of collected vegetation or soil on moist filter paper in a bowl. Sprinkle a light layer of dry oatmeal flakes around the material, and cover the bowl with a glass plate or plastic film. The quick-cooking or "instant" oatmeal does not seem to be satisfactory. If plasmodia or spores are present in the collected materials, eventually a growth of slime mold may appear on the filter paper. Under these culture conditions, other molds may appear on the oatmeal, and the flakes should be replaced as necessary. Store the culture bowl in darkness or in diffused light at 20 to 24°C.

Method 2 (Bold, 1967b). A reliable procedure is to aseptically transfer small portions of plasmodia or sclerotia to petri dishes containing Bold's basal medium with agar (Appendix A). On the organism place several flakes of dry oatmeal that have been autoclaved in a covered petri plate. Within several days, the plasmodia will ingest the food and will spread out over the moist agar. Add fragments of sterile oatmeal as necessary.

When the plasmodia have nearly covered the agar surface, induce formation of sporangia by putting the plates about 10 in below a 100-W lamp. Stop all feeding. Plasmodia will move up and cover the inside surface of the top plate, perhaps in response to the light, where they begin to form fingerlike projections that hang downward. Eventually sporangia and spores develop. When the spores are put into a hanging drop and observed under 100X magnification, students may be able to observe the development of biflagellate amoebae.

To preserve a culture for future use, transfer a portion of the plasmodium to a circle of moist filter paper that has been autoclaved in a covered petri dish. Feed fragments of sterile oat flakes until the plasmodium has covered the paper. Allow

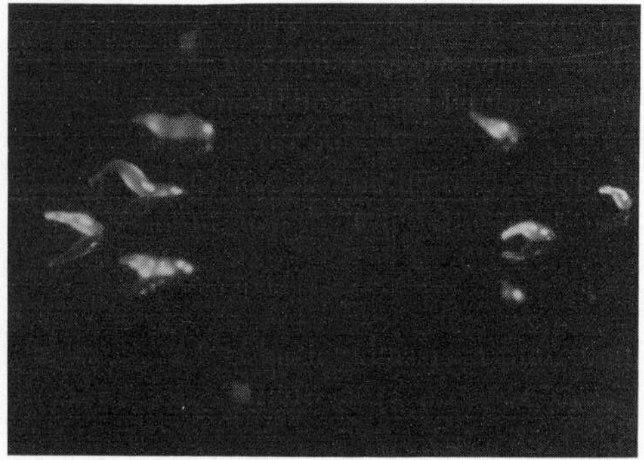

A Courtesy of Carolina Biological Supply Company

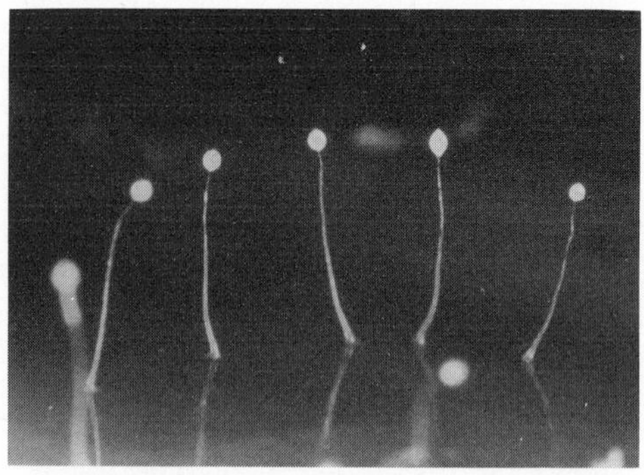

B Courtesy of Carolina Biological Supply Company

Figure 10-3 Other slime molds. **A** *Dictyostelium discoideum*, pseudoplasmodium or slug stage. Each slug is a mass of amoeboid cells, approximately 0.5 mm long, that is migrating over the surface of the agar medium. **B** *D. discoideum*, the sorocarps that grow from pseudoplasmodia and rise about 20 mm above the agar surface. **C** *Fuligo* sp., a large, yellow mass of fruiting bodies, an aethalium, approximately 6 in across.

C Courtesy of Carolina Biological Supply Company

the yellow plasmodium and filter paper to dry gradually. As drying occurs, the plasmodium develops into a sclerotium. Cut the filter paper into 1- to 2-in square portions and store in an envelope. Dried sclerotia may be stored for months and perhaps for 1 or 2 yr. When the paper with dry sclerotia is placed under proper growth conditions, the sclerotia develop into plasmodia.

THE PHYCOMYCETES (ALGA-LIKE MOLDS)

General characteristics of this group are: the mycelium is coenocytic (nonseptate), that is, there are no cross walls and the protoplasm is continuous and multinucleate; or the protoplasm is divided by partial septa (*Allomyces*). Species may be saprophytic or parasitic. The following genera are appropriate for class studies.

Allomyces (Soil Mold; *A. macrogynus* Emerson)

USES: To study the relatively complex life cycle of a relatively simple mold.

DESCRIPTION: Terrestrial; widely distributed in tropical and temperate regions. The mycelium consists of a single hypha that branches dichotomously and is anchored by rhizoids. The hyphae are incompletely septate; are coenocytic and multinucleate (Fig. 10-4). Two types of sporangia are formed: zoosporangia and meiosporangia. A thin-walled, colorless zoosporangium forms at the distal end of a hypha and becomes lateral as a new branch grows beyond it. At maturity, the uniflagellated zoospores pass to the outside through a fissure in the sporangial wall. Each zoospore may germinate into a new diploid "sporophyte." The thick-walled, brown meiosporangia may remain resistant to adverse conditions during long periods. In a favorable environment, haploid zoospores develop into a haploid gamete-producing, "gametophyte" plant from which zygotes and new diploid sporophyte plants develop.

CULTURE: Bold (1967b) describes a method for isolating *Allomyces* from soil on hemp seed in charcoal water. Hemp seed is, of course, from the marijuana plant (*Cannabis sativa* L.); therefore viable seeds cannot be obtained legally except with permission from the state narcotics authorities. Nonviable seeds are as satisfactory for mold culture and may be purchased from Carolina Biological Supply Co. and possibly from other companies. Nonviable and treated seeds can be obtained from pet stores in bird feed, although these are not always satisfactory, perhaps because of special treatment. Other kinds of seeds may be substituted for hemp seed, for example, boiled rice or wheat grain. Small dead insects, such as *Drosophila*, may also be used.

Figure 10-4 *Allomyces* sp. **A** The mold on submerged hemp seeds; each colony is about 10 mm across. **B** Close-up of **A**.

A

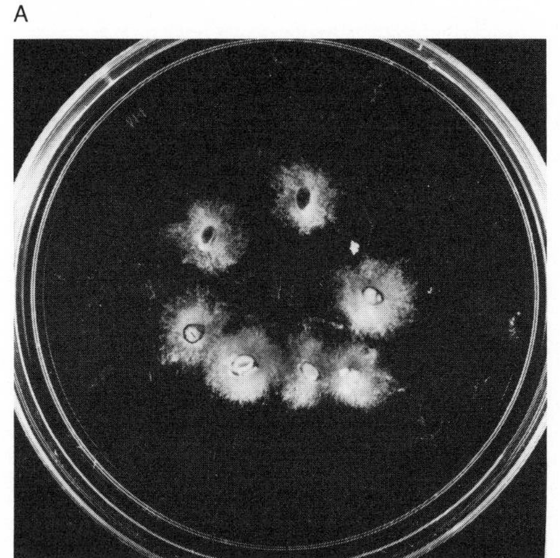

B

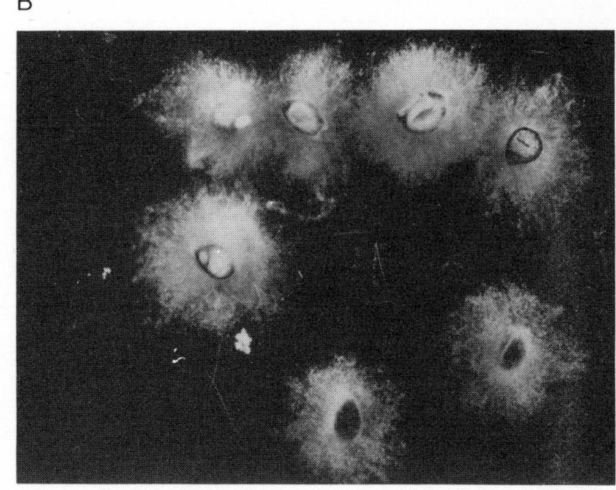

Prepare the charcoal water by placing several 1- to 2-in lumps of activated charcoal in a liter of distilled water. Shake the water vigorously and filter. Pour 100 to 200 ml of the filtered water into each of several small culture bowls. Add 20 to 30 g of soil. When the soil has settled, add four halves of boiled hemp seed (or two boiled rice or wheat grains, or a dead insect, etc.). Generally, the white fuzzy mycelium of *Allomyces*, and other fungi as well, will appear around the seed within 48 to 72 h (Fig. 10-4). To observe the mold, transfer a half-hemp seed or a rice or wheat grain, with mold, to several drops of charcoal water on a microscope slide. Carefully cut away the seed, disturbing the mold as little as possible.

To isolate *Allomyces* from bacteria and other molds that grow on the seed, cut off tips of hyphae (under a stereomicroscope), and transfer the tips to a plate of sterile agar medium consisting of 3 g maltose, 1 g peptone, 20 g agar, and 1,000 ml of distilled water. Within several days, when the mold has grown over the agar, select hyphae of *Allomyces* that are located in a spot removed from the other mold. Use a flamed scalpel to cut loose square-centimeter blocks of agar containing hyphal tips. Transfer the blocks to an empty, sterile petri dish. Add sterile charcoal water to the dish to bring the water to a level at the top edge of the agar blocks. Place a boiled half-seed of hemp on each agar block. Incubate at room temperature for 48 h, and then add sufficient sterile charcoal water to submerge the seed. Usually the mold will develop sporangia within several days.

The ATCC maintains *Allomyces* on hemp seed medium prepared by placing 20 g of crushed hemp seed in 500 ml of tap water. Boil the water for 1 to 2 min. Filter and add enough water to make 1 l volume. For a solid medium, add 20 g of plain agar. Autoclave. Bold (1967b) states that he has grown both the gametophyte and sporophyte stages of *Allomyces* on yeast starch agar and on cornmeal agar. Yeast starch agar is recommended by Alexopoulos and Beneke (1962) as particularly suitable for *Allomyces*. (See Appendix A for media.)

Pilobolus (Dung Mold; *P. crystallinus* (Wiggers) Tode)

USES: (1) To observe the positive phototropism of sporangiophores. (2) To observe the explosive ejection of spores, which may be expelled to a distance of 5 or 6 ft.

DESCRIPTION: Hyphae are coenocytic. Sporangiophores arise from the subsurface mycelium. A hard-walled sporangium develops at the tip of a hypha, followed by a swelling of the hypha immediately below the sporangium (Fig. 10-5). When spores have matured, the subsporangial swelling bursts with an explosive force that ejects the entire hard-walled sporangium. This type of spore dispersal differs from that of other molds, in which the thin-walled sporangium breaks open, releasing the spores. The development of a swelled portion below the sporangium becomes, then, an interesting evolutionary device for removal of the tough-walled sporangium. Also the fact that sporangiophores turn toward a light source presents an additional interesting adaptation.

According to Bold (1967a) the sporangiophores rise above the substrate in the afternoon. During early evening, the terminal spore cases develop, followed later in the night by a swelling of the hypha below the sporangium. In late morning of the next day, the swelled portion bursts and the sporangium is shot from the tip.

CULTURE: Grow *Pilobolus* by placing small portions of fresh horse or cow

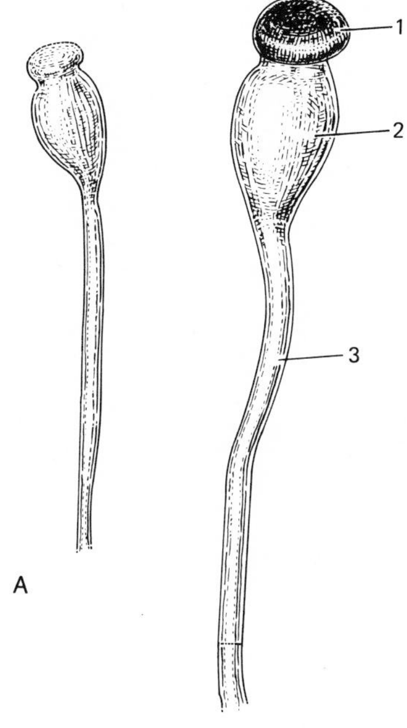

Figure 10-5 *Pilobolus* sp. **A** and **B** Stages in development of sporangium. Mature fruiting body rises 3 to 5 mm above surface of substratum. 1, Sporangium; 2, subsporangial swelling; 3, sporangiophore.

dung (24 to 48 h old) in a culture bowl. Add drops of water to barely cover the bottom surface of the bowl. Cover with glass or plastic film. Incubate at room temperature. Usually the mold will appear within 48 to 72 h. To demonstrate positive phototropism, place a light at one side, or put the bowl in a window or in a box with a small window cut out of one side. Sporangiophores will turn toward the light, and students can observe the expulsion of sporangia in that direction.

Rabbit dung agar. The ATCC maintains *Pilobolus* on agar prepared with rabbit fecal pellets. Sterilize the pellets in plugged test tubes, with three or four fecal pellets per tube. After the first sterilization, to each tube add 4 ml of 1.5% plain agar medium (1.5 g agar in 100 ml distilled water). Sterilize tubes a second time, after which slant the tubes in a position to allow the pellets to protrude above the agar. When agar has solidified, transfer *Pilobolus* to the surface of the pellets from a culture obtained on the horse or cow dung. Since rabbit pellets are easily obtained in an animal room, this method allows the continuation of a culture without the necessity of obtaining fresh dung elsewhere.

Rhizopus (Bread Mold; *R. stolonifer* Lind = *R. nigricans*)

USES: (1) To culture and observe the life history of the mold. (2) To observe protoplasmic flow in the hyphae at 100 to 400X magnifications.

DESCRIPTION: *Rhizopus* forms a mass of white mycelium on and in bread or other organic materials. Stolons (horizontal hyphae) grow over the medium surface, putting down rhizoids at intervals. Groups of erect hyphae (sporangiophores) rise above the rhizoids. A thin-walled sporangium develops at the tip of each erect hypha (Fig. 10-6). Each sporangium contains a sterile central portion, the columnella, and an outer layer of sporogenous protoplasm which segments into small, black-walled spores. The outer, thin wall of the sporangium deteriorates and the mature spores are thereby released, forming a powdery black film over the mycelium and the substratum.

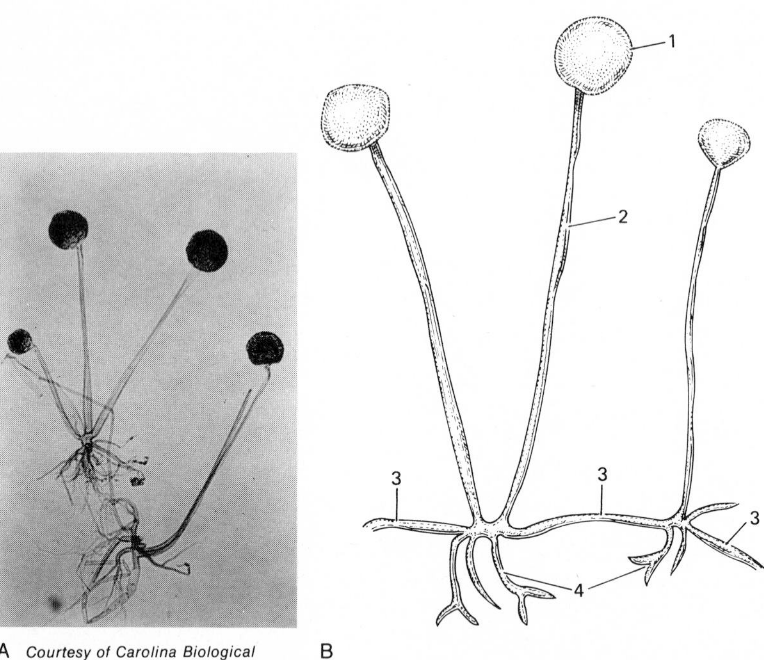

A *Courtesy of Carolina Biological Supply Company*

B

Figure 10-6 *Rhizopus stolonifer.* **A** Habit of growth. **B** Diagram of the mold; 1, sporangium; 2, sporangiophore; 3, stolen; 4, rhizoids. **C** Sexual reproduction; immature zygote at top, left of center; mature zygote at bottom center.

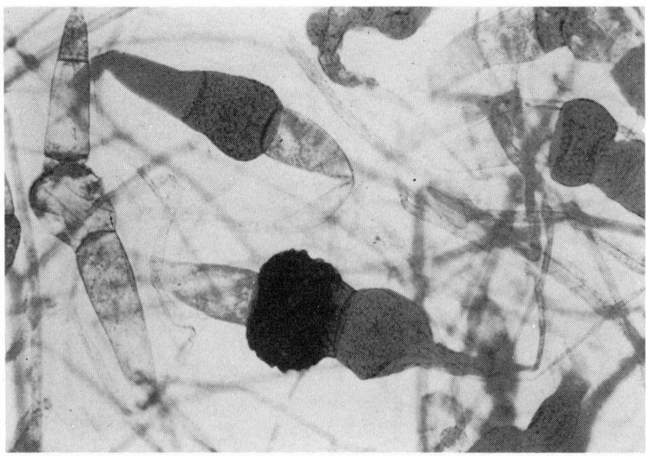

C *Courtesy of Carolina Biological Supply Company*

Sexual reproduction occurs when the hyphae of two mating strains come into contact. The hyphal tips enlarge, and in each a wall grows between the tip and the rest of the hypha, forming two adjacent gametangia. The gametangia fuse, and the many enclosed nuclei unite in pairs. A thick, dark wall forms over the fused gametangia, which then become a zygote containing diploid nuclei (Fig. 10-6). In the laboratory, the resistant zygotes (zygospores) remain dormant for 1 to 3 months (Gauger, 1961). When conditions are favorable, the zygote breaks open and a sporangiophore develops. Meiosis of the diploid nuclei occurs just before zygote germination.

CULTURE: The simplest method of culture is to dampen one-half slice of bread and to rub the bread over a dusty area on the top of a door or window frame or in a dusty corner. Place the bread in a covered bowl at room temperature. Usually *Rhizopus* appears on the bread within 2 or 3 days, and spores mature within another 1 or 2 days. Other molds may also appear, including *Neurospora* (pink mold, an ascomycete), *Mucor*, a black mold much like *Rhizopus*, and *Aspergillus*, a gray-green mold of the Fungi Imperfecti.

To obtain pure cultures of *Rhizopus*, make aseptic transfers of spores to plates or slant tubes of cornmeal agar with dextrose (Appendix A), or purchase starting cultures from a biological supply company.

Prepare a plate of mold for microscopic studies by gently flooding the plate with either 70% alcohol or a weak detergent solution. The mold tends to trap a film of air when put in water, and the other solutions are used to dispel the air pockets. Transfer small blocks of agar containing the mold to several drops of the detergent solution on the slide. Add a coverglass, smash the agar gently, and examine at 100 to 400X magnification.

To demonstrate sexual stages, transfer plus and minus strains to cornmeal agar with dextrose, with each strain placed at one side of the plate about 1½ in apart. Five to seven days later, zygotes may be faintly visible in the middle of the plate. As the zygotes develop, a dark line forms at the point of juncture of the two strains.

The related mold, *Mucor*, can be cultivated by the same methods described for *Rhizopus*. *Rhizopus* and the related molds are readily grown on medium other than cornmeal agar, including oatflake agar, prune agar, potato dextrose agar, and Sabouraud's agar (see Appendix A).

Saprolegnia and Other Water Molds

USES: (1) To demonstrate the life history of a water mold. (2) To observe the unique proliferation of sporangia.

DESCRIPTION: *Saprolegnia* often appears as a white cottony mass on dead insects in water (Fig. 10-7). Most species are saprophytic; a few are parasitic on fish and other aquatic animals. The branched hyphae are nonseptate and multinucleate. Rhizoids serve to anchor the mold to substratum and to absorb nutrient. Primary zoospores, each with two flagella, develop in long sporangia at the tip of hyphae (Fig. 10-7). Mature, pear-shaped zoospores emerge to the outside water through a terminal pore of the sporangium. After a motile period, each zoospore encysts. Each cyst develops into a secondary kidney-shaped zoospore with two lateral flagella; the secondary zoospores then go through a motile period and

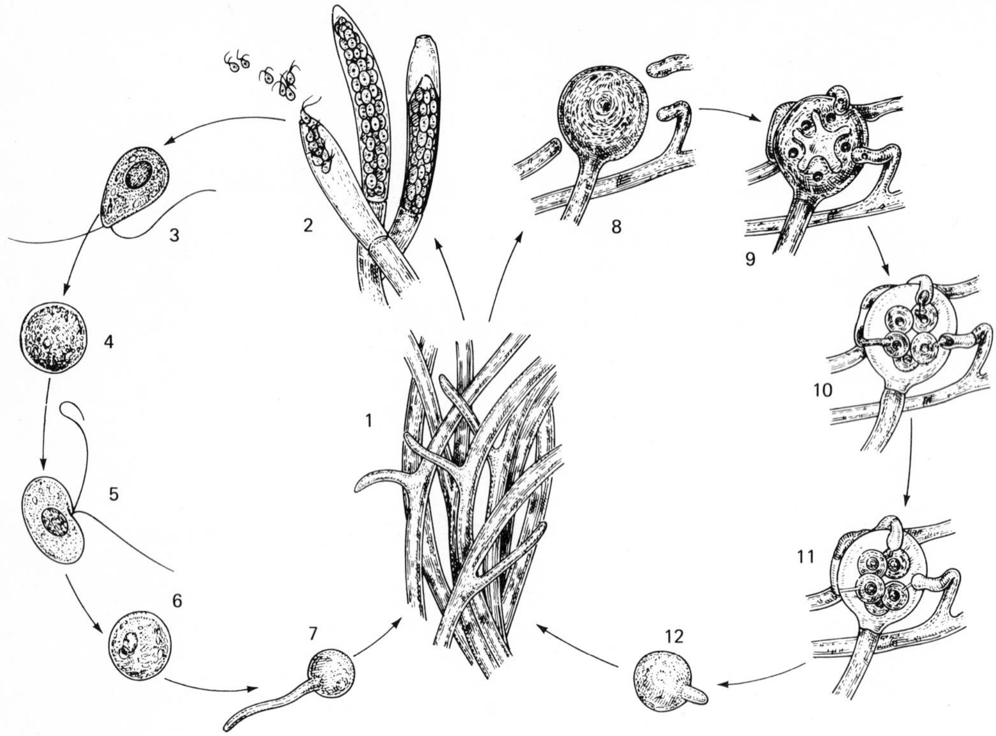

Legend

1. Mycelium
2. Zoosporangium
3. Primary zoospore
4. Encysted primary zoospore
5. Secondary zoospore
6. Encysted secondary zoospore
7. Cyst with germ tube
8 and 9. Oogonium surrounded by developing antheridia
10. A fertilization tube has grown to each oosphere inside oogonium
11. Oogonium with fertilized oospheres
12. An oospore with a germ tube

A *Courtesy of Carolina Biological Supply Company*

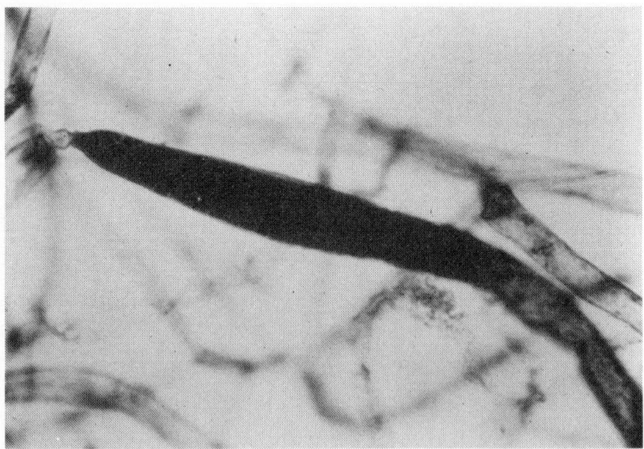

B *Courtesy of Carolina Biological Supply Company*

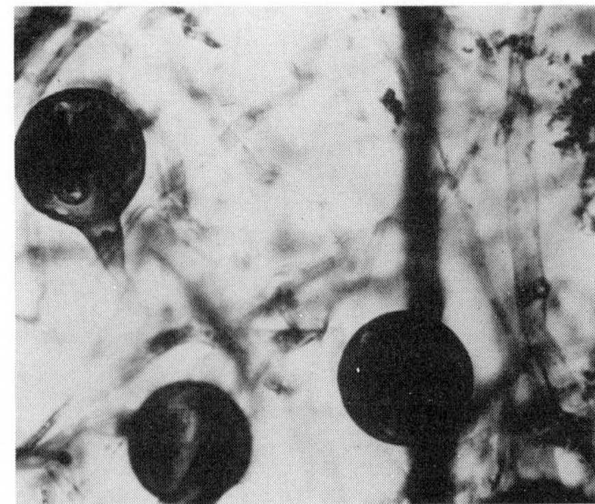

C *Courtesy of Carolina Biological Supply Company*

Figure 10-7 *Saprolegnia* sp., a water mold. **A** Life cycle. **B** Sporangium filled with zoospores; X 100. **C** Oogonia containing fertilized oospheres (egg cells).

eventually encyst again. The secondary cyst germinates, producing a germ tube that develops into mycelium.

Sporangial proliferation is interesting to observe in *Saprolegnia*. When a first sporangium is empty of zoospores, a second sporangium develops at the base of the first. The second sporangium grows through the first, now an empty case, and out beyond it. Additional sporangia are formed in the same manner, each succeeding sporangium growing through the empty case wall of the preceding sporangium. Usually all the empty sporangia remain attached to the hypha.

Sexual reproduction occurs when a thick-walled, globular oogonium becomes surrounded by one or more smaller antheridia, that develop on the same hypha with the oogonium or on a different hypha. A fertilization tube grows from each antheridium through the oogonial wall and to an oosphere (egg cell), of which generally several develop in an oogonium. An antheridial nucleus migrates to the oosphere nucleus and fuses, forming a diploid zygote. Each zygote develops a thick wall and remains dormant as an oospore for a prolonged time. Eventually the oogonial wall disintegrates, and the oospore germinates producing a germ tube that develops into mycelium (Fig. 10-7).

The related water mold, *Achlya*, is much like *Saprolegnia* in structure and life history, and is often studied in place of or with *Saprolegnia*.

CULTURE: *Saprolegnia* and other water molds are widely distributed and are among the easiest of fungi to isolate and culture. Obtain the mold by collecting water with leaves, twigs, dead insects, or bottom soil from a pond or a lake. Place collections in separate culture bowls with the pond water. For bait, add three or four dead flies to each 500 ml of pond water. Other materials may be used for bait, for example, boiled split-hemp seeds, boiled wheat grains, or corn grains. Within 48 to 72 h, a white fuzzy growth should appear on the bait or on the pond debris. It may be necessary to collect water samples from several sources, although this is not generally the case. Occasionally, *Saprolegnia*, *Achlya* and other aquatic molds may also be isolated from soil by the technique described earlier for *Allomyces*.

To stimulate the production of zoosporangia, examine a seed or fly with mold under the microscope. Remove and transfer portions of the desired mold to charcoal water. (See *Allomyces* for preparation of charcoal water.) Bait with boiled split-hemp seed or wheat or corn grain, as described above. Maintain at room temperature. Generally, the transfer to fresh medium stimulates the production of zoosporangia within several days. For class observations, transfer a half-seed of hemp or grain with mold to several drops of charcoal water on a slide. Ordinarily, the release of zoospores occurs shortly after the transfer and can be observed on the slide under 100 to 440X magnification. Sexual reproduction may occur within 4 days in charcoal-water cultures.

THE ASCOMYCETES (Sac Fungi)

Fungi included in this group are the yeasts, some of the black molds and green molds, the mildews, the cup fungi, and the morels and truffles. Techniques for handling the first three types are described here.

The special characteristic that separates Ascomycetes from other fungi is the presence of saclike structures, the asci (ascus, singular), in which a definite number of ascospores, generally eight, develops as the result of sexual reproduction. Ascospore formation involves two meiotic divisions and a mitotic division. Most Ascomycetes also reproduce asexually by conidia (spores), as well as by fragmentation. Yeasts reproduce asexually by division and budding, and more rarely by fusion of haploid cells. In contrast to the Phycomycetes, the mycelium of Ascomycetes is septate, and no flagellated spores are formed.

Aspergillus (Black or Green Mold; *A. niger* van Tiegh.)

USES: (1) To observe conidiophores with conidia. (2) To observe mycelial growth. (3) For various tests that demonstrate growth requirements.

DESCRIPTION: *Aspergillus* forms an extensive, branched, and septate mycelium on various foodstuffs and other organic materials, such as leather and cotton fabrics. The mature spores (conidia) are black, dark green, yellow, or brown, according to species. Conidia are ubiquitous in air and soil and are common contaminants in microbial cultures.

Conidia are formed on erect hyphae called conidiophores. Each conidiophore is an extension of a single hyphal cell, or foot cell, of the mycelium (Fig. 10-8). The tip of the conidiophore enlarges into a swollen, rounded head (a vesicle), which becomes covered with short, tubelike cells, called sterigmata (sterigma, sing.). In

Figure 10-8 *Aspergillus niger.* **A** Culture on agar medium. **B** Conidiophore with conidia. Only a few sterigmata are shown; the head is generally covered with sterigmata and conidia. 1, Conidia; 2, sterigma; 3, vesicle; 4, conidiophore, 5, foot cell.

A

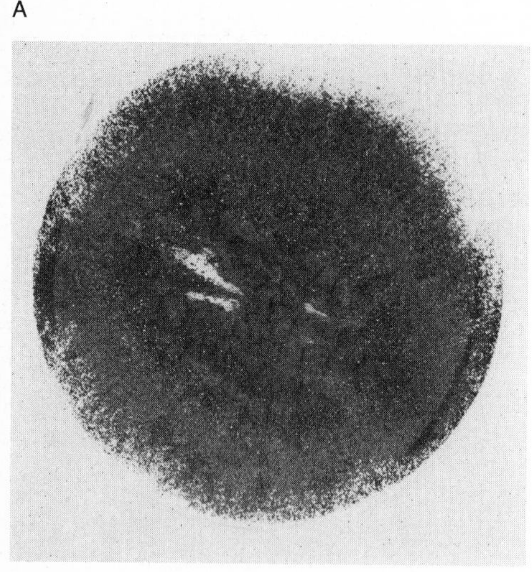

B

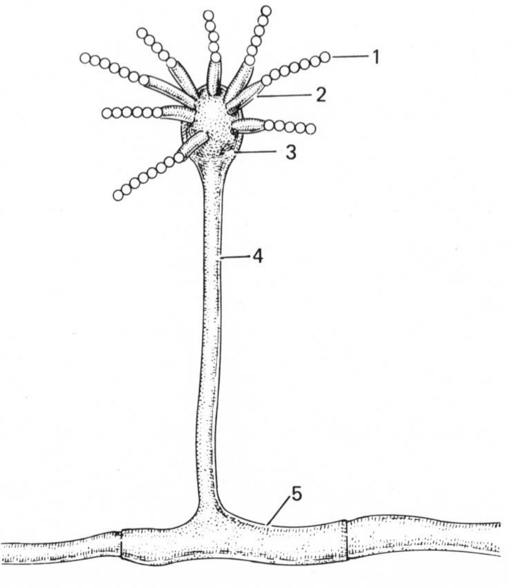

some a second layer of sterigmata is formed. The sterigmata then proliferate at their tips to form chains of small, round spores, the conidia. An abundance of conidiophores is formed, each with many conidia. The ripened conidia break loose and make a powdery covering of characteristic color over the mycelium. Sexual reproduction by ascospores, the perfect stage, has been observed in only a few species. For this reason most *Aspergillus* species are grouped with the imperfect fungi, the Deuteromycetes.

CULTURE: *Method 1.* Expose moist bread, jelly, or cooked fruit to the air for several hours. Cover the container and leave it at room temperature. Generally *Aspergillus* and other molds will appear within 4 to 6 days. *Aspergillus* can be recognized by its powdery film and characteristic color of the ripened conidia. Although *Rhizopus* and *Mucor* form similarly colored spores, they are distinguishable from *Aspergillus* because their mycelium is a loose, somewhat springy material, whereas *Aspergillus*, when mature or old, forms a more flattened film, for example, the scumlike gray-green film that grows in jars of jelly.

Method 2. Obtain pure cultures from biological suppliers or ATCC. Transfer portions to plates of cornmeal agar, or to other media indicated for fungi in Appendix A. Incubate at room temperature for 6 to 8 days, or until the colored spores have formed.

Microscopic examinations. To examine the mold, carefully transfer a small portion of the substratum to a glass slide. Flood with 70% alcohol or a weak detergent solution. Add a coverglass and gently press it against the substratum. Generally conidiophores with intact heads of conidia can be easily observed on mold that is only several days old.

Neurospora (Pink or Red Mold; *N. crassa* Shear and Dodge)

USES: (1) To examine conidia that branch directly from hyphae. (2) To observe ascospores that form in a perithecium. (3) For various genetic and physiological tests.

DESCRIPTION: *Neurospora* is a common contaminant in bakeries and is difficult to eradicate. If not handled carefully in the laboratory, *Neurospora* conidia are easily spread over the laboratory where they remain as insidious contaminants of other microbial cultures. For this reason, transfers of *Neurospora* should be done in still air and with a minimum of movement. Often microbiologists prefer to work with *Neurospora* in an isolated location, removed from other laboratories. Occasionally, a microbiologist will refuse to allow *Neurospora* in his laboratory if he is performing research with other fungi. Therefore once *Neurospora* has been used in a teaching laboratory, most often it can be easily obtained thereafter by exposing food or agar plates to dust or air in the laboratory. The genetics and physiology of *Neurospora crassa* have been studied intensively, and much literature has been published on these subjects.

The branched, septate mycelium quickly forms a dense mat. Hyphal cells are multinucleated. Conidiophores are not distinctly differentiated from the hyphae, as is true for *Aspergillus*. Chains of conidia branch directly from hyphae, as shown in Fig. 10-9A. Mature conidia are pigmented, thus accounting for the pink or red color of the mold. Two types of conidia are formed, small uninucleated

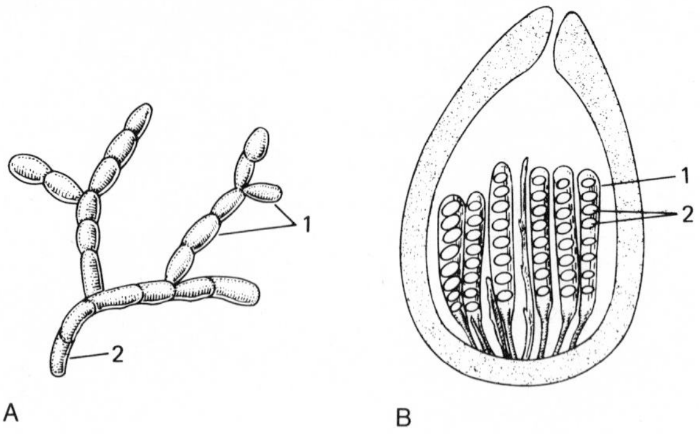

Figure 10-9 *Neurospora* sp. **A** Asexual reproduction; hypha with branching conidia; 1, conidia; 2, hypha. **B** Sexual reproduction, perithecium with asci and ascospores; 1, ascus; 2, ascospores.

microconidia and the multinucleated macroconidia. Each type may develop into mycelium.

As a result of sexual reproduction, flasklike perithecia (perithecium, singular) develop in the mycelium. Female hyphae (trichogynes) grow from the protoperithecium, the precursor of the perithecium, and fuse with macroconidia or microconidia or vegetative hyphae of a compatible strain. Thus a differentiated male cell is not required for sexual reproduction. When a perithecium has matured, it contains many elongated, saclike structures, the asci. Each ascus contains eight dark-colored ascospores, the products of two meiotic divisions and one mitotic division. The mature ascospores are liberated; each may produce new mycelium (Fig. 10-9B).

CULTURE: *Method 1.* Expose moist bread to the air for several hours. Cover the container and leave at room temperature for 4 to 6 days. If *Neurospora* conidia have fallen onto the bread, pink mold will be evident by that time.

Method 2. Obtain compatible strains from a biological supplier or from ATCC. Maintain on cornmeal agar with dextrose or other fungal media (Appendix A). Transfer mating strains to opposite sides of an agar plate. Perithecia form slowly, and ascospores are relatively small. For demonstrating the development of ascospores, the author prefers to use Sordaria (described later) because of its large perithecia and the large, colored ascospores.

Penicillium (Blue-green Mold; *P. notatum* Westling)

USES: To examine the mycelial growth pattern and the conidiophores with conidia.

DESCRIPTION: These are the blue-green molds, which often appear on citrus and other fruits, cloth, leather, and other organic materials. The airborne spores are common contaminants in the laboratory. Various species of *Penicillium* are important commercially: in the cheese industry for flavoring Roquefort and Camembert cheeses; in the production of certain organic acids; and for the production of the antibiotic, penicillin, from *P. notatum* and *P. chrysogenum*.

The mycelium grows in a characteristic circular pattern on the substratum (Fig. 10-10), with the newly formed white mycelium showing at the edges, and with

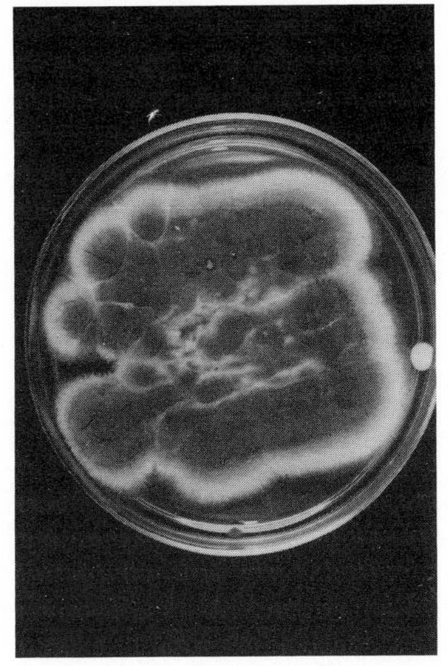

A

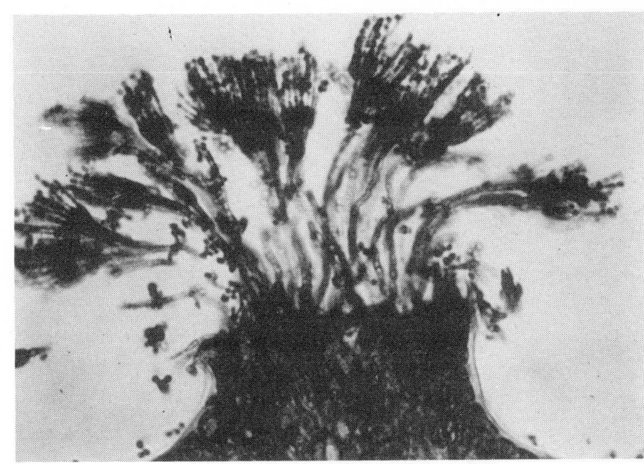

B Courtesy of Carolina Biological Supply Company

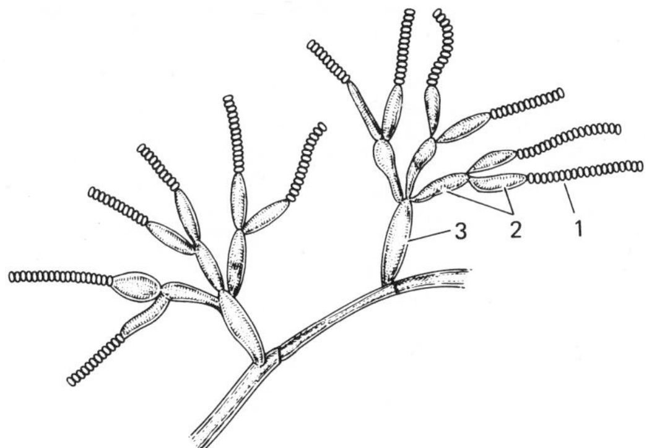

C

Figure 10-10 *Penicillium notatum.* **A** A culture on agar medium showing characteristic circular pattern with newly formed white mycelium at edges and with a cover of mature blue-green conidia over the center mycelium. **B** Photomicrograph showing mass of mycelium at bottom and conidiophores with conidia at top; X 400. **C** Diagram of conidiophores with conidia; 1, conidia; 2, sterigmata; 3, conidiophore.

mature blue-green conidia on mycelium toward the center area. Long, erect conidiophores rise from the mycelium to produce terminal branches of tubelike sterigmata, from which chains of conidia develop (Fig. 10-10). At maturity, the conidia break loose and become dispersed by air currents. Because sexual reproduction, the perfect stage, has been observed in only a few species, *Penicillium* is often grouped with the imperfect fungi, the Deuteromycetes.

CULTURE: *Method 1.* The simplest method is to obtain molded lemons or oranges at a vegetable counter. Transfer a portion of the mold to a glass slide and flood with several drops of alcohol or a weak detergent. Add a coverglass. Examine the septate mycelium and the conidiophores with tufts of stigmata and conidial chains. Young cultures are most satisfactory for study, since the numerous conidia of older cultures are easily shed and tend to cover and obscure other structures in a slide preparation.

Method 2. Obtain cultures from biological suppliers or from ATCC. Transfer to cornmeal agar with dextrose or other agar media described for fungi in Appendix A. In place of purchasing pure cultures, satisfactory cultures are usually obtained by making transfers from molded fruit to agar plates. Within 6 or 7 days, transfer a small portion of agar with mold to a glass slide. Flood with 70% alcohol or weak detergent solution. Add a coverglass and examine. The fresh cultures will give better slide preparation than the old mold taken directly from a fruit to a slide.

Saccharomyces (Brewer's Yeast; *S. cerevisiae* Hansen) and Other Yeasts

USES: (1) To study the structure and reproduction of yeast cells. (2) To use the yeast for demonstrating the rate of gas production, growth curves, population densities, and other physiological and ecological processes.

DESCRIPTION: In contrast to other fungi brewer's yeast is quite simple in structure. A scanty mycelium may develop under certain conditions. Because yeasts ferment carbohydrates to carbon dioxide and alcohol, they are employed widely in the baking and brewing industries. Asexual reproduction is by budding in *Saccharomyces* and by fission in *Schizosaccharomyces* (the fission yeast). See Fig. 10-11. Sexual reproduction occurs when two haploid yeast cells, now called gametes, fuse; or when two ascospores fuse. In either case a zygote is formed which functions directly as an ascus. The nucleus then divides meiotically and four haploid ascospores are formed (*A. cerevisiae*); or the nucleus divides by meiosis followed by mitosis, resulting in eight haploid ascospores (*Schizosaccharomyces*). Ascospores are contained within the zygote wall, now called an ascus.

CULTURE: *Method 1.* Perhaps the simplest method that almost always gives satisfactory results is to add ½ package of dried baker's yeast and 10 ml of molas-

Figure 10-11 Yeasts. **A** *Saccharomyces cerevisiae*, budding cells at right of center; X 600. **B** *Schizosaccharomyces* sp., wild yeast taken from grapes. Many cells are in fission as seen in large cells at left of center; X 400.

A

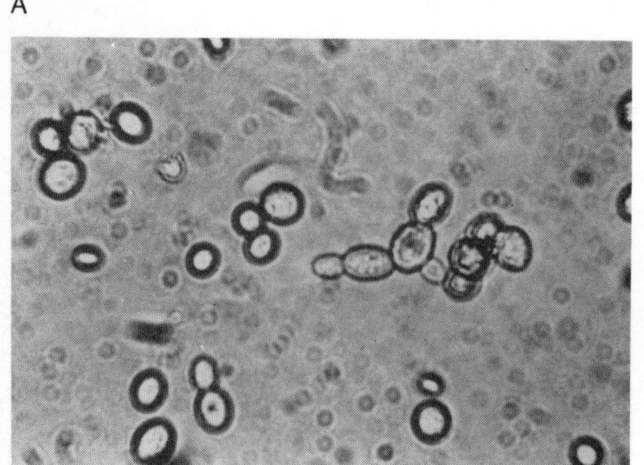

B

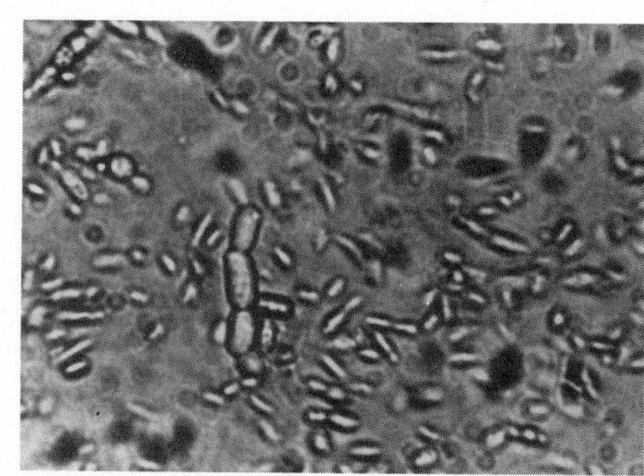

ses to 500 ml of distilled water in a 1-l flask. The flask of yeast culture is left at room temperature. Generally within 24 h the growth and activity of the yeast is evidenced by a considerable foaming. Abundant budding cells will be seen under the microscope.

Method 2. For more exact techniques, involving, for example, a population count, use the following medium:

Yeast extract	2.5 g
Monobasic potassium phosphate	2.0 g
Dextrose	40.0 g
Peptone	5.0 g
Distilled water	1,000 ml

Mix materials and dissolve them slowly over a low heat. The medium should be clear and slightly yellow in color when properly prepared. Put 10 ml of medium into each of the required number of test tubes, and plug with cotton. Sterilize in the autoclave at 15 lb pressure for 15 min. When the medium has cooled, add a measured amount of baker's yeast to each tube, for example, 0.1 g.

Method 3 (Bold, 1967b). Obtain pure cultures of *Saccharomyces cerevisiae* and *Schizosaccharomyces octosporus* from a biological supply company or from ATCC. Maintain the cultures in slant tubes of cornmeal agar with dextrose or on other fungal media described in Appendix A.

For sexual reproduction and formation of ascospores, transfer *Schizosaccharomyces octosporus* to cornmeal agar plates. After 7 or 8 days, examine a loop of yeast on a glass slide. Look for (1) dividing cells; (2) the pairing of the somatic cells (each now called a gamete); and for (3) zygotes (two fused cells) containing eight ascospores. For better observation of the ascospores, stain the slide with acetocarmine (Appendix A) to see the residual cytoplasm around the ascospores.

For budding cells, cover a slant agar culture of *Saccharomyces cerevisiae* with 5% sucrose solution to which have been added 3 or 4 drops of fresh orange juice. Incubate at 30 to 37°C for 24 to 48 h. Transfer several loops of solution to a glass slide for microscopic studies. Stain with methylene blue.

Method 4. Crush several grapes in a small culture bowl. Add a small amount of water to cover the bottom of the bowl. Cover the bowl and leave at room temperature for 4 or 5 days. Almost without exception, a film of wild yeast will form over the crushed grapes. Transfer a loop of film to a drop of water on a slide. Add a coverglass and examine under 100 to 440X magnification. The cell shape varies among species of wild yeast. Often the fission yeast of wild species are rectangular in form and make a chain of cells during their division.

Sordaria (*S. fimicola* Cesati and de Notaris)

USES: (1) *Sordaria fimicola* is an excellent mold for demonstrating segregation of ascospores from two meiotic and one mitotic division. (2) For a wide range of tests including the effects of differences in temperature, pH, and nutritive substances on growth and development of ascospores. See Bretzloff (1954), and Esser and Kuenen (1967).

DESCRIPTION: The mold occurs naturally on dung or on decaying plant parts. Strangely, *Sordaria* does not produce conidia as do the closely related *Neurospora, Aspergillus,* and *Penicillium.* Therefore, the mold presents much less danger as a laboratory contaminant. Although much study has been made of *Sordaria,* the details are not clear about how sexual reproduction occurs or how the perithecia develop. Reports have been published of fusion between two special reproductive structures (Greis, 1942), and also of fusion between two hyphae (Carr and Olive, 1958).

After 7 to 9 days' growth, perithecia can be observed without magnification as small black dots scattered throughout the mycelium (Fig. 10-12). Within each perithecium, 25 to 50 or more tubelike asci are formed, each containing eight mature ascospores in linear order (Fig. 10-12). Ascospores of the wild strain are black, and the mold is homothallic, or self-fertilizing. By using ultraviolet radiation, L. S. Olive produced mutant strains with gray or yellow ascospores (Olive, 1956). When a mutant strain is placed in culture with the wild strain, the asci will contain various combinations of colored ascospores (Fig. 10-12). Since the ascospores can be tabulated, this technique becomes extremely useful for demonstrating the results of hybridization and the segregation of genes during meiosis. As well, crossing-over occurs regularly, and thus the organism is useful for demonstrating the results of this process and for determining the frequency of crossing over. See Cassell and Mertens (1968).

CULTURE: Obtain wild and mutant strains of *Sordaria* from ATCC (Cat. No. 14517 – wild type, and Cat. No. 14518 – gray mutant), or from other biological suppliers. Prepare petri plates of cornmeal dextrose agar, with 0.1% yeast extract added (Appendix A). Inoculate the plate by placing a loop of mycelium with spores from each strain on the agar surface, about 1 cm apart. Leave at room temperature or incubate at 25°C. In 8 to 10 days, perithecia will appear over the plate as small black dots. Since ascospores are shed within 1 or 2 days, the perithecia should be examined as soon as the small black dots appear.

For examination, use a sterile loop to transfer a small portion of agar with mold to a drop of water or 70% alcohol on a glass slide. Gently press a coverglass over the agar to crush the perithecia. Examine under 60 to 440X magnification.

THE CUP FUNGI

The cup fungi are interesting because of their fleshy and often brightly colored cups (apothecia). They may be collected in moist areas on decaying wood, humus soil, dung, and rotting fruit. Only one genus, *Pyronema,* a small inconspicuous fungus, has received much study in culture; apparently other cup fungi are not as easily maintained.

The group is mentioned here because of the large, brightly colored species which are excellent and extremely attractive for terrarium displays, where the fungi may be maintained in fresh condition for 1 or 2 months or longer. Most

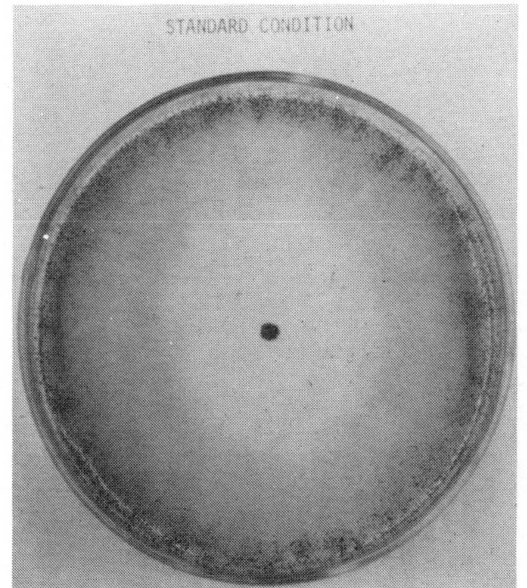

A

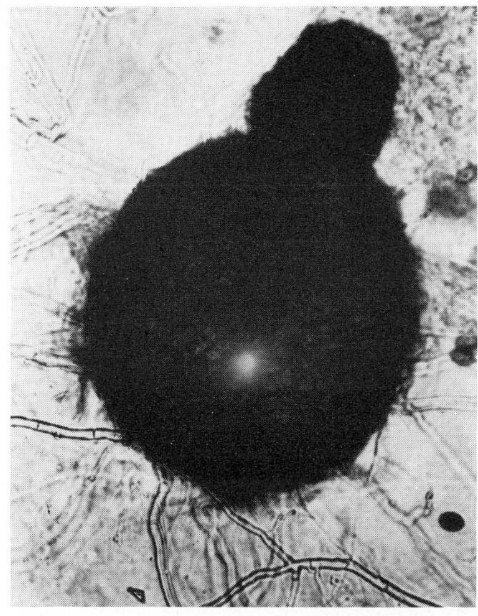

B

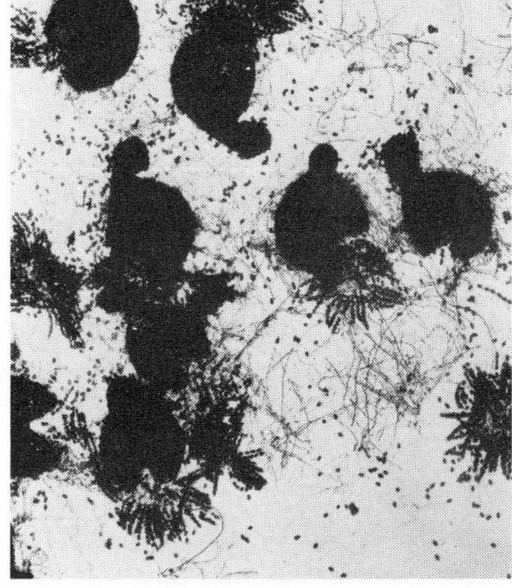

C

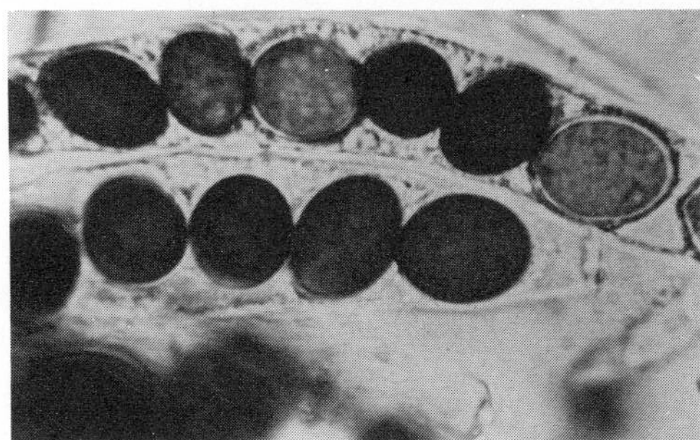

E

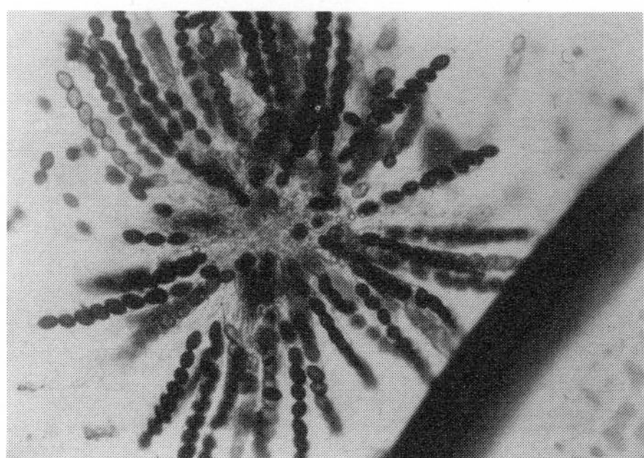

D

Figure 10-12 *Sordaria fimicola.* **A** A culture on agar medium. Note the perithecia that appear as black dots near the rim of the plate. X 0.5. **B** A single perithecium surrounded by hyphal threads. X 100. **C** A group of perithecia that have been crushed under a coverglass. **D** A cluster of asci from a single, crushed perithecium, showing eight ascospores in each ascus. The ascospores are the result of a cross between a wild strain with black ascospores and a mutant strain with gray ascospores. Most of these asci contain only black ascospores; several contain only gray ascospores; and several others can be seen with various linear arrangements of four gray and four black ascospores. **E** Two asci with ascospores. X 400. In the upper ascus the linear arrangement of ascospores is two black, two gray, two black, two gray. (*A Courtesy of Robert T. Pollock. D and E Courtesy of Donald E. Fulton.*)

A

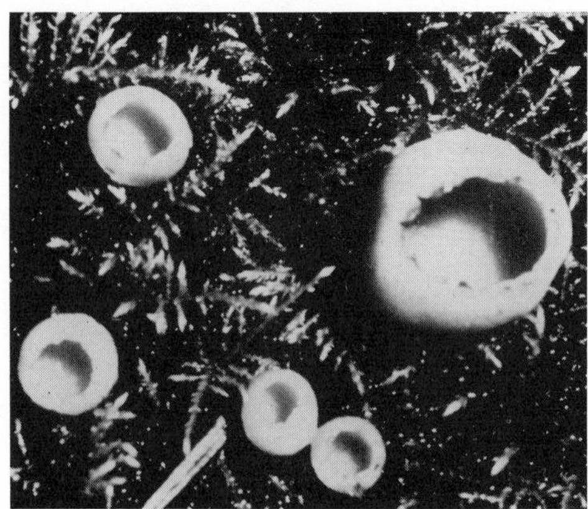

B *Courtesy of Carolina Biological Supply Company*

Figure 10-13 Cup fungi. **A** *Sarcoscypha coccinea*, "Scarlet Cup," or "Fairy Cup." The cup (apothecium) is bright red and about 25 mm across. **B** *Peziza* sp., with beige cups (apothecia). The largest cup is approximately 40 mm across. *(A Courtesy of Ward's Natural Science Establishment, Rochester, N.Y.)*

species can also be dried and retained indefinitely on display or in storage boxes. Many, if not all, retain the color of cups for at least several years when dried.

Their common names add to their attractiveness, for example, the "Scarlet Cup" or "Fairy Cup" *(Sarcoscypha coccinea)* which develops clusters of beautiful scarlet cups, each 15 to 25 mm in diameter. The species is found growing on buried or partially submerged sticks in woodland areas during middle and late spring months. *Scutellinia scutellata*, with bright red cups, 2 to 12 mm in diameter, is fairly common in woodland areas, growing on rotting wood throughout late spring and early summer. Other cup fungi with brown or black apothecia are as common as the more brilliantly colored species, although they are often overlooked because of their dull colors. *Peziza vesiculosa* forms clusters of large white- or beige-colored cups on manure and on rich soil in gardens and greenhouses. Probably all species can be dried and retained indefinitely on display or stored in boxes. See the cup fungi in Fig. 10-13.

BASIDIOMYCETES (BASIDIAL FUNGI)

Basidiomycetes are distinguished from other fungi by their basidia from which basidiospores develop (Figs. 10-14 through 10-16). Basidia develop in layers called hymenia (hymenium, singular). Most Basidiomycetes develop basidia within a fruiting body called a basidiocarp. The fruiting bodies of mushrooms, shelf or bracket fungi, the puffballs and stinkhorns, and others, are all examples of basidiocarps that rise from an extensive mycelium, often buried in substratum where it is unseen and unnoticed. Other Basidiomycetes, including most of the rusts and smuts, develop basidiospores on basidia that arise from germinating spores. Techniques are described in this section for cultivating mushrooms.

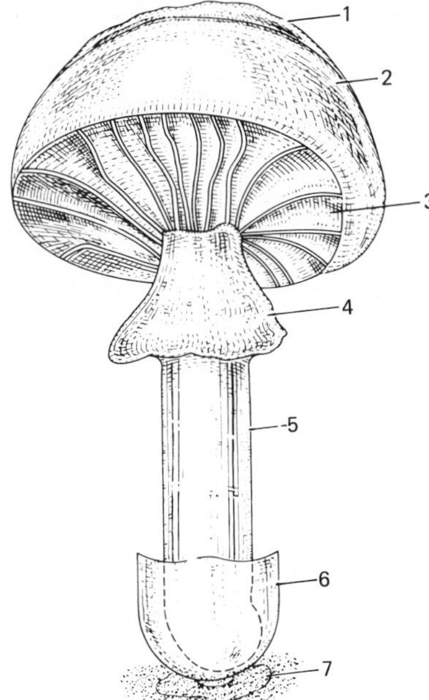

Figure 10-14 The mushroom. Generalized diagram of a basidiocarp (a fruiting body); 1, remnant of universal veil; 2, cap or pileus; 3, gills; 4, ring or annulus; 5, stalk or stipe; 6, cup or volva; 7, underground mycelium.

The Mushrooms

Growing mushrooms can become a fascinating project that is easily accomplished within a laboratory or plant room. Several biological suppliers furnish the "spawn" of *Agaricus bisporus*, a commercial, edible mushroom, along with culture chambers and complete instructions for cultivation. Occasionally, local plant

Figure 10-15 *Coprinus* sp. **A** A portion of a cross section through gills and stalk. The dark edges of the gills are the hymenia, consisting of basidia and basidiospores. **B** Enlarged portion of hymenium; 1, basidiospore; 2, sterigma; 3, basidium.

A *Courtesy of Carolina Biological Supply Company*

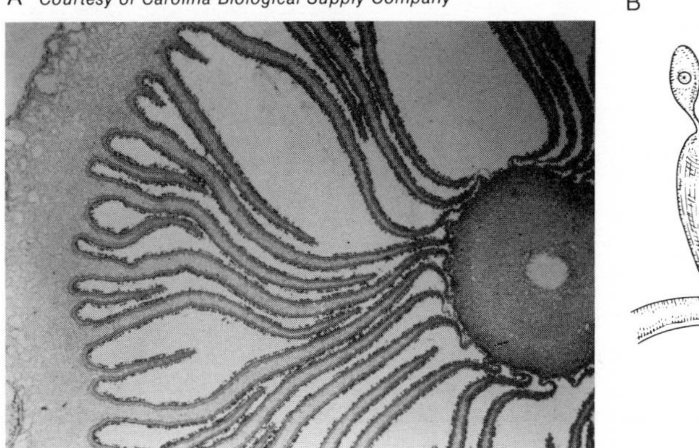

B

A

B *Courtesy of Carolina Biological Supply Company*

C *Courtesy of Carolina Biological Supply Company*

D

E

F

Figure 10-16 Some common basidial fungi. **A** *Laccaria* sp., a common woodland mushroom. The cap is about 4 cm across. **B** *Amanita muscaria,* young fruiting bodies; a poisonous mushroom. Note remnants of volva at base of stalk. **C** *Amanita muscaria,* mature fruiting body. **D** *Pleurotus ostreatus,* oyster mushroom; an edible mushroom. Caps are 8 to 12 in wide. **E** *Pleurotus ostreatus,* close-up of gills in a dried specimen. **F** *Fomes* sp., a bracket or shelf fungus on a log. *(D Courtesy of Joe R. Neel.)*

nurseries or seed stores sell trays of mushroom compost with spawn, prepared for home use. When these prepared chambers or trays are used, the production of mushrooms in the laboratory becomes a simple matter. Alternatively, a growth medium may be prepared by the following procedures from Holden and Wallner (1971).

1. Prepare the following medium:

Dextrose	15 g
Casein hydrolysate	2.5 g
Brewer's yeast	1.0 g
$CaCO_3$	3.0 g
1 *M* KH_2PO_4 (136 g/l H_2O)	10.0 ml

Add 500 ml water to 500 ml of Hoagland's solution, modified (Appendix A). Add enough of this one-half strength Hoagland's solution to the above materials to make 1 l of final medium.

2. Prepare a growth chamber by putting 600 ml of vermiculite or perlite in a 1-l beaker. Add 450 ml of the final medium prepared in step 1.

3. Cover the beaker, first, with a paper towel and, second, with a sheet of aluminum foil. Fasten the two sheets in place with a string around the top of the beaker.

4. Sterilize the covered container in an autoclave at 15 lb pressure for 20 min.

5. When the growth chamber has cooled, transfer portions of the mushroom spawn to the chamber with a transfer loop. This is done with aseptic techniques and by slightly raising one side of the two-layer cover to jab the spawn into the medium to a depth of about 2.5 cm.

6. Again fasten the cover securely to retard loss of moisture. Place the growth chamber in the dark at 18 to 20°C.

Within 3 or 4 weeks, the white mycelium will have spread throughout the medium (Fig. 10-17). To induce fruiting, add to the chamber a layer of moist, fine-textured, unsterilized loam. Replace the cover, and return the chamber to the same location as before. Generally, small white fruiting bodies begin to appear within the following 7 to 10 days. During this time, keep the loam moist by sprinkling with water every 2 days, or more often if necessary. The next stages are easily observed of white buttons that eventually enlarge and open, displaying the gills with spores (Fig. 10-17). Commercial mushrooms are usually harvested at the button stage for table use.

Microscopic study of basidia and basidiospores. Crush a small portion of a gill in a drop of water on a glass slide. Add a coverglass and observe for basidia and basidiospores at 100 to 440X magnification (see Fig. 10-15B).

Making spore prints. Not only are spore prints interesting for class work, but

Figure 10-17 A method for growing mushrooms. (From D. J. Holden and S. J. Wallner.) **A** Beaker is ready for inoculation. Aluminum foil is turned back to show the toweling. **B** The culture after 3 or 4 weeks. The mycelium has grown through the vermiculite and has been topped with a 2.5 cm layer of soil. Note the small fruiting bodies in the top layer. **C** The button stage. The white fruiting bodies are picked at this stage for table use. **D** The results of tests performed by Holden and Wallner show that growth differs with (*a*) natural soil, (*b*) sterile soil, (*c*) natural sand, (*d*) sterile sand. The fruiting bodies are collected at this stage to start a fresh supply of spawn. *(A to D Courtesy of D. J. Holden and S. J. Wallner.)*

the prints are useful for taxonomic purposes. Remove the pileus (cap) from the stipe of a mushroom. Lay the cap, gill-side down, on paper of a color that contrasts to the spore color as indicated by the general color of the gills. Cover the preparation with a glass bowl and leave for 12 to 24 h. The spores will drop onto the paper, thus showing both the color of spores and the pattern of the gills. Although spores are easily blown away, reportedly a permanent record can be obtained by gently spraying from above with several thin coats of artist's fixative used for pencil and charcoal drawings. Allow each coat to dry before adding another. For herbarium storage, place the spore print in the storage box containing the dried mushroom specimen. Since the spray may alter the spore color, it is desirable to also store a vial of untreated spores.

Germination of spores. Used for (1) observing spore germination in class and (2) preparing spawn of wild mushrooms that are to be cultured in the growth chamber described above.

Place the cap of a mature mushroom in a sterile container, such as a covered glass bowl, for 24 to 48 h. Using aseptic techniques, transfer spores with an artist's brush, previously sterilized in alcohol and allowed to dry, to petri plates containing a very thin layer of cornmeal agar, potato dextrose agar, or other fungal agar, all described in Appendix A. If spore germination is to be observed in class, prepare the plate with spores several hours before class time. At the time of observation, invert the plate and observe spores directly through the plate and agar, under a stereoscopic microscope at 30 to 60X magnification.

If spawn is to be developed from wild mushrooms (often an interesting project for individual students), pour a plate about one-third full of agar. After spores begin to germinate, aseptically transfer single germinating spores with mycelium to a growth chamber, according to the techniques described earlier for the preparation of a growth chamber.

Poisonous mushrooms. This topic is too extensive for a simple treatment here. Unless a wild mushroom is identified by a specialist, the only safe procedure is to consider all wild mushrooms as inedible. Caution is required even with "safe" mushrooms, as persons may react differently to the same species. A mushroom that is safe for one person may prove disastrous for another. Much literature is available, for example, the guide book of Smith (1963).

DEUTEROMYCETES (IMPERFECT FUNGI)

For many fungi the sexual stage has not been identified, and it may be lacking. Therefore these fungi are grouped together in the class Deuteromycetes, or the Fungi Imperfecti. Nevertheless, because of certain similarities, some of these fungi have been placed tentatively with other groups. For example, *Penicillium* and *Aspergillus*, because of their conidial reproductive structures, are often grouped with the Ascomycetes that normally develop both the asexual conidia and the sexual ascospores. As more information is gained, it seems probable that members of the large heterogeneous group of Deuteromycetes will be transferred to their appropriate, or at least suspected, categories of fungi. Figure 10-18 shows several kinds of imperfect fungi that often appear in hay infusions or in contaminated mold cultures. *Helminthosporium* and *Alternaria* are carried in house dust and often occur in old cultures of other microorganisms that have been exposed to the air. *Rhizoctonia solani* that causes damping-off on seedlings (mentioned in Chap. 6) is also an imperfect fungus that is found in most soils.

In this section, we will devote attention to one group only, the nematode-trapping or predacious fungi that display unusually interesting traits and that often can be obtained from soil cultures.

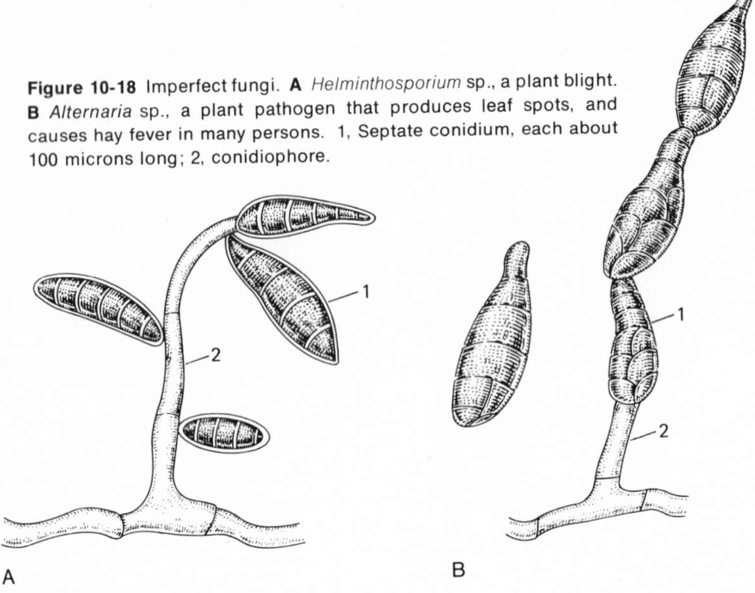

Figure 10-18 Imperfect fungi. **A** *Helminthosporium* sp., a plant blight. **B** *Alternaria* sp., a plant pathogen that produces leaf spots, and causes hay fever in many persons. 1, Septate conidium, each about 100 microns long; 2, conidiophore.

Predacious Fungi

These fungi are unique in that they develop certain structures used for trapping microscopic organisms, such as amoebas, rotifers, and nematodes. In particular, the nematode-trappers are interesting. They may be cultured from soil, although generally some patience and careful observations are required to find them. Pramer (1964), and Witters (1969) have described procedures for obtaining and observing this unusual phenomenon of nature (also see Pramer and Kuyawa, 1963). The following procedures represent a combination and modification of techniques described by the first two authors.

Witters states that his best source for these fungi is in activated sludge from a sewage treatment plant. He suggests that the black muck soil from a marsh or bog may also be productive. Pramer suggests using small portions of partially decomposed organic matter, from: (1) moist, partially decayed wood that is in contact with the soil, (2) freshly collected roots, (3) a mat of moss, or (4) decomposed leaves or grass.

1. Prepare petri plates with cornmeal agar or Czapek Dox agar (Appendix A). Transfer 5 ml of liquid sludge or a 2-cm^2 portion of solid materials to the plate.

2. Incubate at room temperature.

3. At weekly intervals for several weeks, observe under the microscope portions of the mold growth or film that form on the agar. As a rule, a sequence of organisms will develop, comparable to the sequence in a hay infusion, of bacteria, mold, protozoans, rotifers, and eventually nematodes. If no nematodes are present

within 2 to 3 weeks, they may be added from cultures obtained by methods described in Chap. 2 under *Soil Nematodes*.

4. When nematodes are established in the plate, carefully follow the development of the molds with daily observations. According to Pramer, the best way to find the trapping structures of the mold is to look for dead nematodes in material that is transferred from the plate to drops of water on a slide. With careful observations at 100 to 440X magnification it may be possible to locate the mold hyphae on the body of a dead nematode.

Various types of structures used to trap nematodes are shown in Fig. 10-19. *Arthrobotrys conoides* forms mucilaginous hyphal loops at the end of short stalks, in which a worm may become entrapped and preyed upon when hyphae grow into the worm tissue. *Dactylella dreschsleri* forms adhesive pads on short stalks, which adhere to a worm and subsequently invade the worm tissue. *Arthrobotrys dactyloides* forms a closed ring in a hypha, which constricts and holds a worm that enters the ring. Several of the species may be obtained from ATCC for demonstrating this unusual predator-prey relationship.

LICHENS

Lichens consist of a fungus and an alga that grow together to form what may be called a new or different organism. In a sense, lichens are taxonomic "orphans," since they are unassigned to any plant division, although most often they are studied with the fungi. See Hale (1961) for collecting and identifying lichens.

In their symbiotic organization, the alga (or sometimes several species of algae) and the fungus of a lichen form a distinctive thallus, unlike the structure of either component. Approximately 15,000 "species" of lichens have been identified. In reality, each alga and fungus that makes up a lichen maintains a genetic independence. Thus, the traditional classification of lichens into taxa represents an artificial system.

The algal component of a lichen is known as the phycobiont, and the fungal portion as the mycobiont. At least 30 different genera of algae have been found in lichens, of which the majority are the green alga, *Trebouxia* (Fig. 10-20). The fungus of most genera of lichens is a member of the Ascomycetes, although some of the Deuteromycetes and the Basidiomycetes have also been identified as mycobionts of certain lichens. The fungal component is generally a typical hyphal growth that determines the shape and appearance of the thallus.

Three general growth forms are recognized (Fig. 10-21): (1) the *crustose*, the simplest form, which makes a closely attached crust on the substratum; (2) the *foliose*, or leaflike thallus, not as closely attached to substratum as crustose forms, and lobed or circular in pattern; (3) the *fruticose*, an erect or hanging group of radially symmetrical and cylindrical or slightly flattened forms. Intermediate forms are also found between crustose and foliose, and between foliose and fruticose.

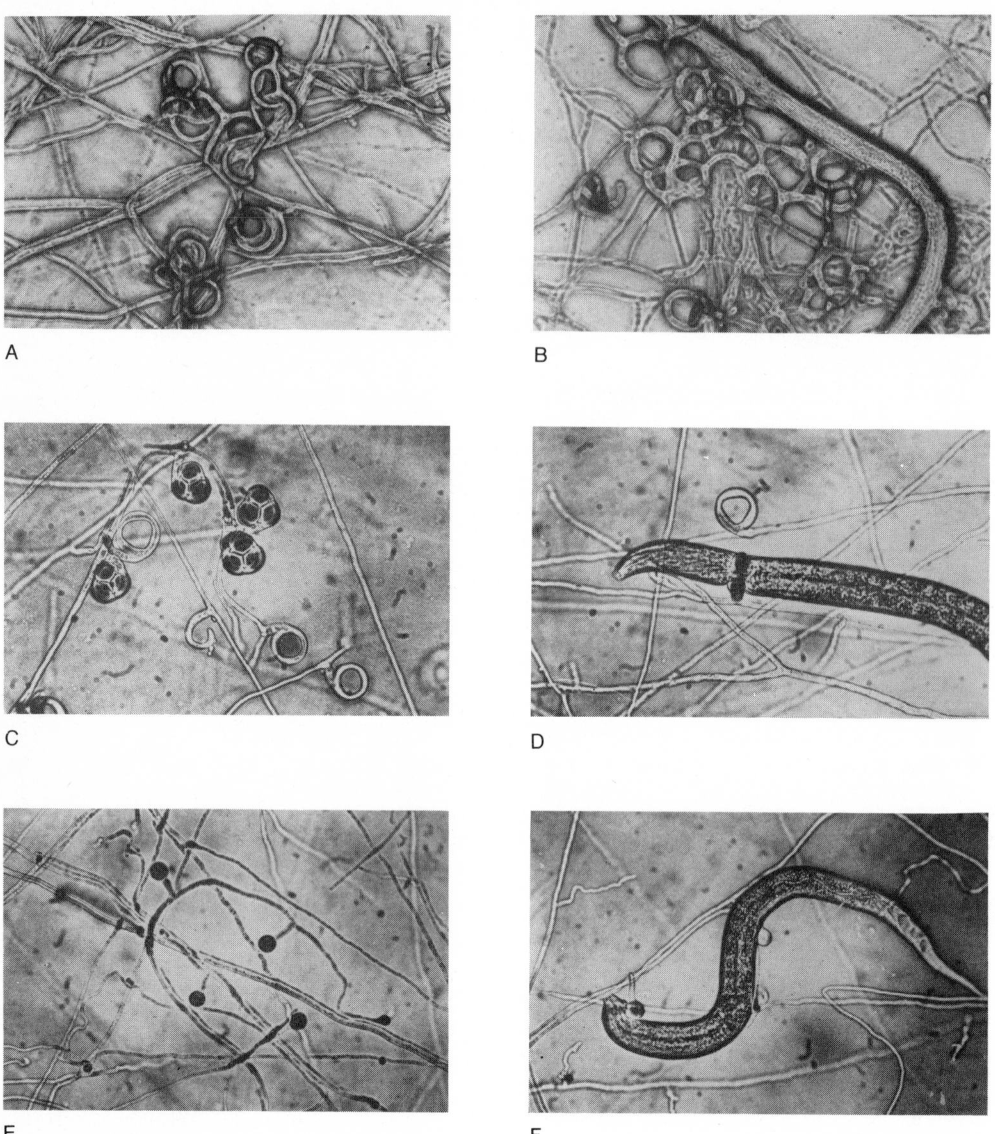

Figure 10-19 Predaceous fungi: nematode trappers. **A** *Arthrobotrys conoides,* X 200. Loops of hyphae, at center, are coated with mucilage. **B** Nematode trapped in loops of *A. conoides.* **C** *Arthrobotrys dactyloides,* equipped with constricting hyphal rings. **D** *A. dactyloides* with a nematode trapped in a constricted ring. **E** *Dactylella dreschsleri* with adhesive pads on short stalks. **F** *D. dreschsleri* with a nematode trapped on adhesive pads. *(A–F Courtesy of David Pramer.)*

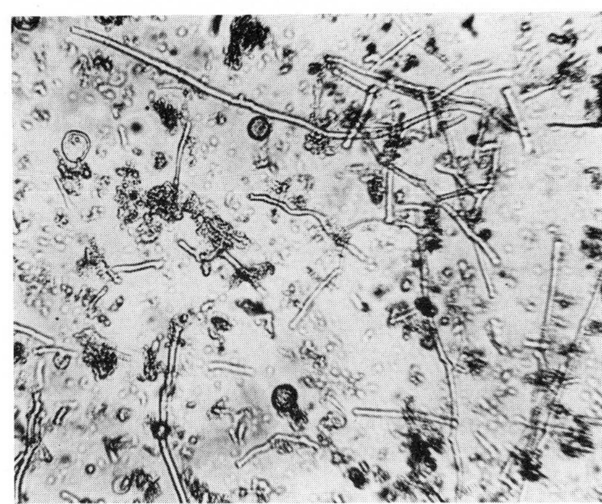

Figure 10-20 Algal and fungal components of a crushed lichen thallus. X 200. The spherical, single cells are algal, possibly *Trebouxia*. The threads are the crushed fungal mycelium.

Much study has been done in cultivating the isolated alga and fungus of a lichen. The techniques are relatively simple and are generally successful. The reverse procedure, of placing an alga and a fungus together to "synthesize" a lichen, is a much more difficult process. An excellent account of isolation techniques and synthesis experiments is given by Ahmadjian (1967).

Isolation of Algae from Lichen

1. Collect fresh lichen, or use those that have been stored for no longer than several months. Wash a thallus under running tap water for about 15 min to remove foreign, external materials, including foreign algae.

2. Scrape away the top layer (the cortex region), exposing the underlying layer of algae.

3. Remove a strip of the algal layer, and wash the portion through three changes of sterile water.

4. Using aseptic techniques, transfer the algal portion to a 250-ml flask containing Bold's basal medium (Appendix A). Place the flask under a 100-W incandescent lamp or in a south or east window at room temperature. Although portions of the fungus will be intermixed with the algal component, the alga should outgrow the fungus within 2 or 3 weeks.

Isolation of a Fungus from a Lichen

1. Collect fresh lichens that contain fruiting bodies (see Fig. 10-21). Often specimens stored for several weeks or even several months can be used successfully.

2. Soak a thallus containing fruiting bodies in cold water for 15 to 20 min.

3. Cut loose a portion of the thallus with one or more fruiting bodies.

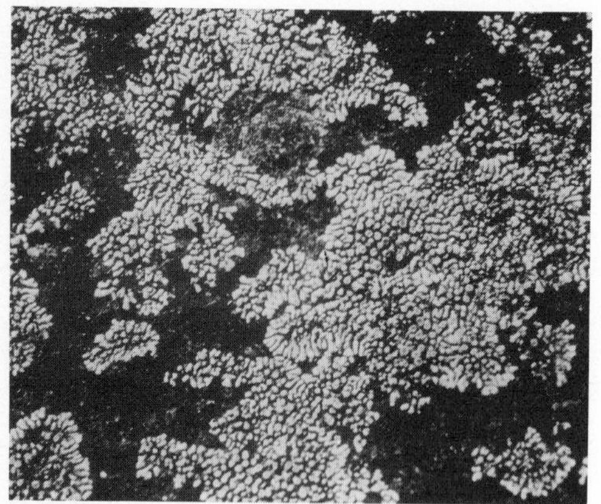

A *Courtesy of Carolina Biological Supply Company*

B

C *Courtesy of Carolina Biological Supply Company*

D

Figure 10-21 Common lichens and typical growth forms. **A** *Caloplaca murorum,* a crustose form with flat, orange thalli on rock. **B** *Parmelia* sp., a foliose form on tree trunk. The fruiting bodies, apothecia, have a chocolate-color lining and a beige outer coat. Largest apothecia are 10 to 15 mm across. Thalli are gray. **C** *Cladonia bacillaris,* a fruticose form on a tree trunk. Many tube-shaped podetia can be seen, topped with bright-red fruiting bodies. **D** *Usnea* sp. Both species are fruticose in growth habit. Upper species consists of yellow-green pendulous thalli; lower species is a bright-green, shrubby type. Both were removed from fir trees on North Pacific coast. At time of photograph both still showed a natural color after 5 years of display in a laboratory.

4. Using aseptic techniques as much as possible, quickly place a small mound of petroleum jelly or Vaseline in the center of the inside surface of a petri plate cover. Add the portion of thallus to the jelly in such a way that the thallus adheres to the jelly and remains in place.

5. Invert the plate cover over a lower petri plate that contains a thin layer of 2% plain agar, or a layer of soil-extract agar medium (Appendix A).

As the fruiting bodies dry, the spores are released onto the agar medium. Because of the low concentration of nutrients in the medium, contaminating fungi do not grow as quickly as the more abundant lichen spores. The amount of spores released and the relative spore germination, as well as the time required for germination, vary among lichen genera. *Cladonia*, a common lichen (Fig. 10-21), may discharge spores within several minutes, and spores may generate within several hours. With other lichens the time of spore germination may range from 24 h to several weeks. Ahmadjian states that from a random collection of lichens only about one-half of the species produce spores that germinate. The reasons for this are not known. The situation points to more research concerning environmental conditions of nutrients, light, and temperature, and could provide excellent studies for students.

To observe spore emission and spore germination, examine an inverted plate under 100X magnification.

Collecting and Preserving Lichens

Lichens can be found in practically all areas, both urban and rural, growing on trees, soil, and rocks. Thus the organism furnishes a ready source for fresh or dried laboratory materials. Use a knife, if necessary, to remove the growth along with a portion of the substratum. A geologist's pick can be used to remove parts of rock with attached lichen.

Dry preservation is the simplest and perhaps the most satisfactory method, since colors are retained quite well. Place the dried lichen on display, or store specimens in labeled and closed boxes containing paradichlorobenzene crystals to prevent insect infestation. Before studying dried lichens, soak them in cold water for several hours. The specimens will regain a surprising, living appearance. Specimens can be repeatedly soaked for class work and dried again for storage.

Microscopic Examination of Lichens

Crush or grind a small portion of a thallus in water. Transfer a drop to a glass slide, where the algal and fungal components can be observed easily at 100X magnification (Fig. 10-20).

REFERENCES

Ahmadjian, V.: 1967, *The Lichen Symbiosis*, Blaisdell Publishing Company, Waltham, Mass., 152 pp.

Alexopoulos, C.J.: 1962, *Introductory Mycology*, 2d ed., John Wiley & Sons, Inc., New York, 613 pp.

Alexopoulos, C.J., and E.S. Beneke: 1962, *Laboratory Manual for Introductory Mycology*, Burgess Publishing Company, Minneapolis, 199 pp.

Alexopoulos, C.J.: 1963, "The Myxomycetes. II," *Bot. Rev.*, **29** (1): 1–78.

Alexopoulos, C.J., and J. Koevenig: 1964, *Slime Molds and Research*, D. C. Heath and Company, Boston, 36 pp.

Bold, H.C.: 1967a, *Morphology of Plants*, 2d ed., Harper & Row, Publishers, Incorporated, New York, 541 pp. (3d ed. is in press.)

Bold, H.C.: 1967b, *A Laboratory Manual for Plant Morphology*, Harper & Row, Publishers, Incorporated, New York, 123 pp. (3d ed. is in press.)

Bretzloff, C.W., Jr.: 1954, "The Growth and Fruiting of *Sordaria fimicola*," *Am. J. Botan.*, **41:** 58–67.

Carmichael, J.W.: 1956, "Frozen Storage for Stock Cultures of Fungi," *Mycologia*, **48:** 378–381.

Carr, A.J.H., and L.S. Olive: 1958, "Genetics of *Sordaria fimicola*. II. Cytology," *Am. J. Botan.*, **45:** 142–150.

Cassell, P., and T.R. Mertens: 1968, "A Laboratory Exercise on the Genetics of Ascospore Color in *Sordaria fimicola*," *Am. Biol. Teacher*, **30** (5): 367–372.

Collins, O.R.: 1966, "Plasmodial Compatibility in Heterothallic and Homothallic Isolates of *Didymium iridis*," *Mycologia*, **58** (3): 362–372.

Collins, O.R.: 1969, "Experiments on the Genetics of a Slime Mold, *Didymium iridis*," *Am. Biol. Teacher*, **31** (1): 33–36.

Esser, K., and R. Kuenen: 1967, *Genetics of Fungi*, trans. Erich Steiner, Springer Verlag New York, Inc., New York, 500 pp.

Gauger, W.L.: 1961, "The Germination of Zygospores of *Rhizopus stolonifer*," *Am. J. Botan.*, **48:** 427–429.

Greis, H.: 1942, "Mutations – und Isolationsversuche zur Beeinflussung des Geschlechtes von *Sordaria fimicola* (Rob.)," *Z. Botan.*, **37:** 1–116.

Hale, Mason E., Jr.: 1961, *Lichen Handbook, A Guide to the Lichens of Eastern North America*, Smithsonian Institution Press, Washington, D.C., 178 pp.

Holden, D.J., and S.J. Wallner: 1971, "Experimental Mushroom-Growing," *Am. Biol. Teacher*, **33** (2): 91–93.

Kerr, N.S.: 1965, "A Simple Method of Lyophilyzation for the Long Term Storage of Slime Molds and Small Soil Amoebae," *BioScience*, **15** (7): 469.

Lilly, V.G., and H.L. Barnett: 1947, "The Influence of pH and Certain Growth Factors on Mycelial Growth and Perithecial Formation by *Sordaria fimicola*," *Am. J. Botan.*, **34** (3): 131–138.

Olive, L.S.: 1956, "Genetics of *Sordaria fimicola*. I. Ascospore Color Mutants," *Am. J. Botan.*, **43:** 97–106.

Pramer, D., and S. Kuyawa: 1963, "The Nematode Trapping Fungi," *Bacteriol. Rev.*, **27:** 282.

Pramer, D.: 1964, *Life in the Soil*, BSCS Laboratory Block, D.C. Heath and Company, Boston. Student Manual, 62 pp.; Teacher's Supplement, 38 pp.

Raper, K.B., and D.F. Alexander: 1945, "Preservation of Molds by the Lyophil Process," *Mycologia*, **37:** 499–525.

Smith, A.H.: 1963, *The Mushroom Hunter's Field Guide*, 2d ed., The University of Michigan Press, Ann Arbor, 264 pp. and 187 color plates.

Vogel, L.H.: 1969, "Slime Mold for BSCS Lab Exercises," *Amer. Biol. Teacher*, **31** (1): 29–32.

Witters, W.L.: 1969, "An Investigation of Predacious Fungi," *Am. Biol. Teacher*, **31** (1): 37–39.

Other Literature

Bonner, J.T.: 1963, "How Slime Molds Communicate," *Sci. Am.*, **209** (2): 84–93.

Culberson, C.F.: 1969, *Chemical and Botanical Guide to Lichen Products*, The University of North Carolina Press, Chapel Hill, 628 pp.

Gorden, R.W.: 1972, *Field and Laboratory Microbial Ecology*, Wm. C. Brown Company Publishers, Dubuque, Iowa, 161 pp.

Gray, W.D., and C.J. Alexopoulos: 1968, *Biology of the Myxomycetes,* The Ronald Press Company, New York, 288 pp.

Hale, M.E., Jr., and W.L. Culberson: 1966, "A Third Checklist of the Lichens of the Continental U.S. and Canada," *Bryologist,* **69:** 141–182.

Lambert, E.B.: 1967, *Mushroom Growing in the United States,* USDA, Farmer's Bulletin 1875, U.S. Govt. Printing Office, Washington, D.C., 12 pp. 10 cents.

Nardi, R.: 1966, *Foto-Atlas Der Pilze,* Dokumente Verlag, Offenburg/Baden, 303 pp. Contains 500 photographs of mushrooms, puffballs, and cup fungi. Distributed in U.S. by LEW's, 2510 Van Ness Ave., San Francisco.

Shushan, Sam, and R.A. Anderson: 1969, "Catalog of the Lichens of Colorado," *Bryologist,* **72** (4): 451–483.

Sprague, W.R., and F.A. Einhellig: 1971, "Technique for Obtaining Slime Molds," *Am. Biol. Teacher,* **33** (6): 359–360.

THE ALGAE

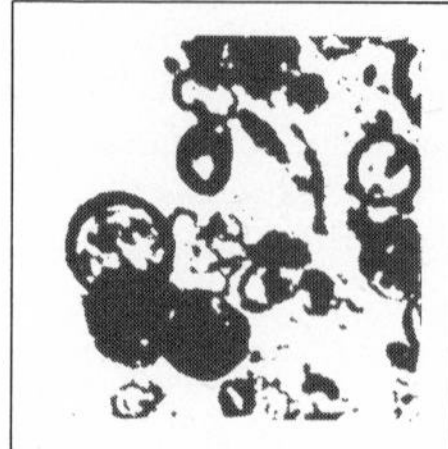

With the increasing concern for water pollution, the term "algae" has become a household word, often spoken with a derogatory overtone which implies, in a way, that algae themselves are to blame for the fouled water. Although the present discussion does not exclude ecological problems, the chapter is more concerned with the matters of what are algae, how can they be collected and identified, and how can they be cultured in the laboratory. Perhaps with this knowledge, the basic ecological problems can be better analyzed and understood.

Despite their role in water pollution, the microscopic beauty of these organisms and their intriguing, complex life patterns present to the student new reasons for an admiration, even reverence, toward the total complementarity of organisms to their environment and to each other. Perhaps, then, the discussions of algae in this chapter, and of microorganisms in other chapters, will help the teacher and the student to see more clearly the beauty of a well-ordered life system, at least within the microscopic world. Finally, however, the microscopic world affects all of us who dwell within the macroscopic world.

FRESHWATER ALGAE

Selected genera of freshwater blue-green, green, golden, and yellow-green algae are described later in the chapter. Although the flagellated organisms of the divisions Chlorophycophyta and Euglenophycophyta display both animallike and plantlike characteristics, here they are arbitrarily placed with the algae. The nomenclature of this chapter is based primarily on Bold (1967a). Much of the description and culture of the various species are based on Bold's publications, and on those of Prescott (1964), and Starr (1964).

Sources

Most genera described in the chapter are widely distributed and easily collected. When large numbers of organisms are needed in pure unialgal culture, the best method may be to purchase a starting culture from a biological supply company. A large group of unialgal cultures, useful for teaching, was acquired in 1969 by the Carolina Biological Supply Company from the collection at Indiana University. Those cultures are available from the Carolina Biological laboratories at Burlington, North Carolina.

Since algae grow in almost any moist situation, they may be found in roadside ditches, bird baths, on tree bark, on soil and rocks, on the backs of water turtles, as well as in other places. Floating algae may be collected with a plankton net as described for planktonic animals in Chap. 1. A dip net may be used at the water bank, or a collecting vessel may be submerged barely below the water near the algae, thus allowing the algae and the water to flow into the container. Small rocks and other substratum with attached algae can be transferred from shallow water to containers of suitable size, to which is added enough water to cover the material. Often a blade is required to scrape algae from large submerged objects. The bottom sediment may be dredged with a bottom sampler or with a container attached to one end of a long pole. Collecting equipment and vessels should be cleaned thoroughly before use. See Chap. 1 for greater details on techniques and equipment.

Collecting algae on submerged glass slides. A simple method for collecting microalgae, with interesting results, is to suspend glass slides in a slowly moving body of water or in other habitats, for example, bogs, swamps, and seawater. The slides are fitted tightly into notches of a frame so that water passes over them. The frame is constructed of wood or rust-resistant metal, or cut from a block of Styrofoam. The glass slides are fitted into grooves on the inner side of the frame. Bonang (1971) describes a slide-rack collector made by cutting a window in the bottom and top of a hinged, plastic slide box, leaving an edge around the windows to hold slides in the box (Fig. 11-1). The frame must be weighted to hold it at the desired depth. As a rule, the first algae accumulate on the slide within a few days and increase for several weeks. Protozoans and small invertebrates will also accumulate on the slide.

Upon removing the frame from the water, transfer it to a flat vessel, with the

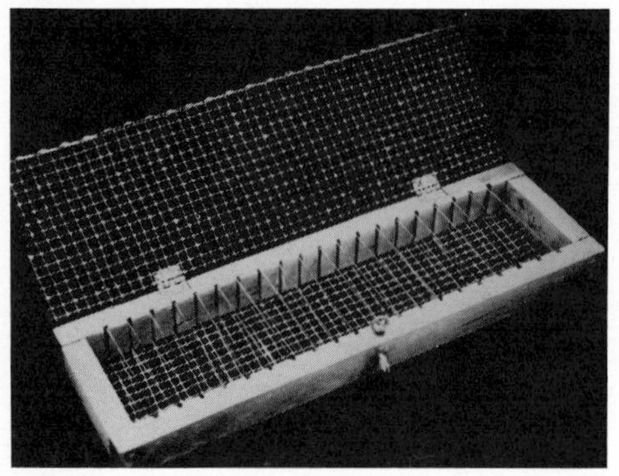

A

B

Figure 11-1 Slide rack for collecting aquatic organisms. **A** Notched wood frame that holds glass slides. The top eyescrew on the front board holds the wire-mesh cover in place when the screw is run through a mesh and turned a half circle. The rack is suspended in the water by a wire looped through the eyescrew on the front edge of the wood strip. **B** Photomicrograph of a portion of a glass slide from the rack that was suspended in a pond for ten days. At center is a group of moving Vorticella; diatoms are scattered through the sediment on the slide.

slides still in place. Gently add pond water to cover the frame and place the set-up in a north window or under moderate light. Slow aeration will help to prolong the culture. At the time of observations, remove a slide, wipe one side clean, and observe the other surface. Alternatively, immerse the slide in source water in a petri dish, and place the latter on the stage of the microscope. When the observations have been completed, the slide may be returned to the frame. In this way, a small ecosystem can be maintained in a growing condition.

General Techniques for Culturing Freshwater Algae

Upon receiving a culture in the laboratory, immediately remove or loosen the cover of the container, except for those of pure cultures that are stoppered so as to admit oxygen. Place the cultures in a north window or under moderate light of 200 to 250 ft-c. After about 10 days' growth, transfer the culture to a lower intensity of 100 to 150 ft-c. Often ordinary room light, away from a window, is sufficient. A temperature of 18 to 20°C is favorable for most algae, although freshwater algae can be maintained in a wider temperature range. Frequency of transfers depends on species, culture media, and, especially, on the rate of desiccation of media. Agar cultures are transferred about every 2 months, although certain species will require more frequent transfers as indicated by their growth rate. Most species in stoppered tubes of liquid media require transfers about every 6 months if kept under low light and a cool temperature.

Collections of mixed algae are often maintained by transferring the collection to flat bowls that are placed in a north window at a cool temperature. The algae must not be crowded in the vessel, and enough collecting water should be added to cover and float the algae. Pond water, dechlorinated tap water, or distilled water, in that order of preference, may be added to replace the water lost in evaporation. Often these collections will exhibit a series of changes in both algal and protozoan forms when retained for several weeks. Cover the container with a glass sheet or plastic film, or transfer small portions of the collection to stacked culture bowls.

To obtain a pure culture from a collection of mixed algae, follow the techniques described for protozoans in Chap. 12. Special media for unialgal cultures are indicated later in the chapter with the descriptions of genera.

SOIL ALGAE

Soil algae are found mostly in the top 6 in of soil, although diatoms and some of the cyanophytes have been found at depths of several feet. Top soil, removed from various habitats, e.g., garden soil, forest soil, or a dry stream bed, generally offers a ready source of algae. Dried crusts of algae are particularly suitable. Store the dry materials in loosely covered containers in low light, or in the dark, at a cool temperature. For later observations, cover the soil with water or Bold's basal medium, or other inorganic medium (Appendix A) in the proportion of 1 g of soil for each 100 ml of medium. Place the culture in a north window at a cool temperature. Usually several species of algae appear within 1 or 2 weeks. Typical soil algae are illustrated in Fig. 11-2.

MARINE ALGAE

Collecting in the intertidal zone is described in Chap. 1. The same precautions must be observed with respect to clothing and equipment, the dangers of becoming stranded by the returning tide, and, especially, the dangers of collecting alone at night.

Remove sessile algae by carefully cutting under the holdfasts to separate algae from rocks. A geologist's pick or a chisel and hammer are useful in removing encrusted algae with small portions of the rock. Place delicate forms in separate plastic bags or vials of seawater. Wrap tougher algae in damp newspapers for transfer to the laboratory. Take only the quantity that is needed, disturbing the site as little as possible. Plankton collecting is done with a plankton net in the manner described for animals in Chap. 1.

Beach sand often contains an abundance of unique forms. Dig through the top, dry layer of sand in an untrammeled area, back away from the water. If algae are present, they will appear in dark-green layers within the lower, moist sand. Little exploration has been done on these forms, which offer an easy method of collecting and a broad area of discovery for students. Place the moist, dark-green sand in a clean vessel and cover with seawater. Gently stir the sand to dislodge and float the algae. Allow the sand to settle, and carefully decant the water containing the floating algae. Several washings in this manner may be required. Concentrate the algae by centrifuging the decanted water in 10 to 15 ml aliquots, at 3,000 rpm for several minutes. Pour off and discard the top water from the centrifuge tubes, and

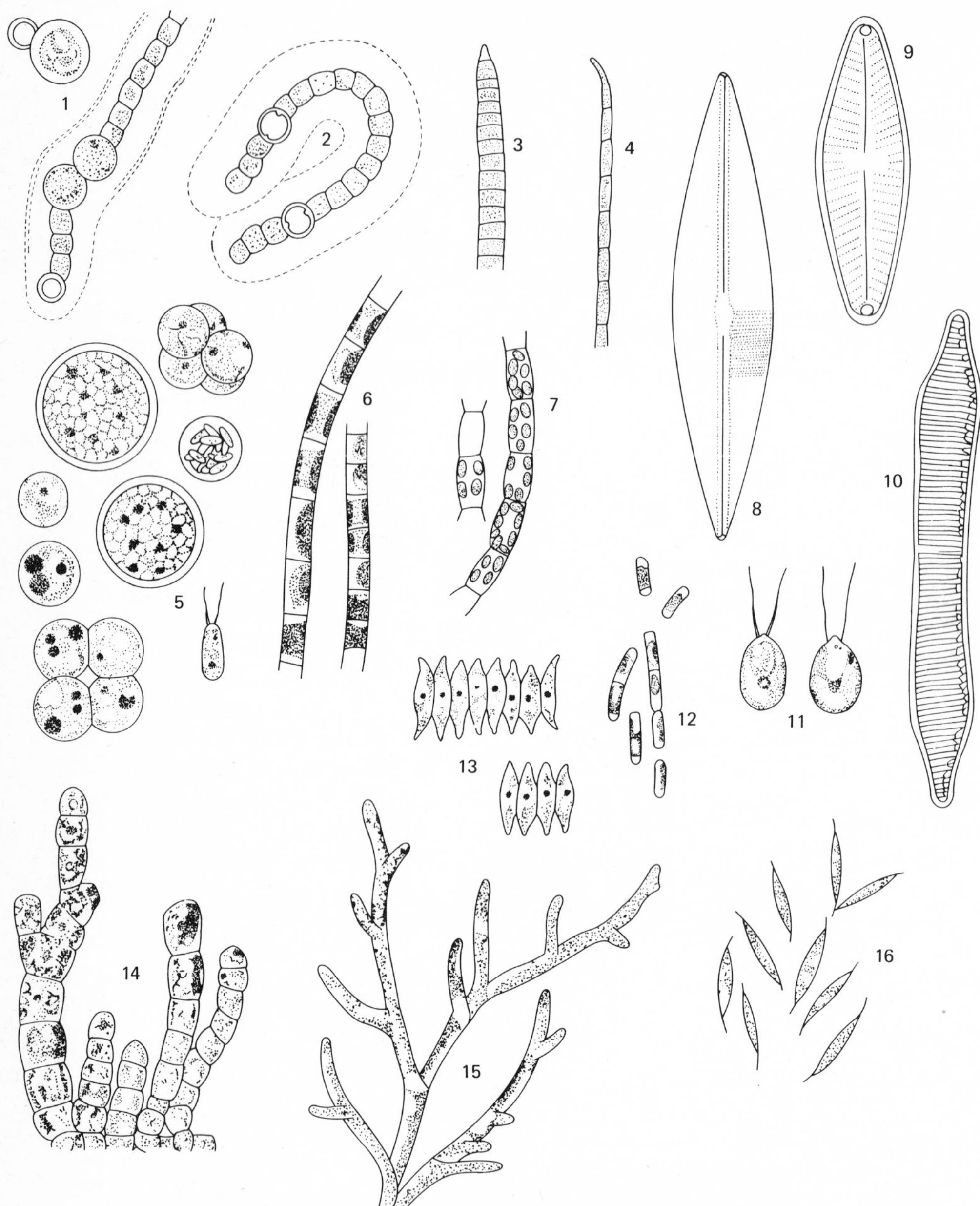

Figure 11-2 Algae of the soil: 1,2, *Nostoc;* 3,4, *Phormidium;* 5, *Chlorococcum;* 6, *Ulothrix;* 7, *Bumilleria;* 8,9, *Navicula;* 10, *Hantzschia;* 11, *Chlamydomonas;* 12, *Stichococcus;* 13, *Scenedesmus;* 14, *Gongrosira;* 15, *Microthamnion;* 16, *Dactylococcus. (From Algae the Grass of Many Waters, by L. H. Tiffany, after Lowe and Moyse. Copyright 1938 by Charles C Thomas, Springfield, Ill.)*

place the lower clumps of algae together in a bowl containing natural or artificial seawater (Appendix A) or one of the media described below.

Culturing Marine Species

Figure 11-3 illustrates marine algae that have been successfully maintained in the laboratory. Although many marine species of blue-green, green, brown, and red algae have been cultured in the laboratory, probably the red alga, *Porphyridium*, and the marine diatoms have been cultured more extensively than others. The media for *Porphyridium* and diatoms are described in Appendix A under *Porphyridium* agar and Diatom seawater agar. Unicellular species, including the marine diatoms, and some of the smaller filamentous forms have been successfully cultured in Erdschreiber solution (Appendix A).

A pure culture of a single species may be isolated from collected water samples with a capillary tube transfer, as described in Chap. 12 for protozoans. Also, unialgal cultures may be obtained from a biological supply company or a marine supply company. Containers must be thoroughly cleaned and autoclaved; cotton-plugged tubes or flasks are often used. Transfer a small portion of the algae into 150 to 200 ml of solution. Requirements of light intensity and temperature are variable but generally fall within the range of 200 to 300 ft-c and a cool temperature of 12 to 18°C. In about 10 days, when growth has started, reduce the light intensity to 100 to 150 ft-c for storage of cultures. Transfer organisms to fresh media every 6 to 8 weeks.

Preserving Algae

Liquid and dry preservation are described here. Techniques for permanent slide preparations will be found in Johansen (1940), Sass (1951), Smith (1951), and others.

Liquid Preservation of Freshwater Algae

1. *Formalin-chrome alum (Klein and Klein, 1970)*

Potassium chrome alum	10 g
4% formalin	5 ml
Water	500 ml

The preservative is recommended for all except the most delicate unicellular forms. The authors state that the natural color of algae is preserved.

2. *Formalin-glacial acetic acid (Sass, 1951)*

Water	72 ml
Formalin (40% formaldehyde)	5 ml
Glycerin	20 ml
Glacial acetic acid	3 ml

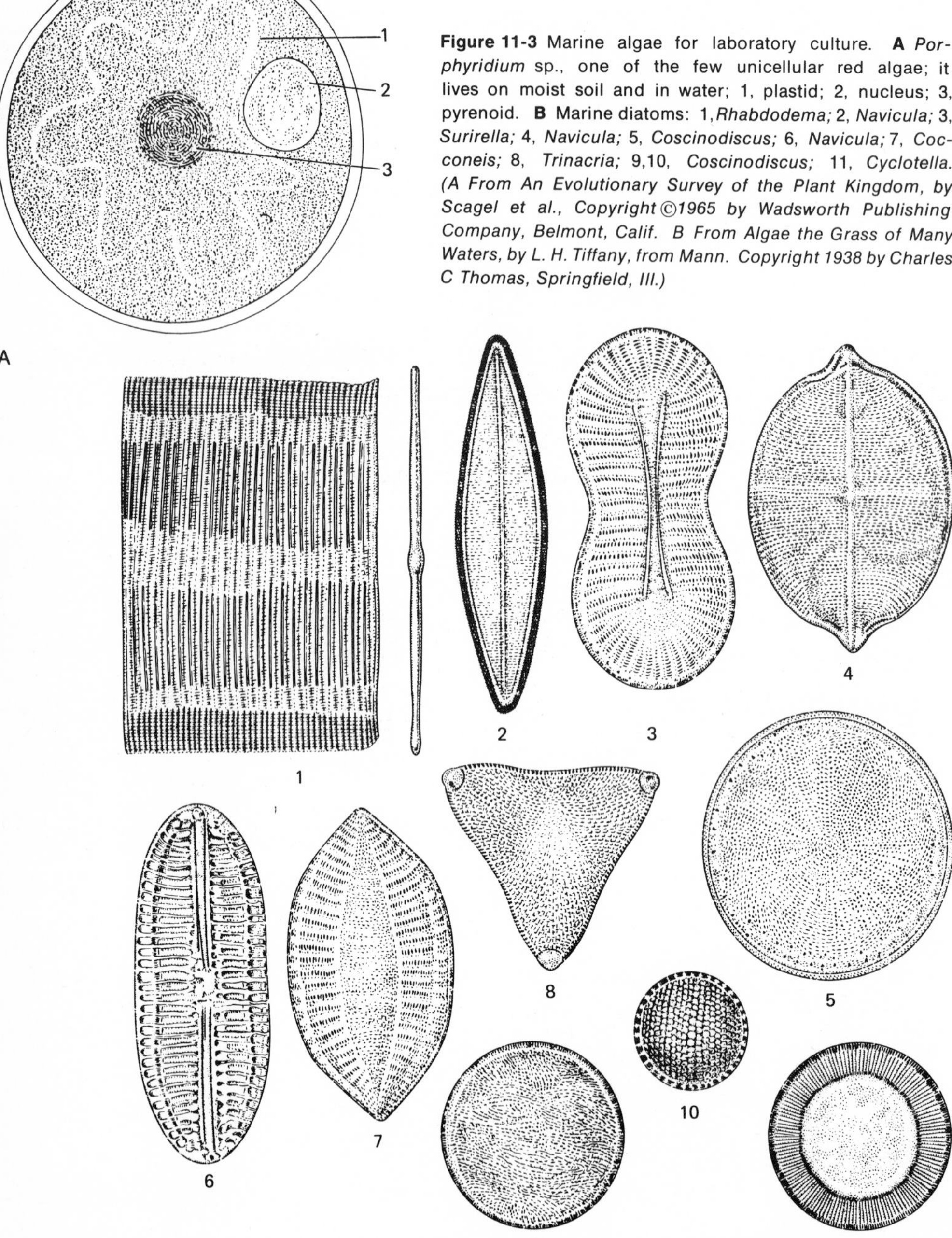

Figure 11-3 Marine algae for laboratory culture. **A** *Porphyridium* sp., one of the few unicellular red algae; it lives on moist soil and in water; 1, plastid; 2, nucleus; 3, pyrenoid. **B** Marine diatoms: 1, *Rhabdodema;* 2, *Navicula;* 3, *Surirella;* 4, *Navicula;* 5, *Coscinodiscus;* 6, *Navicula;* 7, *Cocconeis;* 8, *Trinacria;* 9,10, *Coscinodiscus;* 11, *Cyclotella.* *(A From An Evolutionary Survey of the Plant Kingdom, by Scagel et al., Copyright ©1965 by Wadsworth Publishing Company, Belmont, Calif. B From Algae the Grass of Many Waters, by L. H. Tiffany, from Mann. Copyright 1938 by Charles C Thomas, Springfield, Ill.)*

3. *Transeau's solution (Tiffany, 1938)*

Water	60 ml
Ethyl alcohol (95%)	30 ml
Formalin (40% formaldehyde)	10 ml
Glycerin	5 ml

Adding a small amount of copper sulfate may help to retain color in the green algae.

Liquid Preservation of Marine Algae

1. *Green algae*[1]

Formalin (40% formaldehyde)	25 ml
Saturated solution of cupric acetate	25 ml
Seawater	350 ml

Use to fix color in green algae.

2. *Formalin, 3% (Dawson, 1966)*

Formalin (40% formaldehyde)	3 ml
Seawater	97 ml

With large algae, such as kelp, use 5 ml of formalin.

3. *Ethyl alcohol (Dawson, 1966)*

Ethyl alcohol	70 ml
Seawater	30 ml

Dawson recommends an alcohol solution of 65 to 70% as a better preservative than formalin when specimens are to be stored for a long period of time.

Dry Preparations of Marine and Freshwater Algae

Dry preparations are better for color retention than liquid preservatives.

1. *Calcareous forms.* Soak calcareous algae for several days in 3% formalin (above) to which is added 20 ml glycerin for each 100 ml of the formalin solution. Dry and store in small boxes or vials. The glycerin increases the flexibility of the algae and decreases fragmentation.

2. *Crustose algae on substratum of rocks or wood.* Air dry the algae and substratum. Store in boxes or vials.

---

[1]From Victor Caprilles, Interamerican University, Puerto Rico, by personal communication, July 1969.

3. *Large marine algae.* Kill and fix leathery algae like kelps in 3% formalin (above) for 3 to 4 days. If collections are to be transported during one or more days, begin this procedure at the collecting site by transferring large algae to closed plastic or enamel containers of the solution. Following the killing and fixing period in the 3% formalin, soak the algae for 2 or 3 additional days in 50% glycerin solution (1:1, glycerine: seawater) with a small amount of phenol. Then hang the algae outdoors to dry, after which store them, rolled if necessary, in plastic bags. The glycerin produces a flexibility that allows the algae to be unrolled for observation.

4. *Herbarium preparations.* This procedure is appropriate for fragile, filamentous algae of either fresh- or salt-water forms. It is an interesting and easy method, often providing beautiful displays of red, brown, green, and yellow-green algae that are useful for laboratory studies. Standard 11½ by 16½ herbarium paper, or smaller sheets, and a plant press are used.

Coarse algae may be arranged between sheets of newspaper and pressed in the same manner as for ferns and seed plants. Float the more delicate, filamentous algae in a shallow enamel or plastic pan (1 to 2 in deep) containing fresh water or seawater, according to the type of algae. The pan must be larger than the herbarium paper. Use an artist's brush or other instrument to separate and spread the filaments. Cut and remove portions from very thick sections. Slip a sheet of herbarium paper under the floating algae, and slowly raise the paper to allow the water to run off while the algae remain in a spread position on the paper. Remove the paper and algae from the water, and straighten the algae again with the brush to produce a desirable configuration. Prop the paper against a flat object (tray or sheet of metal) placed at a slight angle to allow the water to drain from the paper. As a rule, the wet algae adhere to the paper. Cover the algae with a sheet of waxed paper, plastic film, or unbleached muslin. Then place each sheet between newspaper, and transfer to a plant press in the usual manner, with a blotter and a layer of corrugated cardboard on each side of the newspaper. Twelve or more layers of herbarium sheets, in newspaper folds with blotters and corrugated cardboard, can be stacked and tightened in the press. Change blotters daily; the algae should be dry within 3 to 7 days depending upon their thickness. Placing the press in moving air, for example, under a fume hood or a fan, will hasten the drying. Specimens must be completely dry before removing them from the press; otherwise the filaments curl and wrinkle, and thick specimens may decay.

If not handled roughly, many algae remain fastened to the paper without further treatment (Fig. 11-4). For other forms, apply spots of standard herbarium paste to the underside of the algae, and press the specimens against the paper. Heavy, coarse algae are often fastened to the paper with narrow strips of herbarium, gummed cloth. All supplies and equipment are available from biological companies.

CLASSIFICATION SYSTEM

The balance of the chapter is concerned with four large divisions of freshwater algae: Cyanophycophyta (blue-green algae); Chlorophycophyta (green algae);

A

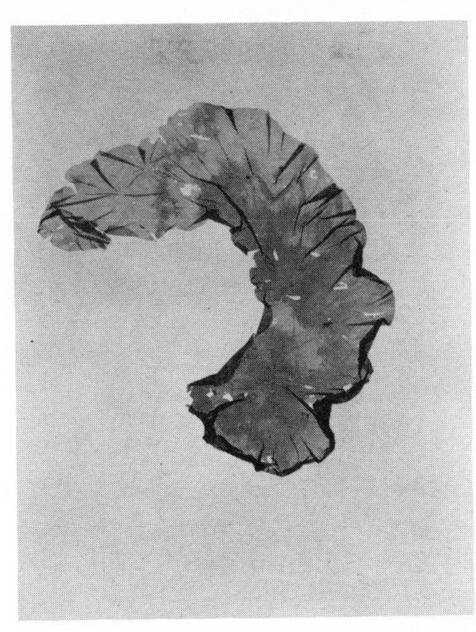

B

Figure 11-4 Herbarium preparation of marine algae. **A** *Sargassum muticum.* **B** *Ulva* sp. Both specimens were taken from water on to the herbarium paper by the procedure described in the text.

Euglenophycophyta (euglenoid algae); and Chrysophycophyta (yellow-green and golden algae). Because biology teachers and students often are more familiar with generic names than with family and order groupings, the algae of each division are arranged in alphabetical sequence according to generic name. Although the alphabetical system allows an easier and quicker reference to a particular organism, it has the severe drawback of not showing the evolutionary relationships among genera. Therefore, the following outline should be used regularly when studying any genus described later.

Division Cyanophycophyta
- *Chroococcus*
- *Microcystis*
- *Oscillatoria*
- *Anabaena*
- *Nostoc*

Division Chlorophycophyta
- *Chlamydomonas*
- *Pandorina*
- *Volvox*
- *Hydrodictyon*
- *Chlorella*
- *Ankistrodesmus*
- *Protococcus*
- *Scenedesmus*
- *Ulothrix*
- *Oedogonium*
- *Cladophora*
- *Spirogyra*
- *Zygnema*
- *Netrium*
- *Cosmarium*
- *Closterium*
- *Micrasterias*

Division Euglenophycophyta
- *Euglena*

Division Chrysophycophyta
- The Diatoms
- *Vaucheria*

THE BLUE-GREEN ALGAE (CYANOPHYCOPHYTA)

A large proportion of blue-green species grow in fresh water, although many species occur also in marine and brackish waters, as well as on moist soil. Cells may be solitary, united in a long chain, or in a mass, all types often within a gelatinous sheath. A chain of cells is sometimes called a trichome; the trichome and outer gelatinous sheath, together, form a filament. Although these algae are called "blue-green," various combinations of pigments produce olive-green, gray-green, brown, and purplish colors, as well as blue-green. Pigments are distributed throughout the cytoplasm, although generally they are most dense in peripheral regions. Nuclear material is in threads or granules of chromatin, scattered throughout the cell.

Reproduction is by cell fission and by fragmentation. There is no sexual reproduction and no swimming cells. A vegetative cell may enlarge and become a heterocyst, which often appears transparent under magnification. Generally the heterocyst produces a weakened site in the trichome that helps to promote fragmentation. Vegetative cells of complex, or "higher," cyanophytes may develop into enlarged, thickwalled spores called akinetes. Research has shown that certain species of *Nostoc, Anabaena,* and others with heterocysts, are able to fix nitrogen from the air when nitrates are lacking in the soil. Several common freshwater genera are described below.

Anabaena

USES: (1) For field and laboratory investigations of an alga that contributes to river and lake pollution by producing a toxin. (2) To study the morphology of a common, blue-green alga.

DESCRIPTION: Trichomes are made up of cylindrical or barrel-shaped cells; the threads may be single or in a tangled mass within a mucilaginous sheath. Heterocysts and akinetes may be present (Fig. 11-5). The genus is easily mistaken for *Nostoc*, but the mucilage of *Anabaena* is usually soft, whereas the gelatinous sheath of *Nostoc* is more firm and skinlike. The algae will be found floating on water, or mixed with other vegetation at the bottom of shallow water, or on moist soil. Gas pockets (pseudovacuoles) within the cells seemingly provide buoyancy to the floating species.

Anabaena is one of the most toxic of algae, playing an important role in water pollution when blooms of the algae are dense. At least six species of *Anabaena* are known to secrete toxins that produce illness and often death in livestock and birds that drink the polluted waters.

CULTURE: To maintain a collection of mixed algae, see the general techniques given earlier in the chapter. For unialgal culture try any of the following media (Appendix A): (1) Bold's basal medium; (2) Bold's basal medium with agar; (3) Cyanophyceae agar; (4) Pringsheim's soil water; (5) Kantz's medium, modified.

Chroococcus

USES: To observe a simple, blue-green alga that may be easily collected within a greenhouse or on moist soils and rocks in nature.

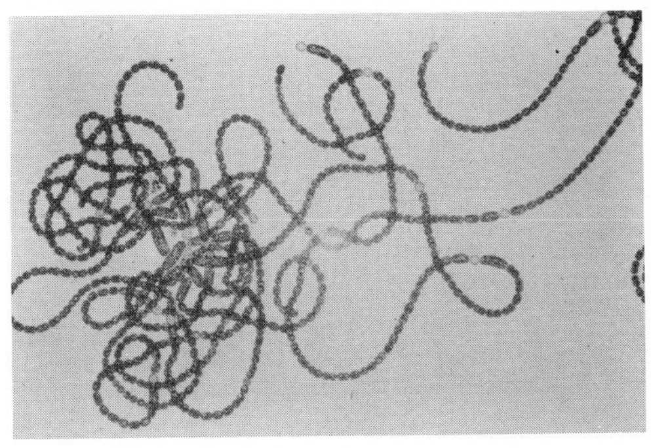

A

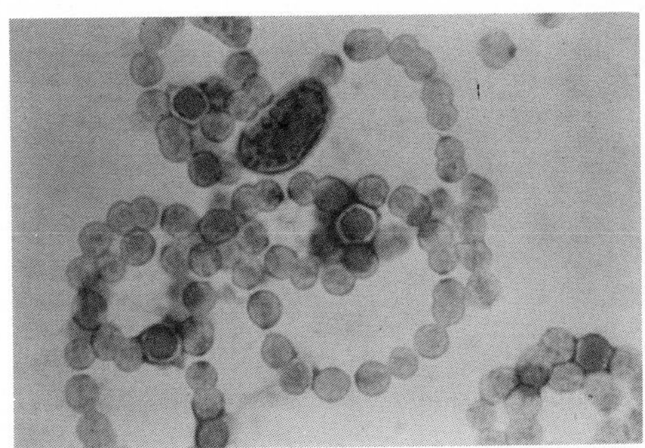

B

Figure 11-5 *Anabaena* sp. **A** Mass of filaments in which heterocysts, the clear, round cells, can be seen. **B** Close-up of several stained filaments. A large, oval akinete is seen near top center; the spherical cells, heavily stained, are heterocysts. *(A Courtesy of Dean Blinn. B Courtesy of Ward's Natural Science Establishment, Inc., Rochester, N.Y.)*

DESCRIPTION: The organism consists of spherical cells, occasionally solitary, but usually in groups of two or four, enclosed in concentric, mucilaginous sheaths (Fig. 11-6). The alga grows in the bottom sediment of fresh water, and on moist rocks or soil. Often it can be collected along with *Gleocapsa* from soil and moist clay pots in a greenhouse.

CULTURE: For culturing collections of mixed algae, see the general techniques given earlier in the chapter. For unialgal cultures use any of the following media (Appendix A): (1) Pringsheim's soil water; (2) Kantz's medium, modified; (3) Shen-Chu solution; (4) Cyanophyceae agar.

Microcystis

USES: (1) For field or laboratory tests related to the toxins produced by *Microcystis*. (2) To test or demonstrate, in the laboratory, environmental condi-

Figure 11-6 Comparison of the cell arrangement of *Chroococcus* and *Gleocapsa*. **A** Diagrams of the two algae: left, *Chroococcus;* right, *Gleocapsa*. **B** *Gleocapsa* sp., living organisms. *(B Courtesy of Ward's Natural Science Establishment, Inc., Rochester, N.Y.)*

A

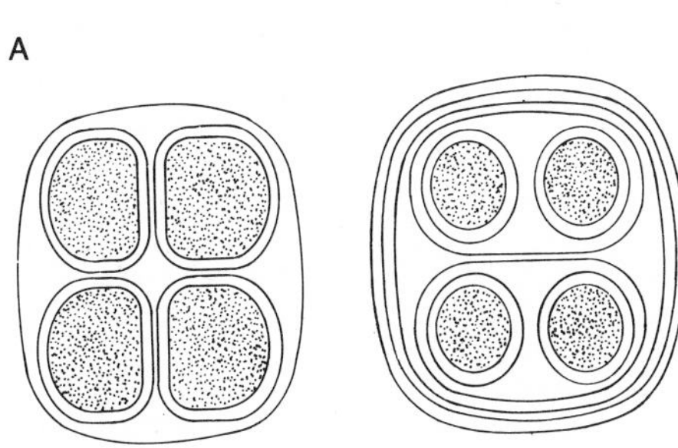

B

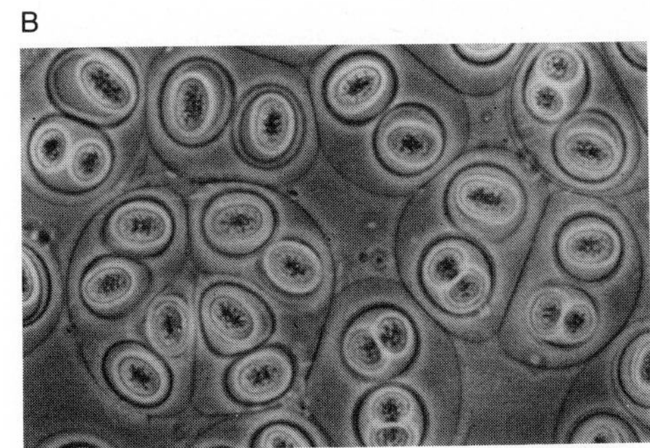

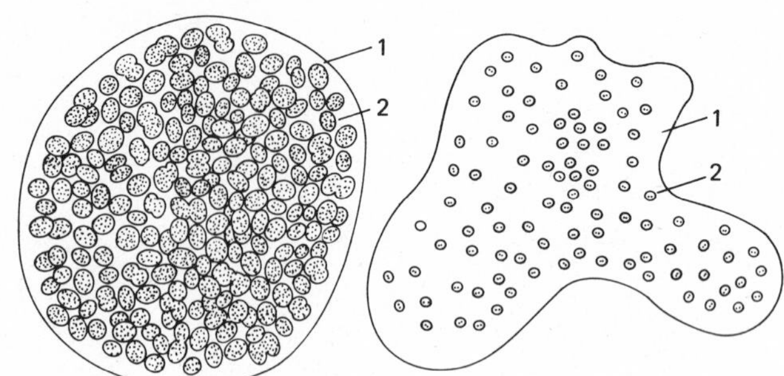

Figure 11-7 *Microcystis* sp. Two species are shown; 1, mucilaginous sheath; 2, cells.

tions that promote the growth and profuse blooming of this alga. (3) Possibly for individual student studies related to the control of the growth of *Microcystis*. (4) To observe the morphology of a simple alga.

DESCRIPTION: An organism consists of a colony of hundreds to thousands of small, spherical cells in an irregular mass of mucilage (Fig. 11-7). Many species contain gas pockets (pseudovacuoles) in the cells that provide buoyancy, and thus the organism floats near the water surface. At least two species, *M. toxica* and *M. aeruginosa*, secrete lethal toxins that kill fish in the water and livestock and birds that drink the polluted water. Apparently *Microcystis* inhibits the development of other algae and perhaps kills other growth in the water, thus producing massive, vegetative death and decomposition. The alga is receiving much study regarding its role in water pollution, methods for controlling its growth, and the nature of the toxins. A starting culture of *M. aeruginosa* may be purchased from the American Type Culture Collection (Appendix B).

CULTURE: For collections of mixed algae see the general techniques given earlier in the chapter. For unialgal cultures try any of the following media (Appendix A): (1) Bold's basal medium; (2) Pringsheim's soil water; (3) Kantz's medium, modified; (4) Shen-Chu solution; (5) Cyanophyceae agar.

Nostoc

USES: To study the interesting morphology of *Nostoc* trichomes and their conglomeration in *Nostoc* balls.

DESCRIPTION: The organism consists of a long chain (a trichome) of beadlike cells. Many trichomes are twisted together in a gelatinous matrix enclosed in an outer, tough membrane. *Nostoc pruniforme* Ag., a common species in lakes and granitic quarries, forms globular colonies that are the typical *Nostoc* balls. The *Nostoc* balls vary in size from less than a millimeter to 7 or 8 cm (Fig. 11-8). Sometimes *Nostoc* forms a conspicuous layer of floating, blue-green balls at the edge of a pond or lake. *Nostoc commune* Vaucher, common in the Arctic tundra, makes tough flat layers on wet limestone soil or on bottom sediment of water. Heterocysts and akinetes may develop regularly.

CULTURE: For collections of mixed cultures see the general techniques given earlier in the chapter. For unialgal cultures use either of the following media (Appendix A): (1) Pringsheim's soil water; (2) Kantz's medium, modified.

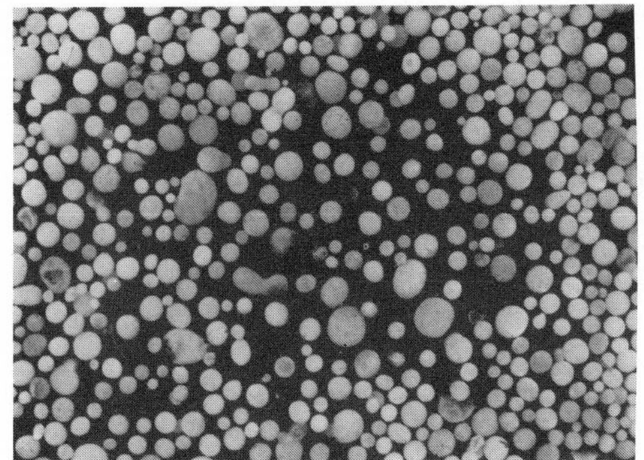

A

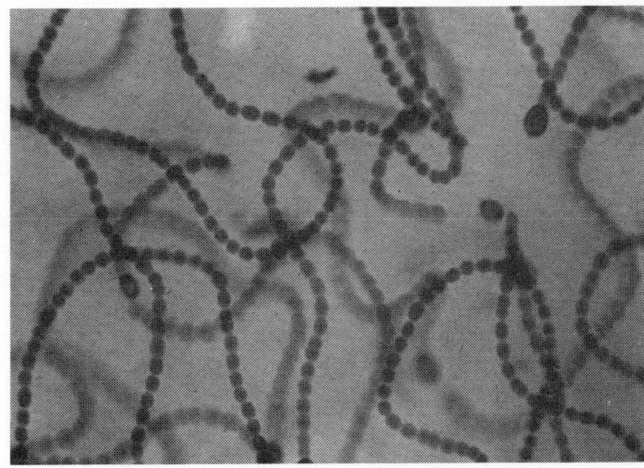

B Courtesy of Carolina Biological Supply Company

Figure 11-8 *Nostoc* sp. **A** Colonies that formed a blue-green cover over water at edge of lake; diameter of the *Nostoc* balls is 2 to 10 mm. **B** Filaments of vegetative cells and larger heterocysts.

Oscillatoria

USES: To observe a simple blue-green alga that is easily obtained in old hay infusions or old collections of pond water.

DESCRIPTION: The alga consists of unbranched trichomes (threads) of short and relatively wide cells. The trichomes may be solitary or in tangled masses (Fig. 11-9). A thin, inconspicuous mucilage may be present around each trichome. The threads move by gliding, rotating, or oscillating motions, the latter motion accounting for the generic name. The forces of movement are not understood. Reproduction is by cell division and fragmentation. Both freshwater and marine species have been described. The alga often appears in old collections of pond water, in old hay infusions, or in pools of water within the greenhouse.

CULTURE: For collections of mixed algae see the procedures described earlier in the chapter. For unialgal cultures use any of the following media (Appendix A): (1) Pringsheim's soil water; (2) Kantz's medium, modified; (3) Cyanophyceae agar.

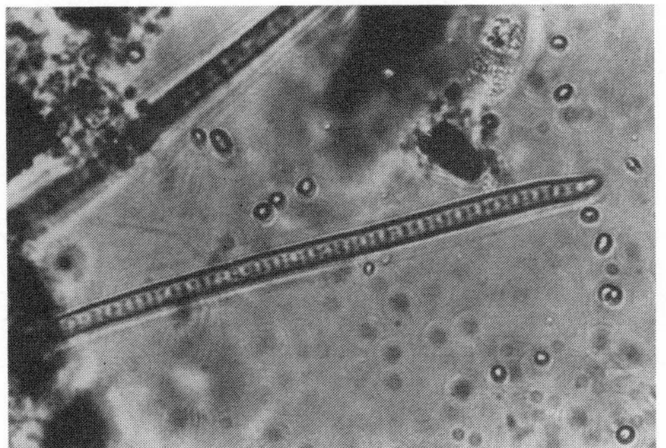

A

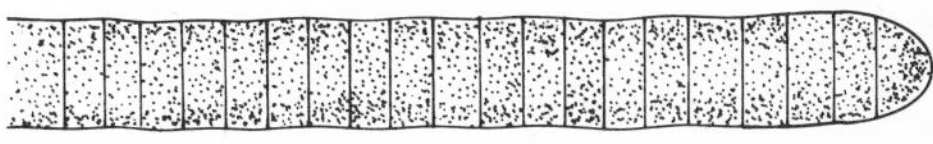

B

Figure 11-9 *Oscillatoria* sp. **A** An oscillating filament. The outer sheath can be seen around the trichome of cells. **B** Diagram of filament. The sheath is not always apparent.

THE GREEN ALGAE (CHLOROPHYCOPHYTA)

The green algae consist of approximately 7,000 species in 450 genera that are chiefly freshwater, although many species are marine or terrestrial. In general, the marine forms are larger and more conspicuous than the freshwater species.

The cells of many green algae are covered by a firm cellulose wall, coated with an outer pectic layer. In contrast to cyanophytes with no definite nuclei, the nuclear membrane of chlorophytes consists of a double-layer, and the nucleus may contain one or more nucleoli. Additionally, the pigments are contained within definite bodies, the chloroplasts, and are not scattered throughout the cytoplasm as in the blue-green algae. Generally the cell contains a large central vacuole that pushes the cytoplasm into a peripheral position. Mitochondria, Golgi bodies, and endoplasmic reticulum, all not present in cyanophytes, are found in the chlorophytes. Some of the more common freshwater and terrestrial genera are described below.

Ankistrodesmus

USES: To identify and observe a unicellular, green alga that often appears in freshwater collections or on soil. The unusual crescent shape of *Ankistrodesmus* and related genera is generally interesting to students.

DESCRIPTION: The organism consists of a single cell that is straight or slightly curved (Fig. 11-10). The cells occur singly or in loose clusters. Occasionally they form a dense green, almost pure growth in an aquarium. The genus is collected rather commonly in fresh water, mixed with other floating algae. A closely related genus, *Selenastrum* (Fig. 11-11) resembles *Ankistrodesmus* but is more curved. The two are often collected together. A starting culture may be obtained from the American Type Culture Collection (Appendix B).

CULTURE: For collections of mixed algae see the general techniques described earlier in the chapter. For unialgal cultures try any of the following media (Appendix A): (1) Bold's basal medium; (2) Bold's basal medium with agar; (3) Pringsheim's soil water.

Chlamydomonas

USES: (1) To study the life cycle, including both the asexual and sexual stages. (2) For genetic studies. The dominant free-living stage is haploid ($n = 8$ or 10).

DESCRIPTION: *Chlamydomonas* is a small, solitary, free-living flagellate, common in fresh water and in soil. The cell is spherical or ovoid, depending on the species, and is covered by a cellulose membrane; two anterior flagella are present. One large, green chloroplast is usually present, almost filling the cell and containing one or several pyrenoids. There are many freshwater species, measuring from 8 to 22 microns long, and 7 to 12 microns wide.

The dominant stage in the life cycle is the haploid organism, described above, which reproduces both asexually and sexually. Asexual reproduction is by fission in which two to eight biflagellated daughter cells are formed from the single intracellular protoplast. Eventually the cell wall ruptures, and the haploid daughter

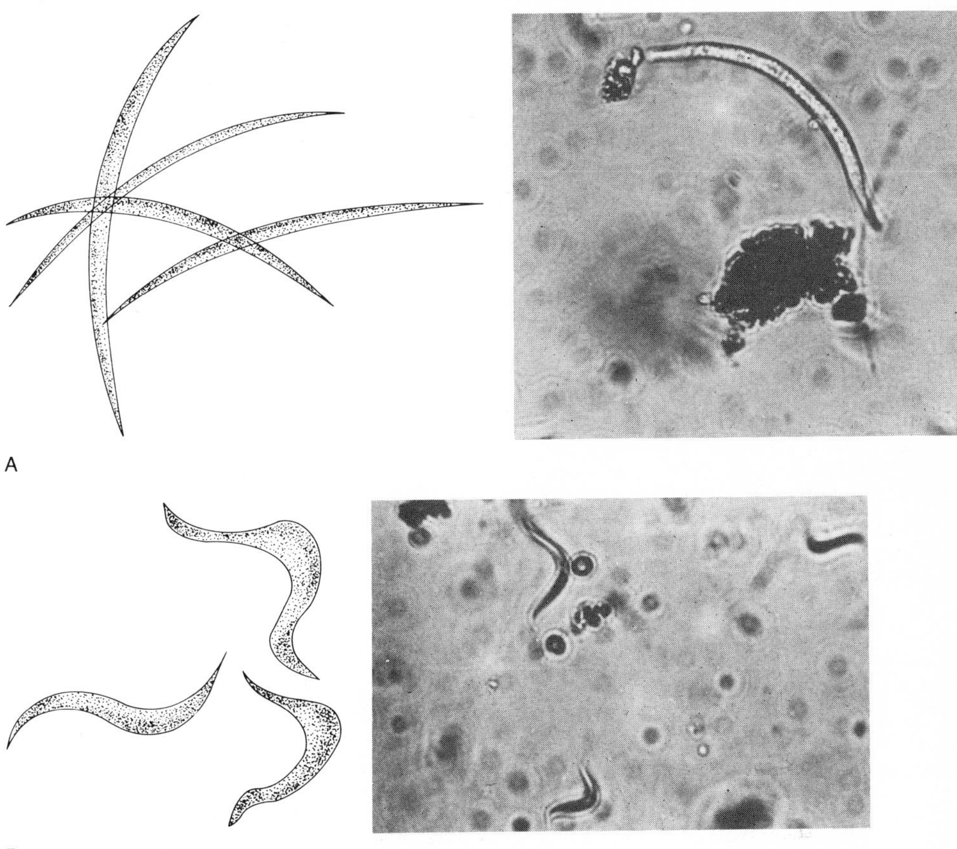

Figure 11-10 *Ankistrodesmus* sp. Two species growing together in a tank of "green soup" water. **A** *A. falcatus:* left, diagram of cell; right, an alga moving in water. **B** *A. convolutus:* left, diagram of cells; right, organisms moving in water.

cells are released. At times, sexual reproduction occurs when morphologically indistinguishable male and female gametes (isogametes) are produced. The haploid gametes fuse to form a diploid zygote that becomes an unflagellated, dormant cyst, or zygote. Under favorable conditions the zygote undergoes meiosis, and forms four or eight biflagellated, haploid organisms. The stages of the life cycle are shown in Fig. 11-12. *Chlamydomonas* may be obtained from most biological supply companies listed in Appendix B.

Much study has been done on the genetics of *Chlamydomonas*, including the work of Sager (1955 and 1960), Lewin (1953), and Sager and Granick (1954).

Technique for demonstrating sexual reproduction in Chlamydomonas. Two mating strains (+ and −) of *C. moewusii* are used for this demonstration. These

Figure 11-11 *Selenastrum* sp., an alga that may contaminate an aquarium and multiply until the water is green.

Figure 11-12 *Chlamydomonas* sp. **A** Haploid cell, the dominant stage. **B** Diagram of cell showing the cuplike chloroplast: 1, flagellum; 2, chloroplast; 3, nucleus; 4, eyespot (stigma); 5, pyrenoid. **C** Life stages. *(C From The Science of Biology, 3d ed., by Paul B. Weisz, 1967. Copyright 1967 by McGraw-Hill, Inc. Used with permission of McGraw-Hill Book Company, New York.)*

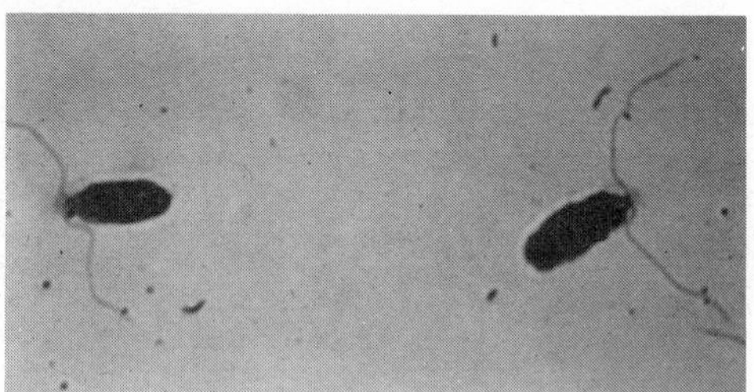

A *Courtesy of Carolina Biological Supply Company*

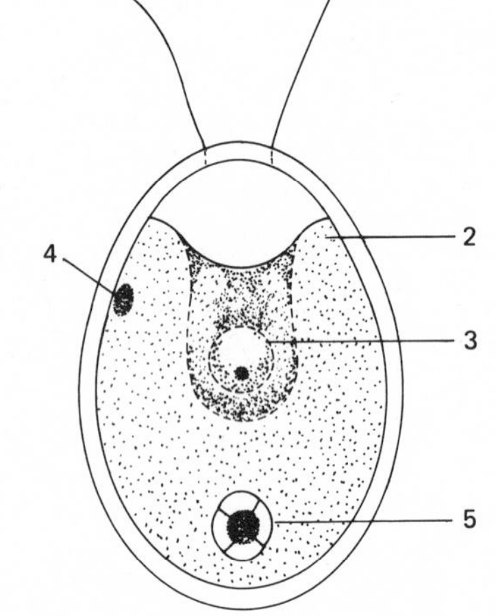

B

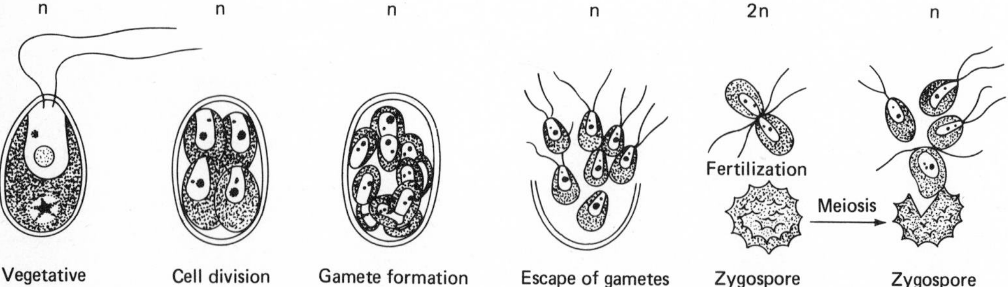

C

strains may be obtained on agar from Carolina Biological Supply Company, Cat. Nos. 15-2034 (+ strain) and 15-2035 (− strain). The company also supplies other mating sets of *Chlamydomonas*. Instructions are included for mating demonstrations.

Prepare petri plates of Bold's basal medium with agar, increased nitrogen (Appendix A). Use a sterile, inoculating loop to transfer the two strains to separate plates, spreading a heavy inoculum on the surface of each plate. Invert the plates and place them under 300 ft-c of light for 12 h daily. Within 5 to 10 days a heavy growth should result. Use the edge of a coverglass to remove the plus and minus cells, separately, by scraping the agar surface while taking care to remove as little agar as possible. Transfer each type of cell to a separate dish, one-half filled with distilled water, by moving the edge of the coverglass slowly back and forth in the water so as to wash the cells into the water. Place the separate plates under light and agitate the water occasionally by rotating each bowl slowly to disperse the cells throughout the water. At the end of 1½ to 2 h, place a drop from each bowl on a slide; mix the two drops together with a toothpick, and observe the cells under the microscope. Very soon, clumping of the two mating types should become apparent. Within a few minutes, the pairs of plus and minus gametes will become evident as they separate from the clumps. If a Vaseline-sealed hanging-drop is made of the two types of cells mixed together, the process can be observed over a period of several hours.

CULTURE: For mixed cultures, see the general techniques given earlier in the chapter. For unialgal cultures use (1) Bold's basal medium with agar or (2) Pringsheim's soil water (Appendix A).

Chlorella

USES: (1) As a producer organism in an aquarium ecosystem. (2) For demonstrations of photosynthetic activity.

DESCRIPTION: The alga consists of small, spherical, bright-green, nonmotile single cells. The cell contains a single cup-shaped chloroplast and a small nucleus in the center of the cytoplasm (Fig. 11-13). Marine, freshwater, and soil species have been described. *Chlorella* often occurs as a symbiont in protozoans, sponges, and *Hydra*, where it may be called *Zoochlorella*. Asexual reproduction is by division of the cell into four or eight nonmotile spores, which are released when the outer cell wall ruptures. Often the organism multiplies rapidly in an aquarium, where it may become a nuisance. *Chlorella* is the organism often used in photosynthetic research.

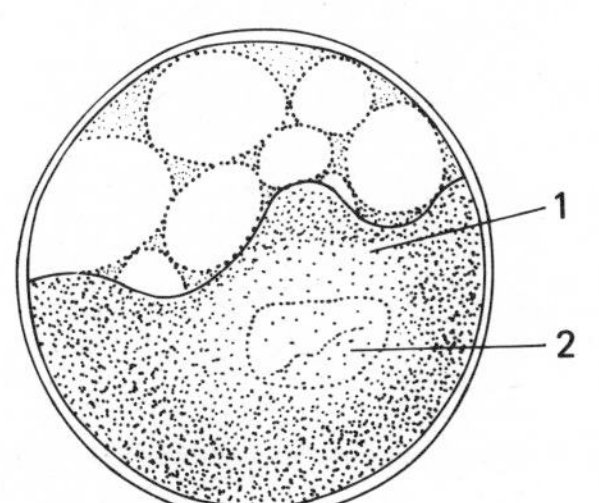

Figure 11-13 *Chlorella* sp., vegetative cell; 1, chloroplast; 2, nucleus.

CULTURE: For collections of mixed algae see the general procedures given earlier in the chapter. For unialgal cultures use any of the following media (Appendix A): (1) Bold's basal medium; (2) Bold's basal medium with agar; (3) Pringsheim's soil water.

Cladophora

USES: (1) For field studies of pond and lake eutrophication. (2) To study the morphology of a branching, filamentous, green alga. (3) To observe, if possible, the development of asexual zoospores, and the fusion of isogametes in sexual reproduction.

DESCRIPTION: The alga consists of branching filaments of cylindrical, multinucleated cells. Thick cell walls give a coarse texture that is easily recognized when handling the filaments. A netlike, often segmented chloroplast occurs in each cell (Fig. 11-14). Many freshwater and marine species have been described. Apparently all species are first attached to substratum by rhizoids, but they may become loosened and free floating at a later stage. Marine forms are abundant in intertidal zones where they may be attached to larger algae. Freshwater species often make dense growths in swift water on rocks and dams, where the plants may extend for several feet with the current. Dense masses of free-floating forms may become a nuisance in ponds and lakes, and may contribute to problems of eutrophication.

Figure 11-14 *Cladophora* sp. **A** Filaments. **B** Close-up of two cells, stained, showing the many nuclei in each cell. *(A and B Courtesy of Ward's Natural Science Establishment, Inc., Rochester, N.Y.)*

A

B

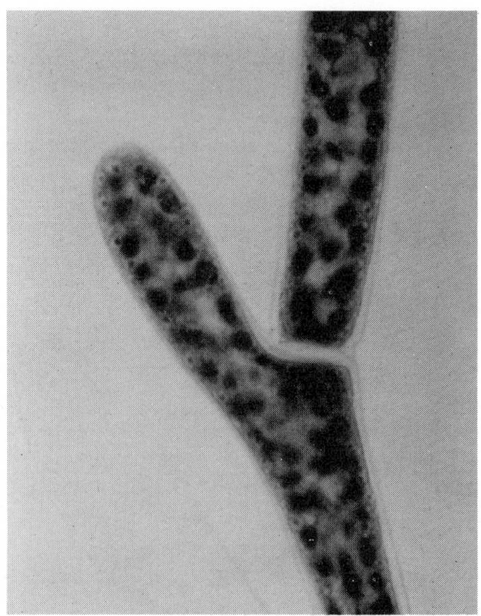

Asexual reproduction occurs when many zoospores, each with one nucleus and four flagella, form in terminal cells of filaments. Zoospores emerge to the outside through a pore in the parental cell wall. Subsequently each zoospore grows into a new plant. Repeated mitotic divisions of the nucleus eventually produce multinucleate cells. Sexual reproduction occurs when biflagellated isogametes form in terminal cells. Isogametes fuse in pairs to form zygotes that develop into new plants. Studies of *Cladophora* reveal that, in at least some species, diploid plants produce haploid zoospores by meiotic divisions; a zoospore then grows into a haploid plant. The haploid plant forms haploid isogametes, and the resulting diploid zygotes grow into diploid plants. The haploid and diploid plants are morphologically indistinguishable.

CULTURE: For collections of mixed cultures see the general procedures given earlier in the chapter. For unialgal cultures use one of the following media (Appendix A): (1) Pringsheim's soil water (for freshwater species); (2) Erdschreiber solution (for saltwater species).

Murphy (1966) describes his method for culturing *Cladophora* in Bristol's solution (Appendix A). He found that growth is more satisfactory when a substratum of agar (1.5 g in 100 ml Bristol's solution) is placed in the culture bowl to anchor the alga. Bristol's solution, without agar, is added above the agar layer.

The Desmids

USES: (1) To study the morphology of these unusually beautiful cell types. (2) To observe the unique terminal vacuoles containing vibrating particles of gypsum. (3) To demonstrate sexual reproduction in mating strains of *Cosmarium*.

DESCRIPTION: The desmids (family Desmidiaceae) comprise a large group of unicellular and filamentous green algae of worldwide distribution. The term "desmid" is derived from the Greek word *desmos* which means "bond." The term well describes most desmids in which the cells are organized into two symmetrical portions connected by a bond or isthmus. Desmids include some of the most beautiful cell types that are known. Four genera commonly found in fresh water are described here. The best collecting season is in late spring and early summer when desmids are most abundant in lakes and ponds.

Closterium. This organism is widely distributed in fresh water that is neutral or slightly acid. The alga is usually associated with other algae. A few species are planktonic. The cell is generally crescent-shaped but may be nearly straight (Fig. 11-15). An elongated, ridged chloroplast is found in each half portion. Each chloroplast contains a row of conspicuous pyrenoids. No cleft appears in the outer wall between the half portions, but the inner cell is clearly divided into two symmetrical semicells. The nucleus is in the isthmus area between the semicells. An interesting feature is the vacuole at each tip, containing vibrating particles of gypsum that may be clearly observed under 100 to 400X magnification. Apparently the gypsum has no physiological function, and perhaps the particles are excretory products. The cell wall consists of an inner cellulose layer and an outer mucilaginous pectic layer. It seems that the secretion of pectin to the outside of the cell produces the gliding motion of many desmids. According to Prescott (1968), desmids move toward light.

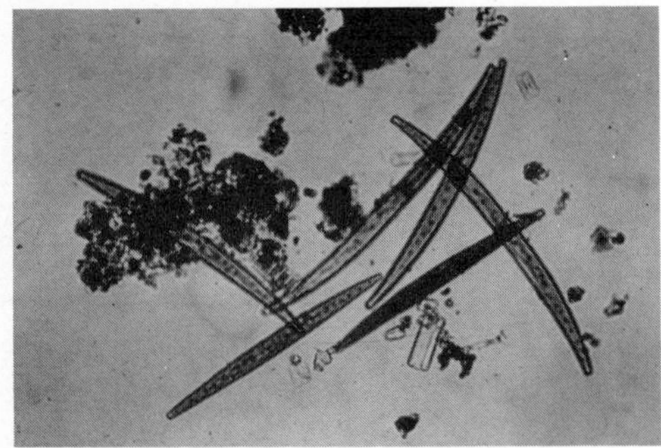

A *Courtesy of Carolina Biological Supply Company*

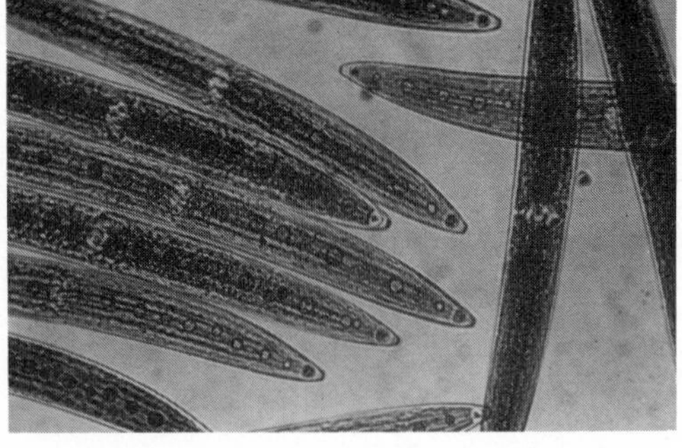

B

Figure 11-15 *Closterium* sp., a desmid. **A** Several vegetative cells. **B** Close-up of portions of cells showing the isthmus, the ridged chloroplast with pyrenoids, and the vacuole containing gypsum at each tip of a cell. *(B Courtesy of Dean Blinn.)*

Asexual reproduction is by mitotic cell division in which the parent cell divides into two cell portions. Each cell portion then regenerates a semicell. Sexual reproduction occurs when two cells become aligned, following which each cell ruptures at the isthmus area. The protoplast of each cell emerges through the opening; the two protoplasts fuse forming a spiny walled zygote between the two empty cells. In favorable conditions the dormant zygote germinates to make two daughter cells.

Cosmarium. A beautiful and interesting unicellular desmid (Greek, *kosmos*, ornament). Many freshwater species are distributed worldwide. A deep incision in the outer walls separates the two half portions. Much variation is found in the shape of semicells and in the texture of the outer wall which may be granular or toothed (Fig. 11-16). In most species the two half portions are rounded or slightly flattened. Asexual and sexual reproduction are similar to those described for *Closterium*. A *Cosmarium* mating set may be obtained from the Carolina Biological Supply Co. Instructions for class studies accompany the set and are similar to

Figure 11-16 *Cosmarium* sp., a desmid. **A** Single vegetative cell. **B** Zygospore at center, and the remains of two conjugating cells. *(A Courtesy of Ward's Natural Science Establishment, Inc., Rochester, N.Y.)*

A

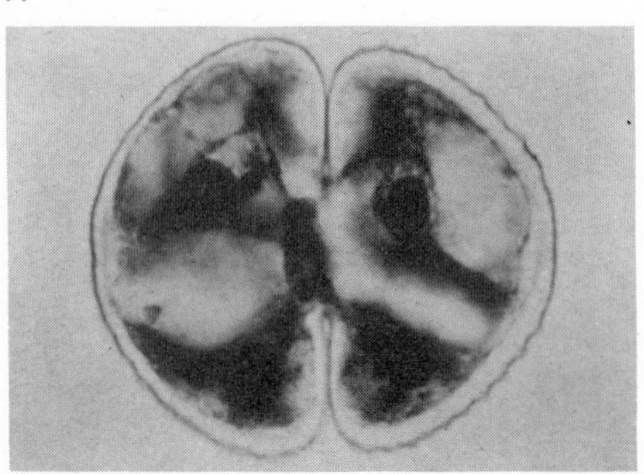

B *Courtesy of Carolina Biological Supply Company*

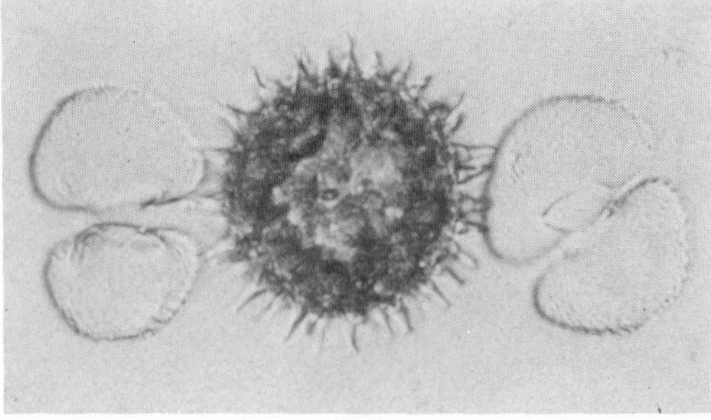

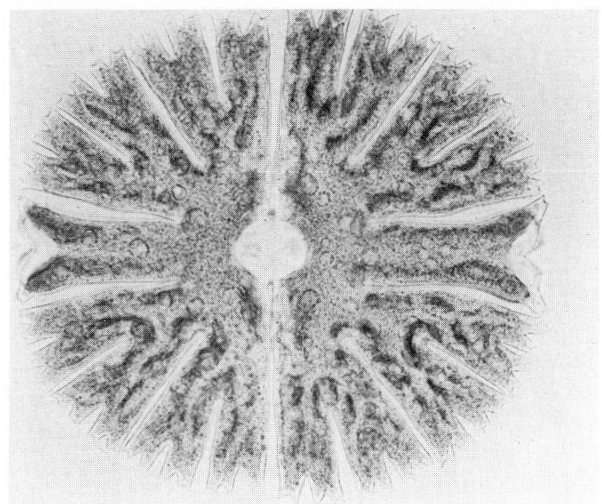

Courtesy of Carolina Biological Supply Company

Figure 11-17 *Micrasterias* sp., a desmid.

techniques described for *Paramecium* (Chap. 12) and for *Chlamydomonas* in this chapter.

Micrasterias. The alga is widely distributed in freshwater ponds and lakes, often floating and mixed with other algae. The single cells are relatively large, flat, and disc-shaped with a deep incision between the semicells. The nucleus is located in the isthmus between the two half portions. The outer wall of the semicells is lobed or deeply cleft, giving a starlike appearance when the cell is viewed from the flat side (Fig. 11-17), and accounting for the generic name (Greek, *micros*, little; *asterias*, star). Occasionally in deeply cleft species, the spines of separate cells become hooked together to form temporary chains or false filaments. Asexual reproduction is similar to that of *Closterium*. Apparently little study has been made of the sexual reproduction.

Netrium. The species *N. digitus* is relatively large, and oval or cucumber-shaped, with little or no indentation between semicells, although the internal cell is clearly divided (Fig. 11-18). The organism occurs in plankton of freshwater lakes and ponds. One or two ridged chloroplasts are present in each semicell. Reproduction is similar to that of *Closterium*.

Courtesy of Carolina Biological Supply Company

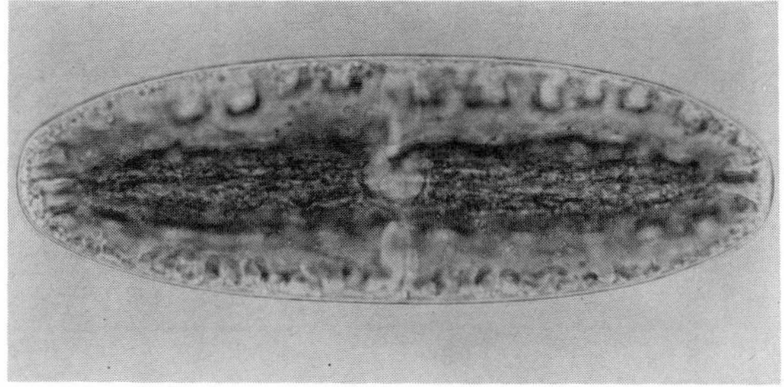

Figure 11-18 *Netrium* sp., a desmid.

CULTURE OF DESMIDS: For mixed cultures see general techniques described earlier in the chapter. For unialgal cultures use either of the following media (Appendix A): (1) Pringsheim's soil water; (2) desmid agar.

Hydrodictyon

USES: (1) To observe the very interesting nets of cells and the development of a new small net within a cell. (2) To demonstrate the induction of asexual and sexual reproduction under different lighting situations.

DESCRIPTION: *Hydrodictyon reticulatum*, commonly called "water net," occurs worldwide in freshwater lakes and ponds where masses of the floating nets may become troublesome. The length of a colony ranges from microscopic to several inches and occasionally to 12 in or more. The cells are cylindrical; each is joined at its end to two other cells, forming a polygonal pattern in the wall of a hollow, tubular net that is closed at both ends (Fig. 11-19). Young cells are uninucleate and become multinucleate as the cell enlarges through mitotic nuclear division. The chloroplast with many pyrenoids forms a network (reticulum) within the mature cell.

The colony is a "coenobium," in that the number of cells does not increase as the young colony matures, and growth occurs by increase in size of each individual cell. Asexual reproduction occurs when the protoplast of a mature cell divides into many small uninucleated and biflagellated zoospores that eventually become joined to form a new net still contained in the outer wall of the parent cell. Thus reproduction is efficient and rapid, in that each cell of a colony may produce a new net containing small, immature cells. Eventually, the old net breaks down and the young nets are released. Each young net may enlarge greatly to form a mature colony. Chapman (1962) says that the asexual zoospores swarm in the parent cell at daybreak, and that zoospore production occurs most abundantly when the

Figure 11-19 *Hydrodictyon reticulatum*, water net. **A** Portion of the tubular net. **B** Several cells magnified to show the junctions of cells that allow the polygonal pattern.

A *Courtesy of Carolina Biological Supply Company*

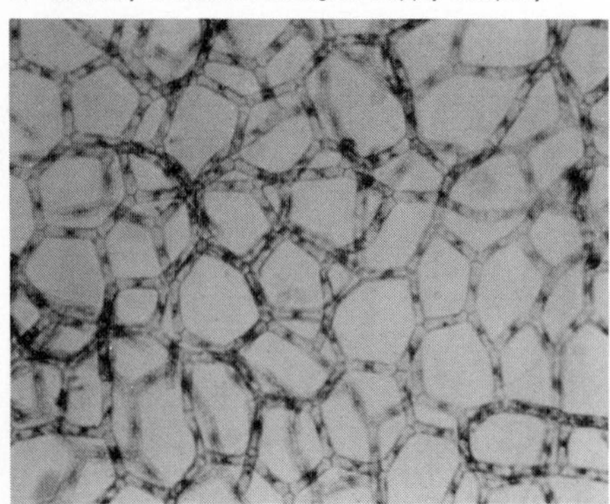

B *Courtesy of Carolina Biological Supply Company*

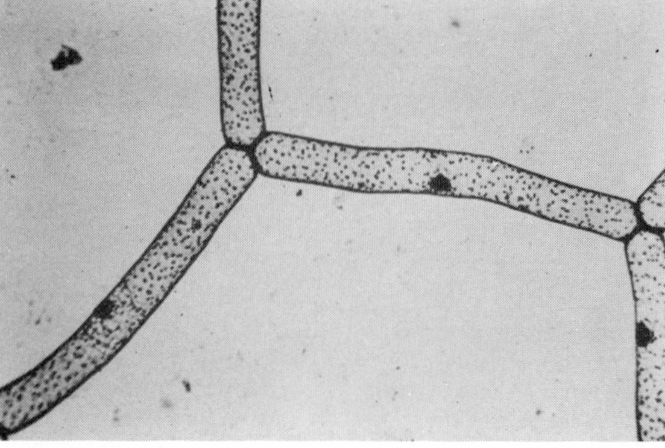

alga is under 10 h of daily illumination of bright light (approximately 1,000 to 1,200 ft-c).

Sexual reproduction occurs when isogametes, morphologically indistinguishable from asexual zoospores, are released from the parent cell. The motile gametes fuse in the water to form zygotes. The zygotes undergo meiosis to form four zoospores, each of which enlarges and becomes a sessile *polyeder*. After later development, each polyeder releases many swimming zoospores in a gelatinous matrix. The zoospores become attached to each other in a polygonal net similar to the adult colony.

CULTURE: For mixed cultures see the general procedures described earlier in the chapter. For unialgal culture use Pringsheim's soil water.

Chapman (1962) describes techniques for inducing asexual and sexual reproduction in *H. reticulatum* by different lighting conditions. He states that if the species is cultured in a weak maltose solution under either bright light or in the dark, and then transferred to distilled water, that zoospores will develop from the algae that were illuminated and gametes from those maintained in the dark.

Oedogonium

USES: (1) To collect "algal paper" from dried ponds and to demonstrate growth from the "paper" when put in water or culture medium. (2) To observe the unique cell division wherein characteristic "annular" rings, or caps, are formed. (3) To observe the development of asexual zoospores and the sexual reproductive organs. (4) To demonstrate mating between two strains of *Oedogonium*.

DESCRIPTION: This is an unbranched filamentous alga occuring usually in fresh water. A few terrestrial species have been identified; no saltwater forms are known. More than 400 species have been identified, primarily by differences in the morphology of sex organs. Young filaments are attached to the substratum (rocks or aquatic angiosperms) by a basal holdfast cell. Mature filaments may remain attached or may become loosened to form floating masses that change with age to a yellow-green or a light yellow. The organism is often found in a dried pond, where the dry *Oedogonium* forms a thin papery layer called "algal paper." *Oedogonium* is often collected from around the base of aquatic plants, where the filaments may be attached to the plants or where layers of the algae may become loosely wrapped around the plants.

Each vegetative cell contains a netlike chloroplast with many pyrenoids and a single nucleus (Fig. 11-20). Cell division is unique in that a thickened ring develops at the upper end of the cell and protrudes toward the cell center. The nucleus moves toward that end and nuclear division occurs. The protoplast of the parent cell divides; the upper portion moves into the ring and protrudes upward to form a new cell, carrying with it a portion of the old outer wall which now becomes the cap, sometimes called an annular ring. Division occurs regularly at the same location, each time forming another cap. Occasionally several rings may be observed at the upper end of a vegetative cell; these become a distinctive trait of this alga.

Asexual reproduction is by zoospores, produced singly in vegetative cells. The

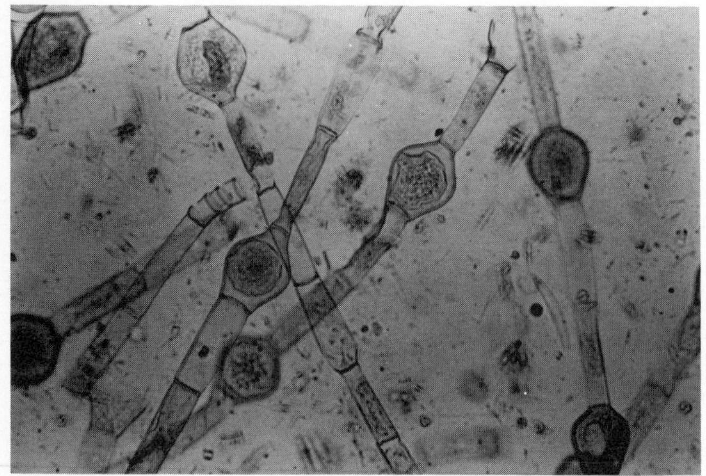

A

B *Courtesy of Carolina Biological Supply Company*

Figure 11-20 *Oedogonium* sp. **A** Filaments with large oogonia. Note the three empty antheridial cells in the filament at left of center. **B** A filament with fertilized oogonia. Note the remains of the dwarf male filament that became epiphytic on the female filament near an oogonium. **C** Diagram of a single cell showing the netlike chloroplast; 1, annular ring; 2, netlike chloroplast; 3, pyrenoid; 4, nucleus. *(A Courtesy of Dean Blinn.)*

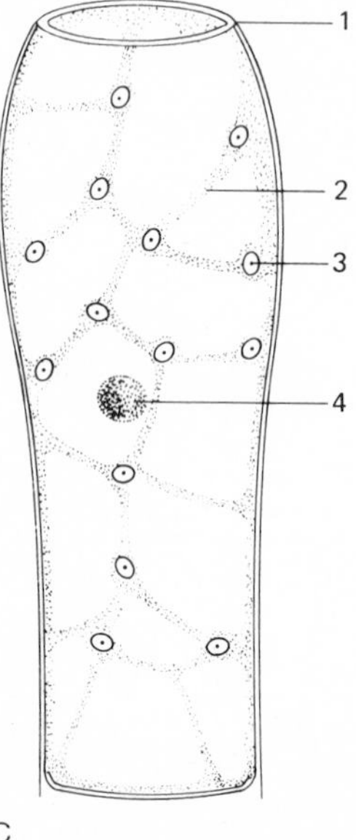

C

zoospores are unique in that they are multiflagellate, each zoospore bearing more than 100 flagella in a crown around the anterior end. Sexual reproduction involves the development of large oogonia, each with one large egg; and the development of special shortened cells, the antheridia, each containing a pair of multiflagellated sperm that are morphologically similar to the asexual zoospores. The oogonia account for the generic name (Greek, *oedos*, swelling; *gonos*, reproductive organ). Species may be monoecious with both kinds of sexual organs developing on a single filament, or dioecious, with separate male and female filaments. In some species, the male filament is dwarfed and becomes attached as an epiphyte on a female filament near a developing oogonium (Fig. 11-20). A mating set of *Oedogonium* can be obtained from Carolina Biological Supply Company. Instructions are sent with the set and are similar to those described earlier in the chapter for *Chlamydomonas*.

CULTURE: For cultures of mixed algae, see the discussion given earlier in the chapter. For unialgal culture use Pringsheim's soil water (Appendix A), or try other media described for other green algae.

Pandorina

USES: (1) To observe a colonial flagellate. (2) To observe sexual reproduction in mating strains.

DESCRIPTION: *Pandorina* is much like *Eudorina* and *Volvox*, and the three genera belong to the same family, Volvocaceae. The alga forms a spherical or slightly flattened colony of 8, 16, 32 (usually 16), biflagellate cells that are closely packed at the center of a firm gelatinous matrix (Fig. 11-21). Individual cells are about 12 microns long; the colony is about 50 microns in diameter, but sometimes much larger, up to 250 microns. Each cell contains one chloroplast, two contractile vacuoles, and a stigma. Asexual reproduction occurs as individual cells divide to form small colonies which are released when the gelatinous matrix breaks open. Sexual reproduction occurs when two isogametes fuse to form a zygote. Mating strains of *Pandorina morum*, + and −, may be obtained from

Figure 11-21 Colonial types of green algae. **A** *Pandorina* sp. **B** *Eudorina* sp. Both are living colonies moving in water. Actual size is approximately 50 and 80 microns, respectively.

A

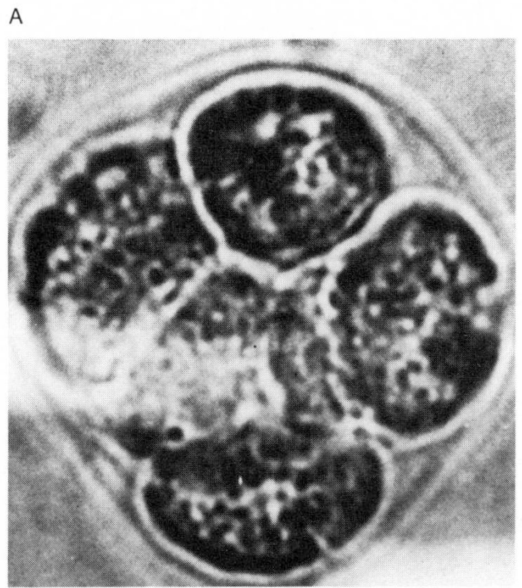

B

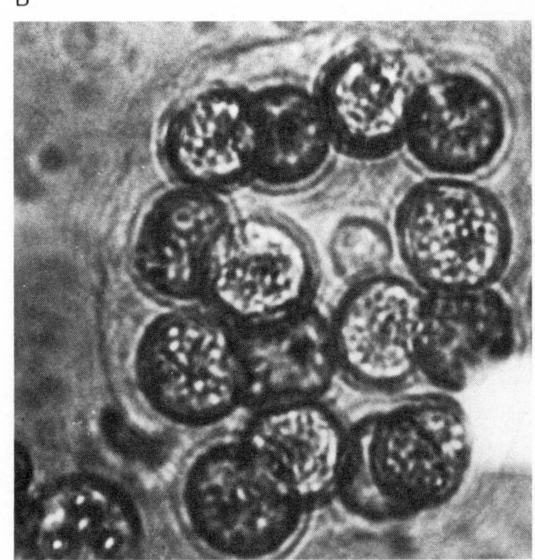

Carolina Biological Supply Company. Instructions for inducing sexual reproduction accompany the shipment.

CULTURE: See *Volvox*.

Protococcus (Pleurococcus)

USES: (1) To observe a simple green alga that is easily collected and stored in a dry condition. (2) To observe the reduced filaments, formed when dividing cells remain joined (see *Pleurococcus* in Scagel et al., 1965).

DESCRIPTION: *Protococcus* is a terrestrial alga found on damp bark of trees, on damp rocks, boards, and other substratum. Generally the round cells are single, although they may remain joined after cell division, a condition considered by some phycologists as a reduced filament (Fig. 11-22). In most regions, the alga is easily obtained by scraping the green growth from trees or rocks. Each cell contains one chloroplast; no pyrenoids are present. The centrally located nucleus can be observed when cells are stained with iodine stain, Lugol's (Appendix A). Reproduction is by cell division. The cells may be carried about by animals and by wind or water.

Protococcus tolerates extreme desiccation. Scrapings can be retained indefinitely in the laboratory when placed in moderate light during the normal daylight hours. At the time of study, moisten a small portion of cells with water. The alga grows readily, sometimes too much so, in the greenhouse on earthen pots and on damp wood.

CULTURE: *Protococcus* is easily maintained in a dry or moist condition as described above.

Scenedesmus

USES: (1) To observe a coenobic alga. (2) To observe the asexual reproduction in which an "autocolony" develops within a cell.

DESCRIPTION: *Scenedesmus* is a freshwater alga that occurs commonly in

Courtesy of Carolina Biological Supply Company

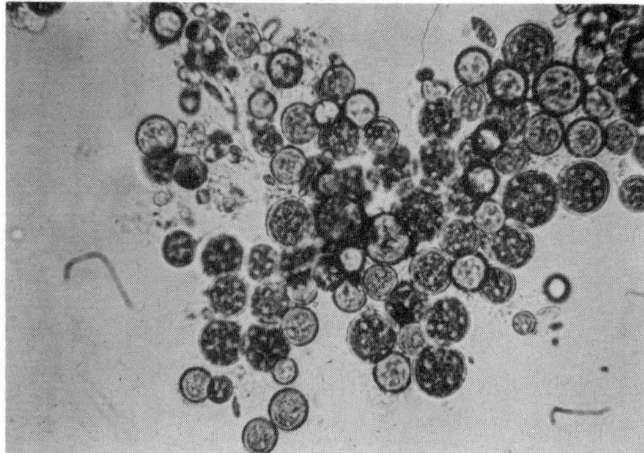

Figure 11-22 *Protococcus* sp.

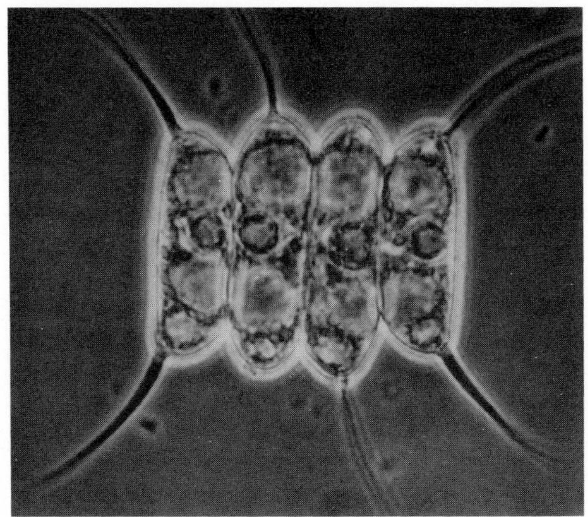

Figure 11-23 *Scenedesmus* sp. *(Courtesy of Ward's Natural Science Establishment, Inc., Rochester, N.Y.)*

open water and at the edge, among weeds and other floating vegetation. Usually four oval or spindle-shaped cells are joined laterally in a single row, as for example, in the common species, *S. quadricauda*. Other species may have double rows of cells, or a greater number of cells in a single row. Spines are often present, particularly on terminal cells. Each cell contains a single nucleus, a chloroplast, and one pyrenoid (Fig. 11-23). The colony is coenobic, in that the number of cells does not increase after the young colony is released from a parent cell. Growth is by an increase in the size of each original cell of a colony. Asexual reproduction occurs in mature colonies when a young colony (an autocolony) develops within each of one or more vegetative cells. The parent cell ruptures and releases the young colony. Occasionally the development and liberation of autocolonies can be observed in cultures several hours after transfer to fresh medium.

CULTURE: For mixed cultures see the general techniques described earlier in the chapter. For unialgal cultures use either of the following media (Appendix A): (1) Pringsheim's soil water; (2) Bold's basal medium with agar (increased nitrogen).

Spirogyra

USES: (1) To study the cellular structure of a beautiful and common freshwater alga. (2) To demonstrate conjugation in the living cultures and to test different media for maximum inducement of conjugation. (3) To observe cytoplasmic streaming with oil immersion lens at 1,000X.

DESCRIPTION: Probably *Spirogyra* is used more often for class studies than any other alga. Microscopically, *Spirogyra* is a beautiful alga because of the ribbonlike chloroplasts that spiral around the cell contents. The alga occurs floating in quiet streams, ponds, and ditches, and is easily identified by its bright-green color and soft, silky texture.

Spirogyra consists of a long, unbranched filament containing a single row of relatively long, cylindrical cells (Fig. 11-24). The chloroplasts vary in number from 1 to 10 or 12 and may have a smooth or serrated margin. Many pyrenoids are present on each chloroplast. To observe the details of cellular organization, filaments with only one or two chloroplasts should be selected. A large vacuole fills the center of the cell. The single nucleus usually appears in a middle position of the cell suspended by strands of cytoplasm in the vacuole; the cytoplasmic threads run from the nucleus to an outer layer of cytoplasm. Cytoplasmic streaming can be observed in the threads when a cell is placed under an oil immersion lens at 960 to 1,000X magnification. When cells are stained with iodine stain

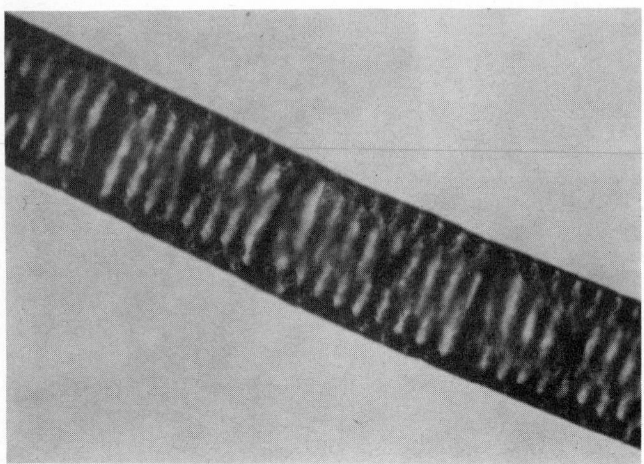

A *Courtesy of Carolina Biological Supply Company*

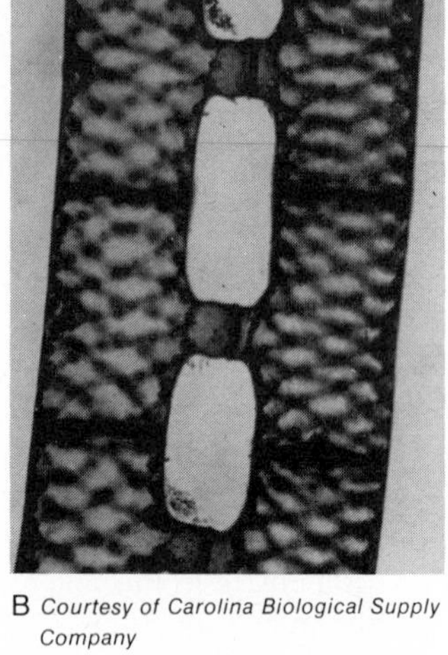

B *Courtesy of Carolina Biological Supply Company*

Figure 11-24 *Spirogyra* sp. **A** Vegetative filament. **B** Conjugation, early stage. **C** Conjugation, later stage. **D** Conjugation at completion; 1, conjugation tube; 2, "male gamete" is passing through tube; 3, "female gamete"; 4, zygote.

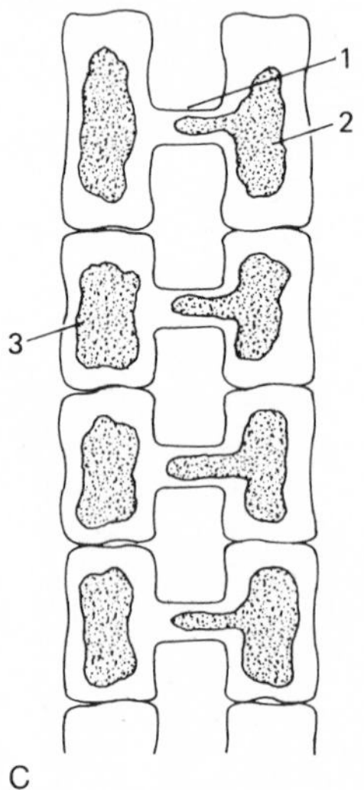

C

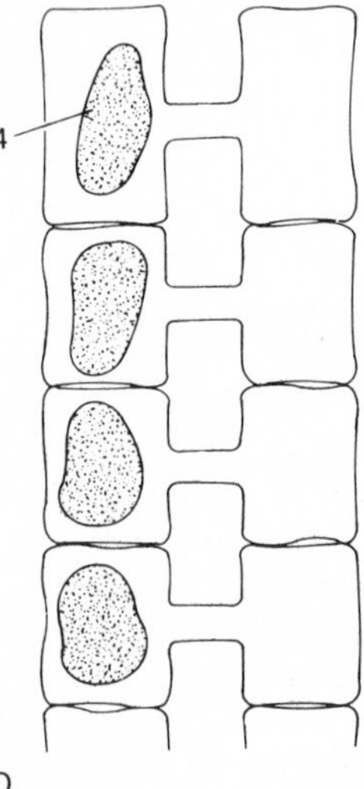

D

(Lugol's)(Appendix A) the cytoplasm and nucleus (in a ring of cytoplasm) are easily observed under about 400X magnification. The filament is covered with a gelatinous coat that accounts for the silky texture of a floating mass.

The filament lengthens as individual cells divide and remain joined. Asexual reproduction is by fragmentation. Sexual reproduction occurs by conjugation when two filaments become aligned opposite to each other. A protuberance forms between opposite cells along the two filaments; the tips of the opposite protrusions touch, and the end walls dissolve forming an open bridge between the two cells. The protoplast of one cell (sometimes called the male gamete) moves through the bridge and unites with the protoplast (female gamete) of the opposite cell. Generally many or all opposite cells of the two filaments conjugate about the same time, thus producing a ladderlike configuration. The outer covering of the fused body hardens, forming a dark-brown, dormant zygote. The zygotes soon are shed into the bottom sediment, and the empty filaments break into pieces and deteriorate. Eventually, each zygote germinates into a new filament. In most regions, *Spirogyra* is abundant in streams and ponds during the spring when zygotes of the previous fall germinate, and again in the fall when some of the spring zygotes germinate.

CULTURE: For mixed cultures see the instructions given earlier in this chapter. For pure cultures use Pringsheim's soil water. Bold (1967b) states that Pringsheim's soil water without $CaCO_3$ is favorable for many species of *Spirogyra*.

On a number of occasions during the spring and fall months we have had *Spirogyra* to conjugate in pond water collections that were placed in shallow glass bowls of pond water in a north window and maintained at room temperature. Bold (1967b) describes the following procedure for inducing conjugation. Transfer portions of rapidly growing *Spirogyra* from Pringsheim's soil water to plates of 1.5% plain agar (1.5 g plain agar in 100 ml of distilled water and autoclaved before pouring into sterile petri plates). Put plates under constant, moderate light (300 to 500 ft-c) and at room temperature (22 to 23°C). Generally conjugation will begin in 3 or 4 days.

Ulothrix

USES: (1) To study the cellular organization, the bandlike chloroplast, and the basal holdfast cell. (2) To observe asexual and sexual reproduction and to test various conditions and media for inducing both types of reproduction.

DESCRIPTION: Both marine and freshwater species have been described. The alga consists of unbranched filaments of cylindrical cells that may be relatively long, or shorter than wide (Fig. 11-25). A basal cell becomes differentiated into a holdfast. A filament may remain attached to substratum or to other plants, or it may break loose and float in a mass with other algae. A single bandlike chloroplast encircles, completely or partially, the cellular content (Fig. 11-25). One or more pyrenoids are present on the chloroplast; the cell contains one nucleus.

Asexual reproduction occurs in mature filaments when one or more quadriflagellate zoospores are released from vegetative cells. Each zoospore may become attached to substratum and develop into a vegetative filament. Sexual reproduc-

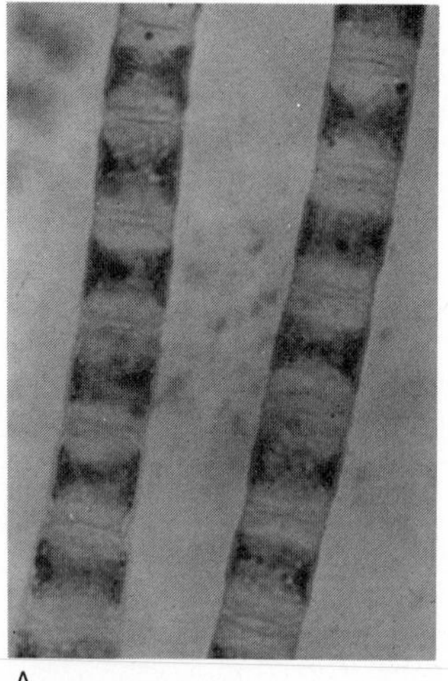
A

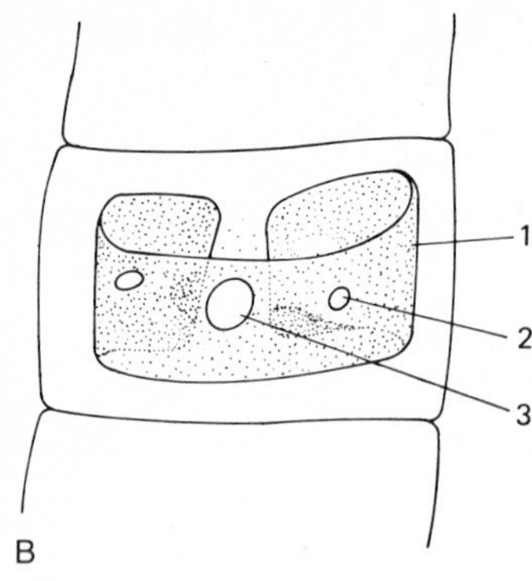

Figure 11-25 *Ulothrix* sp. **A** Filaments showing band-shaped chloroplast, a distinguishing characteristic of this alga. **B** Diagram of a cell: 1, chloroplast; 2, pyrenoid; 3, nucleus. *(A Courtesy of Ward's Natural Science Establishment, Inc., N.Y.)*

tion is by biflagellate isogametes. Species are monoecious or dioecious. Zygotes may remain dormant for several months, following which, meiosis occurs producing four zoospores. Each zoospore may develop into a vegetative filament.

Marine species occur in the intertidal zone, often as epiphytes on other seaweeds. Many freshwater species are found attached or floating in still water. The common *U. zonata* usually occurs in cold, running water, where it makes a silky, green coating on rocks and other substrate. The alga is most abundant during winter and spring months, indicating then that a cool temperature is desirable for culture and maintenance.

CULTURE: For mixed cultures see the procedures described earlier in the chapter. For a pure culture, use Pringsheim's soil water. Maintain cultures at cool temperatures of 15 to 18°C and in a low light of about 150 to 200 ft-c.

Volvox

USES: (1) Used regularly in many general biology courses for studying the colonial form and the asexual and sexual reproduction. (2) For explorations regarding optimum culture conditions and the inducement of asexual and sexual reproduction. See Darden (1966) for tests on sexual differentiation.

DESCRIPTION: *Volvox* is a colonial flagellate, found mostly in fresh water, and similar in structure to *Eudorina* and *Pandorina* (Fig. 11-26). Individual cells of the colony are much like *Chlamydomonas*, with two anterior flagella, a green chloroplast, and a stigma. The large spherical colony of *Volvox globator* is 2 to 3 mm in diameter, and contains more than 5,000 individual cells. Other species may be smaller. Most cells are somatic (vegetative) and are joined by cytoplasmic connections in a gelatinous matrix that surrounds the cells. Asexual reproduction occurs when certain cells divide to form small colonies that are eventually

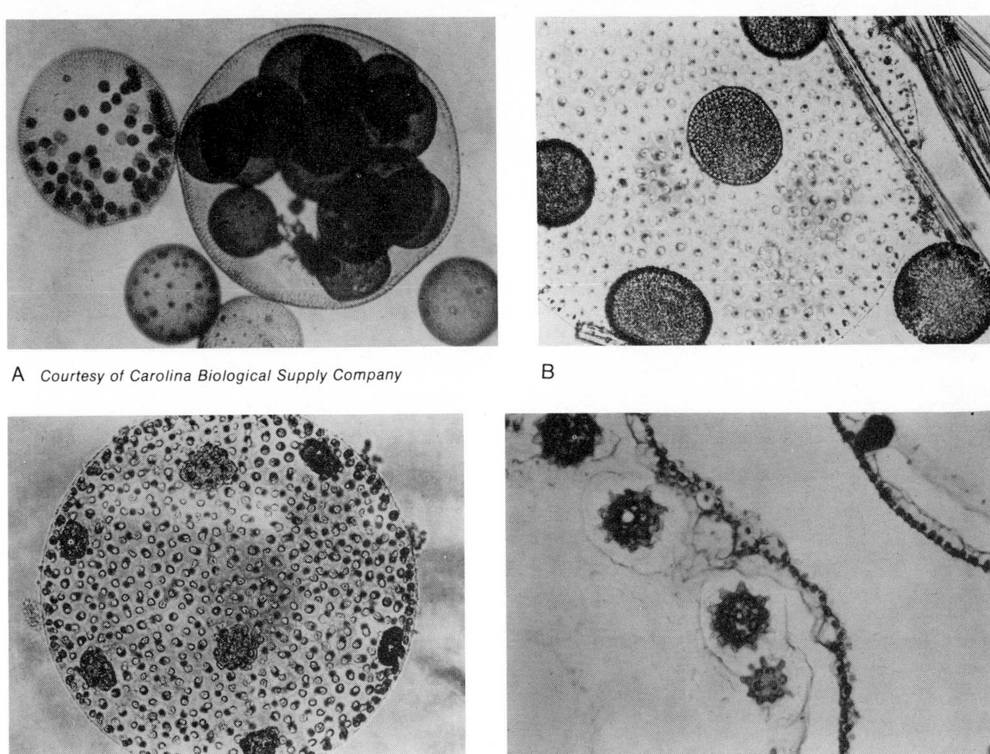

A *Courtesy of Carolina Biological Supply Company*

B

C

D *Courtesy of Carolina Biological Supply Company*

Figure 11-26 *Volvox.* **A** *Volvox aureus.* Left, colony with zygotes. Right, a colony containing large, asexual daughter colonies. **B** *V. aureus,* close-up of a parent colony, showing small vegetative cells and the release of large asexual daughter colonies. **C** *V. aureus,* close-up of parent colony, showing sexual daughter colonies in development. **D** *V. globator,* portion of parent colony containing zygotes. Note spiny covering on zygotes.

released from the parent colony. Some authorities regard the reproductive cells of some species (e.g., *V. aureus*) as gametes that develop without fertilization, and thus they consider the reproduction to be parthenogenetic. Several such parthenogenetic colonies may be present at one time in the parent colony (Fig. 11-26). Occasionally, a "granddaughter" colony can be seen inside a daughter colony before the latter is released. An interesting feature of a young colony is its inversion during development. During the early stage, cells are oriented with the anterior end pointing inward. After the 32-cell stage, inversion occurs through an opening, and the cells become oriented with the anterior end and the flagella pointing outward.

Sexual reproduction occurs in all species and involves the fertilization of large, immotile female gametes (eggs) by small male gametes (sperms) which arise in packets. Depending on the species, colonies are monoecious, with both male and female gametes in the same colony, or they are dioecious, with male and female gametes produced in separate colonies. Male gametes are released and enter the female colonies where fertilization occurs. Eventually the parent colony breaks down, and the dormant zygotes are released. Each zygote may develop into a young colony. Inversion occurs in young, sexually formed colonies in the same manner as in a parthenogenetic colony. The zygote becomes encysted in a tough, often spiny, outer covering (Fig. 11-26). The zygotes sink to the bottom of a pond and remain dormant until favorable conditions allow their germination, generally in the spring.

CULTURE: Collect in fresh water, or obtain from biological suppliers. Carolina Biological Supply Company furnishes a number of species for special studies. *Volvox aureus* and *V. globator* are used most often for routine studies. For unialgal culture, use Pringsheim's soil water (Appendix A). Place cultures in moderate light at temperatures of 20 to 24°C.

Zygnema

Zygnema is closely related to *Spirogyra*. Many freshwater species occur that form bright-green, silky clumps floating in still water. The unbranched filaments consist of uninucleated, cylindrical cells. Each cell contains two large starlike chloroplasts, each with a central pyrenoid (Fig. 11-27). The stellate chloroplasts are a distinctive trait, although occasionally the shape may be obscured by food granules. Staining with Lugol's iodine or with methylene blue (Appendix A) helps in observing the organelles and the outer gelatinous sheath present in several species.

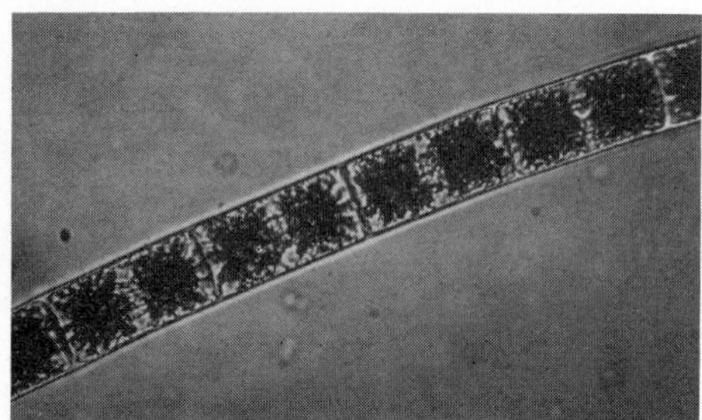

A *Courtesy of Carolina Biological Supply Company*

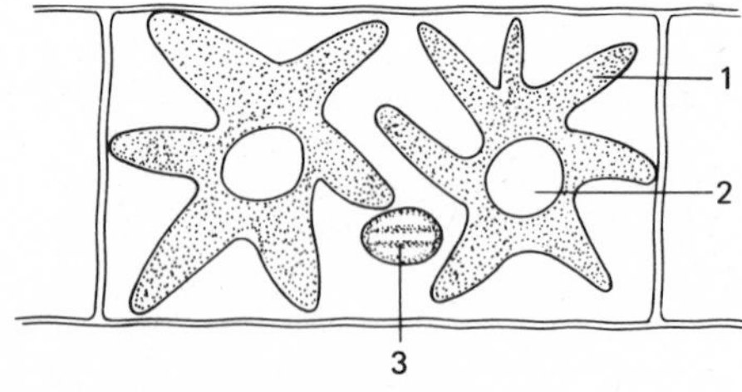

B

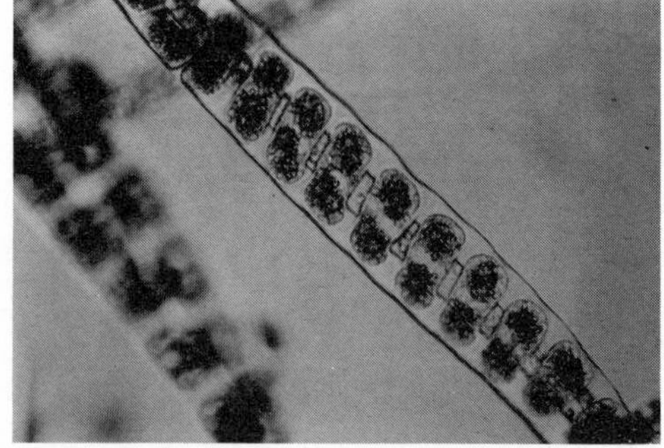

C *Courtesy of Carolina Biological Supply Company*

Figure 11-27 *Zygnema* sp. **A** Filament. **B** Diagram of a single cell showing stellate chloroplasts; 1, chloroplast; 2, pyrenoid; 3, nucleus. **C** Conjugation.

Asexual reproduction is by fragmentation of filaments. Sexual reproduction is similar to conjugation in *Spirogyra*.

CULTURE: For mixed collections, see the procedures described earlier in the chapter. For unialgal culture, use Pringsheim's soil water (Appendix A).

THE EUGLENOIDS (EUGLENOPHYCOPHYTA)

The euglenoids comprise a small group of about 25 genera and 400 species, most of which are uniflagellated and unicellular. Because the organisms display characteristics of both animals and plants, they are often placed in the class Mastigophora of the protozoa. Other classification systems place the chlorophyllous euglenoids with the algae and the nonchlorophyllous genera with the protozoans.

Euglena

USES: (1) To study responses to light. (2) To observe asexual reproduction and encystment.

DESCRIPTION: Species of *Euglena* are common in stagnant ponds and ditches, where the active cells and cysts may form a green film on the water surface, or green spots on very moist soil.

The cell is spindle shaped in several species and ranges in length from 35 to 55 microns (*E. gracilis*), and from 100 to 180 microns (*E. spirogyra*). The pellicle is longitudinally, spirally striated. Some species have a thin, flexible pellicle, and the body shape changes from spindle to ovoid or spherical (Fig. 11-28). The light-sensitive, red or orange stigma (eyespot) is usually anterior. Ten to twenty chloroplasts are present and usually contain pyrenoids. The anterior flagellum arises in the gullet (cytopharynx) and emerges through the mouth (cytostome). The nucleus is usually central. Reserve food is stored in several large bodies in the form of paramylum, a polysaccharide that is distinctive to euglenoids. Asexual reproduction is by longitudinal fission. Most species encyst when conditions become unfavorable to growth. Cysts may survive for several years and may be blown about by wind currents. *Euglena* often excysts in hay infusions.

Euglena gracilis is commonly collected with other freshwater algae, and it is supplied by most biological companies. The organism is easily cultured and has been the subject of much physiological study. For example, when put in total darkness, *E. gracilis* (and certain other species) lose their chlorophyll and then undergo a saprophytic nutrition, at which time protein foods are required. *Euglena* does not ingest solid food.

CULTURE: For collections of mixed algae, see the general procedures described in the introduction of the chapter. For unialgal cultures, use any of the following media (Appendix A): (1) Pringsheim's soil water (to each liter of medium add one grain of wheat, rice, or barley, or one pea); (2) grain or hay infusions. Bold (1967b) recommends a very dilute, split-pea infusion (Appendix A). Place cultures in bright light at 21 to 25°C. *Euglena* is easily maintained when transferred to fresh medium every 2 months.

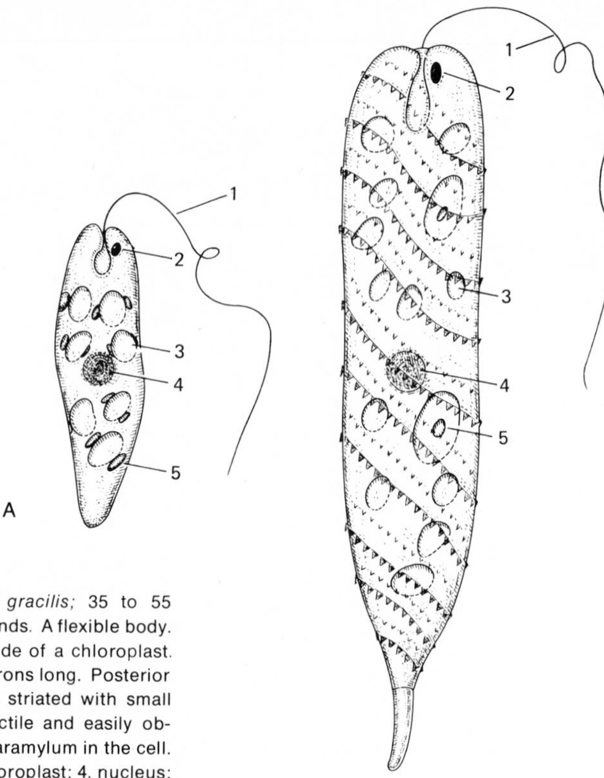

Figure 11-28 *Euglena.* **A** *E. gracilis;* 35 to 55 microns long and rounded at ends. A flexible body. A paramylum body on each side of a chloroplast. **B** *E. spirogyra.* 100 to 180 microns long. Posterior end is a clear spike; body is striated with small conelike peaks that are refractile and easily observed. Two large bodies of paramylum in the cell. 1, Flagellum; 2, eyespot; 3, chloroplast; 4, nucleus; 5, paramylum. *(A and B From T. L. Jahn, 1949.)*

THE YELLOW-GREEN AND GOLDEN ALGAE (CHRYSOPHYCOPHYTA)

The division Chrysophycophyta consists of a large number of algae distinguished by the presence of certain carotenoid and xanthophyll pigments in addition to the chlorophylls, and by their conversion of photosynthetic products to oil and to carbohydrates other than starch.

These algae may be subdivided into three groups: (1) Bacillariophyceae, the diatoms, a large ubiquitous group of marine and freshwater species; (2) the Xanthophyceae, the yellow-green algae that at one time were grouped with the green algae; and (3) Chrysophyceae, the golden-brown algae, which occur rather infrequently in freshwater and marine collections and which are not described further in this chapter. Chrysophyceae may be of considerable interest, however, to persons collecting in cold waters, and further information is available in most of the general listings in the bibliography of this chapter.

The Diatoms

USES: (1) To observe the variations among genera and species in markings and shapes. (2) As a producer organism in an aquarium ecosystem. (3) To develop a small ecosystem on glass slides that have been suspended in fresh or salt water, as described in the introduction to the chapter. (4) To observe fossil diatoms in silver polish or diatomaceous earth, mixed in a drop of water on a glass slide.

DESCRIPTION: Diatoms make up a large group of extremely fascinating organisms. At least 5,000 species have been identified in fresh water, salt water, and soil. Diatoms form an important part of aquatic food chains. In fact, marine diatoms have been called "grass of the sea" because of their extremely vital contribution to the sustenance of marine animals. Diatoms appear in plankton, in layers on the bottom sediment, in beach sand, and as a brown scum on rocks and on other algae. Extensive deposits of diatom shells, both freshwater and marine, are found in several parts of the world and are mined as diatomaceous earth for commercial uses. Perhaps the greatest and most unique collection is the bed of almost pure diatom fossils at Lompoc, California, where the deposit covers an area of about 12 miles square, approximately 1,400 ft deep.

The beauty of diatoms is comparable to, or surpasses, that of the desmids; their variations in shape and pattern far exceed that of any other algal group. Most types are unicellular or colonial; a few are filamentous. Distinctive traits are as follows: (1) The silicon wall (frustule) consisting of two valves, an upper valve (epitheca) that overlaps a lower valve (hypotheca) somewhat like a cover of a box; or with the two valves indirectly joined by connections to overlapping girdles; (2) the wide variety of decorations and markings on the valves, including the beautiful patterns of etched lines, rows of dots, pores and ribs, and hornlike processes; and (3) the variations in shape which can be identified only when observations are made from both a view of the valve surface (upper or lower) and a view of the girdle, or lateral, surface.

Diatoms are grouped into two orders on the basis of symmetry:

1. *Order* Pennales. The cell is elongated in the shape of a spindle, crescent, needle, or wedge (from a valve view), with markings in a bilaterally symmetrical pattern. Although markings vary, each valve, depending on the genus, contains a long groove (raphe) on the midline, or occasionally at one or both sides, and interrupted at midpoint by the central nodule. Usually the cells contain one or two yellow or golden-brown plastids. Species of this order are found primarily in fresh water (Fig. 11-29). Scagel et al. (1965) state that only the pennate diatoms with raphe show movement, and that apparently the motion is a result of cytoplasmic streaming along, or in and out of, the raphe.

2. *Order* Centrales. The cell is circular, triangular, or elongated (from valve view), with markings in a radially symmetrical pattern. Usually the cell contains many yellow or golden-brown chloroplasts. These diatoms have no raphe, but often have spines or horns and are primarily marine (Fig. 11-29).

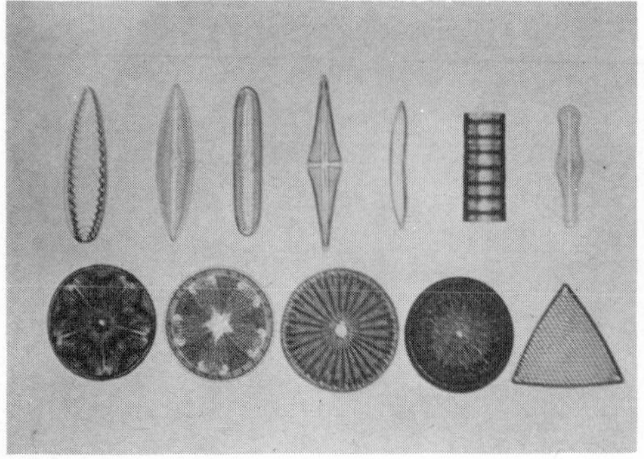

A

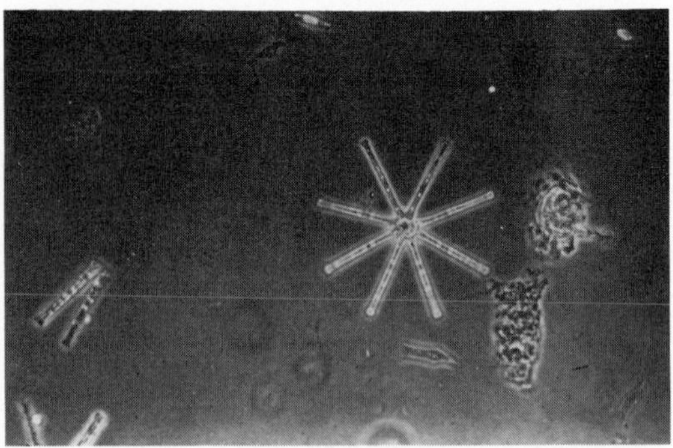

B

Figure 11-29 Diatoms. **A** Top row, pennate forms; bottom row, centric forms. **B** *Asterionella* sp., a colony of cells in girdle view. **C** *Chaetoceros elmorei,* in girdle view; collected in Devil's Lake, North Dakota, a salt lake. *(A Courtesy of Ward's Natural Science Establishment, Inc., Rochester, N.Y. B and C Courtesy of Dean Blinn.)*

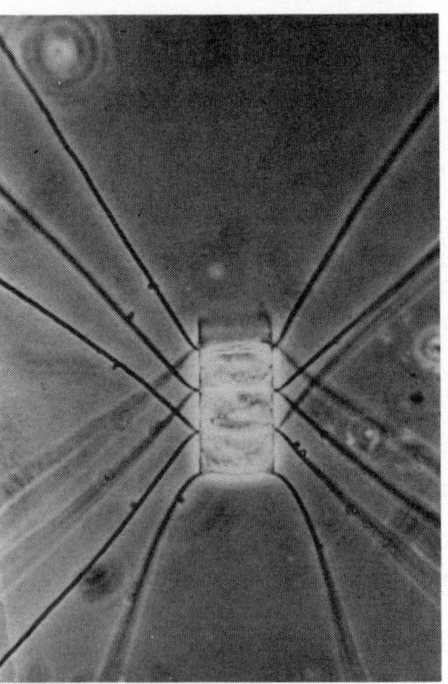

C

Asexual reproduction is by cell division. Each daughter cell receives one valve from the parent which becomes the upper valve of the offspring. A new lower valve is developed by each new offspring. Thus the offspring receiving the lower valve from the parent will be diminished in size, although apparently the size increases with growth. Sexual reproduction varies among species and apparently is not easily induced in laboratory cultures.

Because of the great number of diatoms and their great variations in morphology and ecology, only a relatively brief account is given here. More detailed information can be found in Patric and Reimer (1966), and in many of the general texts listed in the bibliography at the end of this chapter.

CULTURE: For cultures of mixed collections follow the general procedures

A *Courtesy of Carolina Biological Supply Company*

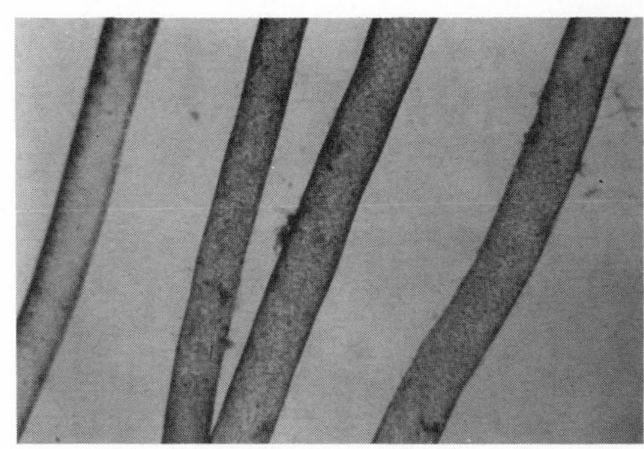

B *Courtesy of Carolina Biological Supply Company*

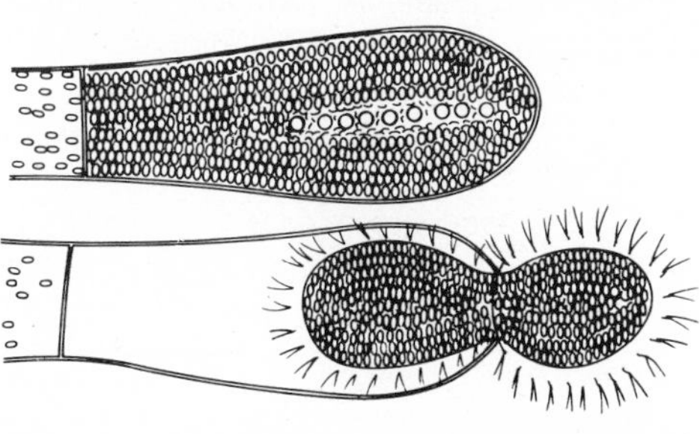

C

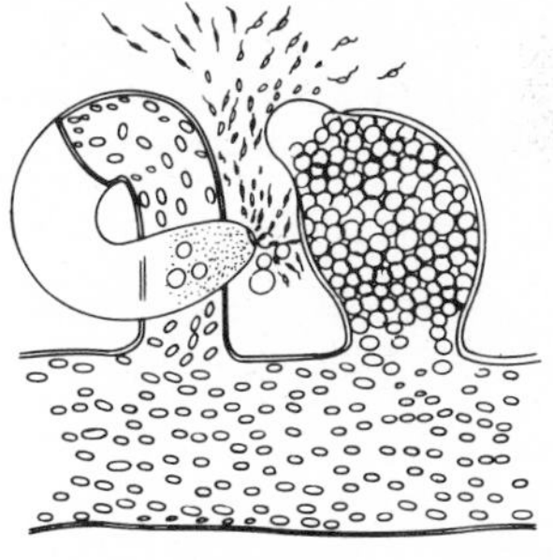

D

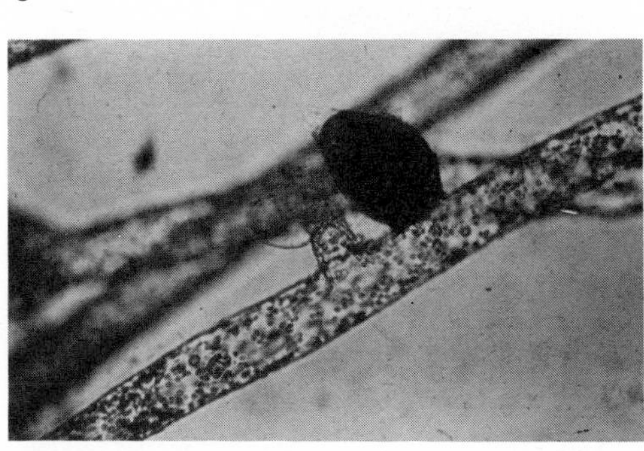

E

Figure 11-30 *Vaucheria* sp. **A** A mass of alga floating on water. **B** Several filaments. **C** Diagram of asexual reproduction; development of a sporangium, at top; diagram of a zoospore, at bottom. **D** Mature antheridium and archegonium. **E** Sexual reproduction in a living alga. A large ovoid archegonium is at center, and an antheridium is on the same filament, below and to left of the archegonium. *(C and D From G. M. Smith after Couch. E Courtesy of Dean Blinn.)*

given earlier in the chapter for freshwater and marine algae. For unialgal cultures use the following media (Appendix A): (1) for freshwater diatoms, Pringsheim's soil water or Bold's basal medium; (2) for marine diatoms, Erdschreiber solution.

Mixed cultures of diatoms can be maintained easily in an aquarium with small fish or invertebrates and aquatic plants. Avoid overcrowding. Many kinds of freshwater species multiply rapidly, often requiring removal of portions to avoid over-density as indicated by the brown or golden color of the water.

Vaucheria (Yellow-green Alga)

USES: (1) To study a common, coenocytic alga. (2) To observe zoosporangia and sexual reproductive organs.

DESCRIPTION: *Vaucheria* is commonly called "water felt" because of its thick, velvetlike mats that float or are submerged in water, or that grow on damp soil. The yellow-green, sparsely branched filaments are tubular, and without cell crosswalls (coenocytic). Thus, the filament is multinucleate (Fig. 11-30).

Asexual reproduction is by the development of a zoosporangium at the end of a filament. Large, multiflagellated and multinucleated zoospores develop within the zoosporangium (Fig. 11-30). After their release, the zoospores swim about for a short while, then lose their flagella, and germinate into a new filament.

Sexual reproduction is by the formation of special sex organs, oogonia and antheridia, either directly on a main filament or on special branches (Fig. 11-30). Most species are monoecious. Each oogonium contains one large egg. Many sperms develop in an antheridium; upon release the mature sperms swim to the eggs, where fertilization occurs. Apparently meiosis takes place when the zygote begins to germinate.

CULTURE: For mixed cultures follow the general procedures given in the introduction of the chapter. For unialgal cultures use Pringsheim's soil water. Zoosporangia and sexual reproductive organs may appear within about 2 weeks.

REFERENCES

Bold, H. C.: 1967a, *Morphology of Plants*, 2d ed., Harper & Row, Publishers, Incorporated, New York, 541 pp.

Bold, H. C.: 1967b, *A Laboratory Manual for Plant Morphology*, Harper & Row, Publishers, Incorporated, New York, 123 pp.

Bonang, C. B.: 1971, "The Slide-Rack Collector," *Am. Biol. Teacher*, **33** (9): 551.

Chapman, V. J.: 1962, *The Algae*, St. Martin's Press, Inc., New York, 472 pp.

Darden, W. H.: 1966, "Sexual Differentiation in *Volvox aureus*," *J. Protozool.*, **13**: 239–255.

Dawson, E. Y.: 1966, *Marine Botany*, Holt, Rinehart and Winston, Inc., New York, 371 pp.

Johansen, D. A.: 1940, *Plant Microtechnique*, McGraw-Hill Book Company, New York, 523 pp.

Klein, R. M., and D. T. Klein: 1970, *Research Methods in Plant Science*, Natural History Press, New York, 756 pp.

Lewin, R.A.: 1953, "The Genetics of *Chlamydomonas moewusii* Gerloff," *J. Genet.*, **51**: 543–560.

Murphy, G. W.: 1966, "For Culturing Algae: A Modification in the Use of Bristol's Solution," *Am. Biol. Teacher*, **28** (2): 122.

Patric, R., and C. W. Reimer: 1966, *The Diatoms of the United States*, vol. I, Monograph 13, Academy of Natural Sciences of Philadelphia.

Prescott, G. W.: 1964, *The Fresh-Water Algae*, Wm. C. Brown Company Publishers, Dubuque, Iowa, 272 pp.

Prescott, G. W.: 1968, *The Algae: A Review*, Houghton Mifflin Company, Boston, 436 pp.

Sager, R. A., and S. Granick: 1954, "Nutritional Control of Sexuality in *Chlamydomonas reinhardi*," *J. Gen. Physiol.*, **37**: 729–742.

Sager, R. A.: 1955, "Inheritance in the Green Alga *Chlamydomonas reinhardi*," *Genetics*, **40**: 476–489.

Sager, R. A.: 1960, "Genetic Systems in *Chlamydomonas*," *Science*, **132**: 1459–1465.

Sass, J. E.: 1951, *Botanical Microtechnique*, 2d ed., The Iowa State University Press, Ames, 228 pp.

Scagel, R. F., R. J. Bandoni, G. R. Rouse, W. B. Schofield, J. R. Stein, and T. M. C. Taylor: 1965, *An Evolutionary Survey of the Plant Kingdom*, Wadsworth Publishing Company, Inc., Belmont, Calif., 658 pp.

Smith, G. M. (ed.): 1951, *Manual of Phycology*, The Ronald Press Company, New York, 373 pp.

Starr, R. C.: 1964, "The Culture Collection of Algae at Indiana University," *Am. J. Botan.*, **51** (9): 1013–1044.

Tiffany, L. H.: 1938, *Algae, The Grass of Many Waters*, Charles C Thomas, Publisher, Springfield, Ill., 171 pp.

Other Literature

Dawson, E. Y.: 1956, *How to Know the Seaweeds*, Wm. C. Brown Company Publishers, Dubuque, Iowa.

Fogg, G. E.: 1965, *Algal Cultures and Phytoplankton Ecology*, The University of Wisconsin Press, Madison.

Hoshaw, R. W.: 1969, "Problems Designed for the Use of Algal Cultures," *Am. Biol. Teacher*, **31** (1): 21–26.

Jackson, D. F. (ed.): 1968, *Algae, Man and the Environment*. Proceedings of an International Symposium, Syracuse University Press, Syracuse, N.Y., 554 pp.

James, D. E.: 1969, "Maintenance and Media for Marine Algae," *Carolina Tips*, **32** (12): 45–46.

James, D. E.: 1969, "Unialgal Cultures," *Carolina Tips*, **32** (9): 33–36.

James, D.E.: 1971, "Isolation and Purification of Algae," *Carolina Tips*, **34** (9): 33–35.

Lissant, E. K.: 1969, "Construction of an Algal Culture Chamber," *Am. Biol. Teacher*, **31** (1): 27–28.

Lund, J. W. G., and J. F. Talling: 1957, "Botanical Limnological Methods with Special Reference to the Algae," *Botan. Rev.*, **23**: 489–583.

Pringsheim, E. G.: 1964, *Pure Cultures of Algae*, Hafner Pub. Company, Inc., New York, 119 pp. (Originally published in 1946 by Cambridge University Press, New York.)

Sohn, B. I.: 1972, "Algae as Pollution Indicators: Analysis Using the Membrane Filter," *Am. Biol. Teacher*, **34** (1): 19–22.

Taft, Clarence E.: 1964, "Collecting Algae for Winter Teaching," *Turtox News*, **42** (10): 268–269.

U.S. Dept. of Interior, 1966, *A Guide to the Common Diatoms of Water Pollution Surveillance System Stations*, Federal Water Pollution Control Administration, 1014 Broadway, Cincinnati, Ohio 45202.

THE PROTOZOA

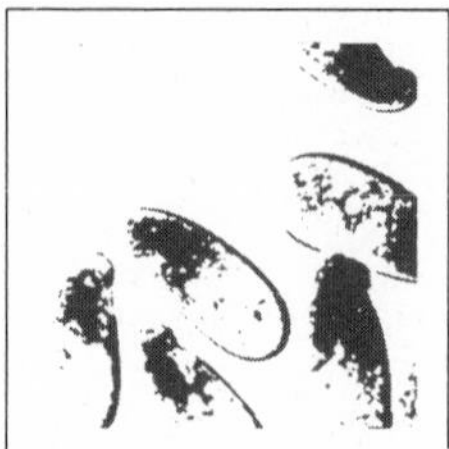

The first portion of the chapter contains general techniques for collecting and handling protozoa. A second portion contains descriptions of particular genera that are appropriate for the general biology laboratory.

COLLECTING PROTOZOA

Protozoa will be found in open surface water (flagellates and ciliates), in the bottom sediment (amoeboid forms and ciliates), on or among vegetation (ciliates, flagellates), and on submerged rocks, stumps, and debris (possibly all forms and particularly the stalked ciliates). In cold-winter regions where ice forms a cover on bodies of water, protozoa can be collected in bottom sediment when a hole is drilled through the ice.

Use a plankton towing net (no. 20 mesh) to collect protozoa in open surface water. A small dip net is satisfactory for collecting at the edge of water. Collect bottom sediment by using a standard bottom sampler or by scraping across the bottom surface with a small can fastened to a pole. To collect floating vegetation, a rich source of protozoa, lower a jar or other container slightly below water surface to allow vegetation and water to flow into the container. Use a blade to scrape scum from rocks and other submerged objects. Transfer collections to small, clean bottles, equipped with a lid. Lids may be tightened while transporting samples to the laboratory, where they should be immediately loosened and laid lightly over the bottle opening. Equipment and techniques are described in greater detail in Chap. 1.

Saltwater Protozoa

Collecting is much the same as for freshwater protozoa and is also described in Chap. 1. Damp beach sand may contain an abundance of protozoa. Little research has been done on these organisms, which could provide worthwhile studies in ecology and physiology as well as taxonomy. Inland salt lakes often contain species not found elsewhere.

Marine protozoa cannot be maintained as easily as those of fresh water. Therefore, the organisms must be studied at once, or they must be preserved, although preservation of marine forms is not always successful. Probably the best preservative is 95% ethyl alcohol, in the proportion of 1 part collecting water with 2 parts of alcohol. If the protozoa are sedimentary, first allow them to sink to the bottom of the container during a 30-min period, and then pour off the top water before adding alcohol. Ciliates and flagellates generally swim to the top of water when a 100-W lamp is placed 10 to 12 in above a collecting bottle. After 30 min, pour off the top water containing the protozoa and add alcohol to the poured-off sample. Often, protozoa may be concentrated with low-speed centrifugation at about 3,000 rpm for 4 or 5 min.

Soil Protozoa

Many protozoan species are found in moist soil where they live in the film of water around soil particles. Most soil protozoa live within the top 1-in layer of soil; rarely are they found below a 12-in depth. Approximately 300 species have been identified, of which the flagellates are in greatest number, followed by amoeboid forms and ciliates, in that order. The motile flagellates and ciliates are easily observed under magnification. Amoebas are more difficult to recognize, because many are small and their slow movement is not easily detected. For better observation of amoeboid forms, soil particles should be well dispersed on the slide, and the microscope light should be reduced. Several types of soil protozoa are illustrated in Fig. 12-1.

A comparison of the number of species and the population counts in different soil types can become an interesting class study. Descriptions for such studies will be found in Pramer (1964) and in Yongue (1966). Pramer suggests that the class select two types of soil from the following soils: (1) an acid forest soil of sandy texture; (2) an agricultural field soil with a silt-loam texture and high organic matter; (3) an agricultural field soil, sandy in texture and relatively low in organic matter; (4) a rich garden or composted soil. Yongue reports a technique for

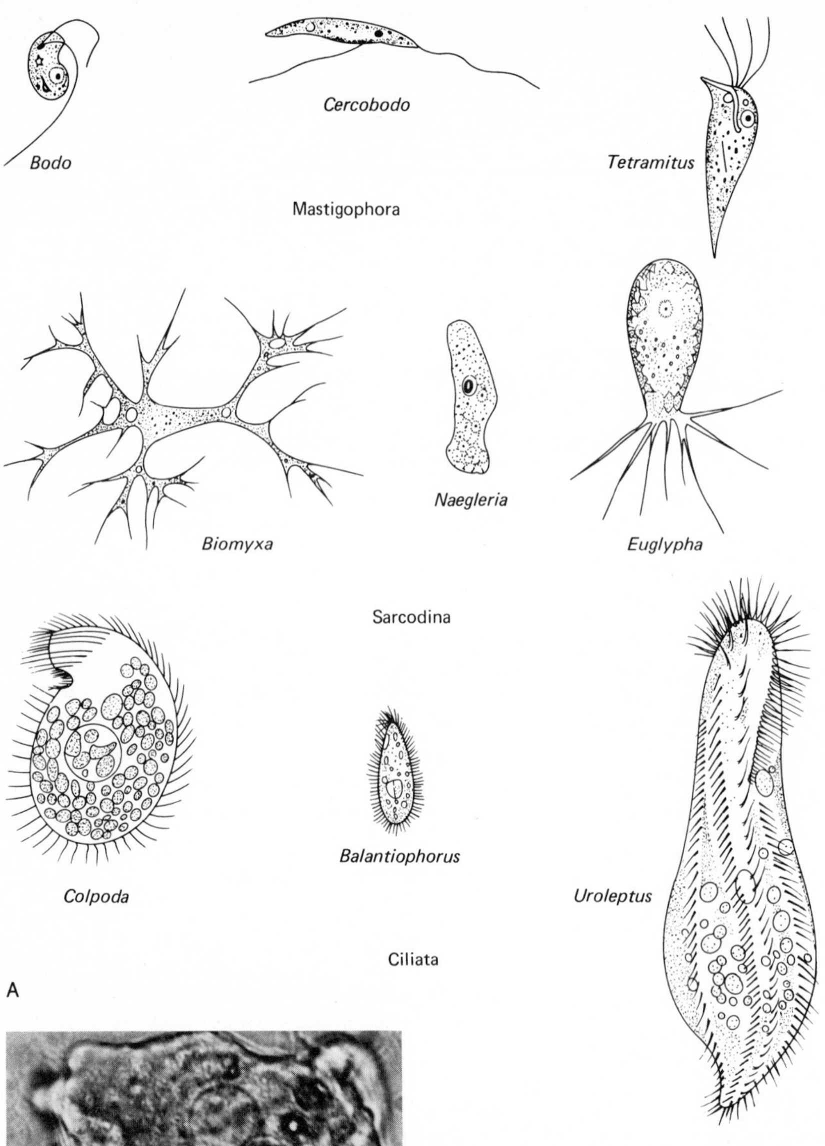

B

Figure 12-1 Soil protozoa. **A** Various protozoa that may be present in a soil sample. **B** A small amoeba observed in soil culture; approximately 75 microns long. *(A From Introduction to Soil Microbiology, by M. Alexander, after H. Sandon, 1927. Copyright 1961 by John Wiley & Sons, Inc., New York.)*

producing soil amoebas in 48 h. He describes the amoebas that he observed in soil samples from: (1) earthworm soil; (2) a rose bed; (3) sand and lake debris of a beach cove; and (4) sand and lake debris from a windswept barren area. Apparently most, if not all, species encyst when the soil becomes dry. These forms excyst when maintained in culture for several days. A variety of media has been described for culturing soil protozoans. Three methods are described here.

Method 1: Hay infusion. Prepare the hay infusion by boiling 0.5 g hay in 100 ml distilled water for 1 or 2 min. After boiling, add enough distilled water to regain the 100-ml volume. Allow the infusion to stand uncovered for 36 to 48 h, during which time bacteria will fall into the infusion and will multiply to become a food source for the protozoans. Add 1 g of soil to the hay infusion. After 48 h, examine drops of sediment under 100 to 400X magnification. Continue to examine drops for at least 6 or 7 days, or longer. As spores excyst, new species may appear, thus providing a series of population changes that may continue during 2 or 3 weeks.

Method 2: Soil extract and agar (Pramer, 1964). Prepare a soil extract by autoclaving (15 lb pressure for 30 min) 325 g of rich garden soil in 1,000 ml of tap water. Filter and refilter until medium is clear. To achieve clearness, it may be necessary to add 0.5 g calcium carbonate and to refilter after 5 min. If final volume is less than 1,000 ml, add enough tap water to regain that volume.

Add 500 ml of the soil extract to the following materials: 1.0 g glucose; 0.5 g K_2HPO_4; 500 ml tap water; 0.5 g yeast extract; 15.0 g agar.

Autoclave for a second time at 15 lb pressure for 15 min. Aseptically pipette 10 ml of the sterilized medium into each of a group of sterile petri plates. When agar has solidified, add 5 ml of sterile tap water to each plate. Then prepare each soil sample by adding 1 g of the sample to 100 ml of sterile water in a sterile flask. Thoroughly mix the soil and water by shaking the flask about 50 times. Use a sterile pipette to transfer 1 ml of the suspension, immediately after shaking the flask, to each plate of medium. Label each plate with the type of soil transferred to it.

After 7 days, use an inoculating loop to transfer a small portion of water from the surface of a plate to a microscope slide. Add a coverglass, and examine under 100 to 400X magnification. Record the number of species and the total number of individuals of each species that are found in five different locations of the drop of water. The counts indicate the relative number of protozoa in each soil type.

Method 3: Rice-cornmeal media. Yongue (1966) describes two media, used successively, that consistently produce amoebas and other protozoans within 48 h to 2 weeks. Our tests have given the same results. Ciliates and flagellates of the soil sample may also appear in either medium.

Rice medium. Put 1 g of soil and six rice grains in a petri dish that contains 15 ml of distilled or tap water. Incubate at 37°C for 24 to 48 h. Amoebas will be found in great numbers (particularly at the end of 48 h) in drops of water taken from around the rice grains. Examination of drops over a period of 1 to 2 weeks will show other protozoa that appear as excystment occurs.

Cornmeal medium. This second medium is used to increase and concentrate the number of amoebas so that they can be located more easily under the microscope. Preparation of the medium is somewhat tedious but well worth the effort for producing dense cultures for class studies. Cook 10 g of cornmeal in 1 l of distilled water at a slow boil for 20 min. Stir the mixture several times each minute to prevent the cornmeal from lumping and sticking to the container. Add 20 g of agar and mix well for several more minutes. Transfer medium in equal proportions to two 500-ml flasks. Plug flasks and autoclave at 15 lb pressure for 15 min. Pour the cooled medium (at about 45°C) into sterile petri plates.

When the medium has hardened, spread over the agar surface of each plate a drop of medium taken from around the rice grains of the first culture (above) at the 24-h stage. Aseptic techniques for transfers are described in Chap. 9. After 24 to 48 h, scrape small portions of the material from the agar surface with a transfer loop. Transfer to a drop of water on a glass slide, and examine under 100 to 400X magnification.

PARASITIC PROTOZOA

A ready source of protozoa is provided by the parasites of most, if not all, laboratory animals. Parasitic protozoans of amphibians, reptiles, the earthworm, cockroach, and grasshopper are described later in the chapter. General techniques are given below for finding parasites in a frog and in an insect. Probably frogs and toads harbor a greater variety of parasites than any other animals. Rarely is an amphibian free of parasites when first captured in the field, although the parasites may be lost within several weeks after transfer to the laboratory. Therefore, freshly collected animals should be examined as soon as possible. The tadpoles of frogs and toads are also excellent sources for parasitic protozoans. The most common protozoan parasites are the ciliates and flagellates in the rectum. Generally other parasites are present in the frog, particularly flukes and nematodes in the intestine, the urinary bladder, and the lungs.

Techniques for Observing Frog Parasites

1. Overanesthetize the frog with chloroform; or, perhaps better, quickly remove the top of the head by inserting one blade of sharp scissors across the mouth opening and making a sharp cut to remove the upper head as far back as possible (Fig. 12-2). Then insert a probe through the spinal canal to destroy the nerve cord and stop reflex actions. Slit the skin and muscle walls on the median ventral line, taking care not to puncture the internal organs. Make transverse cuts at anterior and posterior ends so that the skin and muscle wall can be pinned back to expose the body cavity (Fig. 12-2). Keep the organs moist with a layer of cotton or toweling that has been soaked in 0.7% NaCl solution or in Ringer's solution I (Appendix A).

2. Because the blood quickly clots, it should be examined first. Carefully puncture the heart and remove a drop of blood with a dropper. Put the blood on a glass slide; add a coverglass and examine under 100 to 400X magnification for trypanosomes, described later in the chapter.

3. Then remove the intestine with the rectum and place them in a bowl containing either solution described in Step 1 above. Cut off small portions, and transfer

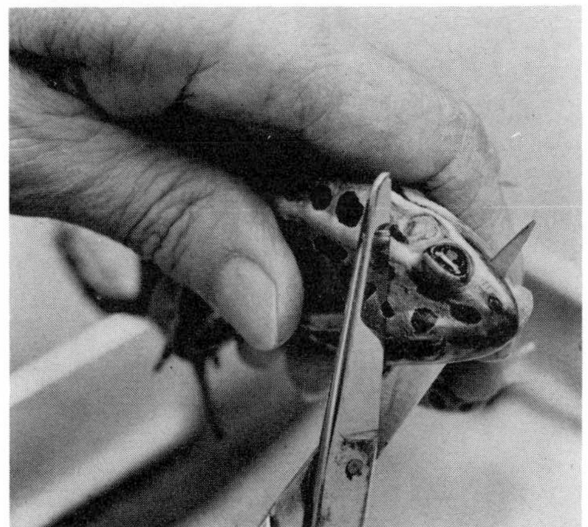

A

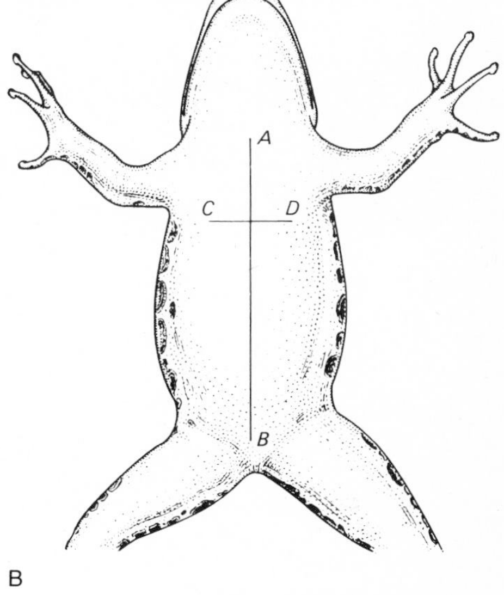

B

Figure 12-2 Preparation of frog for studying parasites. **A** Method for removing upper head. **B** Make incisions through skin and body wall on line *A–B* and then on line *C–D*. *(B Biological Sciences Curriculum Study.)*

one portion at a time to a drop of solution on a slide. Gently tear the tissue apart while observing for protozoans (and parasitic worms) under 100 to 400X magnification.

4. Look for cysts on the liver, and, if present, transfer them with surrounding liver tissue to a solution for microscopic observations. Remove the bladder and lungs, and examine in the same way.

5. See descriptions later in the chapter of *Tritrichomonas*, *Trypanosoma*, *Entamoeba*, and *Opalina*.

6. Use the following stains on the preparations: (1) Iodine stain, (Lugol's), which darkens the cytoplasm causing chromatin material of cysts to appear white against the light brown background of the cytoplasm; (2) *brilliant cresyl blue* to stain living parasites; and (3) *eosin* to stain the background and dead protozoans and to make living, unstained protozoans appear translucent. See Appendix A for stains.

Techniques for Observing Insect Parasites

Prepare the insect by overetherizing or by quickly removing the head with a sharp blade. Dissect out the colon (Fig. 12-3) and transfer it to several drops of 0.7% NaCl solution on a glass slide. Use a needle to carefully tease the tissue apart while examining for parasitic amoebas, flagellates, ciliates, and sporozoans. See descriptions of *Endamoeba*, and *Entamoeba*, later in this chapter.

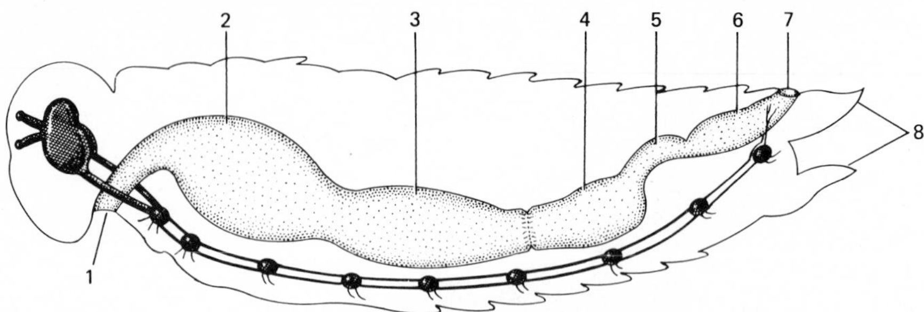

Figure 12-3 Digestive tube of grasshopper: 1, mouth; 2, crop; 3, stomach; 4, ileum; 5, colon; 6, rectum; 7, anus; 8, ovipositor. Brain and ventral nerve cord are shown in black. *(Modified from College Zoology, 7th ed, by R. W. Hegner and K. A. Stiles. Copyright 1959 by the Macmillan Company, New York.)*

GENERAL TECHNIQUES FOR CULTURING PROTOZOA

Appropriate media for each genera are designated later in the chapter with descriptions of individual organisms. General procedures are suggested here.

Agnotobiotic ("impure") cultures. Most often in a teaching laboratory agnotobiotic culture (Greek, *agnostos,* unknown) is practiced, wherein the medium contains unknown microbes in addition to the particular test organism. Often this is called an "impure" culture, for example, a hay infusion. Most protozoa require bacteria or smaller protozoa for food. The smaller protozoa also require live food, generally bacteria that are allowed to accumulate in a medium before protozoa are added. Most often the source for bacteria is from the air, when media are left uncovered for a period of 18 to 24 h before inoculation with protozoa. Although these cultures are "impure," the dominant protozoan type can be maintained by regular transfers to freshly prepared media containing bacteria, and with regular feedings of smaller protozoa when required.

Axenic ("pure") cultures. With this method no species except the test organism is present in a culture. The method is possible only when a protozoan is able to subsist upon organic or inorganic materials and without live food. The method requires aseptic techniques. *Tetrahymena,* described later, is one of the few protozoa that can be maintained, at present, in a synthetic medium without live food. A considerable amount of study is in continuous progress, however, to learn the nutrient requirements of protozoa and to formulate synthetic media that meet these requirements.

With either method, all containers and instruments must be thoroughly cleaned and rinsed at least three times in distilled water. Small culture bowls, 3 to 4 in in diameter, are often used with about 1 in of medium. The culture bowls may be stacked, with the top bowl lightly covered to retard evaporation. Other suitable containers are Pyrex baking dishes, empty food jars, test tubes, or glass vials. It seems that protozoans grow best in diffused light or darkness. Flagellates and ciliates tolerate a wide range of temperatures (18 to 24°C), but amoeboid forms thrive best at a cool temperature of about 20°C.

TRANSFERRING AND ISOLATING A PROTOZOAN

Many types of research require using a culture that is started from one individual. The group that multiplies from the single organism is called a clone.

Techniques

1. Probably a capillary tube is used most often for isolating and transferring a single protozoan to fresh medium. Capillary tubes may be purchased, or they can be made by rotating a 6- or 8-in length of glass tubing within the top of a flame. When the heat has softened the glass, the two ends are slowly pulled apart, thus drawing the softened portion into a fine tube. A triangular file is then used to cut the fine tubing into 3-in lengths, or the fine tubing is cut through the middle, making two tubes, each with an enlarged portion at the outer end. The fine tubing must be examined to be certain that the tip is not sealed by the molten glass.

2. Locate an individual of the desirable type under magnification. While observing through the microscope, place one tip of the capillary tube in the medium near the protozoan. Capillary action will draw a drop of liquid with the protozoan into the tube. The technique is easily learned and requires only a little practice.

3. Place the tip of the tube in a bowl of fresh medium, and gently blow through the other end of the tube to discharge the protozoan. Stopyra (1965) suggests dropping the capillary tube into the fresh container and then breaking the tube into small pieces with a pipette. The pieces of broken glass are left in the culture bowl.

4. Transfers by the above techniques do not produce pure cultures unless aseptic techniques are employed and unless the protozoan is passed through a series of sterile-water washings. As many as ten transfers may be necessary to allow the protozoan to eliminate bacteria and other microbes contained in its vacuoles. Alternatively, in place of washing, penicillin or other antibiotics can be added to the medium. Ordinarily, antibiotics destroy bacteria but do not affect protozoa, although preliminary tests are necessary to determine the concentration of antibiotics.

EXAMINING PROTOZOA WITH THE MICROSCOPE

The lenses, coverglasses, and glass slides must be thoroughly cleaned before use. For careful studies, clean slides and coverglasses by soaking them in concentrated nitric acid for 10 min, followed by rinsing in running water for 10 min. Store slides and coverglasses in 95% alcohol. At time of use, remove them with forceps, or by grasping with thumb and finger at opposite edges. Hold in air until dry.

A better view is obtained when a coverglass is added to a liquid mount. To avoid crushing large protozoa, place several small pieces of a broken coverglass in the drop of liquid to hold the added coverglass above the organisms.

Slowing Protozoa

Fast-moving protozoa must be slowed for close observation. A common procedure is to place small strands of lens paper or other fiber in the drop of liquid; eventually protozoa will become enmeshed and can be examined. Another method is to hold a small piece of paper toweling against one side of the coverglass to draw out the solution. Keep the protozoan in view by moving the slide as water is removed. When the organism is slowed but not completely stopped, remove the paper. If too much water is drawn out, the protozoan will burst. Another disadvantage is that many protozoa are removed with the liquid. To save time, many persons prefer to use a retardant, although a retardant may distort an organism. Several retardants are described below. If stains are to be used, they should be added before the retardant.

1. *Methyl cellulose, 10% (Jahn and Jahn, 1949).* See Appendix A for preparation. Use a toothpick to make a small ring of the retardant on a glass slide; place a drop of protozoa in the center. Gently press a coverglass over the cellulose ring, dispersing the methyl cellulose through the drop of protozoa. Do not use methyl cellulose without a coverglass; a coat of the material on the lenses is extremely difficult to remove and may damage the lenses.

2. *Agar on slides.* An excellent technique for slowing and studying ciliates and perhaps other protozoa is to immobilize the organisms on a layer of 1% plain agar (1 g agar in 100 ml water). Heat the water to dissolve the agar. Pour the warm agar over a microscope slide, and allow it to cool and solidify. Place a small drop of ciliates, e.g., *Paramecium*, on the agar and add a coverglass. The cilia stick to the agar, thus allowing an excellent view of the protozoa.

3. *Polyvinyl alcohol (PVA; Humason, 1967).* Slowly add 15 g of polyvinyl alcohol powder to 100 ml of cold water. Place in water bath at 80°C and heat with continuous stirring until the solution becomes the consistency of thick molasses. Filter through two layers of cheesecloth to remove lumps. Allow the solution to stand for several hours, or longer, while it clears to transparency. Store in capped bottles or vials. The organisms are immediately slowed when a drop of PVA is mixed with a drop of protozoa on a slide.

4. *0.04% Nickel sulfate, 1% copper sulfate, 3% copper acetate, and anesthetizing agents.* The chemicals are toxic to protozoa in concentrated form. Therefore, add 1 small drop to the drop of protozoa. The organisms are immobilized almost immediately, generally without distortion. Fresh solutions are desirable, and they should be prepared no longer than 2 or 3 days before use. Anesthetizing agents are described in Chap. 5.

VITAL STAINS FOR PROTOZOA

Vital stains do not immediately kill an organism, and thus they allow observations of particular organelles in a living protozoan. It seems that basic dyes are less toxic than the acidic ones. Stains may be added to a microscope-slide mount in several ways: (1) Place a small drop of stain on a slide and allow it to dry before adding a drop of protozoans. (2) Place a drop of protozoa on a slide; add a small drop of stain and a coverglass. (3) Observe protozoa in the temporary mount, and then add stain at edge of coverglass. (4) Before microscopic observations, add stain to the culture bowl of protozoa, or to a small portion of the medium.

Methylene blue. Add 0.5 g dry stain to 100 ml 95% ethyl alcohol to make a *stock* solution. At time of use, add 2 ml of stock solution to 98 ml of 95% ethyl alcohol. The stain colors the nucleus, cytoplasmic granules, and cytoplasmic processes. Generally cilia stain a deep blue, and the trichocysts extrude.

Neutral red. Prepare a *stock* solution and dilution in the same manner as for methylene blue. Neutral red becomes concentrated in food vacuoles where it acts as an indicator. A bright red color indicates a pH 7 to 7.5; orange to yellow indicates an alkaline solution; and violet to blue an acid solution. Nuclei may be lightly stained.

Janus green B. Make *stock* solution and dilution as for methylene blue. The dye stains mitochondria a blue-green. In protozoa, mitochondria are sometimes called chondriosomes.

Crystal violet (gentian violet). Prepare a *stock* solution by adding 3 or 4 g of crystal violet to 25 ml of 95% alcohol to make a saturated solution. Allow to stand for several days, stirring 6 to 8 times each day. Filter and store in a capped bottle. At time of use, add 5 ml of stock solution and 5 ml of 95% ethyl alcohol to 40 ml of 1% ammonium oxalate solution (1 g ammonium oxalate in 100 ml water). The stain colors the cilia and cytoplasm a light purple, and causes the extrusion of trichocysts. Bacteria in the culture medium will be stained purple.

Bismarck brown Y. Prepare *stock* solution and dilution by method described for methylene blue. The stain colors the cytoplasm a light brown. The protozoan may retain the stain for 5 or 6 h; apparently the stain does not damage the organism.

Methylene green. Preparation is the same as for methylene blue. Cytoplasm is stained a light green.

Brilliant cresyl blue, 0.1%. Add 0.1 g of powdered stain to 100 ml water. Food vacuoles and perhaps cilia are stained blue.

CLASSIFICATION SYSTEM

Four groups of protozoans are described in the following pages, the Mastigophora, Sarcodina, Sporozoa, and Ciliophora. For easy reference, the organisms are listed alphabetically by generic names within each of the four groups. To understand the relationships among genera, however, the outline below should be consulted regularly. Because literature often refers to a protozoan by an anglicized class or order name, the outline may also be of service when consulting the references contained in the chapter. It must be noted that among different classification systems, the same name-stem may be used for either the class or order rank, and that, in fact, only little conformity is found within the literature. For this reason, the system of Kudo (1966), possibly the most recently published taxonomic system, is used throughout the chapter.

PHYLUM PROTOZOA

Subphylum 1. Plasmodroma

1. Class Mastigophora

Order Cryptomonadida
Chilomonas

Order Phytomonadida
*Chlamydomonas**
*Volvox**
*Pandorina**
*Eudorina**

Order Euglenoidida*
*Euglena**

Order Protomonadida
Trypanosoma

Order Trichomonadida
Trichomonas
Tritrichomonas

2. Class Sarcodina

Order Amoebida
Amoeba
Pelomyxa
Endamoeba
Entamoeba

Order Testacida
Arcella
Difflugia

Order Foraminiferida

3. Class Sporozoa

Order Gregarinida
Monocystis
Gregarina

Subphylum 2. Ciliophora

Class Ciliata

Subclass Holotricha

Order Gymnostomatida
Didinium
Dileptus

Order Hymenostomatida
Tetrahymena
Colpidium
Paramecium

Subclass Spirotricha

Order Heterotrichida
Bursaria
Spirostomum
Blepharisma
Stentor

Order Hypotrichida
Euplotes
Stylonychia

Subclass Peritricha

Order Peritrichida
Vorticella

Order Opalinida
Opalina

*These organisms display both plant and animal characteristics. Although listed here to show their placement, the organisms are described with Algae in Chap. 11.

GENERAL USES FOR PROTOZOA OF THIS CHAPTER

Categories for general uses of protozoa are given below. Specific uses are included with descriptions of genera given later in the chapter.

M = Mastigophora
S = Sarcodina
C = Ciliophora
Sp = Sporozoa

Food Ingestion and Food Vacuoles

Amoeba—S
Arcella—S
Pelomyxa—S
Bursaria—C
Didinium—C
Dileptus—C
Paramecium—C

Food for Other Organisms

Chilomonas—M
Colpidium—C
Paramecium—C
Tetrahymena—C

Genetic Studies

Chlamydomonas—M*
Paramecium—C
Tetrahymena—C

Locomotion and Movement

Amoeba—S
Euglena—M*
Volvox—M*
Difflugia—S
Arcella—S
Didinium—C
Euplotes—C
Paramecium—C
Stentor—C
Vorticella—C
Spirostomum—C

Parasitic Protozoa

Trichomonas—M
Trypanosoma—M
Endamoeba—S
Entamoeba—S
Opalina—C
Monocystis—Sp
Gregarina—Sp

Hyperparasitism: *Entamoeba* in *Opalina* in frog.
Hyper-hyperparasitism: Fungus in *Entamoeba* in *Opalina* in frog.

---

*Described with Algae in Chap. 11.

Pigmented Protozoa

Chlamydomonas—M*
Eudorina—M*
Euglena—M*
Pandorina—M*
Volvox—M*
Blepharisma—C
Stentor—C

Predators and Voracious Feeders

Pelomyxa—S
Bursaria—C
Didinium—C

Observation of Protoplasmic Flow

Amoeba—S
Pelomyxa—S

Regeneration

Spirostomum—C
Stentor—C

Reproduction, Asexual

Eudorina—M*
Pandorina—M*
Volvox—M*
Amoeba—S
Pelomyxa—S
Paramecium—C
Euglena—M*
Vorticella—C

Reproduction, Sexual

Chlamydomonas—M*
Eudorina—M*
Pandorina—M*
Volvox—M*
Paramecium—C
Tetrahymena—C

With Shell or Test

Arcella—S
Difflugia—S
Most of Foraminiferida—S

With Symbionts

Difflugia—S
Paramecium bursaria—C

CLASS MASTIGOPHORA

The Mastigophora, the animallike flagellates, are generally regarded as the most primitive protozoa. All are equipped with one or more flagella, and are found in fresh and salt water, in soil, and as parasites. The pigmented flagellates, claimed by both protozoologists and phycologists, are arbitrarily included with the Algae in Chap. 11.

---

*Described with Algae in Chap. 11.

Chilomonas

USES: Used most often as food for larger protozoa and small invertebrates.

DESCRIPTION: *Chilomonas* is a small, solitary, free-living, freshwater flagellate, with two anterior flagella. It is elliptical in shape with a blunt anterior end and a rounded posterior end (Fig. 12-4). A rigid pellicle (outer layer) is present, and the organism does not change shape. The protozoan feeds on decaying matter and is common in hay infusions and old pond-water collections. It may be purchased from most biological supply companies listed in Appendix B.

CULTURE: General techniques are described earlier in the chapter. The following media are suitable (Appendix A): (1) Grain infusion; (2) hay infusion.

Trichomonas and Tritrichomonas

USES: (1) To observe parasitic and commensal protozoans of the human mouth. (2) To observe parasitic flagellates in amphibians and reptiles.

DESCRIPTION: These are small parasitic or commensal flagellates that are found in most vertebrates and in a few invertebrates. Four freely moving, anterior flagella and a delicate, undulating membrane are present. A long filamentous structure, the axostyle, runs lengthwise from the anterior end through the cytoplasm and extends posteriorally beyond the body. Apparently, the axostyle functions in support and locomotion. Asexual reproduction is by longitudinal fission. Sexual reproduction has not been observed.

Trichomonas tenax (Fig. 12-5) is commensal in man, living in the tartar between teeth and in crevices of the gum. The organism is widespread, and scrapings from around the teeth often show the flagellates when put in 1 or 2 drops of warm Ringer's solution II (Appendix A) at 37°C and examined under magnification. Other protozoa may be present also, particularly if the teeth have not been cleaned regularly. *Trichomonas tenax* has been estimated to occur in at least 10 percent of the human population (Manwell, 1961).

Trichomonas vaginalis infests the female vagina and the male urethra. The infestation produces intense and persistent vaginitis in the female, although symp-

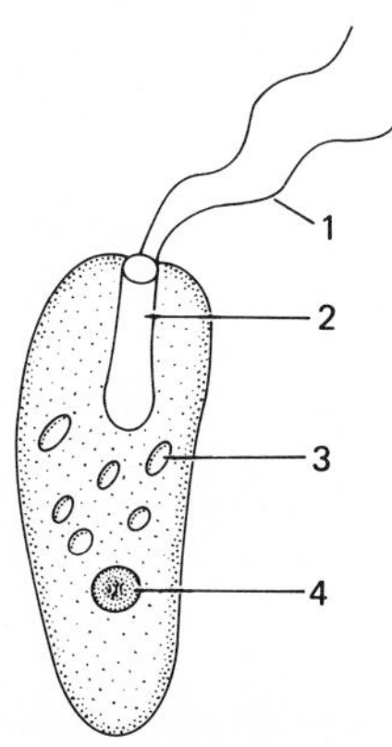

Figure 12-4 *Chilomonas paramecium* 20 to 40 microns long; 1, flagellum; 2, gullet; 3, starch granules; 4, nucleus. *(From T. L. Jahn, 1949.)*

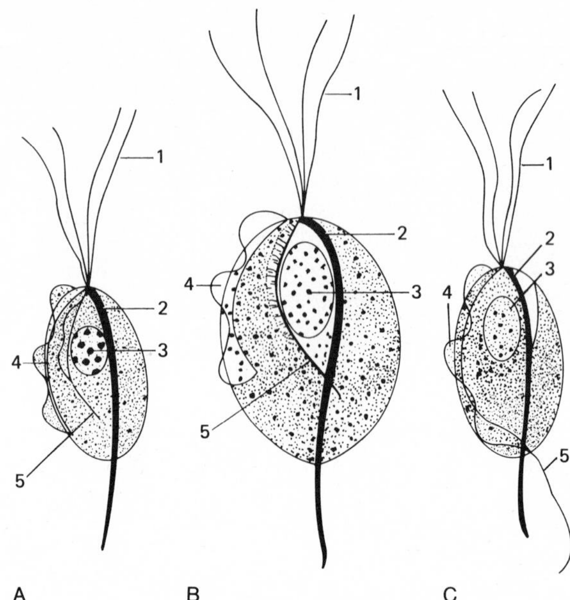

Figure 12-5 *Trichomonas* sp; 10 to 20 microns long. **A** *T. tenax.* **B** *T. vaginalis.* **C** *T. hominis;* 1, anterior flagella; 2, axostyle; 3, nucleus; 4, undulating membrane; 5, flagellum. *(A–C From Kudo, 1966, after Wenrich.)*

toms are rare in the male. Manwell (1961) estimates the incidence in females to be about 25 percent and in males about 4 percent. Probably the infestation should be considered a venereal disease.

Trichomonas hominis is commensal in the colon of man and has been identified with diarrhea. Apparently the incidence in the human is low.

Species of the genus *Tritrichomonas* are similar to the trichomonids and are often studied in the laboratory. The colon parasites, *Tritrichomonas batrachorum* and *T. augusta* are found in perhaps 90 percent or more of amphibians and reptiles (Fig. 12-6). Procedures for studying these flagellates are given in the introduction of the chapter with the techniques for studying parasites in the frog.

CULTURE: Ordinarily these organisms are not cultured in the biology laboratory.

Trypanosoma

USES: To observe a flagellated parasite in the bloodstream of frogs and tadpoles.

DESCRIPTION: Trypanosomes may occur as parasites in the circulatory system of most vertebrates. Several are pathogenic to the human, for example, *T. gambiense* which causes African sleeping sickness, and *T. cruzi* which produces the Chagas disease.

Several species are very common in the amphibian adult and tadpole and are easily observed by the procedures described for frog parasites in the introduction to the chapter. Apparently the trypanosomes are carried from frog to frog by

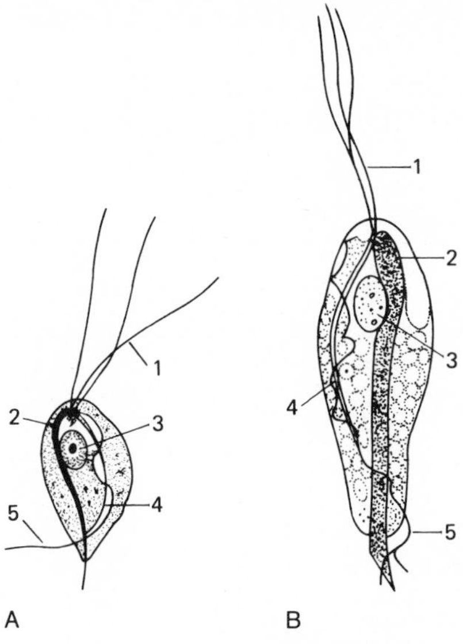

Figure 12-6 *Tritrichomonas* sp; 10 to 20 microns long. **A** *T. batrachorum.* **B** *T. augusta;* 1, flagellum; 2, axostyle; 3, nucleus; 4, undulating membrane; 5, flagellum. *(A From Kudo, 1966, after Bishop. B From Kudo, 1966, after Samuels.)*

bloodsucking leeches in the water. Probably *T. rotatorium* is most common, although a number of similar species have been reported (Fig. 12-7). *Trypanosoma rotatorium* is a rather large species, measuring about 30 microns long in body excluding the flagellum. The undulating membrane is very prominent, making the organism and its waving, swimming motion very evident in a drop of blood. The parasite will be found more often in the blood of internal organs, particularly in the kidney and liver. Although the organism cannot be observed under low magnification, it can be located under 100X magnification by the sweeping and

Figure 12-7 *Trypanosoma* sp., 30 to 35 microns long. **A** *T. rotatorium,* found in blood of frog. **B** *T. lewisi,* in blood of rats and surrounded by red blood cells; 1, nucleus; 2, flagellum; 3, undulating membrane. *(A from T. L. Jahn, 1949.)*

A

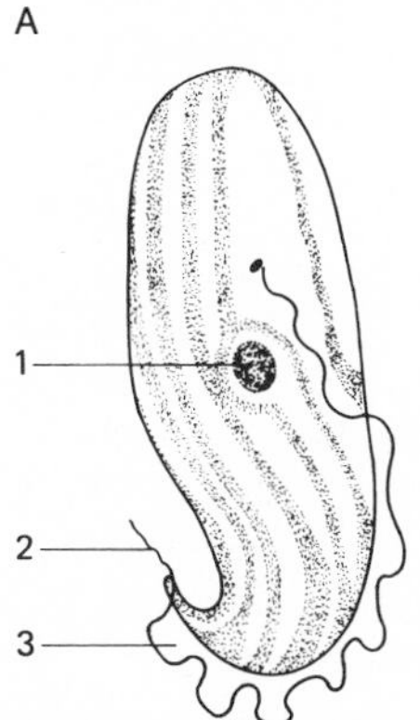

B *Courtesy of Carolina Biological Supply Company*

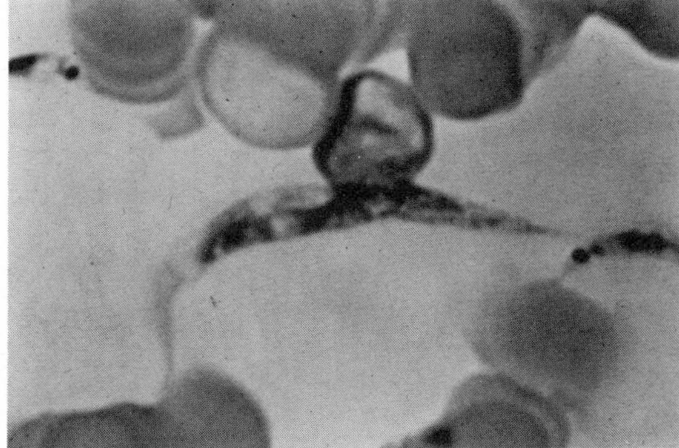

flowing movement it produces in the red blood cells. The organism may then be studied at a higher magnification.

CULTURE: Ordinarily *Trypanosoma* is not cultured in the laboratory, except for its maintenance in host animals.

CLASS SARCODINA

Members of this group lack a rigid pellicle and thus are able to move and obtain food with their flowing cytoplasm. The cytoplasm is usually divided into an outer thin ectoplasm and an inner, more dense endoplasm. The nucleus, food vacuoles, and granules are contained in the endoplasm. Contractile vacuoles are always present in freshwater forms and lacking in saltwater and parasitic species. Asexual reproduction is by binary fission; apparently most species have no sexual reproduction. Encystment is common, and an organism may remain dormant in a cyst for months and occasionally for years.

Sarcodina are found in practically all systems of soil, fresh water, and salt water. Many are parasitic. Although the freshwater *Amoeba* is used most often in biology classes, other forms may be as useful. Freeliving Sarcodina feed on microscopic animals and plants which, then, must be supplied to them when they are in culture.

Amoeba

USES: (1) Used commonly in the teaching laboratory for the study of a protozoan with pseudopodia. (2) To observe food ingestion and binary fission.

DESCRIPTION: The freshwater *Amoeba proteus* is the species commonly obtained from biological suppliers for general biology courses. The protozoan is relatively large, measuring to 500 microns in length (Fig. 12-8). Active organisms produce many unbranched pseudopodia. The outer, clear ectoplasm is easily distinguished from the dense endoplasm when the microscope light is reduced. One nucleus and one large contractile vacuole are present, and many gray or black granules can be seen in the endoplasm. Occasionally a food vacuole will contain a diatom, a small aquatic animal, or other food. In extreme temperatures or when the slide begins to dry, the organism assumes a flat, round form; if the extreme conditions continue, the animal bursts. Occasionally a dividing organism will be located on the slide. Generally, binary fission is completed in about 30 min.

To prepare a slide mount, place a drop of medium with amoebas on a coverglass and leave the glass on the table for several minutes. During this time the amoebas will settle to the bottom of the water and onto the surface of the coverglass. Then invert the coverglass over a concave or deep-well slide. The amoebas will remain on the surface of the coverglass, where they can be more easily located by students. Unless students are given careful assistance, they may mistake sediment for amoebas, a true misfortune, since they then have no opportunity to observe the operations of such a simple organism that is yet so very complex. It seems very

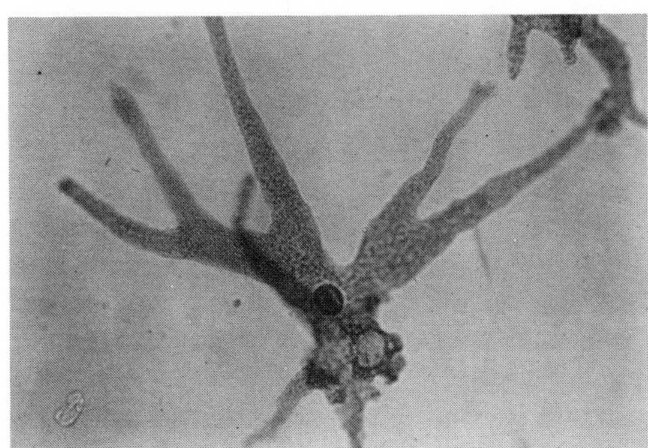

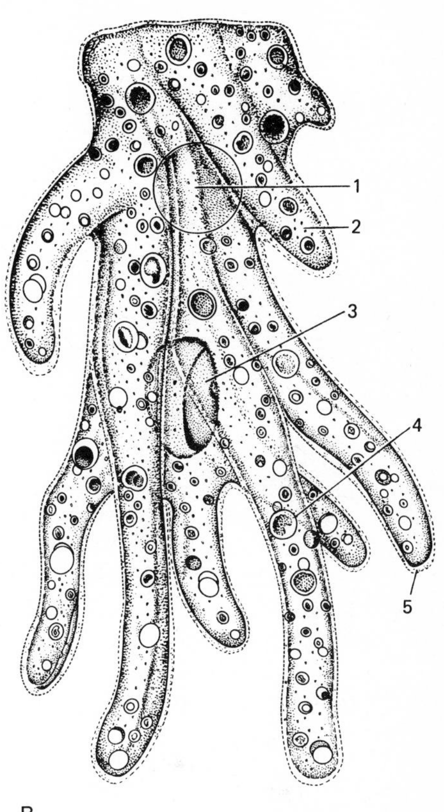

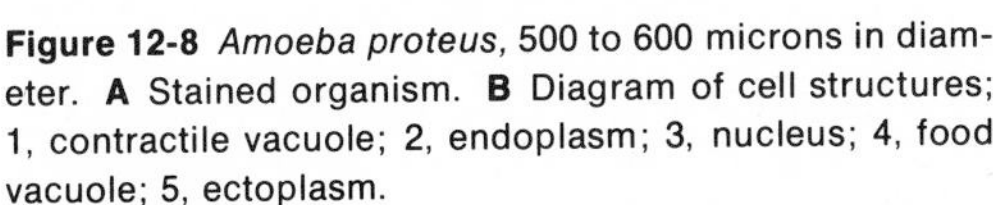

A *Courtesy of Carolina Biological Supply Company*

Figure 12-8 *Amoeba proteus*, 500 to 600 microns in diameter. **A** Stained organism. **B** Diagram of cell structures; 1, contractile vacuole; 2, endoplasm; 3, nucleus; 4, food vacuole; 5, ectoplasm.

necessary to emphasize that the sediment in cultures from biological suppliers is generally yellow or yellowish-green. Amoebas are *dark gray* or *black* globs. The slow cytoplasmic flow becomes apparent after only a little observation.

CULTURE: (1) Add four boiled grains of polished rice to 150 ml of distilled water; to this add 2 ml of *Chilomonas* from a dense culture, and inoculate with about 50 amoebas. (2) Add four boiled wheat grains to 200 ml of Chalkey's medium (Appendix A). To this add 2 ml of *Chilomonas*, and inoculate with about 50 amoebas.

Cover or stack the bowls; store them in the dark or in diffused light at 18 to 22°C. Amoebas should be in abundance around the grains in about 2 weeks. Subculture by dividing the initial culture into several bowls and adding sufficient medium to gain the original amount. Add the boiled grains and *Chilomonas* to each subculture. Mold may form around the rice, but it does not seem to be harmful to the amoebas. Amoebas will thrive if *Chilomonas* does not overmultiply. If *Chilomonas* becomes too abundant, as shown when observed with the microscope, subculture the amoebas.

Arcella

USES: (1) To observe the morphology, movement, and feeding activities of a freshwater protozoan with an outer shell (a test). (2) To observe binary fission.

DESCRIPTION: *Arcella* is a common, freshwater protozoan, characterized by a simple, transparent, chitinlike shell, called a test, which accounts for the order name Testacida (Fig. 12-9). The test becomes yellow or brown with age. From a top view, the test appears to be a disc; a side view shows that the test is a half-sphere. The ventral surface is flat, with a central circular opening that extends inward like an inverted funnel. The protoplasmic body does not fill the shell but is held to the inner wall by cytoplasmic threads. The endoplasm contains two nuclei, two or more contractile vacuoles, and food vacuoles. Several pseudopodia extend through the ventral opening or, at times, are completely withdrawn into test. The pseudopods are used for creeping through mud and sediment. Occasionally, a large gas bubble can be seen inside the test that apparently gives buoyancy to a floating organism. Jahn and Jahn (1949) state that the bubble is formed when the protozoan happens to be turned upside down. The bubble at one side causes that side to rise, enabling the animal to turn over. At times, binary fission may be observed in an established culture. Because of the test, fission is somewhat more complicated than in naked Sarcodina. The two nuclei divide mitotically into four nuclei; the cytoplasm divides into two portions with each portion receiving two nuclei. One portion of cytoplasm extrudes through the opening. Soon a new shell can be seen forming on the extruded cytoplasm, which eventually breaks loose. The remaining cytoplasm and two nuclei retain the old

Figure 12-9 *Arcella* sp; 50 to 250 microns in diameter. **A** Photomicrograph of a test of *A. vulgaris.* **B** Diagram of *A. vulgaris*; side view at left; dorsal view at right with ventral aperture seen through the body. **C** *A. discoides*; 1, test; 2, ventral aperture; 3, vacuole; 4, pseudopodium. *(B from T. L. Jahn, 1949, after Leidy. C From R. R. Kudo, 1966, after Leidy.)*

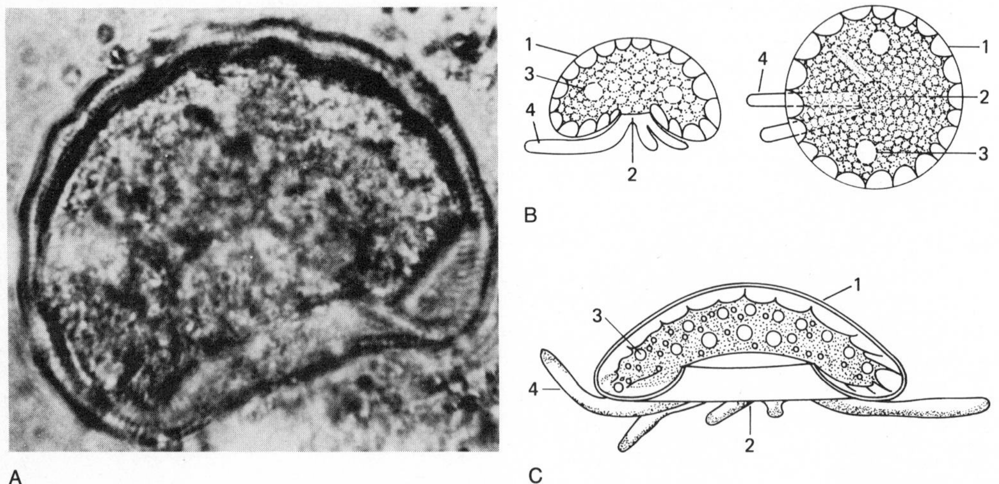

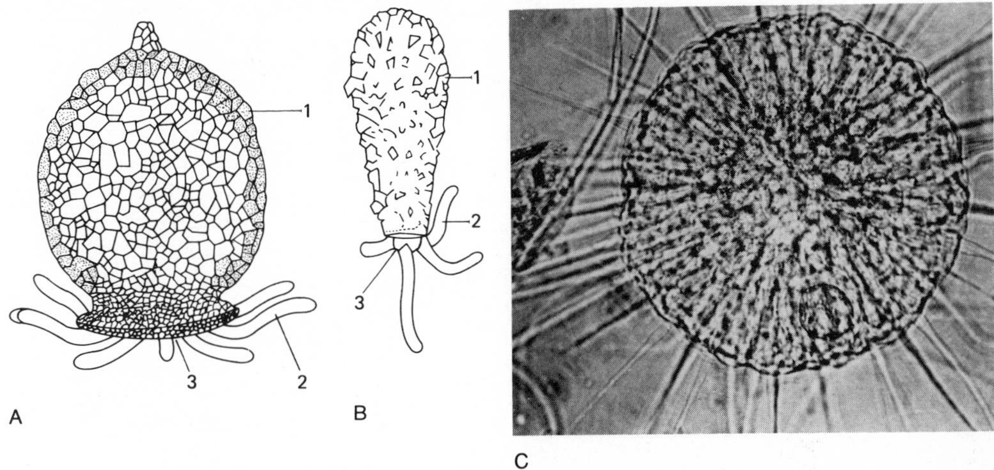

Figure 12-10 Other Sarcodina with tests or with vacuolated ectoplasm. **A** *Difflugia urceolata,* 200 to 250 microns long. **B** *Difflugia oblonga,* 100 to 250 microns long; the test is made of sand. **C** Photomicrograph of *Actinosphaerium* sp.; the body is spherical, 200 to 500 microns in diameter; pseudopodia radiate from endoplasm and through the vacuolated ectoplasm to outside of body. 1, Test; 2, pseudopodium; 3, aperture. *(A From R. R. Kudo, 1966, after Leidy. B From R. R. Kudo, 1966, after Cash.)*

test. Sexual reproduction has been reported for several Testacida but the topic needs more study. Encystment is common, generally by the organism withdrawing pseudopods into the test and, at the same time, drawing foreign material into the opening to form a plug.

Arcella vulgaris and *A. discoides* are supplied by biological companies; both may be collected in sediment of shallow, freshwater ponds. *Arcella vulgaris* is smaller, 50 to 100 microns in diameter on the ventral surface; *Arcella discoides* is somewhat flatter than *A. vulgaris* and is approximately 70 to 250 microns in diameter. The organisms are often found in old pond-water cultures of dead and decaying algae. They may not be numerous, however, and a person may need to search carefully for them in a number of drops of water taken from the bottom of the culture. They are more abundant in an acid environment, for example in peat bogs, than in alkaline water or soil.

To study the protozoans with the microscope, place a drop of medium containing *Arcella* on a coverglass and leave it on the table for 5 or 10 min, during which time the organisms will settle against the glass. When the coverglass is inverted over a concave slide, the *Arcella* will remain attached to the coverglass. The ventral surface of the organism and the movement with pseudopods can be easily observed in this manner.

Difflugia (Fig. 12-10), also a genus of the order Testacida, is sometimes collected in the sediment of fresh water; or it may be obtained from a biological supply company. The test is composed of sand grains, and the organism may be mistaken for a clump of sand on a microscope slide. Culture is the same as for *Arcella* except that a bottom layer of sand, for building the test, should be added to the culture bowl. Many species are green due to symbiotic zoochlorellae.

CULTURE: *Method 1.* Prepare 100 ml of boiled pond water by adding two grains of wheat (or two rice grains) and ½ g of hay in a small culture bowl. Add several droppers of a dense *Chilomonas* culture. After 2 or 3 days, transfer *Arcella* to the culture bowl; cover bowl loosely and store it in darkness or diffused light at approximately 20°C. If available, feed *Paramecia* about twice weekly. *Method 2.* Substitute Chalkey's Medium (Appendix A) for the boiled pond water of method 1. *Method 3.* Davis (1937) described a method for obtaining *Arcella* and a number of other protozoans, including *Stentor, Paramecium, Vorticella,* and *Amoeba.* During the spring months he placed green grass leaves in covered jars that contained wet cotton in the bottom. The grass remained against the side glass surface throughout the next several months. When he needed protozoans, he scraped dead plant tissue from the stems, or mashed the rotten leaves, and incubated the materials for 24 h in a 1% aqueous solution of citric acid.

Endamoeba blattae Butschli

USES: For observing parasitic protozoa in the colon of the cockroach. Use other insects (cricket, grasshopper) to find other amoeboid protozoa in the colons.

DESCRIPTION: Prepare a cockroach by overetherizing or by quickly removing the head with a sharp blade. Dissect out the colon (Fig. 12-3), and transfer it to several drops of 0.7% saline solution on a microscope slide. Use a needle to carefully tease the tissue apart while looking for the round, flat parasites. Probably other parasitic protozoans will be present also, including flagellates, ciliates, and sporozoans. See the discussion of parasitic protozoans in the introduction to this chapter.

The diameter of the trophozoite (active stage) is 15 to 150 microns. The trophozoite moves slowly with several broad pseudopods; the ectoplasm is distinct from the endoplasm (Fig. 12-11). The parasites ingest food from the colon of the roach, and the cellular cytoplasm of the protozoan may contain much ingested food, including bacteria, yeast, and large starch grains. Prepare a slide mount by staining the parasites with iodine stain (Lugol's) (Appendix A) to observe the nucleus and starch grains. Many cysts may also be found in the colon, although it is difficult to distinguish these cysts from those of other parasites and from food

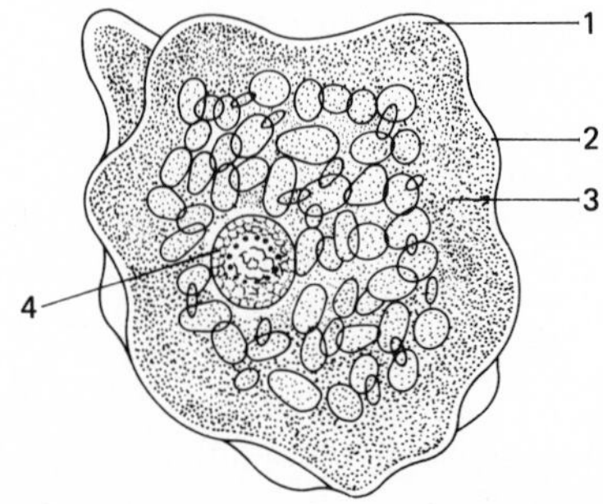

Figure 12-11 *Endamoeba blattae,* 15 to 150 microns in diameter; common in the colon of the cockroach; 1, pseudopodium; 2, ectoplasm; 3, endoplasm; 4, nucleus. *(From R. R. Kudo, 1966.)*

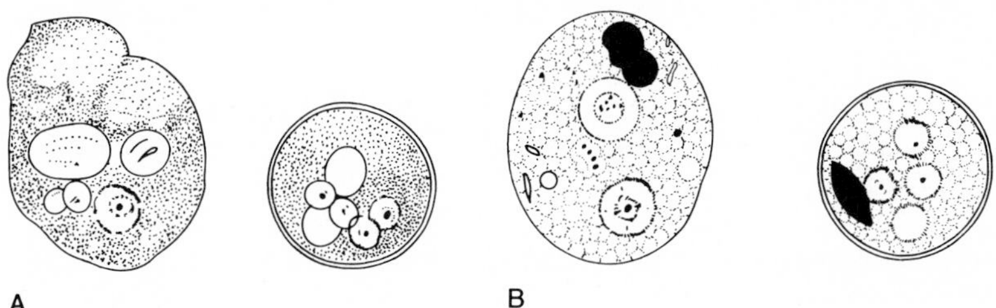

Figure 12-12 *Entamoeba* sp., 10 to 15 microns in diameter. **A** *E. terrapinae;* found in the colon of turtles. Left, trophozoite; right, cyst. **B** Trophozoite and cyst of *E. invadens;* lives in the digestive tube and liver of snakes. Other species named in the text are similar to these. *(A From R. R. Kudo, 1966, after Sanders and Cleveland. B From R. R. Kudo, 1966, after Geimen and Ratcliffe.)*

particles. The life history needs more study. Apparently, the cysts are eliminated with the feces and carried to other roaches in contaminated food and water.

CULTURE: Ordinarily the parasite is not maintained outside its host. See Chap. 3 for culture of the roach.

Entamoeba[1]

USES: To observe amoeboid parasites in the colon of insects and lower vertebrates.

DESCRIPTION: A number of *Entamoeba* species (Fig. 12-12) can be readily observed in the colon of insects and lower vertebrates. Most organisms are small and not easily distinguished from one another. Each of the several species described here has been recorded for a particular animal host. Apparently the amoeboid parasites of the colon, as well as other protozoan parasites, are widespread and can be found in practically all animals including the parasitic worms that are themselves parasitic upon another host.

Entamoeba muris Grassi. The parasite is found in the cecum of rats and mice. The diameter of a trophozoite (active stage) is 10 to 30 microns. Cysts are about 8 to 20 microns in diameter. *Entamoeba gallinarium* Tyzzer. The protozoan lives in the cecum of domestic fowls. The diameter of a trophozoite is 10 to 25 microns; cysts are smaller. *Entamoeba terrapinae* Sanders and Cleveland. The parasite is found in the colon of the turtle, *Chryssemys elegans*, and possibly in other turtles. The trophozoite is about 10 to 15 microns long; cysts are smaller. *Entamoeba invadens* Rodhain. The protozoan lives in the digestive tube and liver of many snakes, where the organism apparently is pathogenic. The trophozoite is approximately 16 microns in diameter; the cyst is somewhat smaller.

CULTURE: Because the organisms can be found in many laboratory animals, maintenance *in vitro* is not essential as a source of supply. The techniques for the isolation and observation of parasitic protozoa are given earlier in the chapter.

---

[1] At one time *Entamoeba* was considered synonymous to *Endamoeba*. In 1954 the International Commission of Zoological Nomenclature ruled to establish the two as distinct genera.

Entamoeba gingivalis Gros

USES: To observe an amoeboid parasite of the human mouth.

DESCRIPTION: This small protozoan is often found in tooth and gum scrapings. The diameter is approximately 10 to 20 microns. Generally the shape is circular and flat when observed in scrapings put in a drop of water on a glass slide (Fig. 12-13). The organism may be quite active on the slide; movement is with several broad pseudopods that can be readily observed under the microscope. Ordinarily, many light-green bodies will be seen in the cytoplasm, which according to Kudo (1966) are probably the nuclei of ingested leucocytes or of other ingested cells of the host. Prepare a slide mount by staining with iodine stain (Lugol's) (Appendix A) to observe the nucleus and other inclusions.

Interestingly, *E. gingivalis* was the first parasitic amoeba reported to be found in man (by Gros, 1849). *Entamoeba gingivalis* is closely related to *E. histolytica*, the organism that produces amoebic dysentery in the human. Much culture and much study of *E. histolytica* have been done, primarily because of its wide incidence as a cause of human illness. Nevertheless, little is known about its life history in the intestine of man.

CULTURE: Apparently few or no attempts have been made to maintain the organism outside the human mouth.

The Foraminiferida

USES: To isolate and observe the fossilized tests (microfossils) as a part of studies in evolution.

DESCRIPTION: The Foraminiferida (or forams) are mostly marine. Many genera have been described. The simplest forms are naked or with a hard cell wall. In more highly evolved genera, the body is enclosed in a test, generally secreted by the amoeboid organism, although a few species form the test by

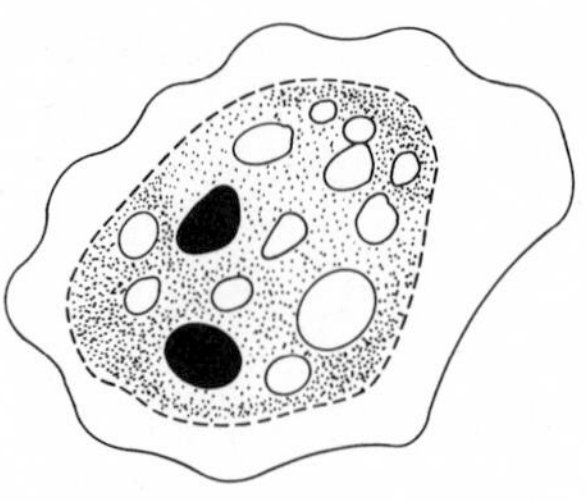

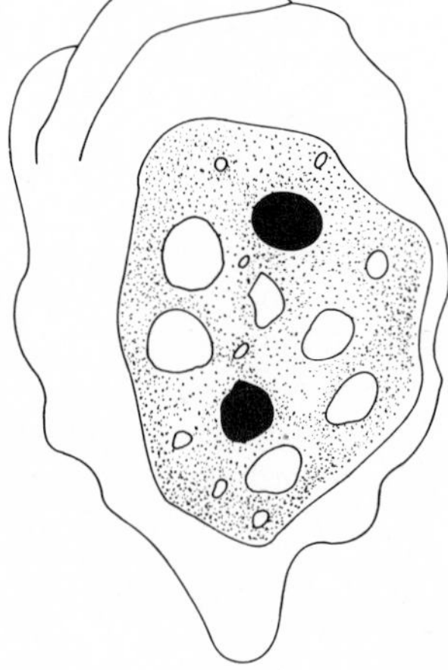

Figure 12-13 A *Entamoeba gingivalis*, 10 to 20 microns in diameter; a parasite of the human mouth. **B** *E. histolytica*, 10 to 20 microns in diameter; a parasite that lives in the large intestine of the human and causes amoebic dysentery. *(A and B From R. R. Kudo, 1966.)*

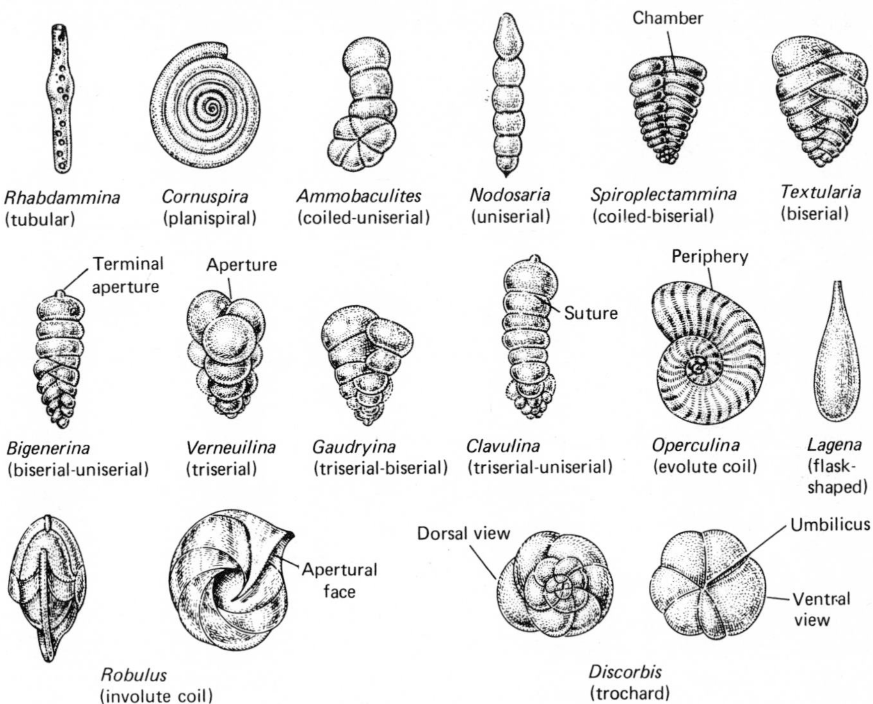

Figure 12-14 Representative genera of Foraminiferida. *(From Invertebrate Fossils, by C. E. Moore et al. Copyright 1952 by McGraw-Hill, Inc. Used with permission of McGraw-Hill Book Company, New York.)*

cementing sand and other particles together. The secreted tests are chitinous, calcareous, or siliceous. A test may contain one chamber, as in the simpler forms, or it may consist of many chambers that are connected by an opening from one chamber to the next. The walls are generally perforated, and in living animals the protoplasmic threads extend through the pores and through the large aperture of the last-formed chamber. The many shapes and arrangements of chambers make these organisms among the most interesting protozoa to study in the laboratory. Various forms are shown in Fig. 12-14.

Living forams may be collected on marine algae in intertidal zones. The microfossils of tests are found in soil samples of clay or limestone from river banks, or in areas where macrofossils are found.

OBSERVATIONS AND TECHNIQUES:[1] Obtain a small sample of soil from a clay bank or a dry river bed. Occasionally a hand lens may expose the larger microfossils, 1 mm or more in diameter, on the surface of rocks or silt, thus helping to avoid collecting barren samples. To separate the fossils from the matrix,

[1]Modified from M. Behringer: 1966, "A Study of Microfossils," *Am. Biol. Teacher*, **28** (4): 282–289.

first dry the clay completely by placing the sample in an oven at about 200°F. Following this, soak the clay for 18 to 24 h in kerosene or a cleaning fluid. Decant the fluid and cover the clay with water. The water completes the disintegration of clay particles. Mix the water and clay; allow the particles to settle to the bottom and pour off the water. Repeat the process until the water is clear after particles have settled out. Finally, gently boil the residue in water for approximately 30 min.

Use a toothpick to make a thin smear of the residue on a microscope slide, and examine under a magnification of 60 to 100X. If particles are still clumped and individual fossils cannot be located easily, the material should be boiled again. The film on the slide will contain sand grains, clay particles, fossil fragments, and other residue, as well as complete microfossils. With a little practice and with careful observations, a person can soon become efficient in locating fossils and in transferring them to a permanent slide preparation. The size of the microfossils varies from 30 microns to 1 mm. Microfossils will be evident among the debris on the slide by the hardness of the shell, by the shape, and by the markings on the shell, such as those shown in Fig. 12-14. The shell markings show more clearly when the fossils are wet. Identification and classification require specialized training and experience. Students may wish to use the references of Cushman (1948), Glaessner (1947), Jones (1956), and Moore et al. (1952).

Microfossils may be preserved by mounting them on standard micropaleontological slides obtained from a geological supply company, or by converting ordinary glass slides to micropaleontological slides as shown in Fig. 12-15. Prepare a thin water suspension of gum of tragacanth, obtained from a drug store. Add several phenol crystals to the gum suspension. Use a small paint brush to spread a thin film of the gum over the black paper of the mounting slide. Locate a microfossil in a smear of the residue, and with a damp sable brush (size 000) carefully transfer the fossil from the smear to the slide. A little practice will be required to make successful transfers. The water on the brush will dampen the fossil sufficiently to cause it to become sealed to the dry film of tragacanth. Care must be taken not to flip a dry fossil off a slide and to lose it. Seal a coverglass or glass slide over the top of the mount, as shown in Fig. 12-15.

Pelomyxa

USES: (1) Interesting for class studies because of the large size, the many nuclei, and the voracious feeding on *Paramecia*, rotifers, and other microbes. (2) To study the reproduction by plasmotomy, in which two or more daughter cells are formed by cytoplasmic division, each offspring receiving an approximately equal number of nuclei from the parent cell. The nuclei do not divide mitotically.

DESCRIPTION: *Pelomyxa carolinensis* is a giant amoeba, sometimes named *Chaos chaos* (Fig. 12-16). The protozoan may become 4 to 5 mm long when moving. It can be easily observed without magnification and isolated by picking up a single organism with an ordinary dropper. Movement is with pseudopodia. Many small nuclei are present, varying in number from less than 100 to 1,000 or more. The cell contains many small contractile vacuoles. Apparently encystment is not common.

Many food vacuoles and crystals can be seen in the cytoplasm. The organism

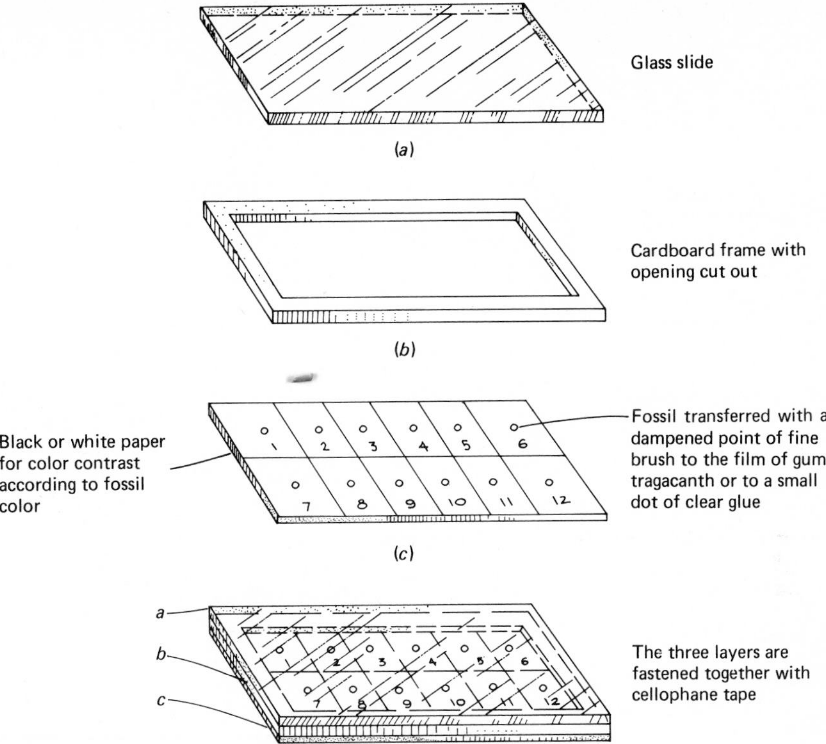

Figure 12-15 A method for mounting microfossils on a glass slide.

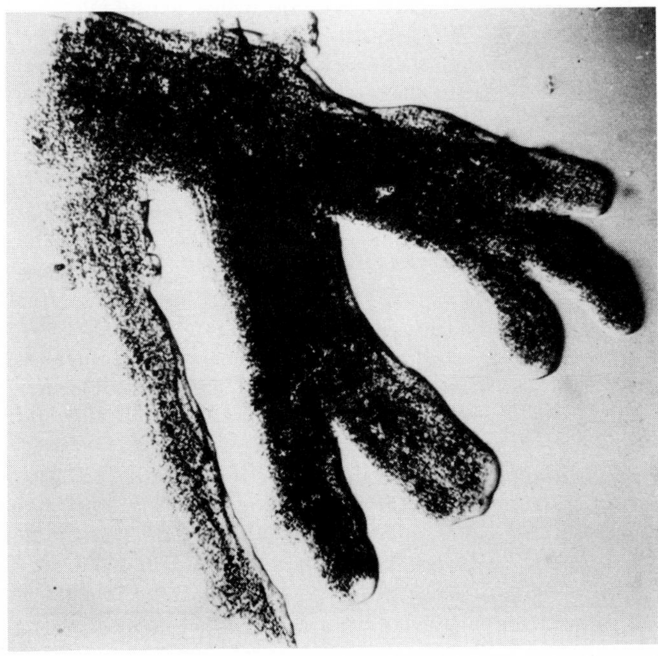

Figure 12-16 *Pelomyxa carolinensis*, a giant amoeba, 4 to 5 mm long when moving. *(Courtesy of CCM: General Biological, Inc., Chicago.)*

feeds on other protozoa and small invertebrates. An interesting class study can be obtained by feeding a drop of a dense culture of *Paramecia* to *Pelomyxa* and watching for a food vacuole to form containing several *Paramecia* all ingested at one time.

Pennak (1953) includes certain species of *Pelomyxa* with a group of organisms commonly found in sewage and called "sewage Protozoa." These may be collected in the sediment of polluted water or on the trickling filters of a sewage disposal plant.

CULTURE: On rare occasions, *Pelomyxa* may be collected in bottom samples from ponds and lakes. A starting culture can be purchased from most biological suppliers listed in Appendix B.

Method 1. Place 200 ml of distilled water and two uncooked rice grains in a culture bowl. After several days, inoculate with *Pelomyxa* and add an abundance of *Paramecia*. *Pelomyxa* requires much food, and *Paramecia* must be added daily, thus necessitating the maintenance of several cultures of *Paramecia*. Maintain culture in diffused light at 20°C and pH 7.0 to 7.5.

Method 2. Chalkey's Solution (Appendix A) may be substituted for the water in method 1.

CLASS SPOROZOA

The Sporozoa constitute a large group of protozoans, all parasitic and all producing spores during one stage of the life cycle. The sporozoans have adapted magnificently to dependency on other organisms for their existence and, like other parasites, have assured the continuity of species with an overly abundant progeny. The extreme variety in life cycles among parasitic species represents some of the most fascinating accounts of evolutionary history. The simplicity of morphology at any one stage is deceiving. In reality, the total series within one life cycle serves to combine a complex system of a division of labor, wherein each stage is beautifully adapted to perform one essential function within one particular ecological niche. For example, the life history, illustrated in Fig. 12-17, displays a total, exceedingly complex morphology, even though each single stage is morphologically simple. An even greater complexity is shown in *Gregarina rigida* (see Jahn and Jahn, 1949).

Essentially, a sporozoan life cycle contains two stages: (1) *schizogony* (asexual stage), and (2) *sporogony* (sexual stage). In the first stage a sporozoite passes through a growth (trophozoite) stage and may then divide to form many small cells (merozoites), which may develop again into trophozoites. This asexual cycle may be repeated again and again, for example, every several days in the human malarial sporozoans.

Eventually some of the merozoites develop into sexual reproductive cells (gametocytes), and the second stage (sporogony) begins. Gametocytes form macro- and microgametes. The two kinds of gametes fuse to form zygotes. From these zygotes many spores are formed, each spore containing sporozoites which begin the first stage again. This outline of the two stages is greatly simplified; most genera display considerable variation in one or both stages.

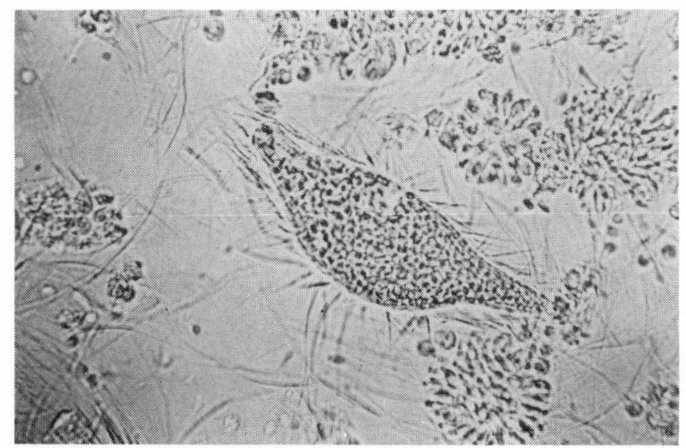

A *Courtesy of Carolina Biological Supply Company*

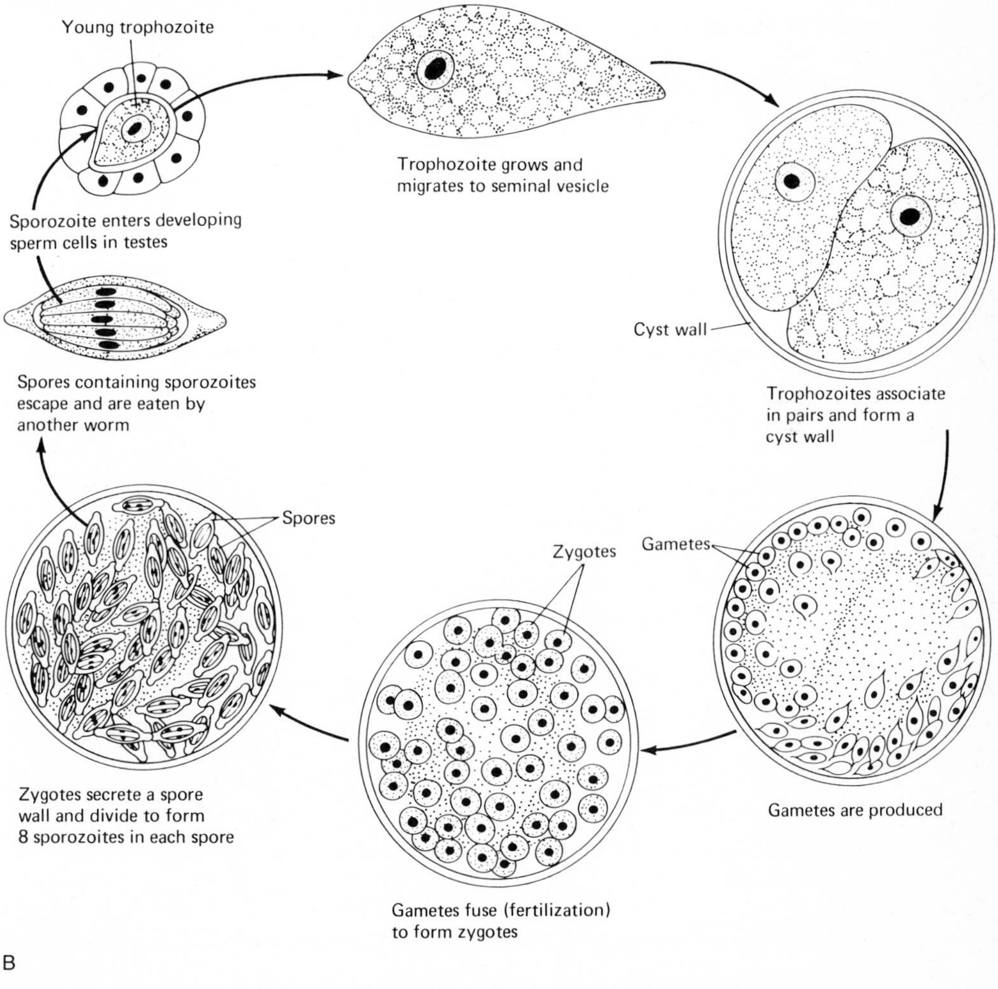

Figure 12-17 *Monocystis lumbrici;* a common parasite of the earthworm. **A** Trophozoite, about 200 microns long, surrounded by sperm in testis of earthworm. **B** Life stages; not drawn to scale. *(B From College Zoology, 7th ed, by R. W. Hegner and K. A. Stiles. Copyright 1959 by the Macmillan Company, New York.)*

Young trophozoite

Trophozoite grows and migrates to seminal vesicle

Sporozoite enters developing sperm cells in testes

Cyst wall

Trophozoites associate in pairs and form a cyst wall

Spores containing sporozoites escape and are eaten by another worm

Spores

Zygotes

Gametes

Zygotes secrete a spore wall and divide to form 8 sporozoites in each spore

Gametes are produced

Gametes fuse (fertilization) to form zygotes

B

The gregarines are a large group of sporozoans that parasitize invertebrates, particularly annelids and arthropods. Descriptions and techniques are given here for studying the life stages of *Monocystis lumbrici* in the seminal vesicles of the earthworm. As an individual project, a student may wish to trace the life history of *Gregarina rigida* in the midgut of grasshoppers, roaches, crickets, *Tenebrio*, and other insects. A good account of *G. rigida* in the grasshopper will be found in Jahn and Jahn (1949).

Monocystis lumbrici Henle

USES: (1) To observe a gregarine sporozoan in the seminal vesicles of the common earthworm. (2) Other gregarines may also be found in the seminal vesicles of the earthworm.

DESCRIPTION: The life history of *Monocystis* is relatively simple when compared to that of many sporozoans. The sporozoan is found almost without exception in all earthworms, especially during spring months. Although live, anesthetized worms are desirable for observing the large, motile trophozoites, preserved worms can be used.

Briefly, the life cycle of *Monocystis* is as follows (Fig. 12-17). The earthworm ingests the spores with soil that is taken into the digestive tube. The eight sporozoites of each spore are released in the intestine from where they move into the bloodstream. Eventually each sporozoite may penetrate a bundle of developing sperm cells within the testes. In this stage the parasite is called a trophozoite. The mature trophozoites (about 200 microns long) move into the seminal vesicles, where they come together in pairs. This association, or gregariousness, accounts for the order name of Gregarinida and the common name of gregarines. Each pair becomes enclosed in a cyst, about 160 microns in diameter. Inside the cyst, the two trophozoites divide by multiple fission to form many macrogametes and microgametes. The two kinds of gametes fuse in pairs to form zygotes. The zygotes change to a spindle shape and secrete a tough wall to become a spore. The nucleus of the spore divides three times to form eight daughter nuclei. Each daughter nucleus, with a portion of cytoplasm, becomes a sporozoite, all encased in the spore. During copulation, the spores may pass into the seminal receptacles of another worm and then into the cocoon, with either subsequent infection of embryos or the release of spores into the soil when the cocoon is broken. Otherwise, the spores may be retained in the adult host and released into the soil upon death and decay of the host worm.

TECHNIQUES AND OBSERVATIONS: To obtain stages of *Monocystis*, overanesthetize an earthworm (see Chap. 5), and slit the worm open on the median dorsal line from segments one to fifteen. The three pairs of cream-colored seminal vesicles will be exposed, lying on either side of the digestive tube (Fig. 12-18). Remove a small portion of a vesicle and tease it well apart in a drop of 0.7% NaCl solution on a microscope slide. Add a coverglass, and examine the tissue under 100X magnification. Usually most stages of Fig. 12-17 can be found: (1) the free, motile trophozoites; (2) cysts, each containing two trophozoites, or each with gametes or zygotes; and (3) the developing spores with sporozoites.

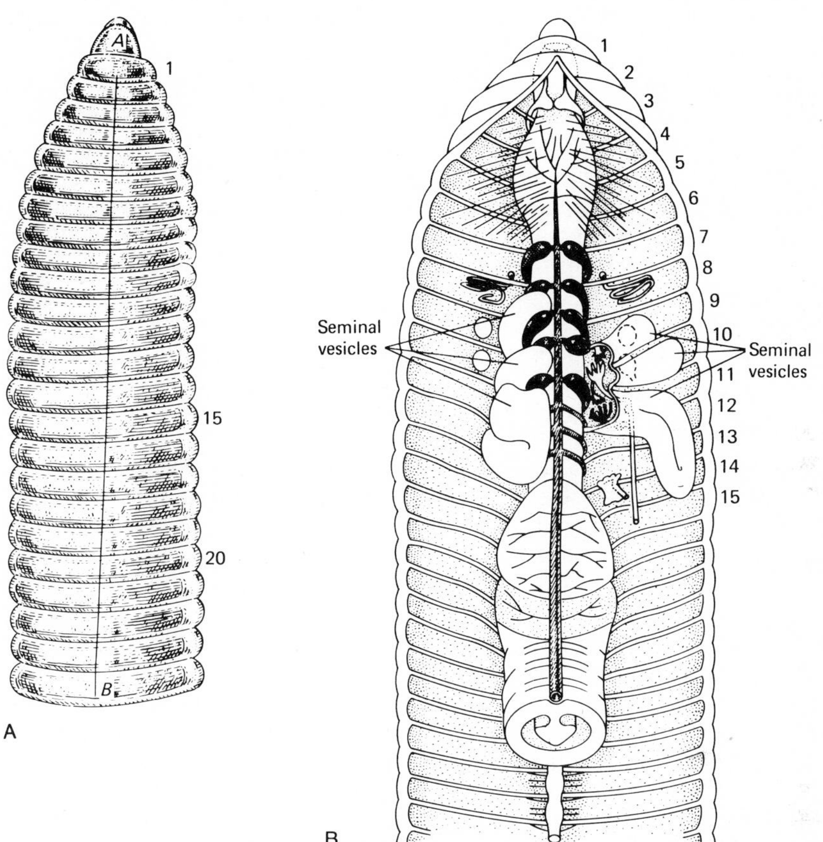

Figure 12-18 Method for opening the earthworm to locate the seminal vesicles. **A** Cut through the dorsal body wall on line *A–B*. Use a needle to tear the internal walls between segments. Pin body walls of worm on wax in pan or cut them away as shown in **B**. **B** The opened earthworm showing the three pairs of seminal vesicles, in place at left, and pulled away from digestive tube at right. *(B Reprinted with permission of the Macmillan Company from College Zoology by R. W. Hegner and K. A. Stiles. Copyright © 1959 by the Macmillan Company.)*

THE CILIOPHORA: CLASS CILIATA

The Ciliophora comprise a large, cosmopolitan group of protozoa that swim with cilia, or creep or walk with cirri (cirrus, singular; a tuft of fused cilia). They are found in fresh water, salt water, and as parasites. The length varies from 10 or 15 microns to 2 or 3 mm. Most species are colorless; a few are brightly colored because of pink, red, or blue chromatophores or because of symbiotic green algae. The form varies from bilateral symmetry (generally free moving, as with *Paramecia* and *Blepharisma*) to radial symmetry (sometimes sessile and attached to substrate, as with *Vorticella*). Most ciliates are solitary; a few live in colonies, particularly as stalked colonies. A prominent mouth (cytostome), in most species, leads through a gullet to the inner cytoplasm. Body regions consist of an endoplasm and an ectoplasm. From the latter are derived an outer, thin covering, the pellicle, and, such structures as cilia, cirri, trichocysts, skeletal plates, and undulating membranes. Contractile threads, myonemes (elementary muscle organelles), as in *Vorticella*, apparently arise from the ectoplasm also.

The location of the cytostome is a distinguishing feature in classifying species. The cytostome is surrounded by an area known as the peristome (or oral groove) that contains cilia which create water currents carrying food into the mouth. In many ciliates, a row or rows of cilia fuse to form a fringe of "membranelles" (adoral zone) along the edge of the peristome. The peristome may wind to the right or left toward the mouth, thus serving also as an important trait in identifying species. Although the arrangement of cilia varies considerably among species, typically, the cilia form rows that begin at the anterior end and continue to the posterior end.

One or two contractile vacuoles are present, even in marine forms, a condition not understood since ordinarily water expulsion is not required in marine organisms. Temporary food vacuoles may be present and vary in number depending on available food. A food vacuole is formed when a food sac is pinched off from the inner gullet and is carried by cytoplasmic streaming around the cell, apparently in a regular course.

One or more macronucleus and micronucleus are present in most species. As a rule, nuclei are visible only in stained organisms. Generally, one micronucleus is present and is closely aligned with, or fused to, the macronucleus, thus making observation difficult. Other inclusions, such as mitochondria and Golgi apparatus, are present but are not discernible with the ordinary laboratory microscope.

Commonly, the bilateral ciliates reproduce asexually by transverse fission, and the radial forms by longitudinal fission. Sexual reproduction (conjugation) occurs when two ciliates fuse and exchange micronuclear material. After separation of the two conjugants, a rapid series of cell division occurs, thus accounting for the reproductive aspect of conjugation. A number of mating types (sometimes called "syngens") have been recognized within certain species.

Blepharisma

USES: (1) For population studies (see Mertens, 1966). (2) To observe taxic responses to light, gravity, and other factors. (3) To test the effects of food, temperature, and pH on reproductive rate. (4) To observe movement with cilia and cirri. (5) To observe the pigmented species.

DESCRIPTION: *Blepharisma* is a relatively large ciliate, 80 to 300 microns long. Both freshwater and saltwater species have been described. The shape is pyriform (pear-shaped), ellipsoid, or spindle, and slightly narrowed at the anterior end (Fig. 12-19). Cilia are uniformly dense over entire body, and a small posterior contractile vacuole is present. The peristome, with a two-layered ciliary membranelle, runs from the anteroventral surface and twists in the right dorsal area where it connects with the cytostome. The two dense layers of cilia are in constant

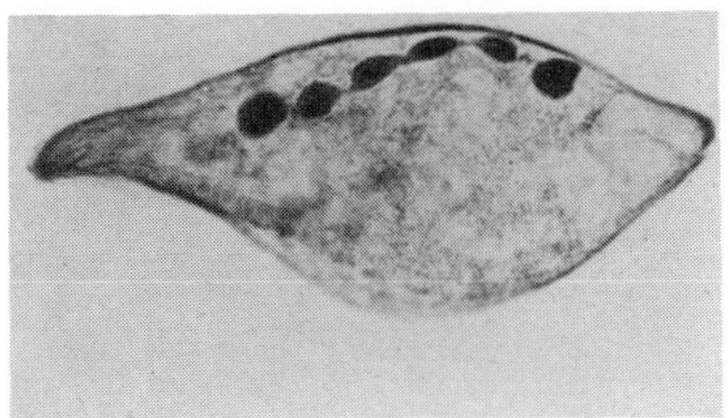

A *Courtesy of Carolina Biological Supply Company*

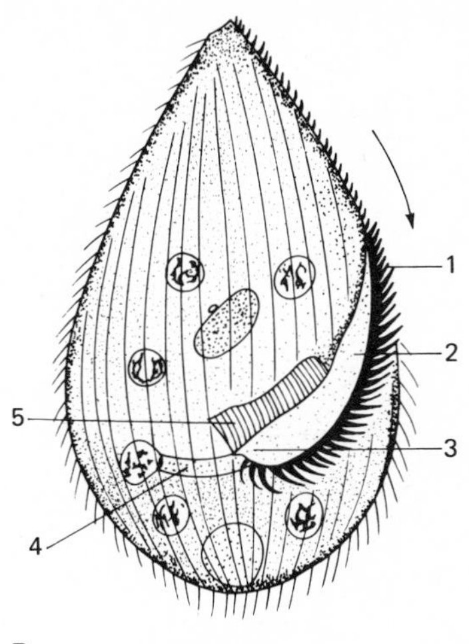

B

Figure 12-19 *Blepharisma* sp., 150 to 300 microns long. **A** *B. undulans americanus,* which contains a light-purple pigment. Organism is stained to show the long macronucleus divided into beadlike parts. **B** Diagram of *B. lateritium,* a pink ciliate; 1, membranelles; 2, oral groove; 3, mouth; 4, gullet; 5, undulating membrane. Arrow indicates the sweep by cilia of water and bacteria into the oral groove. *(From T. L. Jahn, 1949, after Kahl.)*

undulating motion by waves of movement that pass from one end of the layers to the other end.

Several species contain pigment bodies beneath the pellicle, for example, a rose-colored pigment is present in *B. lateritium* (pear-shaped) and *B. steini* (ovoid). Both are found in fresh water among decaying vegetation. Giese (1953) was among the first to study pigment bodies of *B. undulans,* a light-purple organism. He found that the pigment bodies produce a toxin that affects other organisms in the water. Pigment bodies increase in number when *B. undulans* is kept in the dark. The organism loses its color in dim light, and a heavily pigmented organism usually dies when put under bright light. *Blepharisma undulans* is distinguished by its long macronucleus that contains two or more enlarged portions with intervening narrow, stringlike portions.

CULTURE: See general techniques in the introduction of the chapter. Any of the following media (Appendix A) are appropriate: (1) lettuce medium; (2) Chalkey's medium; (3) grain infusion or hay infusion.

In all cases place about 1 in of medium in a culture bowl. Leave bowl uncovered and allow 18 to 24 h for air bacteria to accumulate. Inoculate with *Blepharisma.* Cover bowls lightly, and store in diffused light at 20 to 23°C.

Bursaria truncatella Muller

USES: (1) To observe voracious feeding and digestion in this large protozoan. (2) To study a predatory protozoan.

DESCRIPTION: *Bursaria truncatella* is the only species of the genus. The organism measures 500 to 1,000 microns in length and is one of the largest ciliates.

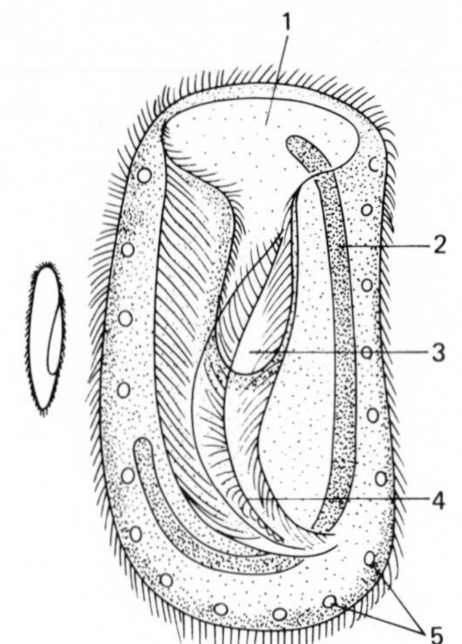

Figure 12-20 *Bursaria truncatella,* 500 to 1,000 microns long. 1, peristome; 2, macronucleus; 3, cytostome; 4, gullet; 5, contractile vacuoles. Note the size of the *Paramecium,* at left, in contrast to the size of the *Bursaria.* *(From T. L. Jahn, 1949, after Kahl.)*

The protozoan is ovoid in shape; the anterior end is truncated (flattened); the posterior end is rounded (Fig. 12-20). The cell is dorsally convex and ventrally flattened. The large peristome begins at the anterior end and extends inward to the cytostome at the center of body. The peristome is divided into two chambers by a longitudinal ridge. At times, *Paramecia* and other ciliates can be seen freely swimming about in the two chambers. Below the mouth, the pharynx (gullet) bends to the left. Cilia cover the entire body. The cell contains a long, rodlike, C-shaped macronucleus that may become coiled. Many micronuclei are present, as well as many contractile vacuoles at the sides and on the ventral surface. The protozoan is a voracious feeder, ingesting large numbers of *Paramecia,* one immediately after the other.

CULTURE: *Bursaria* is not easily maintained in culture, and usually degenerates within a week. Much study is still needed to learn the culture requirements. Any of the media described for other ciliates can be tried. No doubt an *abundance* of *Paramecium* or other protozoans should be supplied for food. Collect in fresh water or purchase from a biological supply company.

Colpidium

USES: (1) Regularly used as a food organism for larger protozoans and for small invertebrates.

DESCRIPTION: *Colpidium* is slightly kidney-shaped and 50 to 150 microns long (Fig. 12-21). A small ventral mouth is located toward the right side and anterior end. A rounded macronucleus, one micronucleus, and one contractile vacuole are present. *Colpidium* closely resembles *Tetrahymena,* another small ciliate, but may be distinguished by the fact that only one row of cilia begins at the mouth in *Colpidium,* whereas two rows begin at the mouth in *Tetrahymena. Colpidium* is widely distributed in fresh and salt water. Encystment is common, and the protozoan often appears in hay infusions.

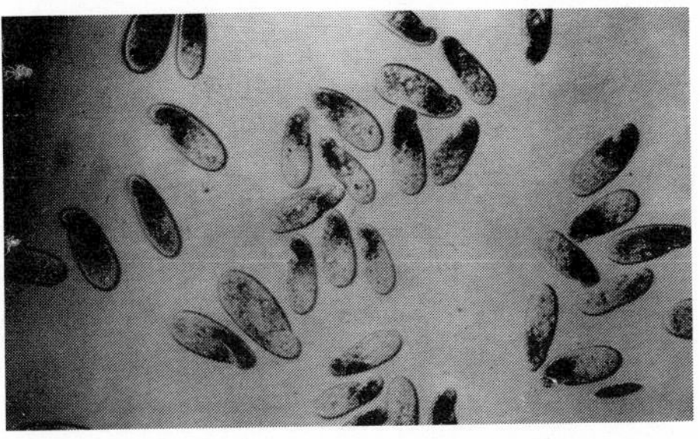

A

B

Figure 12-21 *Colpidium* sp., 50 to 100 microns long. **A** *C. colpoda,* stained organisms. **B** *C. campylum,* diagram of cell; 1, oral groove; 2, mouth; 3, gullet; 4, macronucleus; 5, food vacuole; 6, contractile vacuole. *(A Courtesy of CCM: General Biological, Inc., Chicago. B from T. L. Jahn, 1949.)*

Colpoda is much like *Colpidium* (Fig. 12-22). It has the same uses and may be cultured by the same methods. Although *Colpoda* is collected regularly in pond water, no commercial sources are known.

CULTURE: *Colpidium* is easily cultured in any of the following media (Appendix A): (1) lettuce medium; (2) Chalkey's medium; (3) grain infusion; (4) hay infusion; (5) skim milk medium. Obtain the organism from freshwater collections or from a biological supplier. It is often received with other cultures as a food organism, for example, with *Amoeba,* from which it may be isolated and transferred to fresh media.

Didinium

USES: (1) To observe the ingestion and digestion of *Paramecia.* (2) To observe the general morphology and physiology.

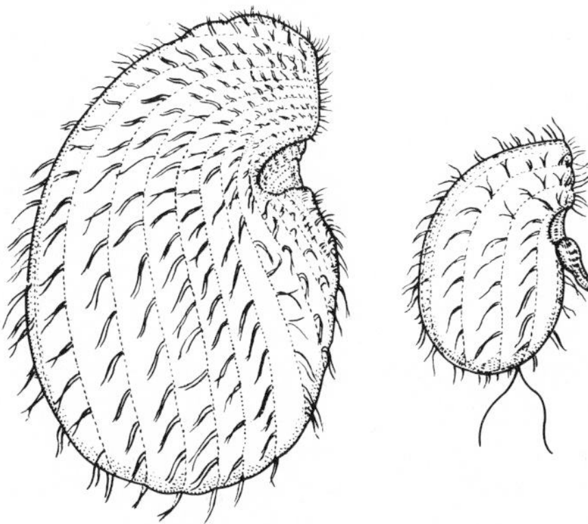

A B

Figure 12-22 *Colpoda* sp., 40 to 110 microns long. **A** *Colpoda cucullus,* a very common species identified by the kidney shape. **B** *C. steini,* a smaller species with fewer rows of cilia and two long posterior cilia. The cilia posterior to the mouth are fused and resemble a beard. *(A and B From T. L. Jahn, 1949, after Burt.)*

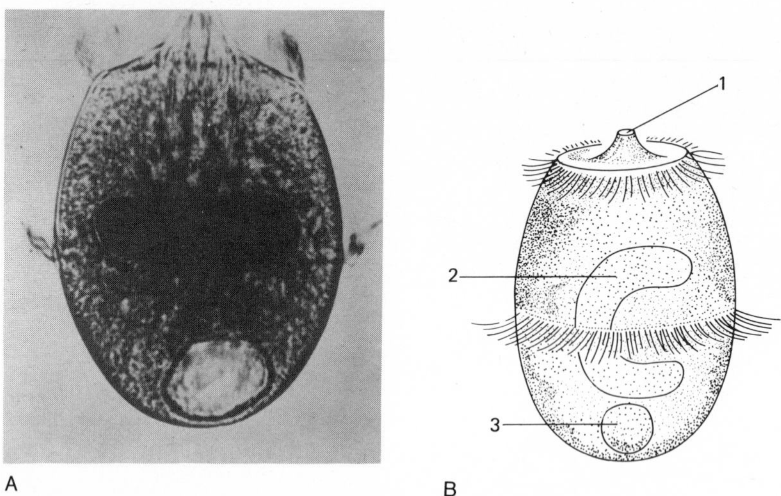

Figure 12-23 *Didinium nasutum,* 80 to 200 microns long. **A** Stained cell showing the two bands of cilia, a bandlike macronucleus, and a posterior vacuole. **B** Diagram of cell: 1, mouth; 2, macronucleus; 3, contractile vacuole. *(A Courtesy of CCM: General Biological, Inc. Chicago. B From T. L. Jahn, 1949.)*

DESCRIPTION: An interesting, relatively large ciliate (80 to 200 microns long), that appears occasionally in hay infusions. The body is barrellike; the cytostome is at the anterior end at the tip of a projecting cone (Fig. 12-23). The cell displays two girdles of cilia, one around the anterior end and one around the middle of body. A U-shaped macronucleus is present, and two to four micronuclei are located near or adherent to the macronucleus. One contractile vacuole is at posterior end.

Didinium nasutum feeds voraciously and almost exclusively on *Paramecia.* The diet limitation is an interesting evolutionary aspect that would appear to be a serious restriction, perhaps accounting for (or at least accompanied by) an encystment within several days when the food supply is exhausted, and a ready excystment when the food supply is replenished. *Paramecia,* often larger than *Didinium,* are seized and sucked into the *Didinium* with its proboscislike mouth, that is extensible and armed with trichocysts. Ingestion is completed in about 1 min, and digestion of the *Paramecium* may be completed in about 20 min (Wessenberg and Antipa, 1970).

CULTURE: Collect on vegetation and in sediment (for cysts) from still, fresh water. Most often the protozoan is collected in the encysted form. Add a dense *Paramecium* culture to the collected sample, and *Didinium,* if present, will excyst within several days. The protozoan may be purchased from most biological supply companies. Cysts can be retained by allowing a culture to dry out in a container. At a later time, a *Paramecium* culture may be added to the container, and *Didinium* will excyst. Continuous cultures of *Paramecium* must be maintained as food. When the supply of *Paramecium* is nearly exhausted in one culture bowl, transfer *Didinium* to a fresh culture of *Paramecium*. *Didinium* is not difficult to maintain as long as an abundance of *Paramecium* is supplied. For culture media, see *Paramecium*.

Dileptus

USES: (1) An interesting protozoan for class study because of the "snout" or "proboscis," an adaptation for searching for food. (2) For research to determine optimum conditions for maintenance.

DESCRIPTION: *Dileptus* is a large ciliate, 200 to 500 microns long. Species are found in both fresh water and salt water. The body is elongated, with a long, slightly curved, anterior neck or snout, and with the posterior end drawn out into a sharp-pointed tail (Fig. 12-24). A round cytostome lies laterally at base of neck. Many trichocysts are present on the ventral surface of the long snout and around the mouth. Cilia are uniform over body. Many contractile vacuoles are contained

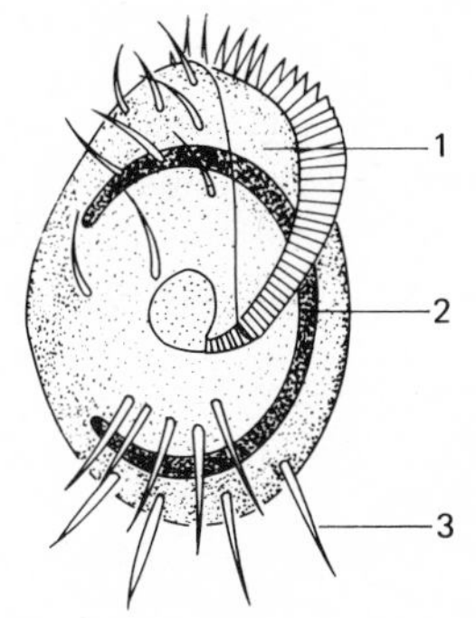

Figure 12-25 *Euplotes patella.* The body is clear, and cirri are easily observed through the body. 1, Oral groove; 2, macronucleus; 3, cirrus. *(From T. L. Jahn, 1949, after Pierson.)*

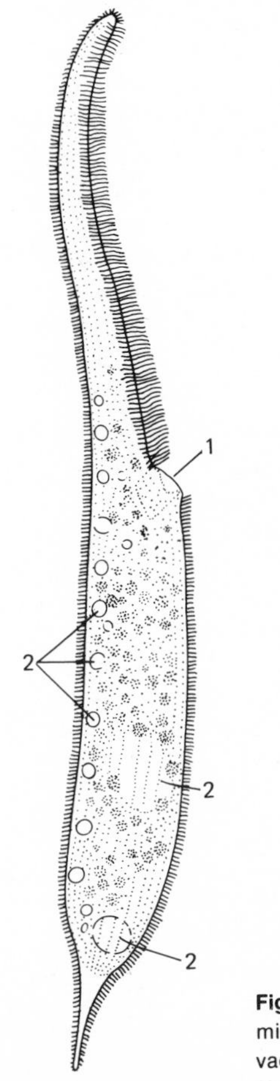

Figure 12-24 *Dileptus anser,* 200 to 500 microns long. 1, mouth; 2, contractile vacuole. *(From Kudo, 1966, after Hayes.)*

in the cell. The macronucleus may be in the form of many small bodies, a string of rounded bodies, or a long band. Many micronuclei are visible in stained organisms. Encystment is common, and *Dileptus* may excyst in hay infusion. See Appendix B for commercial sources.

CULTURE: Apparently little study has been made of the requirements for maintaining *Dileptus* in culture. Any of the media described for *Colpidium* may be appropriate and can be tried.

Euplotes and Other Hypotrichs

USES: To study the large cirri, used for "walking and jumping" and producing a characteristic jerking movement.

DESCRIPTION: Freshwater and saltwater species have been studied somewhat extensively. *Euplotes* and other hypotrichs, for example, *Oxytricha, Urostyla,* and *Stylonychia,* appear in hay infusions and in freshwater collections that contain vegetation and sediment. Cirri and coarse cilia on the flat, ventral surface are used for "walking" on the bottom surface (Fig. 12-25). Rows of short bristles are present on the convex, longitudinally ridged, dorsal surface. The body of *Euplotes* is clear, and the cirri can be seen through the body on the ventral surface. The body is 80 to 200 microns long. A large, triangular-shaped peristome (oral groove) is located on the dorsal surface, and contains ciliated membranelles that wind clockwise to the cytostome at about midlength of the body. A long, band-shaped macronucleus, one micronucleus, and one posterior contractile vacuole are present in the cell.

Asexual reproduction is by transverse division, and sexual reproduction is by conjugation. Encystment is common, and excystment occurs regularly in hay infusions or in sediment from ponds and lakes.

Stylonychia, closely related to *Euplotes,* may also appear in water samples and in hay infusions. *Stylonychia* can be as fascinating as *Euplotes.* Figure 12-26 shows *Stylonychia* "walking" at the bottom of water. Note the marginal cirri.

CULTURE: See the general techniques given in the introduction of the chapter. Use any of the following media (Appendix A): (1) lettuce medium; (2) Chalkey's medium; (3) grain infusion; (4) hay infusion; (5) skim milk medium.

Opalina

USES: (1) To study a large parasitic ciliate in the colon and rectum of amphibians. A few species have been reported in fish, salamanders, and reptiles. (2) To trace the life history of *Opalina* in frogs or tadpoles. (3) To look for parasitic amoebas *(Entamoeba)* in *Opalina,* a condition known as hyperparasitism. (4) To look for a fungus (reportedly *Sphaerita*) that parasitizes the *Entamoeba,* thus producing a condition of hyper-hyperparasitism.

DESCRIPTION: Opalinids are readily found in the cloaca, rectum, and colon of practically all frogs, toads, and their tadpoles. Only rarely is an anuran (tailless Amphibia) without *Opalina* or other related genera. After several weeks in the laboratory, the parasite may disappear from the host animal, although this is not always the case. For preparation of material from the host, see the techniques for

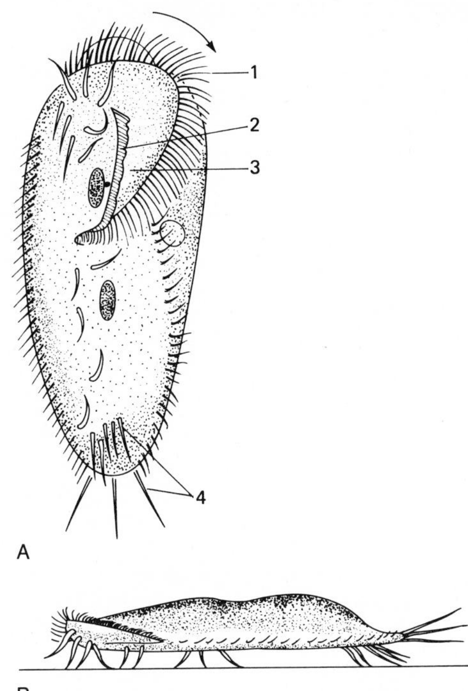

Figure 12-26 *Stylonychia mytilus.* **A** Diagram of a single organism, ventral view: 1, membranelles; 2, undulating membrane; 3, oral groove; 4, cirri. **B** Side view of organism walking on substrate. *(A From T. L. Jahn, 1949, after Stein. B From T. L. Jahn, 1949, after Butschli.)*

studying parasitic protozoans in the introduction of the chapter. Amaro (1967) gives an excellent account of methods for studying *Opalina*.

Opalina is a large protozoan; the average length is approximately 100 to 400 microns, and lengths of 600 to 1,300 microns have been reported. The body is flat, broad and blunt at the anterior end and pointed at the posterior end (Fig. 12-27). The cell is uniformly covered with cilia in oblique longitudinal rows, and swimming is in a spiral motion. A mouth and food vacuoles are absent, and digested food from the host is absorbed through the cell surface. Many small nuclei are present. Asexual reproduction is by oblique fission.

Often *Opalina* will contain numerous small amoebas (a situation of hyperparasitism). The amoebas reflect light and can be seen as refractory bodies within the swimming protozoan. When *Opalina* is stained with iodine solution, the

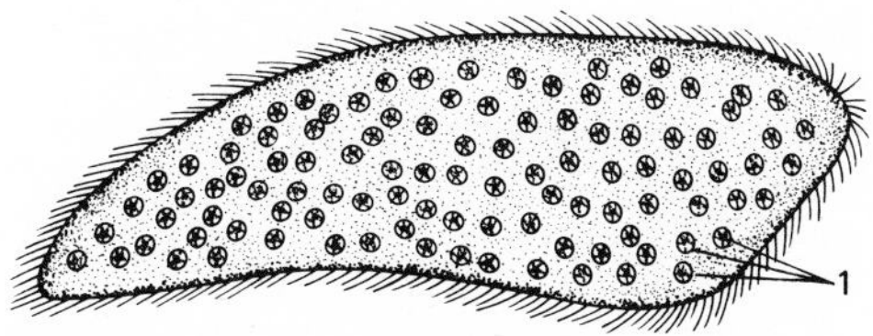

Figure 12-27 *Opalina* sp., 100 to 400 microns long. Many species are parasitic in frogs and toads. 1, micronuclei. *(From T. L. Jahn, 1949.)*

amoebas can be recognized under 400X magnification, with light reduced. The life history is complicated but can, however, be traced in a natural pond inhabited by frogs. Probably *O. ranarum* has been studied more extensively than other species. The parasite's life cycle is remarkably adapted to the frog's reproductive cycle in such a way that the parasite's cysts are released in great quantity within the frog feces just before the time when frogs lay eggs. Thus, the parasite is prepared to perpetuate its species when the young tadpoles ingest the cysts with food from the pond bottom. In the tadpole gut, the excysted opalinids develop into male and female gametes, which fuse to form zygotes that continue to grow and to reproduce asexually by fission.

CULTURE: For routine class studies, *Opalina* are readily obtained from host animals, as described in the introduction of this chapter. The protozoan can be maintained for perhaps 24 to 48 h in Ringer's solution I (Appendix A) or in a 0.7% solution of sodium chloride. Yang (1960) describes a technique for the continued culture of opalinids.

Paramecium

USES: To study: (1) the morphology of a ciliate; (2) the ingestion and digestion of food; (3) taxic responses to light, gravity, electricity, temperature, pH, chemicals, and to contact with solid objects; (4) asexual and sexual reproduction; (5) movement; (6) action of trichocysts; (7) induction and observation of conjugation between mating types of *P. bursaria* or *P. multimicronucleatum*.

DESCRIPTION: A common ciliate, regularly studied in beginning biology courses. Organisms are obtained from freshwater collections or from biological supply companies. At least nine species are recognized, of which the most commonly studied are the four species described below (Fig. 12-28).

Paramecium multimicronucleatum, largest of the four species, is 200 to 350 microns long, with three to seven contractile vacuoles, one macronucleus and four or more micronuclei. Posterior end is somewhat pointed. The species occurs in fresh water and occasionally in hay infusions.

Paramecium caudatum is somewhat smaller than the above species and more pointed at the posterior end. The cell is about 180 to 300 microns long. Two contractile vacuoles are present and one macronucleus with one micronucleus in close association with the larger nucleus. The species occurs in fresh water and often appears in hay infusions.

Paramecium aurelia is much smaller than the two above species and is about 120 to 180 microns long. Both anterior and posterior ends are rounded. Two contractile vacuoles, one macronucleus, and two micronuclei are present. The species is found in fresh water and often appears in hay infusions.

Paramecium bursaria is about the same length as *P. aurelia* but is broader. The species is easily identified by the presence of green, symbiotic zoochlorellae. The cell contains two contractile vacuoles, one macronucleus and one micronucleus. It occurs in fresh water but rarely appears in hay infusion. Collect in pond water or purchase from a biological supply company.

Techniques for observing the feeding apparatus and feeding. In all species, a

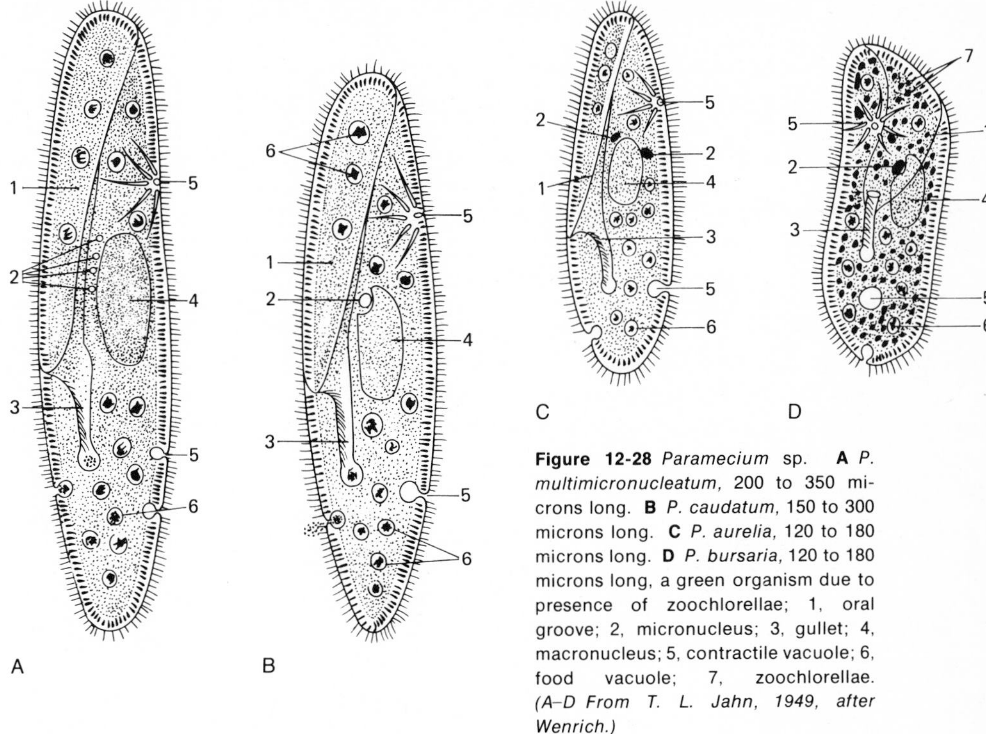

Figure 12-28 *Paramecium* sp. **A** *P. multimicronucleatum,* 200 to 350 microns long. **B** *P. caudatum,* 150 to 300 microns long. **C** *P. aurelia,* 120 to 180 microns long. **D** *P. bursaria,* 120 to 180 microns long, a green organism due to presence of zoochlorellae; 1, oral groove; 2, micronucleus; 3, gullet; 4, macronucleus; 5, contractile vacuole; 6, food vacuole; 7, zoochlorellae. *(A–D From T. L. Jahn, 1949, after Wenrich.)*

long peristome (oral groove) extends from the anterior end toward the middle of body, ending at the cytostome (mouth). The gullet extends from the mouth into the endoplasm. Strong cilia form a lining in the peristome and sweep food and water into the mouth. Food grains move through the gullet, become surrounded with water, and form a food vacuole that floats from the gullet into the endoplasm.

The feeding process can be observed by adding a small portion of carmine powder to a drop of water with *Paramecium.* At 100 to 400X magnification, the red particles can be clearly observed moving through the gullet and accumulating into a food vacuole at the tip of the gullet. Within several minutes many vacuoles of red particles will be seen moving about in the endoplasm. The carmine powder is indigestible, and eventually the protozoan ejects the material from the food vacuole to the outside near the posterior end. *Paramecium* feeds on bacteria and algae; it has been estimated that a large organism may eat several million bacteria in a 24-h period.

Staining and observing the cilia and trichocysts. In all species the ectoplasm contains both cilia and trichocysts distributed over the entire body surface. Apparently, trichocysts (stinging cells) are used defensively and also to paralyze food organisms. The structure of both the cilia and trichocysts is shown in good fashion in the electron micrographs of Jakus and Hall (1946). The trichocysts can be made to discharge by adding iodine solution (Lugol's) (Appendix A) or a weak acid to a drop of *Paramecium* on a microscope slide. Vital stains are described in the first portion of the chapter.

An old standard procedure that almost always produces spectacular staining of cilia and the discharged trichocysts was first described by Halter (1925). Place a small drop of red ink (writing ink for pens) in a drop of *Paramecium* on a glass slide. Allow the mixture to set for 5 min, and then add a coverglass. Examine the drop to be certain the protozoans are still alive and active. If they are dead, prepare another slide with less ink. Then, at the edge of the coverglass, add a small drop of permanent blue-black writing ink. Observe the preparation with the microscope. Generally when the blue-black ink approaches a protozoan, the long trichocysts will be discharged and may become separated from the cell body. Trichocysts will stain blue, the cilia will usually be red, and the cytoplasm pink. Occasionally, the macronucleus will be stained a dark red.

Methods for studying contractile vacuoles. Water-collecting canals (radiating canals) are present in the four species described above. Generally, the contractile vacuoles are near opposite ends; they fill from the radiating canals and then empty, in alternating manner, so that one vacuole is filling while the other is emptying. The action of the canals and water vacuoles is readily observed when the protozoan movement is slowed or stopped. Several methods for slowing protozoans are described in the introduction to this chapter. A number of interesting studies can be done using the rate of vacuole contraction as the determinant for effects of temperature, pH, and tranquilizing or stimulating substances (aspirin, caffein, carbon dioxide in carbonated water, adrenalin, acetylcholine, and other chemicals). Conditions must be standardized for such tests, including food, oxygen, age of culture or clone, and other variables.

Encystment and excystment. The subject of cysts in *Paramecia* has received considerable controversy. Many persons who have studied *Paramecia* intensively and over a long period of time believe that these protozoa do not encyst. Other persons say the cysts of *Paramecia* resemble grains of sand, and therefore, are easily overlooked. The fact that *Paramecia* appear regularly and abundantly in hay infusions, however, attests to their presence in some kind of dormant stage. Thus, the matter of cyst formation in *Paramecia* needs more study for clarification.

Observation of asexual reproduction. Asexual reproduction by transverse fission may be regularly observed in drops of medium from cultures that have been maintained for several days or longer (Fig. 12-29). The division may be completed within 20 to 30 min, or the process may require several hours. In most species, cytoplasmic division is preceded by a mitotic division of the micronucleus (or micronuclei) and by the formation of two new contractile vacuoles. The macronucleus elongates and divides amitotically as the cytoplasm divides. The old mouth and oral groove disappear; new structures are formed in each daughter cell (Wichterman, 1953).

Techniques for inducing conjugation between mating types. Conjugation occurs exclusively in the Ciliata and Suctoria. During conjugation, two protozoans unite and exchange portions of nuclear material. The occurrence of mat-

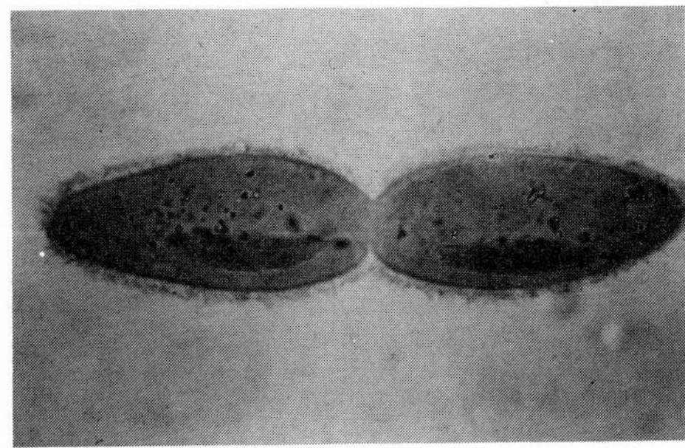

Courtesy of Carolina Biological Supply Company

Figure 12-29 Asexual reproduction in *Paramecium.*

ing types within species of ciliates has been well established by Sonneborn (1939 and 1957), Giese (1939 and 1957), and others. Also it is now known that for conjugation to take place, the partners of mating types must be physiologically different, which means then that mating types might be designated as physiological species. To avoid this difficulty, Sonneborn devised the term "syngen" to designate the mating varieties within a species. Mating types and breeding patterns have been established for a number of ciliates, among which those of *Paramecia, Euplotes,* and *Tetrahymena* are probably the best known. Two or more mating types are recognized within some species; at least 40 are now known for *Tetrahymena pyriformis.*

The mating reaction and conjugation are easily demonstrated between two pure-line cultures obtained from a biological supply company. The two pure-line cultures are maintained separately until the time of demonstration, when drops from each culture are mixed together on a slide and observed under the microscope. Apparently, better results are obtained if the drops are mixed under light, preferably sunlight; and apparently the mating reaction is stronger at noontime. Clumping becomes visible within a few minutes; after several hours large masses can be observed (Fig. 12-30). To observe the process for any length of time, the drops from the two cultures should be mixed in a deep-well slide or in a hanging drop to prevent evaporation. After 16 to 24 h, pairs of conjugants will be seen moving away from a clump. The pairs remain united for approximately 24 to 48 h, after which time they separate, and very few conjugants will be observed. In place of observing drops continuously on a slide, a desirable technique is to mix larger portions of the two mating types in a watch glass or a small culture dish, and to remove drops at intervals for observation. Maintain the cultures under moderate light or in sunlight during the day. Culture dishes should be thoroughly cleaned before use, and, to avoid contamination, separate droppers must be used for each pure-line culture.

CULTURE: *Paramecium* is easily grown in any of the following media (Appendix A): (1) Chalkey's medium; (2) grain infusion; (3) hay infusion; (4) lettuce medium; (5) skim milk medium. See the discussion of *General Techniques* in the introduction of the chapter.

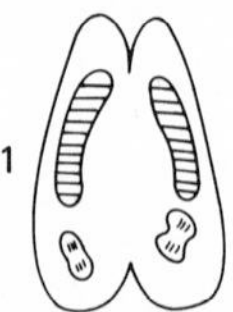

Two conjugants united; micronucleus in each prepares to divide

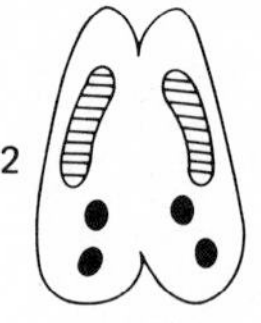

Two micronuclei in each conjugant

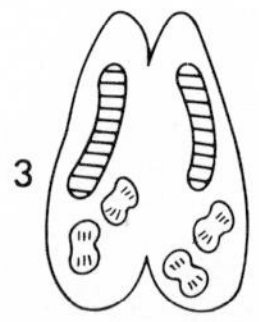

Second division of the micronuclei in each conjugant

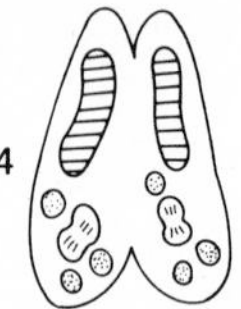

3 of the 4 micronuclei in each conjugant degenerate; the 4th prepares for division

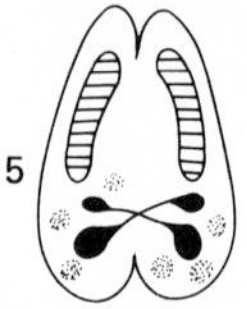

Exchange of migratory nuclei between conjugants

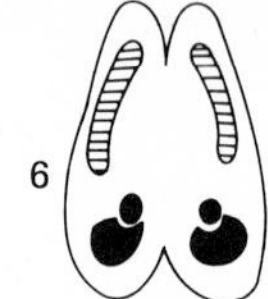

Amphimixis in each conjugant, followed by their separation

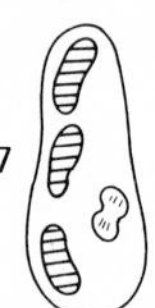

Micronucleus (fusion nucleus) prepares for division; macronucleus fragments

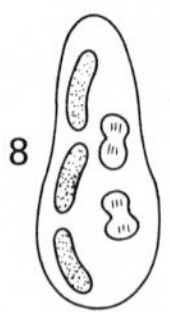

Micronuclei resulting from previous division divide; macronuclear fragments degenerate

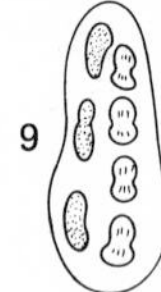

Micronuclei resulting from previous division divide; continued disintegration of macronuclear fragments

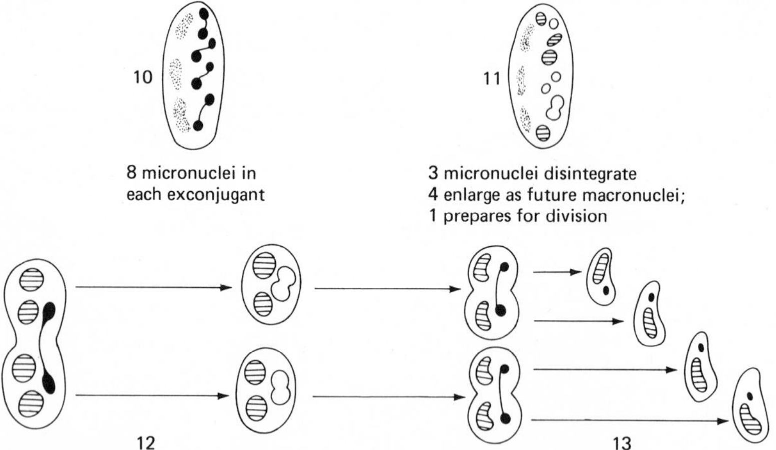

Figure 12-30 Conjugation in *Paramecium. (Reprinted with permission of The Macmillan Company from The Principles of Biology, 2d ed., by M. S. Gardiner and S. C. Flemister, Copyright © 1959 by The Macmillan Company.)*

Spirostomum

USES: (1) A large, interesting ciliate, useful for regular studies in place of other ciliates. It is not abundant in water collections and generally must be purchased. (2) To observe responses due to the particularly sensitive thigmotactic, caudal cilia. (3) To demonstrate regeneration in *Spirostomum ambiguum*.

DESCRIPTION: *Spirostomum* is a large ciliate with a length of 150 to 3,000 microns, and width approximately one-tenth of the length (Fig. 12-31). Species occur in fresh water and salt water. A large contractile vacuole at the posterior end is connected to a long dorsal canal. The macronucleus is ovoid or arranged into a long chain. A long, inconspicuous peristome begins at the anterior end on the flat, ventral surface and extends obliquely to the cytostome at approximately midlength, or more posterior, and at right side. The ectoplasm contains well-developed, longitudinal, contractile threads, called myonemes, that allow contraction to one-fourth the normal length of the body.

Procedures for demonstrating regeneration have been described by Johnson et al. (1966). For the demonstration, place 2 to 3 ml of a dense culture of *S. ambiguum* in a small test tube, and add the broken pieces of several coverglasses. Shake the test tube for several minutes to cut the protozoa into pieces with the broken glass. Transfer drops from the test tube to a glass slide and examine with the microscope. Select cut portions of *Spirostomum*, and use a capillary tube to transfer each piece to a deep-well slide. Place the slides on sections of bent glass

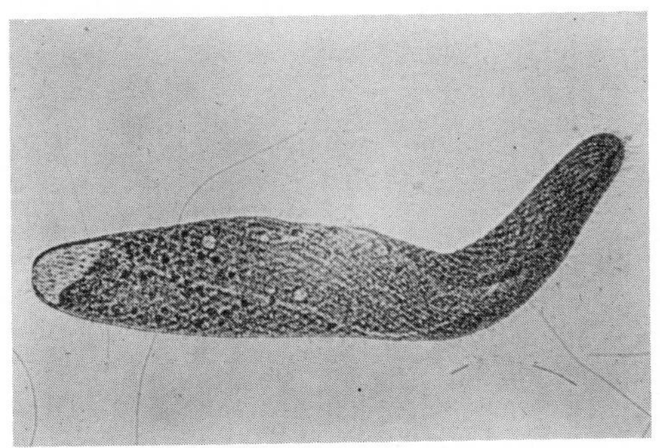

A *Courtesy of Carolina Biological Supply Company*

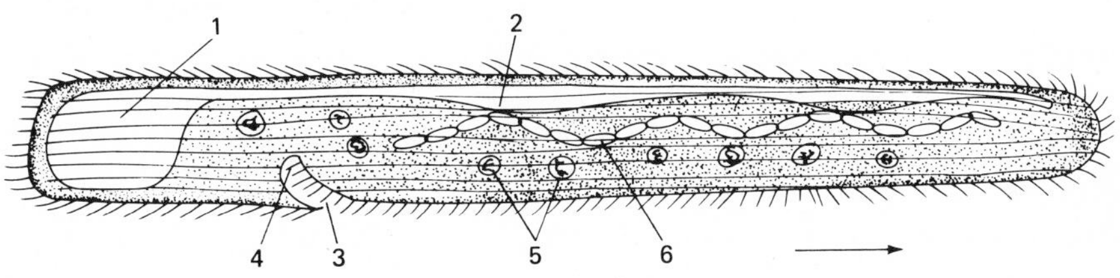

B

Figure 12-31 *Spirostomum* sp., 150 to 3,000 microns. **A** Living organisms. **B** *S. ambiguum,* diagram of cell: 1, contractile vacuole; 2, dorsal canal; 3, oral groove; 4, mouth; 5, food vacuole; 6, macronucleus. *(B From T. L. Jahn, 1949, after Kahl.)*

tubing in a covered moist chamber, e.g., a petri dish or culture bowl with a bottom layer of water. Examine the slides at the end of 24 h. As a rule, all portions of *Spirostomum* that contained a part of the macronucleus will have regenerated into a complete organism.

CULTURE: Collect in water or purchase from a biological supplier. Use any of the following media (Appendix A): (1) lettuce medium; (2) Chalkey's medium; (3) grain infusion; (4) hay infusion. Allow bacteria to accumulate in the medium, and add *Chilomonas* several times weekly as food for the *Spirostomum*. Place culture bowls in diffused light at 21 to 24°C. Cover bowls lightly to retard evaporation. Transfer to fresh cultures every 4 weeks. *Spirostomum* thrives in cultures that contain much bacterial activity.

Stentor

USES: (1) To observe a large and unusually beautiful, pigmented protozoan. (2) To study regeneration of the oral apparatus. (3) To observe conjugation between mating types, if available. No commercial sources are known for mating types.

DESCRIPTION: When extended, *Stentor* displays a long trumpet shape (Fig. 12-32). Several species are beautifully colored, due to blue, yellow, or rose-colored pigments. All contain highly contractile threads, myonemes, that allow the body to be contracted to one-fourth or less of the fully extended size. A conspicuous peristome with strong cilia encircles most of the anterior rim and ends in a small spiral at one edge that encloses the mouth. A short gullet leads into the endoplasm. The body is uniformly covered with cilia. The protozoan is found in fresh water and may be obtained from most biological suppliers.

***Stentor coeruleus* Ehrenberg.** This is a large ciliate; the extended body is 1 to 2 mm long and visible to the unaided eye. The anterior end expands into a large trumpet. The posterior end is greatly narrowed and generally attached to substratum, although an organism may become detached and swim freely. The pellicle contains granules of a dark-blue pigment, stentorin, accounting for the species name. The macronuleus resembles a long chain of beads. A single, relatively small contractile vacuole is toward the anterior end, with a single, long canal extending posteriorally. Asexual reproduction is by transverse division. Webb and Francis (1969) have described mating types and conjugation in *S. coeruleus*.

To demonstrate regeneration, follow the method described for *Spirostomum* (Johnson et al., 1966).

CULTURE: Use any of the following media (Appendix A): (1) lettuce medium; (2) Chalkey's medium; (3) grain infusion; (4) hay infusion; (5) skim milk medium.

Stentor requires a rich culture of mixed protozoans for food. Prepare the media, allow bacteria to accumulate in the exposed media for several days, and then inoculate with *Euplotes*, *Colpidium*, and *Chilomonas*, or another mixture of food protozoans. When the mixed protozoans are well established, add *Stentor*. Place culture bowls in diffused light at 20 to 23°C. Transfer to fresh medium every 2 or 3 weeks. Cover bowls lightly to retard evaporation. Because of the requirement for much food, *Stentor* is not easily maintained.

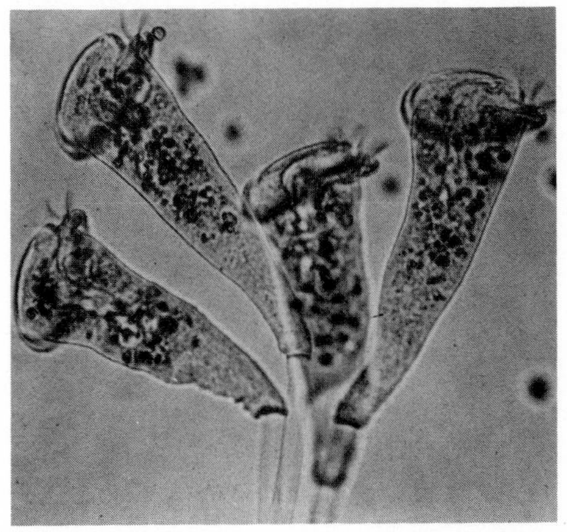

A *Courtesy of Carolina Biological Supply Company*

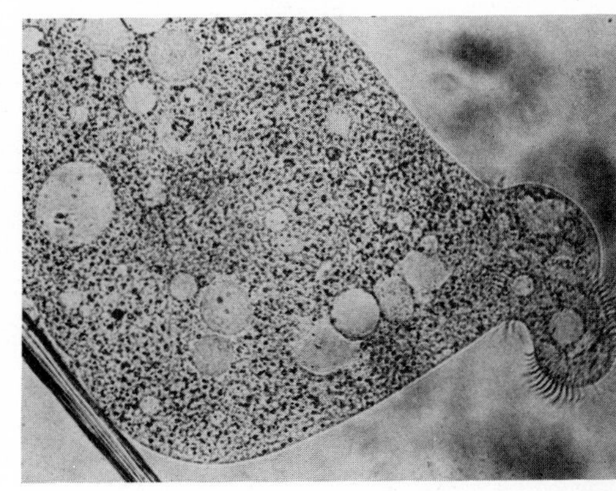

B

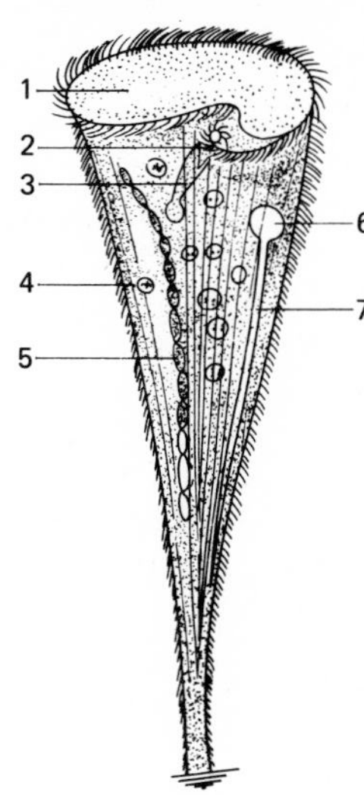

C

Figure 12-32 *Stentor* sp., 1 to 2 mm long when extended. **A** Several living organisms. **B** A living organism crowded on slide and twisted in shape. Anterior cilia and the vacuoles can be seen. **C** *S. coeruleus*, diagram of cell: 1, peristome; 2, mouth; 3, gullet; 4, food vacuole; 5, macronucleus; 6, contractile vacuole; 7, longitudinal canal. *(C From T. L. Jahn, 1949.)*

Tetrahymena

USES: (1) As a food organism (especially *T. pyriformis*) for larger protozoans. (2) For general studies of responses to changes in environment. (3) To demonstrate nutritional requirements in axenic culture. *Tetrahymena* is one of the few protozoa that can be nourished on synthetic media without the addition of bacteria or other living food. (4) To observe conjugation in mating types. Mating types may be purchased from the American Type Culture Collection (see Appendix B).

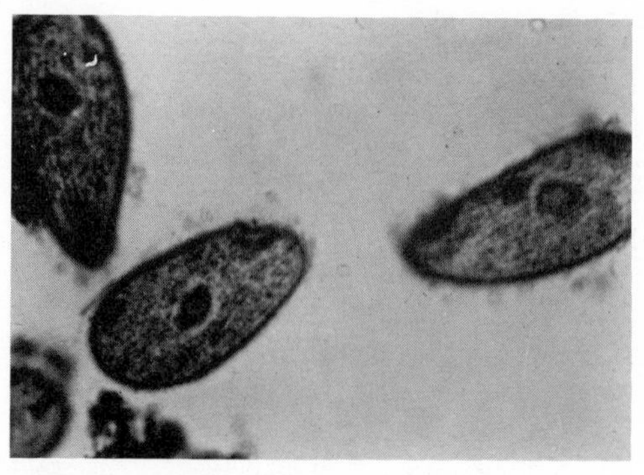

A *Courtesy of Carolina Biological Supply Company*

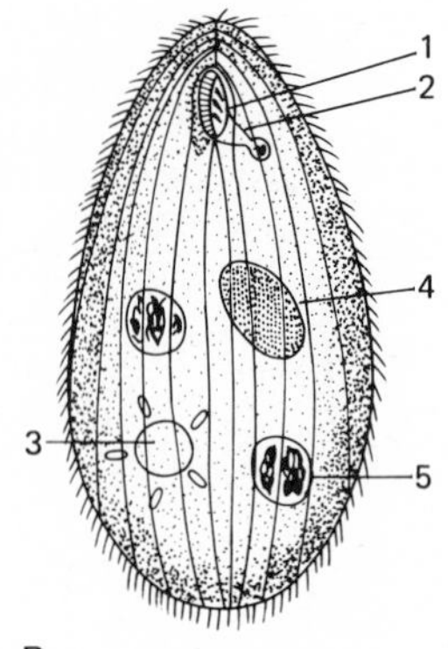

B

Figure 12-33 *Tetrahymena* sp., 30 to 70 microns long. **A** Stained organism showing central nucleus. **B** *T. geleii,* diagram of cell: 1, mouth; 2, gullet; 3, contractile vacuole; 4, macronucleus; 5, food vacuole. *(B From T. L. Jahn, 1949.)*

DESCRIPTION: *Tetrahymena pyriformis,* the species used most often in the laboratory, is 30 to 70 microns long (Fig. 12-33). It is widely distributed in ponds, lakes, small streams, and in rapidly moving water. The protozoan is much used today in physiological, genetic, and cytological research. The shape is generally pyriform (pear-shaped). The cilia are uniformly distributed over the body. A small cytostome is found on the median line near the anterior end; a short cytopharynx leads into the endoplasm. A peristome (oral groove) is not apparent. One contractile vacuole is near the posterior end. The macronucleus is ovoid; a micronucleus is present in some species and lacking in others. Asexual reproduction is by transverse division and may often be observed in hay infusions and in pond water. Conjugation occurs between mating types; forty or more mating types are known. See *Paramecium* for a description of techniques for inducing mating.

CULTURE: For mixed cultures, use any of the following media (Appendix A): (1) lettuce medium; (2) grain infusion; (3) hay infusion; (4) Chalkey's medium; (5) skim milk medium. Allow bacteria to accumulate in the uncovered medium during 1 or 2 days before adding *Tetrahymena.* Place culture bowls in diffused light at 20 to 23°C. Cover lightly to retard evaporation. Transfer to fresh media every 3 or 4 weeks.

A simple medium for axenic culture consists of the following: 5.0 g proteose peptone; 5.0 g tryptone; 0.2 g K_2HPO_4; 1.0 l distilled water. Adjust the pH to 7.2 before autoclaving at 15 lb pressure for 15 min. Use cotton-plugged, culture tubes with 5.0 ml of medium in each tube.

Vorticella

USES: (1) For observation of an unusually interesting protozoan that often occurs in hay infusions and on aquatic vegetation of collected pond water. (2) To

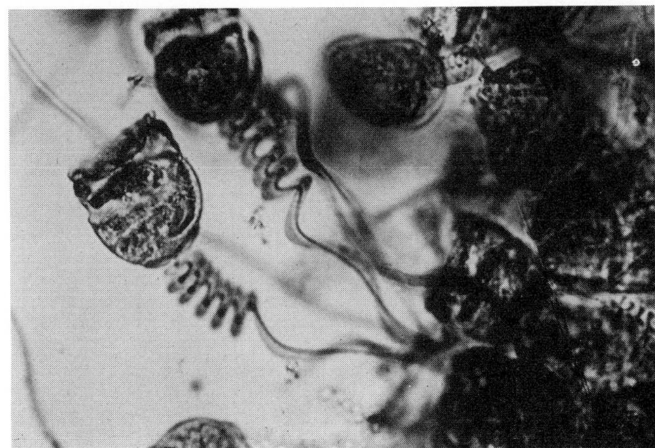

A *Courtesy of Carolina Biological Supply Company*

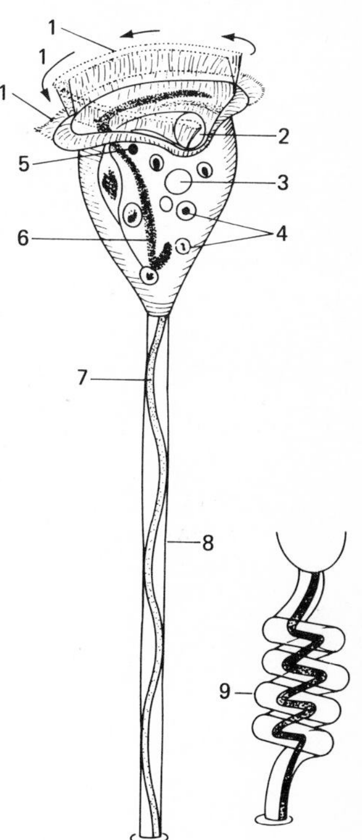

Figure 12-34 *Vorticella* sp. **A** Stained organism showing the myoneme running through the contracted stalks. **B** Diagram of an organism: 1, membranelles; 2, mouth; 3, contractile vacuole; 4, food vacuole; 5, micronucleus; 6, C-shaped macronucleus; 7, myoneme; 8, stalk; 9, contracted stalk with myoneme. *(**B** From T. L. Jahn, 1949, after Noland and Finley.)*

stain and observe the myoneme, a contractile thread that runs through the stalk and branches radially throughout the base of the bell (Fig. 12-34A). (3) So far as is known, the studies that are performed regularly with *Paramecium* and other protozoa have not been performed with *Vorticella*, although the studies may be as interesting and as useful.

DESCRIPTION: Species are found in fresh water and salt water. The bell-shaped ciliate is on a contractile stalk. The organisms are solitary. However, many solitary *Vorticella* may be together in a group, each attached separately to the substratum. Occasionally, an organism becomes detached and swims or floats about in the water with a dangling stalk. The bell portion of the body is 35 to 150 microns long. The immature organism has a ciliary girdle; the mature organism is without body cilia, except for a conspicuous ciliature on the anterior peristome, which winds clockwise to the cytostome. The macronucleus is band- or C-shaped. One micronucleus and generally one contractile vacuole are present.

Asexual reproduction is by longitudinal division. Apparently, only a little study has been made of sexual reproduction in *Vorticella*. (See Finley, 1943; and Finley and Williams, 1955.)

CULTURE: Little information is available concerning the culture of *Vorticella*. Possibly any of the following media will sustain *Vorticella* for several weeks or longer: (1) lettuce medium; (2) Chalkey's medium; (3) grain infusion; (4) hay infusion; (5) skim milk medium. All media are described in Appendix A. Allow bacteria to multiply in a medium before adding *Vorticella*. Most likely, small ciliates (*Chilomonas* or *Colpidium*) should be added as food organisms. These procedures provide many possibilities for explorations.

REFERENCES

Amaro, A.: 1967, "Methods in the General Biology of the Opalinid Protozoa," *Am. Biol. Teacher*, **29** (7): 546–551.

Cushman, J. A.: 1948, *Foraminifera, Their Classification and Economic Use*, 4th ed., Harvard University Press, Cambridge, Mass., 605 pp.

Davis, W. B.: 1937, "A Novel Method of Obtaining Protozoa," in J. G. Needham, Chairman, *Culture Methods for Invertebrate Animals*, Dover Publications, Inc., New York, 1959, p. 134.

Finley, H. E.: 1943, "The Conjugation of *Vorticella microstoma*," *Trans. Am. Microbiol. Soc.*, **62**: 97–121.

Finley, H. E., and H. B. Williams: 1955, "Chromatographic Analysis of the Asexual and Sexual Stages of a Ciliate (*Vorticella microstoma*)," *J. Protozool.*, **2**: 13–18.

Giese, A. C.: 1939, "Mating Types in *Paramecium caudatum*," *Am. Nat.*, **73**: 445.

Giese, A. C.: 1953, "Protozoa in Photobiological Research," *Physiol. Zool.*, **26**: 1.

Giese, A. C.: 1957, "Mating Types in *Paramecium multimicronucleatum*," *J. Protozool.*, **4**: 120.

Glaessner, M. F.: 1947, *Principles of Micropaleontology*, John Wiley & Sons, Inc., New York, 296 pp. (Contains an excellent section on collecting and studying microfossils.)

Gros, G.: 1849, "Fragments d' helminthologie et de physiologie microscopique," *Bull. Soc. Imp. Nat. Moscou*, **22**: 549.

Halter, C. R.: 1925, "Staining *Paramecium* in the Class-room," *Science*, **61**: 90.

Humason, G. L.: 1967, *Animal Tissue Techniques*, 2d ed., W. H. Freeman and Company, San Francisco, 569 pp.

Jahn, T. L., and F. F. Jahn: 1949, *How to Know the Protozoa*, Wm. C. Brown Company Publishers, Dubuque, Iowa, 234 pp.

Jakus, M. A., and C. E. Hall: 1946, "Electron Microscope Observations of the Trichocysts and Cilia of *Paramecium*," *Biol. Bull.*, **91**: 141–144.

Johnson, W. H., R. A. Laubengayer, L. E. DeLanney, and T. A. Cole: 1966, *Laboratory Manual for Biology*, 3d ed., Holt, Rinehart and Winston, Inc., New York, 261 pp.

Jones, D. J.: 1956, *Introduction to Microfossils*, Harper & Brothers, New York, 406 pp. (Contains an excellent discussion on collection and preparation of field samples.)

Kudo, R. R.: 1966, *Protozoology*, 5th ed., Charles C Thomas, Publisher, Springfield, Ill., 1174 pp.

Manwell, R. D.: 1961, *Introduction to Protozoology*, St. Martin's Press, Inc., New York, 642 pp.

Mertens, T. R.: 1966, "Population Explosion in a Test Tube," *Am. Biol. Teacher*, **28** (2): 103–107.

Moore, R. C., C. G. Lalicker, and A. G. Fischer: 1952, *Invertebrate Fossils*, McGraw-Hill Book Company, New York. (A standard reference; for the ambitious student.)

Pennak, R. W.: 1953, *Fresh-Water Invertebrates of the United States*, The Ronald Press Company, New York, 769 pp.

Pramer, D.: 1964, *Life in the Soil*, BSCS Laboratory Block, D. C. Heath and Company, Boston. Student Manual, 62 pp.; Teacher's Supplement, 38 pp.

Sonneborn, T. M.: 1939, "*Paramecium aurelia*: Mating Types and Groups: Lethal Interactions; Determination and Inheritance," *Am. Nat.*, **73**: 390–413.

Sonneborn, T. M.: 1957, "Breeding Systems, Reproductive Methods, and Species Problems in *Protozoa*," in *Species Problem*, Ernst Mayr (ed.), A.A.A.S. Pub. No. 50, pp. 155–324.

Stopyra, T.: 1965, "One Cell," *Am. Biol. Teacher*, **27** (3): 177–184.

Webb, T. L., and D. Francis: 1969, "Mating Types in *Stentor coeruleus*," *J. Protozool.*, **16** (4): 758–763.

Wessenberg, H., and G. Antipa: 1970, "Capture and Ingestion of *Paramecium* by *Didinium nasutum*," *J. Protozool.*, **17** (2): 250–270.

Wichterman, R.: 1953, *The Biology of Paramecium*, McGraw-Hill Book Company, New York, 527 pp.

Yang, W. C. T.: 1960, "On the Continuous Culture of Opalinids," *J. Parasitol.*, **46**: 32.

Yongue, W. H.: 1966, "Amebae in Forty-Eight Hours," *Am. Biol. Teacher*, **28** (2): 108–111.

Other Literature

Chandler, A. C., and C. P. Read: 1961, *Introduction to Parasitology*, 10th ed., John Wiley & Sons, Inc., New York, 822 pp.

Corless, John O.: 1961, *The Ciliated Protozoa*, Pergamon Press, New York, 310 pp.

Gliddon, R.: 1969, "Methods of Studying Movement and Structure in Living Protozoa," *J. Biol. Educ.*, **3** (2): 149–158.

Gregory, W. W., Jr., J. K. Reed, and L. E. Priester, Jr.: 1969, "Accumulation of Parathion and DDT by Some Algae and Protozoa," *J. Protozool.*, **16** (1): 69–71.

Hall, R. P.: 1964, *Protozoa, the Simplest of All Animals*, Holt, Rinehart and Winston, Inc., New York.

Honigberg, B. M., Chairman, The Com. on Taxonomy and Taxonomic Problems of the Soc. of Protozoology, 1964, "A Revised Classification of the Phylum Protozoa," *J. Protozool.*, **11** (1): 7–20.

Noble, E. R., and G. A. Noble: 1964, *Parasitology, The Biology of Animal Parasites*, 2d ed., Lea & Febiger, Philadelphia, 724 pp.

Olsen, O. W.: 1962, *Animal Parasites: Their Biology and Life Cycles*, Burgess Publishing Company, Minneapolis, 346 pp.

Provasoli, L., Chairman: 1958, "A Catalogue of Laboratory Strains of Free-Living and Parasitic Protozoa," *J. Protozool.*, **5** (1): 1–38.

APPENDIX

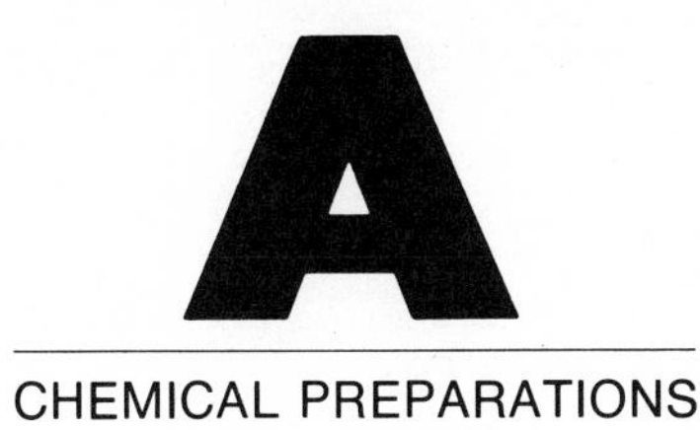

CHEMICAL PREPARATIONS

GENERAL INSTRUCTIONS

1. Unless stated otherwise, the water in a preparation is *distilled* water.
2. Most media, other than inorganic-salt preparations, may be purchased in a premixed form.
3. Before sterilization, all agar media are heated to dissolve the agar.
4. Sterilization of media is generally done in an autoclave or pressure cooker at 15 lb pressure for 15 min.
5. Algal cultures on agar may not show normal morphology until transferred to a liquid medium.
6. *For dilutions of alcohol,* use from the more concentrated alcohol the number of milliliters equal to the desired percentage number. Add enough water to make a volume equal to the original percentage. For example, to make a 50% alcohol solution from 95% alcohol, use 50 ml of the 95% alcohol and add 45 ml of water.
7. See Table A-1 for preparation, pH range, and color changes of common indicators.

TABLE A-1. COLOR CHANGES OF THE INDICATORS OF CLARK AND LUBS AND OF COHEN*

INDICATOR	CONCENTRATION RECOMMENDED,† %	FULL ACID COLOR	FULL ALKALINE COLOR	SENSITIVE pH RANGE
Thymol blue (acid range)	0.04	Red	Yellow	1.2–2.8
Brom phenol blue	0.04	Yellow	Blue	3.0–4.6
Brom chlor phenol blue	0.04	Yellow	Blue	3.0–4.6
Brom cresol green	0.04	Yellow	Blue	3.8–5.4
Methyl red	0.02	Red	Yellow	4.4–6.0
Chlor phenol red	0.04	Yellow	Red	4.8–6.4
Brom cresol purple	0.04	Yellow	Purple	5.2–6.8
Brom phenol red	0.04	Yellow	Red	5.2–6.8
Brom thymol blue	0.04	Yellow	Blue	6.0–7.6
Phenol red	0.02	Yellow	Red	6.8–8.4
Cresol red	0.02	Yellow	Red	7.2–8.8
Thymol blue (alkaline range)	0.04	Yellow	Blue	8.0–9.6
Cresolphthalein	0.04	Colorless	Red	8.2–9.8
Phenolphthalein	0.04	Colorless	Red	8.3–10.0

*Adapted from H.J. Conn, 1961, *Biological Stains*, 7th ed., The Williams & Wilkins Company, Baltimore.
†In 95% ethyl alcohol.

MEDIA AND SOLUTIONS CITED IN THE TEXT

Bold's Basal Medium; BBM (Bold, 1967)

A. Stock Solutions

1. 10 g $NaNO_3$ in 400 ml H_2O
2. 1 g $CaCl_2 \cdot 2H_2O$ in 400 ml H_2O
3. 3 g $MgSO_4 \cdot 7H_2O$ in 400 ml H_2O
4. 3 g K_2HPO_4 in 400 ml H_2O
5. 7 g KH_2PO_4 in 400 ml H_2O
6. 1 g NaCl in 400 ml H_2O

B. Trace-element Solutions

1. 50 g EDTA[1] and 31 g KOH in 1 l H_2O
2. Prepare 1 l of acidified water by adding 1 ml conc. H_2SO_4 *to* 999 ml H_2O. To this add 4.98 g $FeSO_4 \cdot 7H_2O$
3. 11.42 g H_3BO_3 in 1 l H_2O
4. All of following salts together in 1 l of acidified H_2O. (See 2 above.)

$ZnSO_4 \cdot 7H_2O$	8.82 g	$CuSO_4 \cdot 5H_2O$	1.57 g
$MnCl_2 \cdot 4H_2O$	1.44 g	$Co(NO_3)_2 \cdot 6H_2O$	0.49 g
MoO_3	0.71 g		

C. Final Solution

To 940 ml H_2O add 10 ml of each stock solution and 1 ml of each trace-element solution. (Store the final solution in refrigerator.)

Bold's Basal Medium with Agar

Add 15 g plain agar to 1 l of BBM (above).

Bold's Basal Medium with Increased Nitrogen

Modify BBM by using 30 ml of stock soln 1 and 920 ml of water in the final solution, in place of 10 ml of stock soln 1 and 940 ml of H_2O.

Boric Acid Solution (10 ppm)

Stock Solution: Dissolve 1 g of boric acid in 1,000 ml of distilled water.
Add 10 ml of the stock to 990 ml of water.

Bouin's Fixative (Conn et al., 1960)

Picric acid, sat. aqueous soln (about 1.22 g in 100 ml H_2O)	75 ml	Formalin	25 ml
		Acetic acid, glacial	5ml

Bristol's Solution (adapted from Starr, 1964)

A. Stock Solutions

Prepare the six stock solutions of Bold's basal medium.

B. Trace-element Solution

To 1 l of water add:

$ZnSO_4 \cdot 7H_2O$	0.1 g	$MnSO_4 \cdot 4H_2O$	0.15 g
H_3BO_3	0.1 g	$CuSO_4 \cdot 5H_2O$	0.03 g

C. Final Solution

To 940 ml H_2O add 10 ml of each stock solution, 2 ml of the trace-element solution, and *one drop* of a 1% $FeCl_3$ solution.

---

[1]EDTA = ethylene-diamine-tetra-acetic acid.

Buffer, Phosphate (Sorensen) (pH 5.29–8.04)

Stock Solutions:

M/15 dibasic sodium phosphate: 9.465 g made up to 1,000 ml with distilled water.
M/15 potassium acid phosphate: 9.07 g made up to 1,000 ml with distilled water.
For desired pH, mix correct amounts as indicated below.

pH	*M*/15 Dibasic Sodium Phosphate, ml	*M*/15 Potassium Acid Phosphate, ml	pH	*M*/15 Dibasic Sodium Phosphate, ml	*M*/15 Potassium Acid Phosphate, ml
5.29	2.5	97.5	6.81	50.0	50.0
5.59	5.0	95.0	6.98	60.0	40.0
5.91	10.0	90.0	7.17	70.0	30.0
6.24	20.0	80.0	7.38	80.0	20.0
6.47	30.0	70.0	7.73	90.0	10.0
6.64	40.0	60.0	8.04	95.0	5.0

Buffer, Standard (McIlvaine) (pH 2.2–8.0)

Stock Solutions:

1. 0.1 *M* citric acid (anhydrous): 19.212 g made up to 1,000 ml with distilled H_2O.
2. 0.2 *M* disodium phosphate (anhydrous): 28.396 g made up to 1,000 ml with distilled water.

For desired pH, mix correct amounts as indicated below.

pH	Citric Acid, ml	Disodium Phosphate, ml	pH	Citric Acid, ml	Disodium Phosphate, ml
2.2	19.6	0.4	5.2	9.28	10.72
2.4	18.76	1.24	5.4	8.85	11.15
2.6	17.82	2.18	5.6	8.4	11.6
2.8	16.83	3.17	5.8	7.91	12.09
3.0	15.89	4.11	6.0	7.37	12.63
3.2	15.06	4.94	6.2	6.78	13.22
3.4	14.3	5.7	6.4	6.15	13.85
3.6	13.56	6.44	6.6	5.45	14.55
3.8	12.9	7.1	6.8	4.55	15.45
4.0	12.29	7.71	7.0	3.53	16.47

pH	Citric Acid, ml	Disodium Phosphate, ml	pH	Citric Acid, ml	Disodium Phosphate, ml
4.2	11.72	8.28	7.2	2.61	17.39
4.4	11.18	8.82	7.4	1.83	18.17
4.6	10.65	9.35	7.6	1.27	18.73
4.8	10.14	9.86	7.8	0.85	19.15
5.0	9.7	10.3	8.0	0.55	19.45

Chalkey's Medium (Provasoli, 1958)

NaCl	80 mg	$CaCl_2$	4 mg
$NaHCO_3$	4 mg	$CaH_4(PO_4)_2 \cdot H_2O$	1.6 mg
KCl	4 mg	Water	1 l

Add four (for amoebae) or eight (for ciliates) boiled wheat or rice grains for each liter of medium. Allow bacteria to accumulate during 24 to 48 h. Add *Chilomonas*, *Colpidium*, or other food organisms as described in Chap. 12 for particular protozoa.

Colchicine Solution

For inducing polyploidy (Glass, 1965). Dissolve 0.5 g colchicine in 1 l H_2O. To demonstrate polyploidy in root tips, germinate bean or other seeds, or suspend onion bulb in water. When primary root or secondary roots are 2 to 3 cm long, suspend root in colchicine soln for 3 h. Then suspend in clear water for 24 to 48 h. Again suspend roots in colchicine soln for 3 h. Polyploid cells generally form a bulge near tip of root. Excise tip above bulged region and stain by procedure given for acetoorcein or acetocarmine stain.

Cornmeal Agar, with Dextrose (Alexopoulos and Beneke, 1962)

Cornmeal (white)	20 g	Agar	15 g
Peptone	20 g	Water	1 l
Dextrose	20 g		

Mix cornmeal in the water. Place container in a waterbath (or double-boiler) and cook slowly, with periodic stirring, for 1 h. Strain through coarse filter paper (with suction), or allow cornmeal to settle to bottom and pour off top liquid. Add other ingredients to filtrate or decanted liquid, and bring to a volume of 1,000 ml with distilled water. Strain through two layers of cotton on cheesecloth. Disperse into flasks or tubes, and sterilize.

For plain cornmeal agar, omit peptone and dextrose. To culture *Sordaria*, add 1 g of yeast extract to final 1-l volume, before sterilization.

Cyanophyceae Agar (Starr, 1964)

KNO_3	5.0 g	Plain agar	15.0 g
K_2HPO_4	0.1 g	Water	1 l
$MgSO_4 \cdot 7H_2O$	0.05 g		
Ferric ammonium citrate, 1% aqueous soln	10 drops		

Czapek Dox Agar (ATCC, 1970)

Glucose	30.0 g	KCl	0.5 g
$NaNO_3$	3.0 g	$FeSO_4 \cdot 7H_2O$	0.01 g
K_2HPO_4	1.0 g	Agar	15.0 g
$MgSO_4 \cdot 7H_2O$	0.5 g	Water	1 l

Desmid Agar (Starr, 1964)

$MgSO_4 \cdot 7H_2O$, 0.1% aqueous soln	10 ml	Agar	7.5 g
K_2HPO_4, 0.1% aqueous soln	10 ml	Water	1 l
KNO_3, 1.0% aqueous soln	10 ml		

Some desmid strains grow better when supernatant of soil-water extract is added in proportion of 50 ml soil supernatant/liter medium. See Soil-water extract.

Diatom Seawater Agar (Starr, 1964)

Bristol's soln	500 ml	Soil-water supernatant[1]	50 ml
Natural seawater	500 ml	Agar	15 g

Eosin-Methylene Blue (EMB) Agar (Pelczar and Reid, 1972)

Peptone	10.0 g	Eosin Y	0.4 g
Lactose	10.0 g	Methylene blue	0.065 g
K_2HPO_4	2.0 g	Agar	15.0 g
		Water	1 l

Color may disappear during autoclaving, but will reappear when medium cools.

Erdschreiber Solution (modified from Starr, 1964)

$NaNO_3$	2 g in 10 ml H_2O
$Na_2HPO_4 \cdot 12H_2O$	0.3 g in 10 ml H_2O
Seawater (natural)	1 l
Soil-water supernatant (See Soil-water extract)	50 ml

1st Day: Filter seawater through no. 1 filter paper. Heat filtrate to 73°C.

2d Day: Again heat seawater to 73°C. Sterilize the two salt solutions in separate tubes.

3d Day: Add 1 ml of each cold salt solution to the cold soil-water supernatant and the cold seawater. Transfer to sterile tubes or flasks.

Ott's artificial seawater may be substituted for natural seawater (Bold, 1967). Synthetic sea salts in solution should be as effective. (See suppliers of sea salts in Appendix B.)

Bold (1967) suggests the addition of a vitamin solution, prepared as follows:

Biotin	0.1 mg	Thiamin HCl	20.0 mg
B_{12}	0.1 mg	H_2O	100.0 ml

Sterilize, and add 1 ml to each liter of Erdschreiber solution.

---

[1] See Soil-water extract.

Formalin-aceto-alcohol Solution; FAA Solution (Conn, 1960)

70% ethyl alcohol	90 ml
Formalin	5 ml
Glacial acetic acid	5 ml

For delicate materials, use 50% alcohol. For tough materials, e.g., hard, woody tissue, use 3 ml of acetic acid and 7 ml of formalin.

Formalin, Buffered Neutral (Conn, 1960)

Formalin	100 ml	$NaH_2PO_4 \cdot H_2O$	4.0 g
Na_2HPO_4 (anhyd.)	6.5 g	Water	900 ml

Dilute according to instructions given for particular organisms.

Gibberellic Acid Solution (1 ppm)

Stock Solution
Dissolve 100 mg of gibberellic acid in 1 to 2 ml of 95% ethyl alcohol and add 900 ml of water. Warm gently on a hot plate or over a steam bath to evaporate the alcohol. Add water to make 1 l. Store under refrigeration in a flask covered with aluminum foil. Make fresh stock solution every 2 weeks.

Final Solution
Add 10 ml of stock solution to 990 ml of distilled water for a 1-ppm solution.

Gilson's Fixative (Humason, 1967)

Nitric acid, conc.	15 ml	Ethyl alcohol, 60%	100 ml
Glacial acetic acid	4 ml	Water	880 ml
Mercuric chloride crystals	20 g		

Grain Infusion

Add four boiled wheat or rice grains to 100 ml of boiled pond or spring water. Allow bacteria to accumulate during 24 to 48 h. Where indicated for particular protozoa, add *Chilomonas*, *Colpidium*, or other ciliates for food. (Also see Split-pea infusion.)

Hay Infusion (Kudo, 1966)

Add five to ten 2-in lengths of timothy hay (or other hay or dry grass) to 100 ml of distilled or spring water. Boil for 10 min. Add enough distilled water to regain a volume of 100 ml. Allow bacteria to accumulate during 24 to 48 h. When indicated, add *Chilomonas*, *Colpidium*, or other ciliates, as food organisms.

Hoagland's Solution, Modified (Lee, 1963)

A. Stock Solutions. Prepare each of the following in 100 ml H_2O:

1. $Ca(NO_3)_2 \cdot 4H_2O$	11.8 g
2. KNO_3	5.0 g

3. $MgSO_4 \cdot 7H_2O$	1.4 g
4. KH_2PO_4	1.4 g
5. Iron chelate[1]	1.0 g

B. Trace-element Solution

1. $ZnSO_4 \cdot 7H_2O$	2.2 g/100 ml H_2O (Use 10 ml)
2. $CuSO_4 \cdot 5H_2O$	0.8 g/100 ml H_2O (Use 10 ml)
3. $Na_2MoO_4 \cdot 2H_2O$	2.5 g/100 ml H_2O (Use 1 ml)

Final Trace-element Solution:

10 ml, each, of (1) and (2) above	$MnCl_2 \cdot 4H_2O$	1.8 g
1 ml of (3) above	H_3BO_3	2.8 g
	Water	979 ml

C. Final Solution

To 940 ml H_2O add 10 ml of each of the five stock solutions and 10 ml of the final trace-element solution.

Indole Acetic Acid (IAA) Solution (1 ppm)

Use procedure for gibberellic acid.

Kantz's Medium, Modified (Bold, 1967)

A. Stock Solutions

Prepare separate solutions of the following six salts, in the amounts shown:

1. $NaNO_3$	10.0 g in 400 ml H_2O
2. $CaCl_2 \cdot 2H_2O$	1.0 g in 400 ml H_2O
3. $MgSO_4 \cdot 7H_2O$	3.0 g in 400 ml H_2O
4. K_2HPO_4	3.0 g in 400 ml H_2O
5. NaCl	1.0 g in 400 ml H_2O
6. Tris[2]	12.5 g in 250 ml H_2O

B. Transfer 5 ml of each solution into 900 ml H_2O.
C. To this solution, add 10 ml of Eagle vitamin mixture (from Difco Laboratories) and 2 drops of vitamin B_{12} (in concentration of 1 mg B_{12} crystals/ml H_2O).
D. Add enough water to make a total volume of 996 ml. Adjust to pH 7.5 with 1*N* HCl.
E. Add 1 ml of each of the four trace-element solutions of Bold's basal medium.

Lactobacillus Medium (ATCC, 1970)

Skim milk, dry	100.0 g
Tomato juice (filtered, pH 7.0)	100.0 ml
Yeast extract	5.0 g

Add enough water to make 1 l.

---

[1]Iron chelate = iron ethylene-diamine-tetra-acetic acid.
[2]Tris = tris-(hydroxymethyl) aminomethane.

Lactophenol Mounting Medium (Humason, 1967)

Melted phenol (carbolic acid)	3 parts
Lactic acid	1 part
Glycerol	2 parts
Water	1 part

If staining is desired, add 0.05 to 0.1% solution of aniline blue (cotton blue), brilliant green, eosin, or other stain indicated for a particular organism.

Lactose-peptone Broth (Pelczar and Reid, 1972)

Lactose	5 g	Beef extract	3 g
Peptone	5 g	Water	1 l

Lettuce Medium (Provasoli, 1958)

Dried lettuce	1.5 g	$CaCO_3$, slightly in excess of amount that dissolves
Water	1 l	

Aseptic technique: Heat to boiling point. Filter, and dispense 5 to 10 ml to each of required number of tubes. Autoclave. Allow medium to cool and add *Aerobacter aerogenes*. Incubate at room temperature for 18 h. Add food protozoa, if appropriate. Add culture organism.

Nonsterile technique: Nonsterile techniques are satisfactory for most teaching purposes. Combine dried lettuce, water, and a slight excess of $CaCO_3$. Boil for 5 min. Filter hot medium into 250-ml flasks. Plug with cotton and let stand until cool. If flask is sealed tightly with "parafilm," medium can be stored in refrigerator for several weeks. At time of use, put 2 parts of medium and 1 part of distilled water in an open culture bowl. Allow bacteria to accumulate during 24 to 48 h. Add small ciliates as food organisms, if appropriate. Add the culture organism. Cover bowl with plastic film and place in dark or low light at room temperature. Ciliates with symbiotic algae require more light but not direct sunlight.

Dried lettuce may be obtained from Difco, or under the trade name of Cerophyl from Cerophyl Laboratories, Inc., Kansas City, Mo.

The lettuce can be prepared in the laboratory by slowly drying clean leaves in the oven until the leaves become crisp and brown. Discard any black leaves and grind others with a mortar and pestle. The powdered leaves may be stored indefinitely in a glass-stoppered bottle at room temperature.

Mannitol Agar (ATCC, 1970)

Yeast extract	5.0 g	Agar	15.0 g
Peptone	3.0 g	H_2O	1 l
Mannitol	25.0 g		

Mannitol Solution, Nitrogen-free

Mannitol	10.0 g	$MnSO_4 \cdot 4H_2O$	trace
K_2HPO_4	0.5 g	$FeCl_3 \cdot 6H_2O$	trace

$MgSO_4 \cdot 7H_2O$	0.2 g	Water	1 l
NaCl	0.2 g	$CaCO_3$	in excess

Sterilize in flasks. After sterilization, to each flask add slightly more (sterilized) $CaCO_3$ than will dissolve. (Sterilize $CaCO_3$ in a dry oven.)

Methyl Cellulose

Mix 10 g of methyl cellulose with 45 ml of boiling water. Immediately remove from heat and allow mixture to stand for 20 min. Add 45 ml of cool water and mix thoroughly. Place at 10°C (top shelf of refrigerator) until solution becomes transparent. Dispense into dropping bottles. The prepared methyl cellulose is sold under various trade names, for example, Methocel.

Nitrate Sucrose Agar

Sucrose	10.0 g	KCl	0.5 g
$NaNO_3$	2.0 g	$FeSO_4 \cdot 7H_2O$	0.5 g
K_2HPO_4	1.0 g	Agar	15.0 g
$MgSO_4 \cdot 7H_2O$	0.5 g	H_2O	1 l

Nutrient Agar

Peptone	5.0 g	Agar	15.0 g
Beef extract	3.0 g	Water	1 l

Nutrient Broth

Omit agar in nutrient agar, above.

Nutrient Broth, Beef Bouillon Cubes

Dissolve one beef bouillon cube in 500 ml of warm distilled water. Filter to remove fat and sediment. Autoclave.

Nutrient Solution, 1% Agar (Provasoli, 1958)

Proteose-peptone (Difco)	100 mg	$MgSO_4 \cdot 7H_2O$	2 mg
KNO_3	20 mg	Agar	1 g
K_2HPO_4	2 mg	H_2O	100 ml
Bring soln to pH 6.0 to 7.0			

When solidified, agar surface should be wet.

Oatflake Agar (Alexopoulos and Beneke, 1962)

Agar	15 g	Oatflakes, enough for
H_2O	1 l	bottom ½ in of tube

Heat agar in water. Place oatflakes in bottom ½ in of each tube. Pour 10 to 15 ml of agar water over oatflakes of each tube. Sterilize, and slant tubes when cooling.

Photobacterium Agar (Carolina Biological Supply Co.)

NaCl	30.0 g	Agar	20.0 g
Beef extract	3.0 g	Water	1 l
Peptone	5.0 g		

Transfer to fresh medium every 3 weeks. A fresh culture must be started for each demonstration of bioluminescence.

Physiological Salt Solutions

In each case, add the indicated amount of NaCl to 1 l of H_2O.

Frog	6.4 g NaCl
Salamander	8.0 g NaCl
Bird	7.5 g NaCl
Mammals	9.0 g NaCl

Porphyridium Agar (Starr, 1964)

Natural seawater	250.0 ml	Tryptone	0.5 g
Soil-water supernatant (see Soil-water extract)	50.0 ml	Agar	7.5 g
		Glass-distilled H_2O	250.0 ml
Yeast extract	0.5 g		

Potato-dextrose Agar (Alexopoulos and Beneke, 1962)

Dextrose	20.0 g	Agar	15.0 g
Potato infusion	1 l		

Boil, for 1 h, 200 g of peeled and diced white potatoes in 1 l H_2O. Strain through cloth and add H_2O to regain volume of 1 l. Add dextrose and agar; heat and dispense in tubes or flasks. Autoclave.

Pringsheim's Soil Water (Bold, 1967)

1. Select a garden soil that contains a medium amount of humus and that has not been recently fertilized with commercial fertilizer. Soil with a high clay content is not suitable.
2. For culture chambers, use test tubes, 250-ml flasks, or ½-pt milk bottles that have been washed thoroughly with soapy water and rinsed at least five times in distilled water.
3. Place about 0.2 g $CaCO_3$ in bottom of test tube (0.5 g to 1.0 g $CaCO_3$ in flask or bottle). The $CaCO_3$ should be in an amount so that it is visible under soil when steaming of step 5 is completed.
4. Add ¼ to ½ in of soil and enough distilled water to make chamber ¾ full. Plug loosely with cotton.
5. Steam (without pressure) for 1 h on each of three consecutive days.
6. Some algae, e.g., *Euglena*, grow better when one fourth of a pea cotyledon is added before steaming.
7. A few algae, e.g., *Spirogyra*, do not require the $CaCO_3$.

Prune Agar (Alexopoulos and Beneke, 1962)

Dried prunes	40.0 g	Water	1,000 ml
Agar	20.0 g		

Boil prunes until soft in 200 to 300 ml water. Crush fruit, and press through several layers of cheesecloth. Add water to make 1 l. Add agar. Autoclave.

Rhizobium Medium (ATCC, 1970)

Yeast extract	1.0 g	Mannitol	10.0 g
Soil-water extract	200.0 ml	Agar	15.0 g
		Water	800 ml

Ringer's Solution I, For cold-blooded vertebrate tissue (Humason, 1967, Modified)

KCl	0.42 g	NaCl	6.5 g
$CaCl_2$	0.25 g	Water	1,000 ml

Best when prepared fresh. For invertebrate tissue, add 0.20 g of $NaHCO_3$ and use only 6.0 g of NaCl.

Ringer's Solution II, For warm-blooded vertebrate tissue (Humason, 1967)

KCl	0.42 g	NaCl	9.0 g
$CaCl_2$	0.25 g	Water	1,000 ml

Best when prepared fresh.

Sabouraud's Agar (Alexopoulos and Beneke, 1962)

Glucose (or maltose)	40.0 g	Agar	15.0 g
Peptone	10.0 g	Water	1 l

Sea Water, Ott's Artificial (Bold, 1967)

A. Stock Solutions

Salt	g/100 ml H_2O	ml for Final Soln
1. NaCl	25.0	85.0
2. $MgSO_4 \cdot 7H_2O$	10.0	60.0
3. $MgCl_2 \cdot 6H_2O$	10.0	50.0
4. $CaCl_2 \cdot 2H_2O$	10.0	10.0
5. KCl	10.0	8.0
6. NaBr	1.0	10.0
7. $NaHCO_3$	1.0	20.0
8. H_3BO_3	1.0	6.0
9. $Na_2SiO_3 \cdot 9H_2O$	1.0	1.0
10. $Sr(NO_3)_2$	1.0	3.0
11. $NaNO_3$	1.0	20.0
12. Na_2HPO_4	1.0	2.0

B. To 700 ml of glass-distilled H_2O add the amounts of each stock solution shown in the third column above.
C. To this solution add 1 ml from each of EDTA, iron, and boron solutions [trace-element solutions (1), (2), and (3)] of Bold's basal medium.
D. Add enough glass-distilled H_2O to make a final volume of 1 l.

Shen-Chu Solution (Bold, 1967)

A. Stock Solutions. Add each of the following salts to 100 ml of water.

1. $Co(NH_2)_2$	2.0 g	4. Na_2SiO_3	1.0 g
2. $CaCl_2 \cdot 2H_2O$	10.0 g	5. K_2HPO_4	0.055 g
3. $MgSO_4 \cdot 7H_2O$	5.0 g	6. KCl	5.0 g

The Na_2SiO_3 may be omitted except for diatoms.

B. Final Solution

1. 1 ml of each of the above six stock solutions
2. 1 ml of each of the four trace-element solutions in Bold's basal medium
3. Add distilled water to bring the volume to 950 ml
4. Add 0.5 g Tris[1]
5. Adjust to pH 7.2
6. Add distilled water to bring final volume to 1,000 ml

Sterilize in autoclave or pressure cooker.

Skim Milk Medium

Nonfat dried milk	1.0 g	Water	1 l

Skim milk medium (Code # 0032) may be ordered from Difco Laboratories.

Milk agar may be prepared as follows:

Nutrient agar	23.0 g
Skim milk (liquid)	50.0 ml

Add water to make 1,000 ml volume.

Soil-water Extract (ATCC, 1970, Modified)

African violet soil	77.0 g
Na_2CO_3	0.2 g
Water	200.0 ml

Sterilize in autoclave or pressure cooker for 1 h at 15 lb pressure. Filter through paper and resterilize. Obtain African violet soil at a garden store.

Bold (1967) gives the following method: Prepare a mixture of 1 part, each, of garden soil and distilled water. Autoclave for 1 h. Allow mixture to cool and settle. Decant the supernatant, filter, and resterilize.

[1]Tris = tris-(hydroxymethyl)aminomethane.

Split-Pea Infusion (Bold, 1967)

Boil 40 split peas in 1 l of tap water. Allow infusion to cool and settle. Pour off supernatant. Bold recommends the very dilute supernatant for *Euglena gracilis*, that is transferred to fresh medium every 2 months. For class studies, transfer *Euglena* to fresh medium 3 or 4 days before use.

Sporulation Agar (ATCC, 1970, Modified)

Beef extract	1.0 g	Dextrose	10.0 g
Yeast extract	1.0 g	$FeSO_4$	Trace
Tryptose	2.0 g	Agar	15.0 g
		Water	1 l

Trypticase Agar (Pelczar and Reid, 1972)

Trypticase	10.0 g	K_2HPO_4	2.5 g
NaCl	5.0 g	Agar	15.0 g
		Water	1 l

For *Soft Trypticase Agar* use 7.0 g agar in place of 15.0 g.

Trypticase Broth (Pelczar and Reid, 1972)

Trypticase	10.0 g	K_2HPO_4	2.5 g
NaCl	5.0 g	Water	1 l

Tryptone Glucose Extract Agar

Use dehydrated product from Difco Laboratories (Cat. No. 0002).

Yeast-starch Agar (Alexopoulos and Beneke, 1962)

Starch, soluble	15.0 g	$MgSO_4 \cdot 7H_2O$	0.5 g
Yeast extract	4.0 g	Agar	20.0 g
K_2HPO_4	1.0 g	Water	1 l

STAINS CITED IN THE TEXT

Acetocarmine Staining (also see Acetoorcein)

Glacial acetic acid	45 ml
Water	55 ml
Powdered carmine (excess)	0.5 g, approx.

1. Dissolve carmine in acetic acid and add water.
2. Boil for 2 to 4 min in well-ventilated room. The quantity of carmine should be greater than the amount that goes into solution.
3. Cool; filter; add 1 or 2 *drops* of ferric chloride prepared in 50% acetic acid (5 g $FeCl_3$ in 50 ml glacial acetic acid and 50 ml H_2O).

4. Transfer tissue to be stained (e.g., plant sporocytes in sporangia of liverworts, mosses and ferns, and in anther of seed plants; also onion or seedling root tips) to a watch glass containing a mixture of 9 parts of stain with 1 part of 1 *N* HCl.
5. Hold watch glass over a 100-W light bulb or 4 to 5 in above a flame. Heat until steam is visible, but do not boil.
6. Transfer tissue to a drop of stain on a glass slide. Add a coverglass and spread the cells by pressing thumb against the coverglass. Check slide under 100X magnification to determine if cells are spread to a single layer. Press coverglass again, if necessary.

Acetoorcein Staining

Prepare both a 1 and a 2% stain.

1. Dissolve 1 g orcein (1% stain), or 2 g orcein (2% stain), in 45 ml of hot glacial acetic acid (near boiling). Cool and add 55 ml water. Mix thoroughly and filter.
2. Place tissue that is to be stained in a watch glass containing 9 parts of 2% acetoorcein stain and 1 part of 1 *N* HCl.
3. Heat as described in step 5 of acetocarmine staining.
4. Transfer tissue to a drop of 1% acetoorcein stain on a glass slide. Treat as described in step 6 of acetocarmine staining.

Ammonium Oxalate-crystal Violet

Solution A

Dissolve 2 g crystal violet in 20 ml of 95% ethyl alcohol.

Solution B

Dissolve 0.8 g ammonium oxalate in 80 ml water. Combine the two solutions. For bacterial Gram staining.

Aniline Blue

Dissolve 2.0 g dry stain (water soluble) in 100 ml water. For fungal mycelium.

Bismarck Brown Y

Stock Solution

Dissolve 0.5 g dry stain in 100 ml 95% ethyl alcohol.

Final Solution

Add 2 ml stock solution to 98 ml 95% ethyl alcohol. Vital stain for protozoa; a general cellular stain.

Brilliant Cresyl Blue

Add 0.1 g of powdered stain to 100 ml water. A vital stain for protozoa and a general cellular stain.

Carbol Fuchsin, Ziehl's

Dissolve 0.3 g basic fuchsin in 10 ml of 95% ethyl alcohol. Add this to 100 ml of 5% aqueous phenol (5 g phenol in 95 ml H_2O). Bacterial and general cellular stain.

Carbol Rose Bengal

Dissolve 1 g rose Bengal in 100 ml of 5% aqueous phenol. Add 0.01 g of $CaCl_2$. For staining soil bacteria.

Crystal Violet (Tyler's Aqueous Solution)

Dissolve 1 g crystal violet in 100 ml H_2O. General purpose stain for bacteria, protozoa, and cellular staining.

Eosin

Add 0.5 g dry eosin Y stain to 100 ml water. For staining dead protozoa, and to make living, unstained protozoa appear translucent against the red background.

Iodine Stain, Gram's

Dissolve 1 g iodine crystals and 2 g potassium iodide in 300 ml water. For Gram staining of bacteria.

Iodine Stain, Lugol's (Iodine Potassium Iodide)

Iodine crystals	2 g	Water	300 ml
Potassium iodide	3 g		

Dilute in water, as necessary.

Janus Green B

Stock Solution

Dissolve 0.5 g dry stain in 100 ml of 95% ethyl alcohol.

Final Solution

Add 2 ml stock soln to 98 ml 95% ethyl alcohol. Use a very small drop as a vital stain for protozoa. The stain is a hydrogen acceptor, e.g., in mitochondria where the compound is red in a reduced state and green in an oxidized state.

Methylene Blue

Stock Solution

Dissolve 0.5 g dry stain in 100 ml 95% ethyl alcohol. Let stand for 2 to 3 days, stirring occasionally. Filter and store.

Final Solution

Add 10 ml stock soln to 90 ml distilled water. A vital stain for protozoa; a general cellular stain. (Alternative: Add 2 ml stock soln to 98 ml 95% ethyl alcohol.)

Methylene Blue, Alkaline (Loeffler's)

Dissolve 0.3 g dry stain in 30 ml of 95% ethyl alcohol. Add this to 100 ml of 0.01% aqueous solution of KOH (0.01 g of KOH pellets in 100 ml H_2O). Bacterial stain and general cellular stain.

Methylene Green

Stock Solution

Dissolve 0.5 g dry stain in 100 ml 95% ethyl alcohol.

Final Solution

Add 2 ml stock soln to 98 ml 95% ethyl alcohol. Use a very small drop as a vital stain for protozoa. Also a general cellular stain.

Neutral Red

Stock Solution

Dissolve 0.5 g dry stain in 100 ml 95% ethyl alcohol.

Final Solution

Add 2 ml stock soln to 98 ml 95% ethyl alcohol. Use a very small drop as a vital stain for protozoa. Also use as an indicator solution:

bright red color	pH 7–7.5
orange to yellow	alkaline
violet to blue	acid

Nigrosin Stain

Add 10 g nigrosin (water soluble) to 100 ml water in a beaker. Use a wax pencil to mark the top level of the contents. Place beaker in a boiling water bath for 30 min to dissolve stain. Remove beaker; add distilled water to regain original volume. Add 0.5 ml of formalin. Filter twice through double layers of filter paper.

Safranin Stain

Dissolve 0.25 g safranin O in 10 ml 95% ethyl alcohol. Add this to 90 ml water. For Gram stain of bacteria.

REFERENCES

Alexopoulos, C.J., and E.S. Beneke: 1962, *Laboratory Manual for Introductory Mycology*, Burgess Publishing Company, Minneapolis.

ATCC: 1970, *Catalogue of Strains*, 9th ed., The American Type Culture Collection, Rockville, Md.

Bold, H.C.: 1967, *A Laboratory Manual for Plant Morphology*, Harper & Row, Publishers, Incorporated, New York.

Conn, H.J., M.A. Darrow, and V.M. Emmel: 1960, *Staining Procedures*, 2d ed., The Williams & Wilkins Company, Baltimore.

Glass, B.: 1965, *Genetic Continuity*, D.C. Heath and Company, Boston.

Humason, G.L.: 1967, *Animal Tissue Technique*, W.H. Freeman and Company, San Francisco.

Kudo, R.R.: 1966, *Protozoology*, 6th ed., Charles C Thomas, Publisher, Springfield, Ill.

Lee, A.E.: 1963, *Plant Growth and Development*, D.C. Heath and Company, Boston.

Pelczar, M.J., Jr., and R.D. Reid: 1972, *Laboratory Exercises in Microbiology*, 3d ed., McGraw-Hill Book Company, New York.

Provasoli, L. (Chairman): 1958, "A Catalogue of Laboratory Strains of Free-Living and Parasitic Protozoa," *J. Protozool.*, **5** (1): 1–38.

Starr, R.C.: 1964, "The Culture Collection of Algae at Indiana University," *J. Botan.*, **51** (9): 1013–1044.

APPENDIX

B

SUPPLIERS AND MANUFACTURERS

ANIMAL SUPPLIERS

The following tables *cannot* be used for ordering from a supplier, and before placing an order, a person must obtain a supplier's catalog. The tables are adapted from *Animals for Research*, published by the Institute of Laboratory Animal Resources. A more complete listing will be found in that publication.[1]

All names and addresses contained in the tables have been updated by direct correspondence with each supplier; the information in the tables was furnished by the suppliers. The author cannot assume a responsibility for the quality of products, nor does the author endorse or recommend any specific supplier.

[1]Institute of Laboratory Animal Resources, *Animals for Research*, Publication 1678 of the National Academy of Sciences, Washington, D.C., 1968.

TABLE B-1. CHECKLIST FOR SOURCES OF LABORATORY ANIMALS

(See Table B-3 for names of suppliers that correspond to the code numbers given in this table.)

ANIMALS	CODE NUMBER OF SUPPLIER
Porifera	
Freshwater sponge:	20, 22, 27
Saltwater sponge:	20, 39, 40, 78, 84, 114, 118
Cnidaria	
Sea anemone:	20, 39, 40, 78, 84, 114, 118
Coral:	20, 39, 78, 84, 114
Aurelia:	20, 40
Hydra:	17, 20, 21, 22, 27, 66, 68, 101, 102, 104, 113
Platyhelminthes	
Freshwater planarian:	17, 20, 21, 22, 27, 66, 68, 101, 102, 104, 113
Nematoda	
Vinegar eel:	17, 20, 22, 27, 66, 68, 101, 102, 104, 113
Soil nematode:	20, 22
Annelida	
Aeolosoma:	22, 27
Tubifex:	17, 20, 22, 27, 106
Enchytraeus:	20, 22, 27, 66, 106
Lumbricus:	17, 20, 21, 22, 27, 66, 68, 84, 101, 102, 104, 106
Hirudinea:	17, 20, 22, 27, 66, 68, 78, 101, 104
Mollusca	
Freshwater snail:	17, 20, 21, 22, 27, 66, 68, 84, 101, 104, 113
Land snail:	17, 20, 21, 66
Marine slug:	20, 39, 114
Freshwater clam and mussel:	17, 20, 22, 27, 66, 68, 104
Marine gastropod:	20, 39, 78, 84, 114, 118
Marine pelecypod:	20, 39, 78, 84, 114, 118
Arthropoda	
Artemia (eggs):	17, 20, 22, 27, 66, 104, 113
Daphnia:	17, 20, 22, 27, 66, 68, 101, 102, 104, 113
Cyclops:	20, 22, 27, 101, 113
Land isopod:	22
Gammarus:	22, 27, 66, 104, 118
Crayfish:	17, 20, 21, 22, 27, 66, 84, 101, 104, 113
Cockroach:	17, 22, 27, 127
Cricket:	17, 20, 21, 22, 27, 84, 100
Grasshopper:	20, 21, 27
Tenebrio:	17, 20, 22, 27, 84, 100, 101, 104, 106, 108, 113, 122, 127
Tribolium:	22
Butterfly (eggs and pupae):	17, 20, 21
Moth (eggs and pupae):	17, 20, 22, 27, 66, 68, 104
Drosophila:	17, 20, 21, 22, 27, 66, 68, 104, 113
Housefly:	17, 21, 22
Blowfly:	17, 22
Nasonia (Mormoniella):	22
Ant:	20, 27, 132
Honey bee:	20, 22, 27, 52, 66, 101
Scorpion:	17, 82, 84

TABLE B-1 (Continued)

ANIMALS	CODE NUMBER OF SUPPLIER
Spider:	17, 20, 22, 82, 100, 101
Millipede:	17, 20, 22, 82, 84, 101
Centipede:	17, 20, 21, 22, 82, 84
Echinodermata	
Sand dollar:	20, 39, 40, 78, 84
Starfish:	20, 27, 39, 40, 78, 84, 114, 118, 123
Sea urchin:	20, 21, 27, 39, 40, 78, 84, 114, 118, 123
Sea cucumber:	20, 39, 40, 78, 84, 114, 123
Fish	
Goldfish:	20, 21, 22, 66, 68, 84, 104
Stickleback:	20, 40, 66, 68, 104
Japanese Medaka:	22
Betta splendens and other anabantids:	21, 84
Haplochromis and other cichlids:	84
Sunfish (*Lepomis*):	84
Guppy:	20, 31, 84
Minnow (*Gambusia*):	20, 22, 40
Marine fish, aquarium:	39, 40, 78, 84, 114, 118, 128
Other aquarium fish:	31
Amphibia	
Salamanders and newts:	17, 20, 22, 27, 66, 68, 84, 89, 100, 101 103, 104, 122, 127
Hyla:	27, 66, 82, 84, 100, 103, 108, 122, 127
Rana:	12, 17, 20, 22, 27, 66, 68, 82, 84, 89, 91, 100, 101, 104, 108, 113, 121, 122, 127
Bufo:	16, 17, 20, 27, 66, 68, 82, 83, 84, 91, 100, 103, 104, 108, 122, 127
Frog ovulation set:	17, 21, 22, 27, 66, 68, 101, 104, 113
Reptilia	
Turtle:	16, 17, 20, 21, 22, 27, 39, 40, 66, 68, 82, 83, 84, 89, 100, 101, 104, 108, 121, 122, 123, 127
Lizard:	16, 17, 20, 21, 22, 66, 82, 83, 84, 89, 100, 101, 103, 104, 108, 122, 123, 127
Snake (nonpoisonous):	12, 16, 17, 20, 22, 66, 68, 82, 83, 84, 100, 103, 108, 122, 123, 127
Aves	
Bantam chicken:	20, 59, 89, 111, 115, 122
Pigeon, domestic:	17, 20, 27, 37, 59, 66, 79, 84, 89, 91, 104, 111, 115, 122, 125
Japanese quail (*Coturnix*):	89, 115, 122
Chicks, domestic:	66, 115
Mammals	
Mice:	12, 17, 20, 23, 53, 63, 64, 66, 71, 80, 81, 89, 90, 91, 92, 97, 101, 104, 107, 121, 122
Rats:	6, 12, 17, 20, 23, 53, 63, 64, 66, 71, 77, 80, 81, 89, 90, 91, 92, 97, 101, 104, 107, 121, 122, 125
Hamsters:	12, 64, 66, 76, 84, 90, 91, 99, 104, 122, 129

TABLE B-1 (Continued)

ANIMALS	CODE NUMBER OF SUPPLIER	ANIMALS	CODE NUMBER OF SUPPLIER
Guinea pigs:	6, 12, 20, 37, 47, 48, 53, 64, 66, 77, 80, 81, 84, 85, 89, 90, 91, 97, 104, 122, 125	Dutch:	6, 20, 37, 44, 48, 59, 64, 84, 89, 90, 91, 111
Rabbits, strains:		Polish:	44, 122, 125
New Zealand:	6, 12, 20, 37, 47, 48, 53, 59, 64, 80, 81, 84, 89, 90, 91, 111, 121, 122, 125		

TABLE B-2. CHECKLIST FOR SOURCES OF EQUIPMENT AND SUPPLIES*

(See next table (B-3) for names of suppliers that correspond to the code numbers given in this table.)

ITEM	CODE NUMBER OF SUPPLIER	ITEM	CODE NUMBER OF SUPPLIER
Ant nest, observation:	27, 35, 52, 120	Bottles and tubes, drinking:	5, 12, 21, 22, 23, 27, 35, 36, 45, 48, 49, 50, 51, 55, 67, 68, 77, 84, 89, 91, 101, 119, 120, 122
Aquarium, freshwater, and accessories:	15, 21, 22, 27, 29, 35, 36, 49, 50, 60, 66, 68, 84, 89, 101, 104, 129	Brooders, chick:	21, 51, 60, 66, 68
Aquarium, marine, and accessories:	8, 15, 21, 22, 29, 35, 49, 68, 78, 84, 101, 114, 129	Cages:	5, 21, 22, 23, 27, 33, 34, 35, 36, 45, 48, 49, 50, 51, 54, 55, 58, 66, 68, 77, 80, 84, 96, 101, 116, 119, 120, 122
Bedding (cage litter):	5, 12, 23, 36, 48, 50, 51, 77, 84, 89, 91, 116, 122	Chemicals, standard laboratory:	21, 22, 24, 30, 32, 61, 65, 95
Beehive, observation:	22, 27, 35, 52, 66, 129	Feeders:	21, 22, 23, 29, 35, 36, 45, 48, 49, 50, 51, 55, 58, 68, 77, 80, 89, 96, 101, 104, 119, 120
Behavioral apparatus:	21, 50, 55, 58, 68, 89, 96, 105, 119	Food, animal:	2, 22, 27, 34, 35, 36, 48, 66, 74, 77, 84, 88, 98, 100, 101, 106, 109, 115, 120, 122
Biochemicals:	10, 19, 21, 22, 30, 34, 38, 62, 74, 94, 95		

*Many items listed in the table are also furnished by the general biological suppliers whose code numbers are given under the item of "General biological supplies and equipment."

TABLE B-2 (Continued)

ITEM	CODE NUMBER OF SUPPLIER	ITEM	CODE NUMBER OF SUPPLIER
General biological supplies and equipment:	3, 13, 21, 22, 24, 25, 27, 28, 32, 35, 57, 60, 66, 67, 68, 93, 95, 101, 102, 104, 110, 113, 120	Seines:	22, 35, 104, 124
Incubator, egg:	17, 21, 22, 51, 54, 66, 68, 84, 89, 120	Snares, small animal:	21, 33, 50, 51, 127
Microscopes:	4, 11, 21, 22, 35, 93, 95, 120, 126	Stains, biological:	22, 27, 35, 120
Nets:	15, 17, 21, 22, 35, 36, 101, 104, 120, 124	Tags, bands, and tapes (identification):	21, 46, 48, 50, 51, 68, 70, 86, 87, 95, 101, 119
Physiological apparatus:	21, 22, 35, 43, 71, 93, 95, 120	Tattoo equipment:	5, 29, 33, 48, 51, 68, 87, 119
Preserved animals:	14, 20, 21, 22, 27, 35, 66, 67, 93, 102, 104, 113, 120, 121	Traps:	7, 15, 42, 72
Radioisotopes:	1, 9, 18, 73, 117	Veterinary supplies:	29, 33, 36, 48, 51, 68, 70, 87
Sea salts, synthetic:	8, 15, 21, 22, 27, 29, 36, 39, 51, 68, 78, 84, 101, 104	Washers, bottle and equipment:	21, 36, 48, 51, 54, 77, 89, 95
		Watering system, automatic:	41, 80

TABLE B-3. SOURCES FOR ANIMALS, EQUIPMENT, AND GENERAL SUPPLIES

CODE NUMBER (See Tables B-1 and B-2) — SUPPLIER

1. Abbott Laboratories, Oak Ridge Division, Oak Ridge, Tenn. 37830
2. Allied Mills, Inc., 110 N. Wacker, Chicago, Ill. 60606
3. Aloe Scientific, 1831 Olive Street, St. Louis, Mo. 63103
4. American Optical Co., Instrument Division, Buffalo, N.Y. 14215
5. Ancare Corp., 47 Manhasset Ave., Manhasset, N.Y. 11030
6. Animal Resources, Inc., P.O. Box 67, Woodsboro, Md. 21798
7. Animal Trap Co. of America, Lititz, Pa. 17543
8. Aquarium Systems, Inc., 1462 East 289 St., Wickliffe, Ohio 44092
9. Argonne National Laboratories, P.O. Box 299, Radioisotopes Sales Dept., Lemont, Ill. 60439
10. J. T. Baker Chemical Co., Phillipsburg, N.J. 08865
11. Bausch & Lomb, Inc., 635 St. Paul St., Rochester, N. Y. 14602
12. Beaumanor Farms, 1712 Sheridan Rd., Cleveland, Ohio 44121
13. Bico Scientific Co., 2325 So. Michigan Ave., Chicago, Ill. 60616

14. Biological Research Products, 243 West Root St., Stockyards Station, Chicago, Ill. 60609
15. Bio Metal Associates, BioQuip Products, P.O. Box 61, Santa Monica, Calif. 90406
16. Blue Ribbon Pet Farm, 14300 S.W. 86th Ave., Miami, Fla. 33158
17. Blue Spruce Biological Supply, P.O. Box Y, Castle Rock, Colo. 80104
18. CALATOMIC, Div. of CALBIOCHEM, 3625 Medford St., Los Angeles, Calif. 90063
19. CALBIOCHEM, 10933 N. Torrey Pines Rd., LaJolla, Calif. 92037
20. Canadian Breeding Farm and Laboratories, Ltd., 188 Lasalle, St. Constant, Laprairie Co., Quebec, Canada
21. R. F. Carle Co., P.O. Box 3155, Chico, Calif. 95926
22. Carolina Biological Supply Co., Burlington, N. C. 27215, or Powell Laboratories Division, Gladstone, Ore. 97027
23. Carworth, Div. of Becton, Dickinson and Company, 216 Congers Road, New City, N. Y. 10956
24. Central Scientific Co., 2600 So. Kostner Ave., Chicago, Ill. 60623
25. Clay-Adams Co., 141 E. 25th St., New York, N. Y. 10010
26. Combined Scientific Supplies, P.O. Box 125, Rosemead, Calif. 91770 (Scientific specimens of insects and arachnids, with collecting data)
27. Connecticut Valley Biological Supply Co., Inc., Valley Road, Southampton, Mass. 01073
28. W. H. Curtin Co., Box 1546, 4220 Jefferson Ave., Houston, Texas 77023
29. Dayno Sales Co., 678 Washington St., Lynn, Mass. 01901
30. Eastman Organic Chemicals, Eastman Kodak Co., Rochester, N. Y. 14650
31. G. Ertrachter's Tropical Fish Farm, Route 5, P.O. Box 5, Tampa, Fla. 33614
    Faust Scientific Supply Co. (See Mogul-ED)
32. Fischer Scientific Co., 1458 N. Lamon Ave., Chicago, Ill. 60651
33. Fulton Veterinary Supply Co., Inc., 169-09 Jamaica Ave., Jamaica, N. Y. 11432
34. General Biochemicals, Div. of Mogul Corp., 950 Laboratory Park, Chagrin Falls, Ohio 44022
35. General Biological, Inc., 8200 So. Hoyne Ave., Chicago, Ill. 60620
36. General Pet Supply of Ill., Inc., 1310 So. Fourth Ave., Maywood, Ill. 60153
37. Gingrich Animal Supply, Rt. 1, P.O. Box 189, Fredericksburg, Pa. 17026
38. Grand Island Biological Co., Div. of Mogul Corp., 3175 Staley Rd., Grand Island, N. Y. 14072 (media, biochemicals, microbiological equipment)
39. Gulf Specimen Co., Inc., P.O. Box 237, Panacea, Fla. 32346
40. Harborton Marine Lab, Box 11, Harborton, Va. 23389
41. Hardco Scientific, Div. of Fieldstone Corp., 3229 Omni Drive, Cincinnati, Ohio 45245
42. Harvard Apparatus, 150 Dover Road, Millis, Mass. 02054
43. Havahart, P.O. Box 551, Ossining, N. Y. 10562
44. Hel 'C' Bob Rabbitry (Hanson Enterprises), P.O. Box 10177, Santa Ana, Calif. 92711
45. Hoeltge, Inc., 5242 Crookshank Rd., Cincinnati, Ohio 45238
46. Hollister, Inc., 211 E. Chicago Ave., Chicago, Ill. 60611
47. Horton's Lab. Animals, Inc., 20151 Thompson Rd., Los Gatos, Calif. 95030
48. Isaacs Lab Stock, 1011 N. Walnut St., Litchfield, Ill. 62056
49. Jewel Aquarium Co., Inc., 5005 W. Armitage Ave., Chicago, Ill. 60639
50. Keyco Co., Inc., P.O. Box 12, Peach Bottom, Pa. 17563
51. K. O. B. Associates, Inc., 180 Higbie Lane, P.O. Box 91, W. Islip, N. Y. 11795
52. Herman Kolb, Bee Hobbyist, P.O. Box 183, Edmund, Okla. 73034

53. Lab Associates, Inc., 13640 132d Ave. N. E., Kirkland, Wash. 98033
54. Lab-Line Instruments, Inc., 15th & Bloomingdale Aves., Melrose Park, Ill. 60160
55. Lafayette Instrument Co., P.O. Box 1279, Lafayette, Ind. 46202
56. LaMotte Chemical Products Co., Chestertown, Md. 21620 (chemicals and equipment for testing water and soil)
57. LaPine Scientific Co., 6001 S. Knox Ave., Chicago, Ill. 60629.
    The Lemberger Co. (See Mogul-ED)
58. Lenderking Metal Products, Inc., 1000 So. Linwood Ave., Baltimore, Md. 21224
59. George Lomax, Rt. 1, Opdyke, Ill. 62872
60. Macalaster Scientific Co., Rt. 111 and Everett Turnpike, Nashua, N. H. 03060
61. Mallinckrodt Chemical Works, 2d and Mallinckrodt Sts., St. Louis, Mo. 63160
62. Mann Research Laboratories, 136 Liberty St., New York, N. Y. 10006
63. Manor Research, 223 Highway 18, East Brunswick, N. J. 08816
64. Marland Breeding Farms, Inc., P.O. Box 75, Wayne, N.J. 07470
65. Matheson Coleman and Bell, P.O. Box 85, East Rutherford, N.J. 07073
66. Mogul-ED, P.O. Box 482, Oshkosh, Wisc. 54901
67. Nalge Company, Nalgene Labware Division, 75 Panorama Creek Drive, Rochester, N.Y. 14625
68. Nasco, Fort Atkinson, Wisc. 53538
69. National Appliance Co., 3000 Taft St., Hollywood, Fla. 33021 (incubators, ovens, washers)
70. National Band and Tag Co., 721 York St., Newport, Ky. 41072
71. National Laboratory Animal Co., P.O. Box 12939, Creve Coeur, Mo. 63141
72. National Live Trap Corp., P.O. Box 302, Rt. 1, Tomahawk, Wisc. 54487
73. Nuclear Chicago Corporation, 223 West Erie St., Chicago, Ill. 60607
74. Nutritional Biochemicals Corp., 26201 Miles Road, Cleveland, Ohio 44128
75. Ohaus Scale Corp., 29 Hanover Road, Florham Park, N.J. 07932
76. Con Olson Co., Inc., P.O. Box 4021, Madison, Wisc. 53711
77. Osceola Cavies and Small Animal Farm, Inc., P.O. Box 15381, Orlando, Fla. 32808
78. Pacific Bio-Marine Supply Co., P.O. Box 536, Venice, Calif. 90291
79. Palmetto Pigeon Plant, P.O. Box 1585, Sumter, S.C. 29150
80. Pel-Freez Bio-Animals, Inc., P.O. Box 68, Rogers, Ark. 72756
81. Perfection Breeders, Inc., P.O. Box 75, Douglassville, Pa. 19518
82. The Pet Corral, 4146 Oracle Rd., Tucson, Ariz. 85705
83. The Pet Farm, Inc., 3310 N.W. So. River Drive, Miami, Fla. 33142
84. Pets Unlimited, 903 Hennepin Ave., Minneapolis, Minn. 55403
85. C. A. Phillips, Rt. 4, Cortland, N.Y. 13045
86. Products International Company, 2345 West Holly St., Phoenix, Ariz. 85009
87. Professional Veterinary Services, Inc., 683 W. 26th St., Hialeah, Fla.
88. Ralston Purina Company, 835 South 8th St., St. Louis, Mo. 63199
89. Redwood Game Farm and Lab, 1955 No. Redwood Road, Salt Lake City, Utah 84116
90. The Charles River Breeding Laboratories, Inc., 251 Ballardvale Street, Wilmington, Mass. 01887
91. Rockland Farms, Inc., P.O. Box 316, Gilbertsville, Pa. 19525
92. Royalhart Laboratory Animals, Inc., P.O. Box 198, New Hampton, N.Y. 10958
93. Sargent-Welch Scientific Co., 7300 North Linder Ave., Skokie, Ill. 60076
    Schettle Biologicals (See Mogul-ED)
94. Schwarz BioResearch, Inc., Mountain View Ave., Orangeburg, N.Y. 10962

95. Scientific Products, Div. of American Hospital Supply Corp., 1210 Leon Place, Evanston, Ill. 60201
96. Scientific Prototype Mfg. Corp., 615 W. 131st St., New York, N.Y. 10027
97. Simonsen Laboratories, Inc., 5228 Centerville Road, P.O. Box 8586, White Bear Lake, Minn. 55110
98. Skidmore Enterprises, 275 W. Mitchell Ave., Cincinnati, Ohio 45232
99. Gilbert M. Slater, Lakeview Hamster Colony, P.O. Box 85, Newfield, N.J. 08344
100. Snake Farm, P.O. Box 96, Laplace, La. 70068
101. Southern Biological Supply Co., McKenzie, Tenn. 38201
102. Southwestern Biological Supply Co., P.O. Box 4084, Station A, Dallas, Tex. 75208
103. Southwestern Herpetological Research and Sales, P.O. Box 282, Calimesa, Calif. 92320
104. E. G. Steinhilber & Co., Inc., Oshkosh, Wisc. 54901
105. Stoelting Company, 424 North Homan Ave., Chicago, Ill. 60624
106. Sure Live Mealworm Company, P.O. Box 206, Torrance, Calif. 90501
107. Taconic Farms, Inc., Germantown, N.Y. 12526
108. Tarpon Zoo, Inc., P.O. Box 847, Tarpon Springs, Fla. 33589
109. Teklad, Inc., P.O. Box 348, Monmouth, Ill. 61462
110. Arthur H. Thomas, Vine Street at Third, P.O. Box 779, Philadelphia, Pa. 19105
111. Three Springs Kennels Co., Inc., R.D. 1, Zelienople, Pa. 16063
112. Torsion Balance Co., 35 Monhegan, Clifton, N.J. 07013
113. Trans-Mississippi Biological Supply, 892 West County Road B, St. Paul, Minn. 55113
114. Tropical Atlantic Marine Specimens, P.O. Box 62, Big Pine Key, Fla. 33043
115. Truslow Farms, Inc., Chestertown, Md. 21620
116. Unifab Corporation, 5260 Lovers Lane Rd., Kalamazoo, Mich. 49002
117. Union Carbide Corp., P.O. Box 366, Tuxedo, N.Y. 10987
118. Viking Marine Biological Co., P.O. Box 256, Manasquan, N.J. 08736
119. Wahmann Manufacturing Company, P.O. Box 6883, Baltimore, Md. 21204
120. Ward's Natural Science Establishment, Inc., P.O. Box 1712, Rochester, N.Y. 14603, or P.O. Box 1749, Monterey, Calif. 93940

    Welch Scientific Co. (See Sargent-Welch Scientific Co.)
121. West Jersey Biological Supply, South Marion Avenue, Wenonah, N.J. 08090
122. White Animal Farm, Scarboro, Maine 04074
123. Wild Animal Importers, 519 W. 18th, Kennewick, Wash. 99336
124. Wildlife Supply Co., 2200 S. Hamilton St., Saginaw, Mich. 48602
125. World Wide Animal Brokers, Inc., 11 E. Mt. Royal Ave., Baltimore, Md. 21202
126. Carl Zeiss, Inc., 444 Fifth Ave., New York, N.Y. 10018
127. Zoological Center, International, 15W506 W. 63d (Burr Ridge), Hinsdale, Ill. 60521
128. Zoological Fauna, 1526 W. Highland Ave., Chicago, Ill. 60626
129. Zucca's Hamstery, 1541 Allen Ave., P.O. Box 507, Vineland, N.J. 08360

PLANT SUPPLIERS

Because many types of plants and seeds are available locally, the sources listed here are mostly those for unusual plants and unique seeds. Several general suppliers are included, who ship to all parts of the United States and to other countries.

Often a plant specialist may ask a small price for his catalog, and the approximate price, when known, is indicated below. In all cases a person must obtain a supplier's catalog before placing an order. A more extensive list of plant suppliers will be found in the *Plant Buyer's Guide* from the Massachusetts Horticultural Society.[1]

Suppliers of Plants and Seeds

Alberts & Merkel Bros., Inc., P.O. Box 537, Boynton Beach, Fla. 33435. Tropical plants, ferns, euphorbs, cycads, small pineapple plants.

A. E. Allgrove, North Wilmington, Mass. 01887. Specialist in terrarium plants, including mosses, ferns, and insectivorous plants.

Antonelli Brothers, 2545 Capitola Rd., Santa Cruz, Calif. 95060. Specialists in begonias, gloxinias, *Colchicum*, Venus-flytrap. Catalog, 50¢.

Armstrong Associates, Inc., Plant Oddities, Box 127, Basking Ridge, N.J. 07920. Insectivorous plants.

Beahm Gardens, 2686 Paloma St., Pasadena, Calif. 91107. Specialist in epiphyllums and other epiphytic cacti.

Buell's Greenhouses, Eastford, Conn. 06242. Catalog and culture handbook, $1.00. Specialist in gloxinias and African violets.

Burnett Brothers, Inc., 92 Chambers St., New York, N.Y. 10007. Seeds, indoor plants, equipment.

W. Atlee Burpee Co., Philadelphia, Pa. 19132. Seeds, indoor plants, and general equipment.

Cactus Land, 5740 S. 6th Ave., Tucson, Ariz. 85706. Importers and exporters of a wide variety of cacti.

Cactus Ranch, P.O. Box 128 Morristown, Ariz. 85342. Cacti and other succulents.

Carolina Biological Supply Co., Burlington, N.C. 27215

A. Hugh Dial, 7685 Deer Trail, Yucca Valley, Calif. 92284. Cacti, epiphyllums.

Fantastic Gardens, 9550 S.W. 67th Ave., Miami, Fla. 33156. Bromeliads.

Farmer Seed and Nursery Co., Faribault, Minn. 55021. Seeds and indoor plants.

French's, Bulb Importer and Seedsman, P.O. Box 37, Lima, Pa. 19060

Henrietta's Nursery, 1345 N. Brawley, Fresno, Calif. 93705. Cacti and succulents. Catalog, 20¢.

Hilltop Herb Farm, Box 866, Cleveland, Tex. 77327. Mosses, ferns, herbs, indoor plants, and seeds.

The House Plant Corner, Box 810, Oxford, Md. 21654. An extensive listing of indoor equipment. Catalog, 20¢.

P. de Jager & Sons, Inc., 188 Asbury St., So. Hamilton, Maine 01982. Bulb specialist.

Johnson Cactus Garden, 2735 Olive Hill Road, Fallbrook, Calif. 92028. Catalog, 50¢. An extensive listing of cacti and other succulents.

Kartuz Greenhouse, 92 Chestnut St., Wilmington, Mass. 01887. Specialists in begonias and other indoor plants.

Logee's Greenhouses, Danielson, Conn. 06239. Catalog, 50¢. Begonias, geraniums, ferns, and unusual indoor plants.

Merry Gardens, Camden, Maine 04843. Catalog, 25¢. Unusual indoor plants; herbs; geraniums; begonias; ferns, mosses, and bromeliads.

---

[1] H. G. Mattoon, ed., 1958, *Plant Buyer's Guide of Seed and Plant Materials*, 6th ed., The Massachusetts Horticultural Society, Boston. (Supplements to the guide are issued at intervals. Supplement One is now available.)

New Mexico Cactus Research, P.O. Box 787, Belen, N.M. 87002. An extensive listing of seeds of cacti and other succulents. Seed list, 50¢.

Geo. W. Park Seed Co., Inc., Greenwood, S.C. 29646. A large variety of seeds, indoor plants, and indoor equipment.

Peter Pauls Nurseries, R. D. # 4, Canandiagua, N.Y. 14424. Insectivorous plants.

Clyde Robin, P.O. Box 2091, Castro Valley, Calif. 94546. Seeds of wild flowers and wild trees. Catalog, 50¢.

Roehrs Exotic Nurseries, Farmingdale, N.J. 07727. A large listing of indoor plants.

Harry E. Saier, Seedsman, Dimondale, Mich. 48821. Seeds of flowering plants, vegetables, herbs, cacti, and succulents. Bog plants and ferns.

Wilson Brothers, Roachdale, Ind. 46172. Geranium specialists; other indoor plants.

Van Bourgondien Bros., 245 Farmingdale Rd., Rt. 109, Babylon, N.Y. 11702. Bulbs and indoor plants.

Western Seed Testing Service, 439 Pierce St., Twin Falls, Idaho 83301. Many types of seeds for laboratory use, including genetic and light-sensitive seeds.

Manufacturers of Greenhouses

Aluminum Greenhouses, Inc., 14615 Lorain Ave., Cleveland, Ohio 44111. Prefabricated greenhouses and accessories.

Chicopee Mfg. Co., Cornelia, Georgia 30531. Plastic shade fabrics. Ask for names of their distributors.

Lord & Burnham, Greenhouse Specialists, Irvington, N.Y. 10533. Specialists in designing and constructing greenhouses for educational institutions. They also supply prefabricated units, including window units, and accessories.

Redfern's Prefab Greenhouse Mfg. Co., 55 Mt. Hermon Rd., Scott's Valley, Calif. 95060.

Stearns Greenhouses, 98 Taylor St., Neponset, Boston, Mass. 02122. Prefabricated aluminum greenhouses and accessories.

Sturdi-built Mfg. Co., 11304 S.W. Boones Ferry Rd., Portland, Ore. 97219. Prefabricated redwood houses.

Texas Greenhouse Co., Inc., 2717 St. Louis Ave., Fort Worth, Tex. 76110. Prefabricated aluminum and redwood greenhouses, and accessories.

Turner Greenhouses, P.O. Box 1260, Hwy. 117 South, Goldsboro, N.C. 27530. Prefabricated greenhouses and accessories.

Plant Lighting

Floralite Co., 4124 East Oakwood Rd., Oak Creek, Wisc. 53154. Specialist in plant growth stands, carts, light fixtures, and accessories.

General Electric Co., Lamp Division, Nela Park, Cleveland, Ohio 44112

Shoplight Company, Inc., 566 Franklin Ave., Nutley, N.J. 07110. Specialists in plant lighting. Write for a price list of their booklets.

Sylvania Lighting Products, Engineering Dept., Salem, Mass.

Tube Craft, Inc., 1311 W. 80th St., Cleveland, Ohio 44102. Plant growth carts, fluorescent fixtures, accessories.

Westinghouse Electric Corp., Lamp Division, Bloomfield, N.J. 07003

Manufacturers of Growth Chambers

Environmental Growth Chambers, Chagrin Falls, Ohio 44022

Controlled Environments, Inc., 601 Stutsman St., Pembina, N. Dak. 58271. Specialist in plant growth chambers, incubators, controlled environmental rooms, seed germinators, and entomology cabinets.

Percival Refrigeration and Mfg. Co., Inc., Controlled Environment Div., P.O. Box 249, Boone, Iowa 50036. Specialists in all types of environmental control equipment.

MICROORGANISM SUPPLIERS

Sources of Microorganisms

American Type Culture Collection, 12301 Parklawn Drive, Rockville, Md. 20852. For viruses, bacteria, fungi, algae, and protozoa.

Carolina Biological Supply Co., Burlington, N.C. 27215, or Gladstone, Ore. 97027. For viruses, bacteria, fungi, algae, and protozoa.

General Biological, Inc., 8200 So. Hoyne Ave., Chicago, Ill. 60620. For bacteria, fungi, algae, and protozoa.

Indiana University, Algae Culture Collection, Dept. of Botany, Bloomington, Indiana. Use this source for unusual algae or those for research. Many of their algae for educational purposes have been transferred to the Carolina Biological Supply Co.

Also see general biological supply companies listed in Table B-2.

Media and Equipment

Baltimore Biological Laboratory, Inc., 2201 Aisquith St., Baltimore, Md.

Carolina Biological Supply Co., Burlington, N.C. 27215, or Gladstone, Ore. 97027

Difco Laboratories, Inc., Detroit, Mich. 48201

General Biological, Inc., 8200 So. Hoyne Ave., Chicago, Ill. 60620

Grand Island Biological Co., 3175 Staley Rd., Grand Island, N.Y. 14072

Ben Venue Laboratories, Inc., Bedford, Ohio. (For media in a tablet form that is used to prepare small quantities.)

Also see general biological supplies listed in Table B-2.

index

Note: an asterisk beside a page number indicates an illustration.